Methods in Enzymology

Volume 247

NEOGLYCOCONJUGATES

Part B

Biomedical Applications

METHODS IN ENZYMOLOGY

Methods in Enzymology

Volume 247

Neoglycoconjugates

Part B
Biomedical Applications

EDITED BY

Y. C. Lee

Reiko T. Lee

BIOCHEMISTRY DEPARTMENT
THE JOHNS HOPKINS UNIVERSITY
BALTIMORE, MARYLAND

ACADEMIC PRESS
San Diego New York Boston London Sydney Tokyo Toronto

This book is printed on acid-free paper. ∞

Academic Press, Inc.
A Division of Harcourt Brace & Company
525 B Street, Suite 1900, San Diego, California 92101-4495

United Kingdom Edition published by
Academic Press Limited
24-28 Oval Road, London NW1 7DX

International Standard Serial Number: 0076-6879

International Standard Book Number: 0-12-182148-X

PRINTED IN THE UNITED STATES OF AMERICA
94 95 96 97 98 99 EB 9 8 7 6 5 4 3 2 1

Table of Contents

Section I. Oligosaccharide Derivatives and Glycopeptides

Section II. Enzymatic and Affinity Methods

Section III. Binding Site Characterization

Section IV. Biomedical Applications

Contributors to Volume 247

Article numbers are in parentheses following the names of contributors.
Affiliations listed are current.

TOSHIHIRO AKAIKE (30), *Faculty of Bioscience and Technology, Tokyo Institute of Technology, Yokohama 227, Japan*

ELEONORA ALTMAN (17), *Institute for Biological Sciences, National Research Council of Canada, Ottawa, Ontario, Canada K1A 0R6*

SHIN AOKI (13), *Department of Chemistry, The Scripps Research Institute, La Jolla, California 92037*

REINHARD BROSSMER (11, 12), *Institute für Biochemie II, Universität Heidelberg, D-69120 Heidelberg, Germany*

DAVID R. BUNDLE (17, 20), *Department of Chemistry, University of Alberta, Edmonton, Alberta, Canada T6G 2G2*

CHRISTOPHER C. Q. CHIN (18), *Department of Biochemistry and Molecular Biology, University of Texas Medical School, Houston, Texas 77225*

MING H. CHIU (3), *Division of Pharmaceutics and Pharmaceutical Chemistry, College of Pharmacy, Ohio State University, Columbus, Ohio 43210*

M. L. CORRADI DA SILVA (3), *Division of Pharmaceutics and Pharmaceutical Chemistry, College of Pharmacy, Ohio State University, Columbus, Ohio 43210*

STEVEN P. DOUGLAS (10), *Department of Molecular and Medical Genetics, and Protein Engineering Network of Centres of Excellence, University of Toronto, Toronto, Ontario, Canada M5S 1A8*

TAMAO ENDO (15), *Department of Biochemistry, Institute of Medical Science, University of Tokyo, Minato-ku, Tokyo 108, Japan*

MARK A. FINDEIS (23), *TargeTech, Inc., Meriden, Connecticut 06450*

ANDRZEJ GAMIAN (24), *Department of Immunochemistry, Institute of Immunology and Experimental Therapy, Polish Academy of Sciences, 53-114 Wroclaw, Poland*

CORNELIS P. J. GLAUDEMANS (21), *National Institutes of Health, Bethesda, Maryland 20892*

HANS JÜRGEN GROSS (11, 12), *Institute für Biochemie II, Universität Heidelberg, D-69120 Heidelberg, Germany*

SEN-ITIROH HAKOMORI (22), *Department of Pathobiology, University of Washington, Seattle, Washington, 98195*

HITOSHI HASHIMOTO (5), *Faculty of Pharmaceutical Sciences, Mukogawa Women's University, Nishinomiya 663, Japan*

OLE HINDSGAUL (14), *Department of Chemistry, University of Alberta, Edmonton, Alberta, Canada T6G 2G2*

YOSHITAKA ICHIKAWA (7), *Department of Chemistry, The Scripps Research Institute, La Jolla, California 92037*

KATSUJI IKEKUBO (27), *Department of Nuclear Medicine, Kobe City General Hospital, Chuo-ku, Kobe 650, Japan*

KOJI KIMATA (25), *Institute for Molecular Science of Medicine, Aichi Medical University, Nagakute, Aichi 480-11, Japan*

SUMIO KITAHATA (5), *Faculty of Pharmaceutical Sciences, Mukogawa Women's University, Nishinomiya 663, Japan*

AKIRA KOBATA (15), *Tokyo Metropolitan Institute of Gerontology, Itabashi-ku, Tokyo 173, Japan*

AKIRA KOBAYASHI (30), *Faculty of Agriculture, Nagoya University, Chikusa, Nagoya 464-01, Japan*

KAZUKIYO KOBAYASHI (30), *Faculty of Agriculture, Nagoya University, Chikusa, Nagoya 464-01, Japan*

NAOHISA KOCHIBE (15), *Department of Biology, Faculty of Education, Gunma University, Maebashi 371, Japan*

KYOKO KOIZUMI (5), *Faculty of Pharmaceutical Sciences, Mukogawa Women's University, Nishinomiya 663, Japan*

HIROSATO KONDO (13), *Department of Chemistry, The Scripps Research Institute, La Jolla, California 92037*

PAVOL KOVÁČ (21), *National Institutes of Health, Bethesda, Maryland 20892*

JIRI J. KREPINSKY (10), *Department of Molecular and Medical Genetics, and Protein Engineering Network of Centres of Excellence, University of Toronto, Toronto, Ontario, Canada M5S 1A8*

KENNETH A. KROHN (29), *Departments of Radiology and Chemistry, University of Washington, Seattle, Washington 98195*

MASATOSHI KUDO (26, 27, 29), *Division of Gastroenterology, Department of Medicine, Kobe City General Hospital, Chuo-Ku, Kobe 650, Japan*

HORST KUNZ (1), *Institut für Organische Chemie, Johannes Gutenberg-Universität, D-55099 Mainz, Germany*

CRAIG A. LAFERRIÈRE (24), *Department of Chemistry, University of Ottawa, Ottawa, Ontario, Canada K1N 6N5*

JOCHEN LEHMANN (19), *Albert-Ludwigs-Universität Freiburg, Institute für Organische Chemie und Biochemie, 79104 Freiburg, Germany*

BJÖRN LÜNING (6), *Department of Chemistry, Swedish University of Agricultural Science, S-750 07 Uppsala, Sweden*

NICULINA LUPESCU (10), *Department of Molecular and Medical Genetics, and Protein Engineering Network of Centres of Excellence, University of Toronto, Toronto, Ontario, Canada M5S 1A8*

TAMARA MCBROOM (3), *Division of Pharmaceutics and Pharmaceutical Chemistry, College of Pharmacy, Ohio State University, Columbus, Ohio 43210*

FREDERICK L. MOOLTEN (10), *E. N. Rogers Veterans Administration Hospital, Bedford, Massachusetts 01730, and Department of Microbiology, School of Medicine, Boston University, Boston, Massachusetts*

EUGENIA M. NASHED (21), *National Institutes of Health, Bethesda, Maryland 20892*

THOMAS NORBERG (6), *Department of Chemistry, Swedish University of Agricultural Science, S-750 07 Uppsala, Sweden*

SEIICHIRO OGAWA (8, 9), *Department of Applied Chemistry, Faculty of Science and Technology, Keio University, Yokohama 223, Japan*

MONICA M. PALCIC (14), *Department of Chemistry, University of Alberta, Edmonton, Alberta, Canada T6G 2G2*

JAMES C. PAULSON (16), *Cytel Corporation, San Diego, California 92121*

MICHAEL PIERCE (14), *Department of Biochemistry, University of Georgia, Athens, Georgia 30602*

ROBERT A. PON (24), *Department of Chemistry, University of Ottawa, Ottawa, Ontario, Canada K1N 6N5*

DAVID PULLEYBLANK (10), *Department of Biochemistry, University of Toronto, Toronto, Ontario, Canada M5S 1A8*

KEVIN G. RICE (2, 3), *Division of Pharmaceutics and Pharmaceutical Chemistry, College of Pharmacy, Ohio State University, Columbus, Ohio 43210*

RENÉ ROY (24), *Department of Chemistry, University of Ottawa, Ottawa, Ontario, Canada K1N 6N5*

SUBRAMANIAM SABESAN (16), *Central Science and Engineering, The Dupont Company, Wilmington, Delaware 19880*

MARKUS SCHMIDT-SCHUCHARDT (19), *Albert-Ludwigs-Universität Freiburg, Institute für Organische Chemie und Biochemie, 79104 Freiburg, Germany*

MING-CHUAN SHAO (18), *Department of Biochemistry, Shanghai Medical University, Shanghai 200032, China*

BENT W. SIGURSKJOLD (20), *Department of Chemistry, Carlsberg Laboratory, DK-2500 Copenhagen Valby, Denmark*

ROBERT C. STADALNIK (26, 28, 29), *Division of Nuclear Medicine, Department of Radiology, University of California, Davis Medical Center, Sacramento, California 95817*

NOBUO SUGIURA (25), *Institute for Molecular Science of Medicine, Aichi Medical Unviersity, Nagakute, Aichi 480-11, Japan*

TOSHIAKI TAMURA (3), *Division of Pharmaceutics and Pharmaceutical Chemistry, College of Pharmacy, Ohio State University, Columbus, Ohio 43210*

JAN TEJBRANT (6), *Department of Chemistry, Swedish University of Agricultural Science, S-750 07 Uppsala, Sweden*

SARA J. TOPCU (28), *Department of Radiology, University of California, Davis Medical Center, Sacramento, California 95817*

TATSUSHI TOYOKUNI (22), *The Biomembrane Institute and Department of Chemistry, University of Washington, Seattle, Washington 98119*

HIDETOSHI TSUNODA (9), *Department of Applied Chemistry, Faculty of Science and Technology, Keio University, Yokohama 223, Japan*

DAVID R. VERA (26, 28, 29), *Division of Nuclear Medicine, Department of Radiology, University of California, Davis Medical Center, Sacramento, California 95817*

KARSTEN VON DEM BRUCH (1), *Institut für Organische Chemie, Johannes Gutenberg-Universität, D-55099 Mainz, Germany*

MANPREET S. WADHWA (3), *Division of Pharmaceutics and Pharmaceutical Chemistry, College of Pharmacy, Ohio State University, Columbus, Ohio 43210*

RUO WANG (7), *Department of Chemistry, The Scripps Research Institute, La Jolla, California 92037*

KOMEI WASHINO (27), *Research and Development Division, Central Research Laboratory, Nihon Medi-Physics, Chiba 299-02, Japan*

JASMINDER WEINSTEIN (16), *Amgen Center, Thousand Oaks, California 91320*

DENNIS M. WHITFIELD (10), *Department of Molecular and Medical Genetics, and Protein Engineering Network of Centres of Excellence, University of Toronto, Toronto, Ontario, Canada M5S 1A8*

CHI-HUEY WONG (7, 13), *Department of Chemistry, The Scripps Research Institute, La Jolla, California 92037*

CATHERINE H. WU (23), *University of Connecticut Health Center, Division of Gastroenterology, Farmington, Connecticut 06030*

GEORGE Y. WU (23), *University of Connecticut Health Center, Division of Gastroenterology, Farmington, Connecticut 06030*

YOSHIHIRO YAMAMICHI (27), *Research and Development Division, Central Research Laboratory, Nihon Medi-Physics, Chiba 299-02, Japan*

TOMOAKI YOSHIDA (4), *Department of Immunology, Nagoya University School of Medicine, Nagoya 466, Japan*

Preface

Rising interest in glycobiology in recent years has led to similar interest in neoglycoconjugates, a term given to carbohydrates conjugated to a variety of materials, including proteins, lipids, and synthetic polymers. Carbohydrates to be conjugated can be synthetically obtained or isolated from natural products. Such neoglycoconjugates are useful tools in a wide area of disciplines, ranging from clinical medicine to synthetic chemistry. They serve as a chemomimetic of natural glycoconjugates with well-defined structure and provide certain advantages over natural glycoconjugates.

Methods in Enzymology Volumes 242 and 247 contain many practical methods on how to prepare and use neoglycoconjugates, which have been contributed by a group of international experts. This volume (247) deals with biomedical applications and Volume 242 with synthesis. Readers should peruse all sections or utilize the indexes, since many chapters contain both preparation and application.

We were pleased to be able to develop these volumes in the *Methods in Enzymology* series which complement our treatise "Neoglycoconjugates: Preparation and Applications" (1994, Academic Press).

Y. C. Lee
Reiko T. Lee

METHODS IN ENZYMOLOGY

Volume XVII. Metabolism of Amino Acids and Amines (Parts A and B)
Edited by Herbert Tabor and Celia White Tabor

Volume XVIII. Vitamins and Coenzymes (Parts A, B, and C)
Edited by Donald B. McCormick and Lemuel D. Wright

Volume XIX. Proteolytic Enzymes
Edited by Gertrude E. Perlmann and Laszlo Lorand

Volume XX. Nucleic Acids and Protein Synthesis (Part C)
Edited by Kivie Moldave and Lawrence Grossman

Volume XXI. Nucleic Acids (Part D)
Edited by Lawrence Grossman and Kivie Moldave

Volume XXII. Enzyme Purification and Related Techniques
Edited by William B. Jakoby

Volume XXIII. Photosynthesis (Part A)
Edited by Anthony San Pietro

Volume XXIV. Photosynthesis and Nitrogen Fixation (Part B)
Edited by Anthony San Pietro

Volume XXV. Enzyme Structure (Part B)
Edited by C. H. W. Hirs and Serge N. Timasheff

Volume XXVI. Enzyme Structure (Part C)
Edited by C. H. W. Hirs and Serge N. Timasheff

Volume XXVII. Enzyme Structure (Part D)
Edited by C. H. W. Hirs and Serge N. Timasheff

Volume XXVIII. Complex Carbohydrates (Part B)
Edited by Victor Ginsburg

Volume XXIX. Nucleic Acids and Protein Synthesis (Part E)
Edited by Lawrence Grossman and Kivie Moldave

Volume XXX. Nucleic Acids and Protein Synthesis (Part F)
Edited by Kivie Moldave and Lawrence Grossman

Volume XXXI. Biomembranes (Part A)
Edited by Sidney Fleischer and Lester Packer

Volume XXXII. Biomembranes (Part B)
Edited by Sidney Fleischer and Lester Packer

Volume XXXIII. Cumulative Subject Index Volumes I–XXX
Edited by Martha G. Dennis and Edward A. Dennis

Volume XXXIV. Affinity Techniques (Enzyme Purification: Part B)
Edited by William B. Jakoby and Meir Wilchek

Volume XXXV. Lipids (Part B)
Edited by John M. Lowenstein

VOLUME XXXVI. Hormone Action (Part A: Steroid Hormones)
Edited by BERT W. O'MALLEY AND JOEL G. HARDMAN

VOLUME XXXVII. Hormone Action (Part B: Peptide Hormones)
Edited by BERT W. O'MALLEY AND JOEL G. HARDMAN

VOLUME XXXVIII. Hormone Action (Part C: Cyclic Nucleotides)
Edited by JOEL G. HARDMAN AND BERT W. O'MALLEY

VOLUME XXXIX. Hormone Action (Part D: Isolated Cells, Tissues, and Organ Systems)
Edited by JOEL G. HARDMAN AND BERT W. O'MALLEY

VOLUME XL. Hormone Action (Part E: Nuclear Structure and Function)
Edited by BERT W. O'MALLEY AND JOEL G. HARDMAN

VOLUME XLI. Carbohydrate Metabolism (Part B)
Edited by W. A. WOOD

VOLUME XLII. Carbohydrate Metabolism (Part C)
Edited by W. A. WOOD

VOLUME XLIII. Antibiotics
Edited by JOHN H. HASH

VOLUME XLIV. Immobilized Enzymes
Edited by KLAUS MOSBACH

VOLUME XLV. Proteolytic Enzymes (Part B)
Edited by LASZLO LORAND

VOLUME XLVI. Affinity Labeling
Edited by WILLIAM B. JAKOBY AND MEIR WILCHEK

VOLUME XLVII. Enzyme Structure (Part E)
Edited by C. H. W. HIRS AND SERGE N. TIMASHEFF

VOLUME XLVIII. Enzyme Structure (Part F)
Edited by C. H. W. HIRS AND SERGE N. TIMASHEFF

VOLUME XLIX. Enzyme Structure (Part G)
Edited by C. H. W. HIRS AND SERGE N. TIMASHEFF

VOLUME L. Complex Carbohydrates (Part C)
Edited by VICTOR GINSBURG

VOLUME LI. Purine and Pyrimidine Nucleotide Metabolism
Edited by PATRICIA A. HOFFEE AND MARY ELLEN JONES

VOLUME LII. Biomembranes (Part C: Biological Oxidations)
Edited by SIDNEY FLEISCHER AND LESTER PACKER

VOLUME LIII. Biomembranes (Part D: Biological Oxidations)
Edited by SIDNEY FLEISCHER AND LESTER PACKER

VOLUME LIV. Biomembranes (Part E: Biological Oxidations)
Edited by SIDNEY FLEISCHER AND LESTER PACKER

VOLUME LV. Biomembranes (Part F: Bioenergetics)
Edited by SIDNEY FLEISCHER AND LESTER PACKER

VOLUME LVI. Biomembranes (Part G: Bioenergetics)
Edited by SIDNEY FLEISCHER AND LESTER PACKER

VOLUME LVII. Bioluminescence and Chemiluminescence
Edited by MARLENE A. DELUCA

VOLUME LVIII. Cell Culture
Edited by WILLIAM B. JAKOBY AND IRA PASTAN

VOLUME LIX. Nucleic Acids and Protein Synthesis (Part G)
Edited by KIVIE MOLDAVE AND LAWRENCE GROSSMAN

VOLUME LX. Nucleic Acids and Protein Synthesis (Part H)
Edited by KIVIE MOLDAVE AND LAWRENCE GROSSMAN

VOLUME 61. Enzyme Structure (Part H)
Edited by C. H. W. HIRS AND SERGE N. TIMASHEFF

VOLUME 62. Vitamins and Coenzymes (Part D)
Edited by DONALD B. MCCORMICK AND LEMUEL D. WRIGHT

VOLUME 63. Enzyme Kinetics and Mechanism (Part A: Initial Rate and Inhibitor Methods)
Edited by DANIEL L. PURICH

VOLUME 64. Enzyme Kinetics and Mechanism (Part B: Isotopic Probes and Complex Enzyme Systems)
Edited by DANIEL L. PURICH

VOLUME 65. Nucleic Acids (Part I)
Edited by LAWRENCE GROSSMAN AND KIVIE MOLDAVE

VOLUME 66. Vitamins and Coenzymes (Part E)
Edited by DONALD B. MCCORMICK AND LEMUEL D. WRIGHT

VOLUME 67. Vitamins and Coenzymes (Part F)
Edited by DONALD B. MCCORMICK AND LEMUEL D. WRIGHT

VOLUME 68. Recombinant DNA
Edited by RAY WU

VOLUME 69. Photosynthesis and Nitrogen Fixation (Part C)
Edited by ANTHONY SAN PIETRO

VOLUME 70. Immunochemical Techniques (Part A)
Edited by HELEN VAN VUNAKIS AND JOHN J. LANGONE

VOLUME 71. Lipids (Part C)
Edited by JOHN M. LOWENSTEIN

VOLUME 72. Lipids (Part D)
Edited by JOHN M. LOWENSTEIN

VOLUME 73. Immunochemical Techniques (Part B)
Edited by JOHN J. LANGONE AND HELEN VAN VUNAKIS

VOLUME 74. Immunochemical Techniques (Part C)
Edited by JOHN J. LANGONE AND HELEN VAN VUNAKIS

VOLUME 75. Cumulative Subject Index Volumes XXXI, XXXII, XXXIV–LX
Edited by EDWARD A. DENNIS AND MARTHA G. DENNIS

VOLUME 76. Hemoglobins
Edited by ERALDO ANTONINI, LUIGI ROSSI-BERNARDI, AND EMILIA CHIANCONE

VOLUME 77. Detoxication and Drug Metabolism
Edited by WILLIAM B. JAKOBY

VOLUME 78. Interferons (Part A)
Edited by SIDNEY PESTKA

VOLUME 79. Interferons (Part B)
Edited by SIDNEY PESTKA

VOLUME 80. Proteolytic Enzymes (Part C)
Edited by LASZLO LORAND

VOLUME 81. Biomembranes (Part H: Visual Pigments and Purple Membranes, I)
Edited by LESTER PACKER

VOLUME 82. Structural and Contractile Proteins (Part A: Extracellular Matrix)
Edited by LEON W. CUNNINGHAM AND DIXIE W. FREDERIKSEN

VOLUME 83. Complex Carbohydrates (Part D)
Edited by VICTOR GINSBURG

VOLUME 84. Immunochemical Techniques (Part D: Selected Immunoassays)
Edited by JOHN J. LANGONE AND HELEN VAN VUNAKIS

VOLUME 85. Structural and Contractile Proteins (Part B: The Contractile Apparatus and the Cytoskeleton)
Edited by DIXIE W. FREDERIKSEN AND LEON W. CUNNINGHAM

VOLUME 86. Prostaglandins and Arachidonate Metabolites
Edited by WILLIAM E. M. LANDS AND WILLIAM L. SMITH

VOLUME 87. Enzyme Kinetics and Mechanism (Part C: Intermediates, Stereochemistry, and Rate Studies)
Edited by DANIEL L. PURICH

VOLUME 88. Biomembranes (Part I: Visual Pigments and Purple Membranes, II)
Edited by LESTER PACKER

VOLUME 89. Carbohydrate Metabolism (Part D)
Edited by WILLIS A. WOOD

VOLUME 90. Carbohydrate Metabolism (Part E)
Edited by WILLIS A. WOOD

VOLUME 91. Enzyme Structure (Part I)
Edited by C. H. W. HIRS AND SERGE N. TIMASHEFF

VOLUME 92. Immunochemical Techniques (Part E: Monoclonal Antibodies and General Immunoassay Methods)
Edited by JOHN J. LANGONE AND HELEN VAN VUNAKIS

VOLUME 93. Immunochemical Techniques (Part F: Conventional Antibodies, Fc Receptors, and Cytotoxicity)
Edited by JOHN J. LANGONE AND HELEN VAN VUNAKIS

VOLUME 94. Polyamines
Edited by HERBERT TABOR AND CELIA WHITE TABOR

VOLUME 95. Cumulative Subject Index Volumes 61–74, 76–80
Edited by EDWARD A. DENNIS AND MARTHA G. DENNIS

VOLUME 96. Biomembranes [Part J: Membrane Biogenesis: Assembly and Targeting (General Methods; Eukaryotes)]
Edited by SIDNEY FLEISCHER AND BECCA FLEISCHER

VOLUME 97. Biomembranes [Part K: Membrane Biogenesis: Assembly and Targeting (Prokaryotes, Mitochondria, and Chloroplasts)]
Edited by SIDNEY FLEISCHER AND BECCA FLEISCHER

VOLUME 98. Biomembranes (Part L: Membrane Biogenesis: Processing and Recycling)
Edited by SIDNEY FLEISCHER AND BECCA FLEISCHER

VOLUME 99. Hormone Action (Part F: Protein Kinases)
Edited by JACKIE D. CORBIN AND JOEL G. HARDMAN

VOLUME 100. Recombinant DNA (Part B)
Edited by RAY WU, LAWRENCE GROSSMAN, AND KIVIE MOLDAVE

VOLUME 101. Recombinant DNA (Part C)
Edited by RAY WU, LAWRENCE GROSSMAN, AND KIVIE MOLDAVE

VOLUME 102. Hormone Action (Part G: Calmodulin and Calcium-Binding Proteins)
Edited by ANTHONY R. MEANS AND BERT W. O'MALLEY

VOLUME 103. Hormone Action (Part H: Neuroendocrine Peptides)
Edited by P. MICHAEL CONN

VOLUME 104. Enzyme Purification and Related Techniques (Part C)
Edited by WILLIAM B. JAKOBY

VOLUME 105. Oxygen Radicals in Biological Systems
Edited by LESTER PACKER

VOLUME 106. Posttranslational Modifications (Part A)
Edited by FINN WOLD AND KIVIE MOLDAVE

VOLUME 107. Posttranslational Modifications (Part B)
Edited by FINN WOLD AND KIVIE MOLDAVE

VOLUME 108. Immunochemical Techniques (Part G: Separation and Characterization of Lymphoid Cells)
Edited by GIOVANNI DI SABATO, JOHN J. LANGONE, AND HELEN VAN VUNAKIS

VOLUME 109. Hormone Action (Part I: Peptide Hormones)
Edited by LUTZ BIRNBAUMER AND BERT W. O'MALLEY

VOLUME 110. Steroids and Isoprenoids (Part A)
Edited by JOHN H. LAW AND HANS C. RILLING

VOLUME 111. Steroids and Isoprenoids (Part B)
Edited by JOHN H. LAW AND HANS C. RILLING

VOLUME 112. Drug and Enzyme Targeting (Part A)
Edited by KENNETH J. WIDDER AND RALPH GREEN

VOLUME 113. Glutamate, Glutamine, Glutathione, and Related Compounds
Edited by ALTON MEISTER

VOLUME 114. Diffraction Methods for Biological Macromolecules (Part A)
Edited by HAROLD W. WYCKOFF, C. H. W. HIRS, AND SERGE N. TIMASHEFF

VOLUME 115. Diffraction Methods for Biological Macromolecules (Part B)
Edited by HAROLD W. WYCKOFF, C. H. W. HIRS, AND SERGE N. TIMASHEFF

VOLUME 116. Immunochemical Techniques (Part H: Effectors and Mediators of Lymphoid Cell Functions)
Edited by GIOVANNI DI SABATO, JOHN J. LANGONE, AND HELEN VAN VUNAKIS

VOLUME 117. Enzyme Structure (Part J)
Edited by C. H. W. HIRS AND SERGE N. TIMASHEFF

VOLUME 118. Plant Molecular Biology
Edited by ARTHUR WEISSBACH AND HERBERT WEISSBACH

VOLUME 119. Interferons (Part C)
Edited by SIDNEY PESTKA

VOLUME 120. Cumulative Subject Index Volumes 81–94, 96–101

VOLUME 121. Immunochemical Techniques (Part I: Hybridoma Technology and Monoclonal Antibodies)
Edited by JOHN J. LANGONE AND HELEN VAN VUNAKIS

VOLUME 122. Vitamins and Coenzymes (Part G)
Edited by FRANK CHYTIL AND DONALD B. MCCORMICK

VOLUME 123. Vitamins and Coenzymes (Part H)
Edited by FRANK CHYTIL AND DONALD B. MCCORMICK

VOLUME 124. Hormone Action (Part J: Neuroendocrine Peptides)
Edited by P. MICHAEL CONN

VOLUME 125. Biomembranes (Part M: Transport in Bacteria, Mitochondria, and Chloroplasts: General Approaches and Transport Systems)
Edited by SIDNEY FLEISCHER AND BECCA FLEISCHER

VOLUME 126. Biomembranes (Part N: Transport in Bacteria, Mitochondria, and Chloroplasts: Protonmotive Force)
Edited by SIDNEY FLEISCHER AND BECCA FLEISCHER

VOLUME 127. Biomembranes (Part O: Protons and Water: Structure and Translocation)
Edited by LESTER PACKER

VOLUME 128. Plasma Lipoproteins (Part A: Preparation, Structure, and Molecular Biology)
Edited by JERE P. SEGREST AND JOHN J. ALBERS

VOLUME 129. Plasma Lipoproteins (Part B: Characterization, Cell Biology, and Metabolism)
Edited by JOHN J. ALBERS AND JERE P. SEGREST

VOLUME 130. Enzyme Structure (Part K)
Edited by C. H. W. HIRS AND SERGE N. TIMASHEFF

VOLUME 131. Enzyme Structure (Part L)
Edited by C. H. W. HIRS AND SERGE N. TIMASHEFF

VOLUME 132. Immunochemical Techniques (Part J: Phagocytosis and Cell-Mediated Cytotoxicity)
Edited by GIOVANNI DI SABATO AND JOHANNES EVERSE

VOLUME 133. Bioluminescence and Chemiluminescence (Part B)
Edited by MARLENE DELUCA AND WILLIAM D. MCELROY

VOLUME 134. Structural and Contractile Proteins (Part C: The Contractile Apparatus and the Cytoskeleton)
Edited by RICHARD B. VALLEE

VOLUME 135. Immobilized Enzymes and Cells (Part B)
Edited by KLAUS MOSBACH

VOLUME 136. Immobilized Enzymes and Cells (Part C)
Edited by KLAUS MOSBACH

VOLUME 137. Immobilized Enzymes and Cells (Part D)
Edited by KLAUS MOSBACH

VOLUME 138. Complex Carbohydrates (Part E)
Edited by VICTOR GINSBURG

VOLUME 139. Cellular Regulators (Part A: Calcium- and Calmodulin-Binding Proteins)
Edited by ANTHONY R. MEANS AND P. MICHAEL CONN

VOLUME 140. Cumulative Subject Index Volumes 102–119, 121–134

VOLUME 141. Cellular Regulators (Part B: Calcium and Lipids)
Edited by P. MICHAEL CONN AND ANTHONY R. MEANS

VOLUME 142. Metabolism of Aromatic Amino Acids and Amines
Edited by SEYMOUR KAUFMAN

VOLUME 143. Sulfur and Sulfur Amino Acids
Edited by WILLIAM B. JAKOBY AND OWEN GRIFFITH

VOLUME 144. Structural and Contractile Proteins (Part D: Extracellular Matrix)
Edited by LEON W. CUNNINGHAM

VOLUME 145. Structural and Contractile Proteins (Part E: Extracellular Matrix)
Edited by LEON W. CUNNINGHAM

VOLUME 146. Peptide Growth Factors (Part A)
Edited by DAVID BARNES AND DAVID A. SIRBASKU

VOLUME 147. Peptide Growth Factors (Part B)
Edited by DAVID BARNES AND DAVID A. SIRBASKU

VOLUME 148. Plant Cell Membranes
Edited by LESTER PACKER AND ROLAND DOUCE

VOLUME 149. Drug and Enzyme Targeting (Part B)
Edited by RALPH GREEN AND KENNETH J. WIDDER

VOLUME 150. Immunochemical Techniques (Part K: *In Vitro* Models of B and T Cell Functions and Lymphoid Cell Receptors)
Edited by GIOVANNI DI SABATO

VOLUME 151. Molecular Genetics of Mammalian Cells
Edited by MICHAEL M. GOTTESMAN

VOLUME 152. Guide to Molecular Cloning Techniques
Edited by SHELBY L. BERGER AND ALAN R. KIMMEL

VOLUME 153. Recombinant DNA (Part D)
Edited by RAY WU AND LAWRENCE GROSSMAN

VOLUME 154. Recombinant DNA (Part E)
Edited by RAY WU AND LAWRENCE GROSSMAN

VOLUME 155. Recombinant DNA (Part F)
Edited by RAY WU

VOLUME 156. Biomembranes (Part P: ATP-Driven Pumps and Related Transport: The Na,K-Pump)
Edited by SIDNEY FLEISCHER AND BECCA FLEISCHER

VOLUME 157. Biomembranes (Part Q: ATP-Driven Pumps and Related Transport: Calcium, Proton, and Potassium Pumps)
Edited by SIDNEY FLEISCHER AND BECCA FLEISCHER

VOLUME 158. Metalloproteins (Part A)
Edited by JAMES F. RIORDAN AND BERT L. VALLEE

VOLUME 159. Initiation and Termination of Cyclic Nucleotide Action
Edited by JACKIE D. CORBIN AND ROGER A. JOHNSON

VOLUME 160. Biomass (Part A: Cellulose and Hemicellulose)
Edited by WILLIS A. WOOD AND SCOTT T. KELLOGG

VOLUME 161. Biomass (Part B: Lignin, Pectin, and Chitin)
Edited by WILLIS A. WOOD AND SCOTT T. KELLOGG

VOLUME 162. Immunochemical Techniques (Part L: Chemotaxis and Inflammation)
Edited by GIOVANNI DI SABATO

VOLUME 163. Immunochemical Techniques (Part M: Chemotaxis and Inflammation)
Edited by GIOVANNI DI SABATO

VOLUME 164. Ribosomes
Edited by HARRY F. NOLLER, JR., AND KIVIE MOLDAVE

VOLUME 165. Microbial Toxins: Tools for Enzymology
Edited by SIDNEY HARSHMAN

VOLUME 166. Branched-Chain Amino Acids
Edited by ROBERT HARRIS AND JOHN R. SOKATCH

VOLUME 167. Cyanobacteria
Edited by LESTER PACKER AND ALEXANDER N. GLAZER

VOLUME 168. Hormone Action (Part K: Neuroendocrine Peptides)
Edited by P. MICHAEL CONN

VOLUME 169. Platelets: Receptors, Adhesion, Secretion (Part A)
Edited by JACEK HAWIGER

VOLUME 170. Nucleosomes
Edited by PAUL M. WASSARMAN AND ROGER D. KORNBERG

VOLUME 171. Biomembranes (Part R: Transport Theory: Cells and Model Membranes)
Edited by SIDNEY FLEISCHER AND BECCA FLEISCHER

VOLUME 172. Biomembranes (Part S: Transport: Membrane Isolation and Characterization)
Edited by SIDNEY FLEISCHER AND BECCA FLEISCHER

VOLUME 173. Biomembranes [Part T: Cellular and Subcellular Transport: Eukaryotic (Nonepithelial) Cells]
Edited by SIDNEY FLEISCHER AND BECCA FLEISCHER

VOLUME 174. Biomembranes [Part U: Cellular and Subcellular Transport: Eukaryotic (Nonepithelial) Cells]
Edited by SIDNEY FLEISCHER AND BECCA FLEISCHER

VOLUME 175. Cumulative Subject Index Volumes 135–139, 141–167

VOLUME 176. Nuclear Magnetic Resonance (Part A: Spectral Techniques and Dynamics)
Edited by NORMAN J. OPPENHEIMER AND THOMAS L. JAMES

VOLUME 177. Nuclear Magnetic Resonance (Part B: Structure and Mechanism)
Edited by NORMAN J. OPPENHEIMER AND THOMAS L. JAMES

VOLUME 178. Antibodies, Antigens, and Molecular Mimicry
Edited by JOHN J. LANGONE

VOLUME 179. Complex Carbohydrates (Part F)
Edited by VICTOR GINSBURG

VOLUME 214. Carotenoids (Part B: Metabolism, Genetics, and Biosynthesis)
Edited by LESTER PACKER

VOLUME 215. Platelets: Receptors, Adhesion, Secretion (Part B)
Edited by JACEK J. HAWIGER

VOLUME 216. Recombinant DNA (Part G)
Edited by RAY WU

VOLUME 217. Recombinant DNA (Part H)
Edited by RAY WU

VOLUME 218. Recombinant DNA (Part I)
Edited by RAY WU

VOLUME 219. Reconstitution of Intracellular Transport
Edited by JAMES E. ROTHMAN

VOLUME 220. Membrane Fusion Techniques (Part A)
Edited by NEJAT DÜZGÜNEŞ

VOLUME 221. Membrane Fusion Techniques (Part B)
Edited by NEJAT DÜZGÜNEŞ

VOLUME 222. Proteolytic Enzymes in Coagulation, Fibrinolysis, and Complement Activation (Part A: Mammalian Blood Coagulation Factors and Inhibitors)
Edited by LASZLO LORAND AND KENNETH G. MANN

VOLUME 223. Proteolytic Enzymes in Coagulation, Fibrinolysis, and Complement Activation (Part B: Complement Activation, Fibrinolysis, and Nonmammalian Blood Coagulation Factors)
Edited by LASZLO LORAND AND KENNETH G. MANN

VOLUME 224. Molecular Evolution: Producing the Biochemical Data
Edited by ELIZABETH ANNE ZIMMER, THOMAS J. WHITE, REBECCA L. CANN, AND ALLAN C. WILSON

VOLUME 225. Guide to Techniques in Mouse Development
Edited by PAUL M. WASSARMAN AND MELVIN L. DEPAMPHILIS

VOLUME 226. Metallobiochemistry (Part C: Spectroscopic and Physical Methods for Probing Metal Ion Environments in Metalloenzymes and Metalloproteins)
Edited by JAMES F. RIORDAN AND BERT L. VALLEE

VOLUME 227. Metallobiochemistry (Part D: Physical and Spectroscopic Methods for Probing Metal Ion Environments in Metalloproteins)
Edited by JAMES F. RIORDAN AND BERT L. VALLEE

VOLUME 228. Aqueous Two-Phase Systems
Edited by HARRY WALTER AND GÖTE JOHANSSON

VOLUME 229. Cumulative Subject Index Volumes 195–198, 200–227 (in preparation)

VOLUME 230. Guide to Techniques in Glycobiology
Edited by WILLIAM J. LENNARZ AND GERALD W. HART

VOLUME 231. Hemoglobins (Part B: Biochemical and Analytical Methods)
Edited by JOHANNES EVERSE, KIM D. VANDEGRIFF AND ROBERT M. WINSLOW

VOLUME 232. Hemoglobins (Part C: Biophysical Methods)
Edited by JOHANNES EVERSE, KIM D. VANDEGRIFF AND ROBERT M. WINSLOW

VOLUME 233. Oxygen Radicals in Biological Systems (Part C)
Edited by LESTER PACKER

VOLUME 234. Oxygen Radicals in Biological Systems (Part D)
Edited by LESTER PACKER

VOLUME 235. Bacterial Pathogenesis (Part A: Identification and Regulation of Virulence Factors)
Edited by VIRGINIA L. CLARK AND PATRIK M. BAVOIL

VOLUME 236. Bacterial Pathogenesis (Part B: Integration of Pathogenic Bacteria with Host Cells)
Edited by VIRGINIA L. CLARK AND PATRIK M. BAVOIL

VOLUME 237. Heterotrimeric G Proteins
Edited by RAVI IYENGAR

VOLUME 238. Heterotrimeric G-Protein Effectors
Edited by RAVI IYENGAR

VOLUME 239. Nuclear Magnetic Resonance (Part C)
Edited by THOMAS L. JAMES AND NORMAN J. OPPENHEIMER

VOLUME 240. Numerical Computer Methods (Part B)
Edited by MICHAEL L. JOHNSON AND LUDWIG BRAND

VOLUME 241. Retroviral Proteases
Edited by LAWRENCE C. KUO AND JULES A. SHAFER

VOLUME 242. Neoglycoconjugates (Part A)
Edited by Y. C. LEE AND REIKO T. LEE

VOLUME 243. Inorganic Microbial Sulfur Metabolism
Edited by HARRY D. PECK, JR., AND JEAN LEGALL

VOLUME 244. Proteolytic Enzymes: Serine and Cysteine Peptidases
Edited by ALAN J. BARRETT

VOLUME 245. Extracellular Matrix Components (in preparation)
Edited by E. RUOSLAHTI AND E. ENGVALL

VOLUME 246. Biochemical Spectroscopy (in preparation)
Edited by KENNETH SAUER

VOLUME 247. Neoglycoconjugates (Part B: Biomedical Applications)
Edited by Y. C. LEE AND REIKO T. LEE

VOLUME 248. Proteolytic Enzymes: Aspartic and Metallo Peptidases (in preparation)
Edited by ALAN J. BARRETT

VOLUME 249. Enzyme Kinetics and Mechanisms (Part D) (in preparation)
Edited by DANIEL L. PURICH

VOLUME 250. Lipid Modifications of Proteins (in preparation)
Edited by PATRICK J. CASEY AND JANICE E. BUSS

VOLUME 251. Biothiols (in preparation)
Edited by LESTER PACKER

Section I

Oligosaccharide Derivatives and Glycopeptides

[1] Neoglycoproteins from Synthetic Glycopeptides

By HORST KUNZ and KARSTEN VON DEM BRUCH

Introduction

Saccharide side chains of glycoproteins not only influence the physicochemical properties of these biomacromolecules and their stability against proteolytic degradation, but also play important roles as ligands in biological recognition[1] and in the organized distribution of these compounds within multicellular organisms. Carbohydrate–lectin interactions are important, for example, in viral infections[2] and for the recruitment and invasion of leukocytes into injured tissues.[3] Although in a number of processes carbohydrates were revealed to be decisive recognition labels,[4] in other biological selections peptide sequences proved to be the recognized areas.[5] However, for glycoproteins it seems inappropriate to separate rigorously the role of peptides and carbohydrates since both parts carry a number of polar functionalities predestinated for interacting with one another. Therefore, it is highly presumable that both parts, carbohydrates and peptide sequences, can contribute to a complex recognition label or epitope of a natural glycoprotein. Logically, model compounds useful for the investigation of such recognition phenomena should have exactly specified structures in both the carbohydrate and the peptide portion. Furthermore, the linkage region also should be of clearly defined stereochemistry. To meet these requirements, the synthesis of glycopeptides[6,7] has to be considered an indispensible precondition.

Synthetic T Antigen Glycoproteins

Springer[8] reported that glycoproteins with the Thomsen–Friedensreich antigen structure (T antigen) occur in membranes of a number of epithelial tumors, whereas they are absent in normal cells. The monoclonal antibod-

[1] See, for example, K. Drickamer and J. Carver, *Curr. Opin. Struct. Biol.* **2,** 653 (1992).
[2] See, for example, K.-A. Karlsson, *Annu. Rev. Biochem.* **58,** 309 (1989).
[3] See, for example, M. J. Polley, M. L. Philips, E. Wagner, E. Nudelman, A. K. Shinghal, S.-I. Hakomori, and J. C. Paulson, *Proc. Natl. Acad. Sci. U.S.A.* **88,** 6224 (1991).
[4] G. Ashwell, *Trends Biochem. Sci.* **2,** 186 (1977).
[5] M. E. Hemler, *Annu. Rev. Immunol.* **8,** 365 (1990).
[6] H. Kunz, *Angew. Chem., Int. Ed. Engl.* **26,** 294 (1987).
[7] H.Kunz, *Pure Appl. Chem.* **65,** 1223 (1993).
[8] G. F. Springer, *Science* **224,** 1198 (1984).

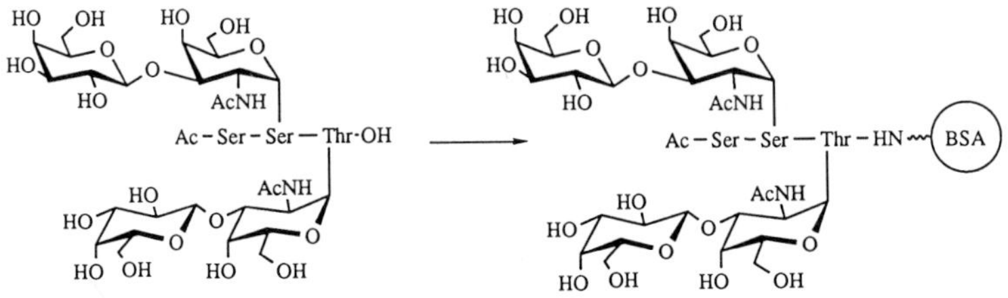

SCHEME 1

ies obtained from these tumor-associated T antigen glycoproteins were found to be cross-reactive with asialoglycophorin.[8] We have synthesized the N-terminal glycopeptide of human glycophorin **1** containing two T antigen disaccharide side chains and have linked this synthetic glycopeptide to bovine serum albumin (BSA) in order to furnish a synthetic antigen **2**[9,10] (Scheme 1).

The synthesis of the glycopeptide is realized starting from 9-fluorenylmethoxycarbonyl (Fmoc) serine or threonine benzyl ester, respectively.[9,10] As the glycosidic bonds are potentially sensitive to acids and, in particular, *O*-glycosylserine and -threonine derivatives are prone to base-catalyzed β-elimination, the removal of the protecting groups from such conjugates requires a mild and selective methodology. We have found that for β-xylosylserine derivatives removal of the Fmoc group can be achieved with complete selectivity and without affecting the *O*-glycosylserine linkage by application of the weak base morpholine.[11] The intrinsic basicity of morpholine is sufficient to cause β-elimination on the Fmoc group, but it is too weak to initiate β-elimination of the carbohydrate portion from the serine and threonine conjugates. This procedure for the N-terminal deblocking in combination with selective hydrogenolysis of the benzyl ester[6] provides a general strategy for the synthesis of glycopeptides. It has enabled us to synthesize the protected N-terminal glycotripeptide **1** of human glycophorin with M blood group specificity and the T antigen side chains. As the synthetic T antigen glycopeptide should be applied for induction of antibodies, we have aimed at a coupling of **1** to the carrier protein without applying an artificial linker which can cause strongly immunogenetic, but unnatural epitope structures. With this

[9] H. Kunz and S. Birnbach, *Angew. Chem., Int. Ed. Engl.* **25,** 360 (1986).

[10] H. Kunz, S. Birnbach, and P. Wernig, *Carbohydr. Res.* **202,** 207 (1990).

[11] P. Schultheiss-Reimann and H. Kunz, *Angew. Chem., Int. Ed. Engl.* **22,** 62 (1983); *Angew. Chem. Suppl.* 39 (1983).

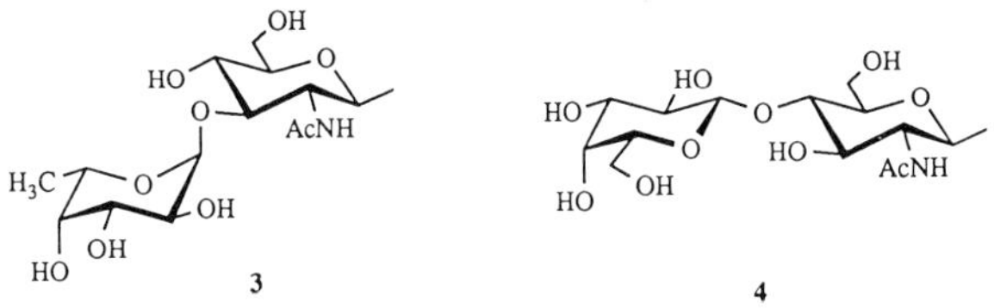

3 4

SCHEME 2

aim we have carried out the attachment of **1** to BSA in water at pH 6 and 4° by using the water soluble 1-ethyl-3-(3-dimethylaminopropyl)carbodiimide (EDC)[12] in combination with 1-hydroxybenzotriazole (HOBt).[13] According to carbohydrate analysis,[14] **2** contains on average 38 molecules of **1** per molecule protein.[10]

Preliminary immunological results[15] show that the monoclonal antibody obtained with the synthetic antigen **2** reacts with all epithelial tumors tested, but also binds to normal cells of the same tissues. However, further investigations have revealed that the antibody assumed to recognize the T antigen disaccharides differentiates between asialoglycophorin of M and N blood group specificity differing in the N-terminal peptide sequence. Thus, it has to be concluded that the antibody especially recognizes an epitope formed from both the saccharide and the peptide part of **1.**

Synthetic N-Glycoproteins Containing Fucosylglucosamine and Lactosamine Side Chains

Model glycopeptides are not only interesting for immunological studies but also as ligands for investigations of recognition phenomena and enzyme specificities. With these intentions we are interested in the synthesis of glycopeptides which carry partial structures of the Lewis^{x} saccharide as side chains, namely, the α-fucosyl(1-3)-*N*-acetylglucosamine **3** and the β-galactosyl(1-4)-*N*-acetylglucosamine **4** (Scheme 2). As a rule, multivalent ligands exhibit dramatically enhanced activity in biological recognition.[16] Therefore, we have synthesized glycopeptides of this type which contain two of these saccharide side chains separated by a distance proper for the simulation of a biantennary structure.

[12] J. C. Sheehan, J. Preston, and P. A. Cruickshank, *J. Am. Chem. Soc.* **87,** 2492 (1965).

[13] W. König and R. Geiger, *Chem. Ber.* **103,** 788 (1970).

[14] M. Dubois, K. A. Gilles, J. K. Hamilton, P. A. Rebers, and F. Smith, *Anal. Chem.* **28,** 350 (1956).

[15] W. Dippold, A. Steinborn, and K. H. Meyer zum Büschenfelde, I. Med. Klinik, Univ. Mainz, Germany, unpublished results.

[16] R. T. Lee and Y. C. Lee, *Glycoconjugate J.* **4,** 317 (1987).

Synthesis of Disaccharides

In the synthesis of the disaccharide units **3** and **4** we apply the azido group as the anomeric protection of the glucosamine, which can also serve as the precursor of the anomeric amino group required later for the construction of the N-glycosidic linkage to asparagine.[17] To this end, 2-acetamido-2-deoxy-3,4,6-tri-*O*-acetyl-β-D-glucopyranosyl azide (**5**)[18] is de-O-acetylated using Amberlyst A 26 (OH^- form) in methanol,[19] and the product is subjected to acetalization with α,α-dimethoxytoluene to give the key intermediate **6.** Reaction of **6** with 2,3,4-tri-*O*-(4-methoxybenzyl)-α-L-fucopyranosyl bromide[20] according to the *in situ* anomerization procedure[21] stereoselectively yields the α-fucosylglucosaminylazide **7.** To provide an acid stability of the fucoside bond sufficient for the planned glycopeptide syntheses,[17] the 4-methoxybenzyl (Mpm) protection of **7** is replaced with *O*-acetyl protection. Oxidative removal of the Mpm group using ceric ammonium nitrate[22] proceeds without affecting the anomeric azide. Subsequent acetylation gives **8.** The stabilizing effect of the *O*-acetyl protection on the fucoside linkage[17] is illustrated by the acidolytic removal of the 4,6-benzylidene group, which is carried out with tetrafluorohydroboric acid in aqueous acetonitrile to give **9** (Scheme 3).

It should be noticed that an analogous treatment of **7** completely destroys both the fucoside bond and the Mpm protections. Acetylation of the hydroxyl groups of **9** furnished the saccharide azide **10,** which is transformed to the disaccharide amine **11** by hydrogenolysis. For the synthesis of the lactosamine derivatives (Scheme 4), **6** is acetylated at the 3-position. Acidolytic removal of the benzylidene group and subsequent regioselective benzoylation carefully carried out with benzoyl chloride/pyridine in acetonitrile at 0° give the glucosamine azide **12** having a selectively unblocked 4-hydroxyl group.

Glycosylation of **12** with *O*-acetyl-protected galactosyl bromide **13**[23] promoted by silver triflate resulted in the formation of the lactosamine azide as an α/β mixture (α/β 1 : 4). The β anomer **14** is isolated by flash chromatography (71%). Hydrogenolysis catalyzed by Raney nickel furnished the desired lactosaminyl amine **15.**

[17] H. Kunz and C. Unverzagt, *Angew. Chem., Int. Ed. Engl.* **27,** 1697 (1988).
[18] The compound was synthesized via phase-transfer catalysis; see C. Unverzagt and H. Kunz, *J. Prakt. Chem.* **334,** 570 (1992).
[19] L. Szilagyi and Z. Györgydeak, *Carbohydr. Res.* **143,** 21 (1985).
[20] J. März and H. Kunz, *Synlett,* 589 (1992).
[21] R.U. Lemieux and H. Diguez, *J. Am. Chem. Soc.* **97,** 4063 (1975).
[22] R. Johansson and B. Samuelson, *J. Chem. Soc., Perkin Trans 1,* 2371 (1984).
[23] H. Ohle, W. Marecek, and W. Bourjan, *Ber. Dtsch. Chem. Ges.* **62,** 833 (1929).

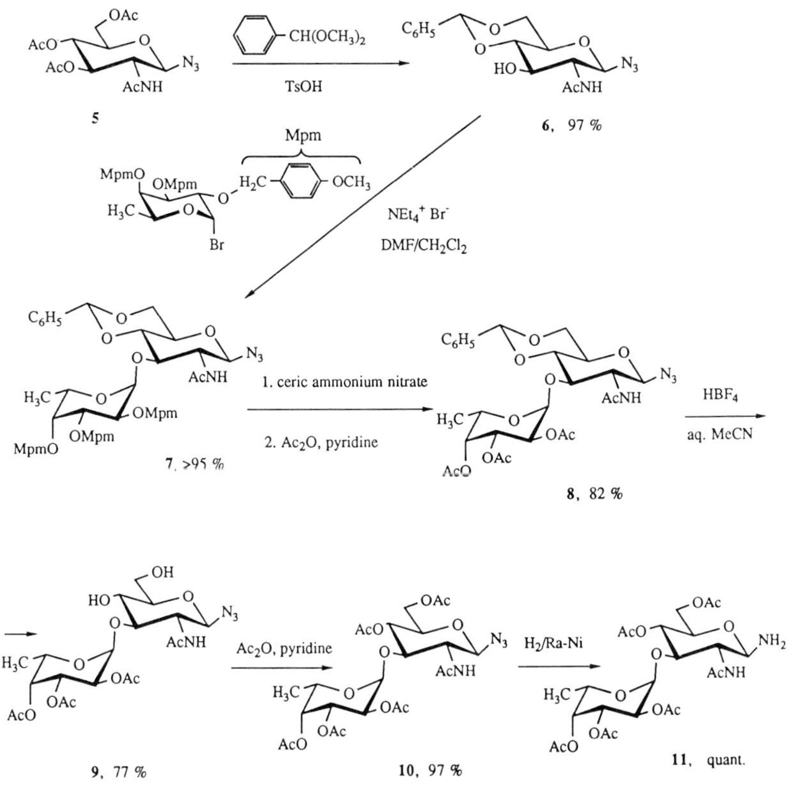

SCHEME 3

Synthesis of Disaccharide Asparagine Conjugates

N-Glycoconjugates of asparagine have been obtained by Garg and Jeanloz[24] via reaction of *N*-benzyloxycarbonyl(*Z*)-aspartic acid anhydride with O-protected glycosylamines of *N*-acetylglucosamine and chitobiose. As this procedure results in the formation of two regioisomers, we have synthesized the *N*-glycosidic conjugates from glycosylamines and asparatic derivatives selectively deblocked at the β-carboxyl function.[17,20,25,26] A general method consists of the regioselective formation of asparatic acid β-benzyl ester **16** from asparatic acid and benzyl alcohol catalyzed with tetrafluorohydroboric acid[27] and subsequent selective introductions

[24] H. G. Garg and R. W. Jeanloz, *J. Org. Chem.* **41,** 2480 (1976).
[25] H. Waldmann and H. Kunz, *Liebigs Ann. Chem.* 1712 (1983).
[26] W. Günther and H. Kunz, *Angew. Chem., Int. Ed. Engl.* **29,** 1050 (1990).
[27] R. Albert, J. Danklmaier, H. Hönig, and H. Kandolf, *Synthesis,* 635 (1987).

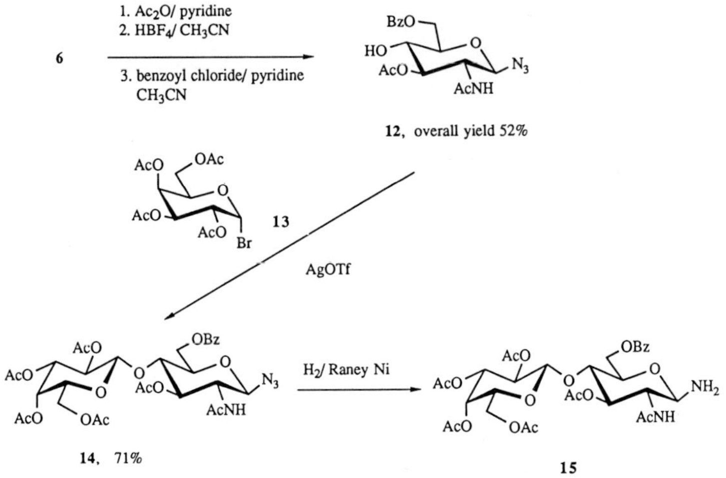

SCHEME 4

of either the amino protecting group or the α-carboxylic ester. Proton-catalyzed reaction of **16** with isobutene gives the asparatic acid diester **17,** which is transformed to the *N*-Fmoc derivative **18.** Alternatively, **18** is obtained from **16** by first introducing the Fmoc group and subsequently forming the *tert*-butyl ester. Hydrogenolysis of the β-benzyl ester of **18** gave the desired *N*-protected aspartic acid α-monoester **19** (Scheme 5).

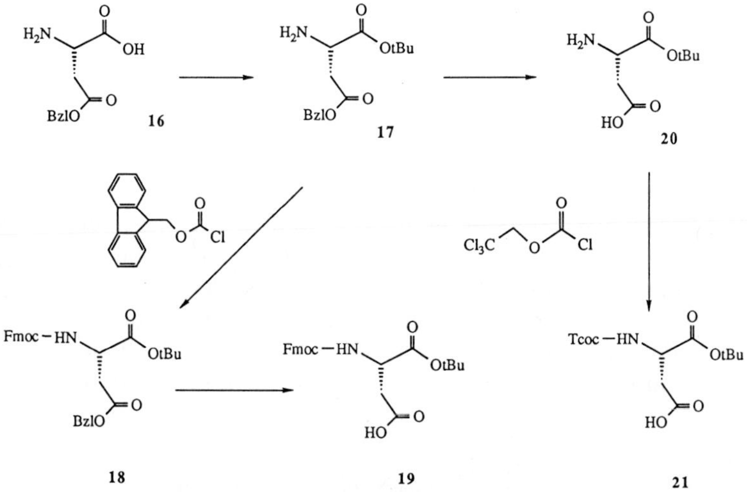

SCHEME 5

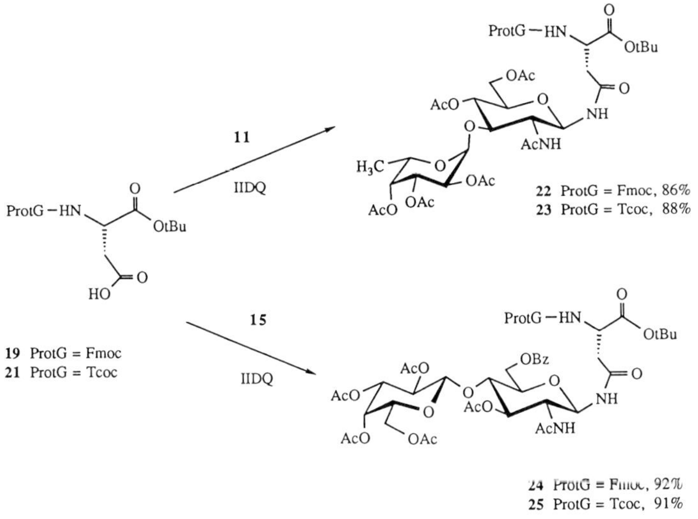

ProtG−HN, OtBu, HO, O

19 ProtG = Fmoc
21 ProtG = Tcoc

11, IIDQ

22 ProtG = Fmoc, 86%
23 ProtG = Tcoc, 88%

15, IIDQ

24 ProtG = Fmoc, 92%
25 ProtG = Tcoc, 91%

SCHEME 6

A variation of this methodology is based on the hydrogenolytic cleavage of the β-benzyl ester of **17** to yield the aspartic acid α-monoester **20,** which is then equipped with suitable N-protection, for example, with the 2,2,2-trichloroethoxycarbonyl (Tcoc) group[20,28] in **21.** Condensation of **19** or **21** with the disaccharide amines **11** or **15** by using isobutyl-2-isobutoxy-3,4-dihydroquinoline 1-carboxylate (IIDQ)[29] furnishes the disaccharide asparagine conjugates **22** and **23** or **24** and **25**, respectively (Scheme 6).

Synthesis of Glycopeptides Containing Disaccharide Side Chains

The selective removal of the Tcoc group from the amino function of the glycosyl asparagine derivatives is exemplified for **23** and **25.** The reductive elimination using zinc in acetic acid proceeds quantitatively and with complete selectivity to give **26** or **27,** respectively. Likewise, the Fmoc group can be removed, for example, from the lactosamine conjugate **24,** to yield **27** using morpholine/dichloromethane with complete selectivity. Alternatively, the acetolytic cleavage of the *tert*-butyl esters (e.g., of **23** and **25**), is accomplished selectively by using trifluoroacetic acid. As *O*-acetylgroups of the saccharide exhibit a marked indirect protection,[17]

[28] T. B. Windholz and D. B. R. Johnston, *Tetrahedron Lett.* 2555 (1967).
[29] Y. Kiso, Y. Kai, and H. Yajima, *Chem. Pharm. Bull.* **21,** 2507 (1973).

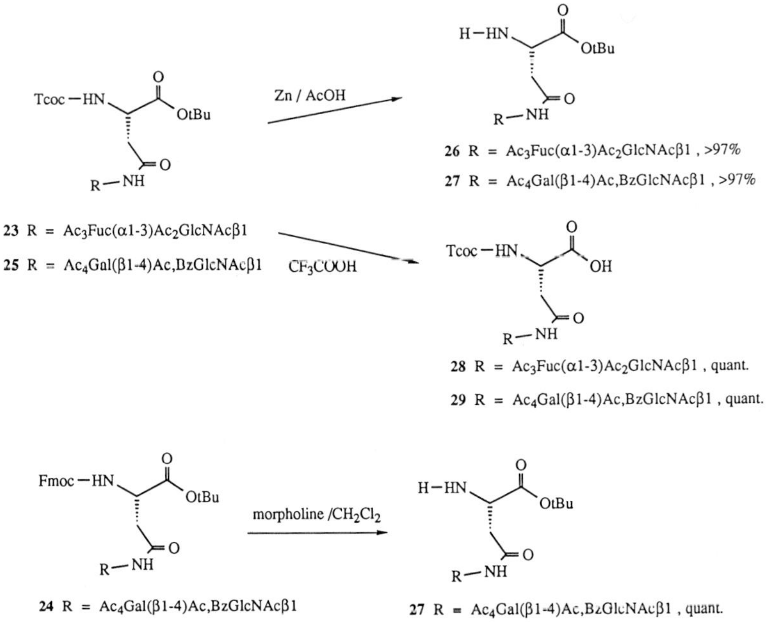

Tcoc—HN, OtBu, R—NH

Zn / AcOH

H—HN, OtBu, R—NH

26 R = $Ac_3Fuc(\alpha 1\text{-}3)Ac_2GlcNAc\beta 1$, >97%

27 R = $Ac_4Gal(\beta 1\text{-}4)Ac,BzGlcNAc\beta 1$, >97%

23 R = $Ac_3Fuc(\alpha 1\text{-}3)Ac_2GlcNAc\beta 1$

25 R = $Ac_4Gal(\beta 1\text{-}4)Ac,BzGlcNAc\beta 1$

CF_3COOH

Tcoc—HN, OH, R—NH

28 R = $Ac_3Fuc(\alpha 1\text{-}3)Ac_2GlcNAc\beta 1$, quant.

29 R = $Ac_4Gal(\beta 1\text{-}4)Ac,BzGlcNAc\beta 1$, quant.

Fmoc—HN, OtBu, R—NH

morpholine /CH_2Cl_2

H—HN, OtBu, R—NH

24 R = $Ac_4Gal(\beta 1\text{-}4)Ac,BzGlcNAc\beta 1$

27 R = $Ac_4Gal(\beta 1\text{-}4)Ac,BzGlcNAc\beta 1$, quant.

SCHEME 7

the intersaccharide bonds of compounds **28** and **29** remained unaffected (Scheme 7).

Applying this methodology, we have synthesized glycopeptides which contain two saccharide side chains. Condensation of either **26** or **27** with Tcoc glycine gives the glycodipeptide **30** or **31,** respectively. Removal of the Tcoc group with zinc/acetic acid forms the amino-deblocked compounds **32** and **33.** Condensation of **32** with **28** furnishes the glycotripeptide **34,** from which the Tcoc group is removed once more by reductive elimination. N-acetylation of the product **35** gives **36** which is C-terminally deblocked by acidolysis using formic acid. Condensation of the obtained carboxyl component **37** with the conventionally synthesized tripeptide *tert*-butyl ester **38** by applying *O*(1-benzotriazolyl)-*N,N,N',N'*-tetramethyluronium tetrafluoroborate (TBTU)[30] yields the protected glycohexapeptide **39** which carries two α-fucosyl(1-3)-glucosamine side chains (Scheme 8).

Analogously, the amino-deblocked lactosamine dipeptide ester **33** is

[30] R. Knorr, A. Trzeciak, W. Bannwarth, and D. Gillessen, *Tetrahedron Lett.* **30,** 1927 (1989).

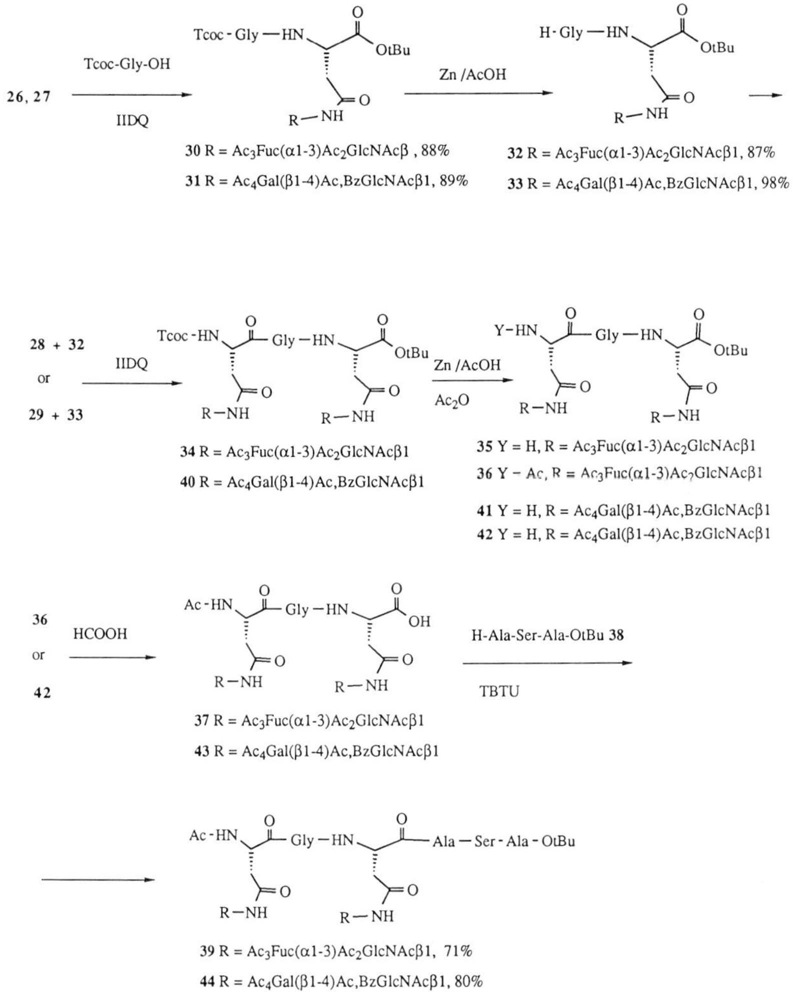

26, 27 → (Tcoc-Gly-OH, IIDQ) → Tcoc-Gly—HN…OtBu → (Zn/AcOH) → H-Gly—HN…OtBu →

30 R = $Ac_3Fuc(\alpha 1\text{-}3)Ac_2GlcNAc\beta$, 88%

31 R = $Ac_4Gal(\beta 1\text{-}4)Ac,BzGlcNAc\beta 1$, 89%

32 R = $Ac_3Fuc(\alpha 1\text{-}3)Ac_2GlcNAc\beta 1$, 87%

33 R = $Ac_4Gal(\beta 1\text{-}4)Ac,BzGlcNAc\beta 1$, 98%

28 + **32** or **29** + **33** → (IIDQ) → Tcoc-HN…Gly—HN…OtBu → (Zn/AcOH, Ac_2O) → Y-HN…Gly—HN…OtBu

34 R = $Ac_3Fuc(\alpha 1\text{-}3)Ac_2GlcNAc\beta 1$

40 R = $Ac_4Gal(\beta 1\text{-}4)Ac,BzGlcNAc\beta 1$

35 Y = H, R = $Ac_3Fuc(\alpha 1\text{-}3)Ac_2GlcNAc\beta 1$

36 Y = Ac, R = $Ac_3Fuc(\alpha 1\text{-}3)Ac_2GlcNAc\beta 1$

41 Y = H, R = $Ac_4Gal(\beta 1\text{-}4)Ac,BzGlcNAc\beta 1$

42 Y = H, R = $Ac_4Gal(\beta 1\text{-}4)Ac,BzGlcNAc\beta 1$

36 or **42** → (HCOOH) → Ac-HN…Gly—HN…OH → (H-Ala-Ser-Ala-OtBu **38**, TBTU) →

37 R = $Ac_3Fuc(\alpha 1\text{-}3)Ac_2GlcNAc\beta 1$

43 R = $Ac_4Gal(\beta 1\text{-}4)Ac,BzGlcNAc\beta 1$

→ Ac-HN…Gly—HN…Ala—Ser-Ala-OtBu

39 R = $Ac_3Fuc(\alpha 1\text{-}3)Ac_2GlcNAc\beta 1$, 71%

44 R = $Ac_4Gal(\beta 1\text{-}4)Ac,BzGlcNAc\beta 1$, 80%

SCHEME 8

condensed with the α-carboxyl deblocked asparagine conjugate **29** to yield the glycotripeptide **40.** Reductive removal of the Tcoc group from **40** and subsequent N-terminal acetylation of **41** deliver compound **42.** Cleavage of the *tert*-butyl ester by formic acid and condensation of the obtained compound **43** with the tripeptide ester **38** furnish the glycohexapeptide **44** containing two neolactolactosamine side chains.

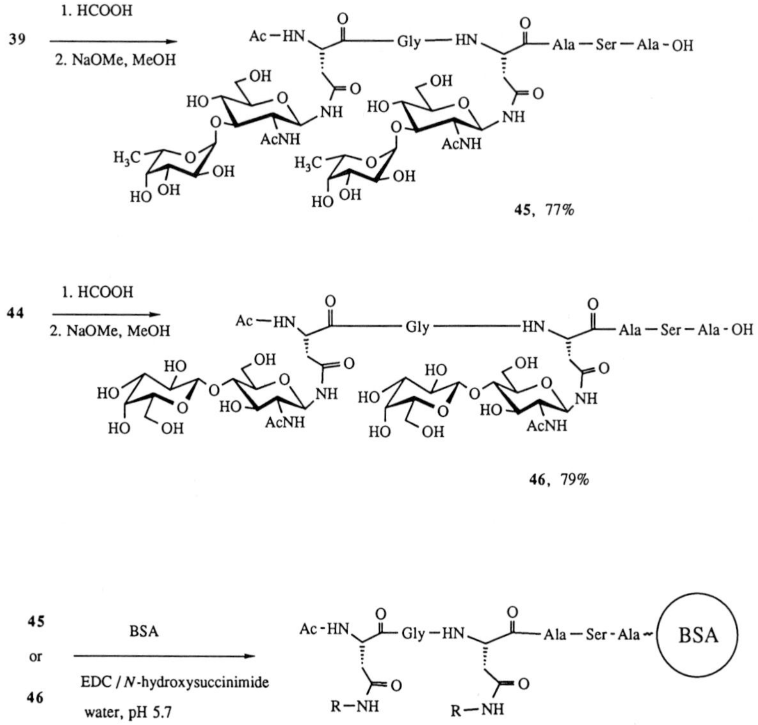

39 → (1. HCOOH; 2. NaOMe, MeOH) → 45, 77%

44 → (1. HCOOH; 2. NaOMe, MeOH) → 46, 79%

45 or 46 → (BSA; EDC / *N*-hydroxysuccinimide; water, pH 5.7) → Ac-HN-Asn(R-NH)-Gly-HN-Asn(R-NH)-Ala-Ser-Ala-BSA

47 R = Fuc(α1-3)GlcNAcβ1, 24 molecules/ molecule BSA

48 R = Gal(β1-4)GlcNAcβ1, 27 molecules/ molecule BSA

SCHEME 9

Deblocking of Glycopeptides and Coupling to Bovine Serum Albumin

Deblocking of the synthesized model glycopeptides **39** and **44** is carried out by first cleaving the C-terminal *tert*-butyl ester and subsequently using base-catalyzed methanolysis of the ester groups of the carbohydrate portions to yield **45** and **46** (Scheme 9).

Coupling of the synthetic glycopeptides to BSA is carried out in water at room temperature by application of methods from peptide chemistry.[31,32] The carboxylic group of **45** or **46**, respectively, is activated using EDC[12]

[31] C. P. Stowell and Y. C. Lee, *Adv. Carbohydr. Chem. Biochem.* **37,** 225 (1980).
[32] B. F. Erlanger, this series, Vol. 70, p. 85.

in the presence of *N*-hydroxysuccinimide.[33] The solution is kept at pH 5.7–5.8. After a reaction time of about 1 day the solution is subjected to dialysis. The formed neoglycoproteins, **47** containing the fucosylglucosamine glycopeptide **45** and **48** containing the lactosamine glycopeptide **46,** are isolated by lyophilization. Photometric determination of the carbohydrate content of the synthetic glycoproteins according to the phenol–sulfuric acid method[14] reveals that **47** contains on average 24 molecules of glycopeptide (**45**) per molecule BSA and **48** carries on average 27 molecules of glycopeptide (**46**) per molecule of protein. The O-deacetylated disaccharides obtained from **10** or **14,** respectively, serve as standards in the photometrical analysis.

The glycopeptides **45** and **46** have been obtained as pure compounds. The structures have been elucidated by high-field nuclear magnetic resonance (NMR) spectra and by fast atom bombardment-mass spectrometry (FAB MS). Therefore, the neoglycoproteins obtained from these compounds are useful models for the investigation of their immunological and other biological recognition reactions.

Experimental

General Instrumentation

Thin-layer chromatography (TLC) is carried out on silica gel (E. Merck, Darmstadt, Germany), using three detection systems: (a) UV light (λ 254 nm); (b) ninyhydrin (0.3%) in methanol containing 3% acetic acid; and (c) 1 : 1 mixture of 1 *M* sulfuric acid and a solution of 3-methoxyphenol (0.2%) in ethanol. As solvents are applied either 2 : 1 (v/v) ethyl acetate/light petroleum ether (A) or 50 : 1 (v/v) ethyl acetate/acetic acid (B) unless indicated otherwise. Flash chromatography is performed on silica gel 60 (0.04–0.063 mm; E. Merck). For high-performance liquid chromatography (HPLC), a reversed-phase (RP) silica gel Spherisorb ODS II column (Bischoff, Leonberg, Germany) is used. The NMR spectra are recorded with Bruker AC-200 (200 MHz for ^{1}H; 50.3 MHz for ^{13}C) and Bruker AM 400 (400 MHz for ^{1}H; 100.6 MHz for ^{13}C) spectrometers. Optical rotation values at the sodium D line (589.5 nm) are measured on a Perkin-Elmer (Norwalk, CT) polarimeter, Model 241. The FAB mass spectra are recorded on a MAT 95 mass spectrometer (Finnigan, Bremen, Germany) by application of 3-nitrobenzyl alcohol (3-NOBA) or glycerol as the matrix. Melting points are uncorrected.

[33] E. Wünsch and F. Drees, *Chem. Ber.* **99,** 110 (1966).

Synthetic Procedures

2-Acetamido-4,6-O-benzylidene-2-deoxy-β-D-glucopyranosylazide (**6**). To a solution of 2-acetamido-2-deoxy-3,4,6-tri-*O*-acetyl-β-D-glucopyranosylazide (18.9 g, 50 mmol), obtained from the corresponding glycosyl chloride and sodium azide under phase-transfer catalysis,[18] in dry methanol (150 ml) is added ion-exchange resin Amberlyst A26 (OH^- form, 5 ml). The mixture is shaken for 4 hr and filtered. The resin is washed with methanol. The combined methanolic solutions are evaporated to dryness. The remaining 2-acetamido-2-deoxy-β-D-glucopyranosyl azide is recrystallized from methanol–ethyl acetate: yield, quantitative; mp 149°–150°; $[\alpha]_D^{20}$ −26.9° (*c* 1, water); literature[19] mp 146°–147°; $[\alpha]_D^{20}$ −45.0° (*c* 0.6, water). To a stirred mixture of this product (12.3 g, 50 mmol) and *p*-toluene sulfonic acid (0.95 g, 5 mmol) is added benzaldehyde dimethyl acetal (38 ml, 0.25 mol). After 10–15 min the mixture congeals to a gel. After 1 hr, 1 : 1 (v/v) light petroleum ether/diethyl ether (200 ml) is added, and the solid is suspended. Triethylamine (0.7 ml, 50 mmol) is added to the suspension. After filtration, the solid is washed with diethyl ether, dried *in vacuo*, dissolved in methanol, and added to silica gel (0.04–0.2 mm, 50 g). The loaded silica gel is dried under high vacuum applied to a column filled with another 50 g of silica gel, extensively washed with 1 : 1 (v/v) light petroleum ether/ethanol, and eluted with 1 : 1 (v/v) ethyl acetate/ethanol. The yield is 16.3 g (97%); mp 216° (literature[34] mp 214°–216°, $[\alpha]_D^{20}$ −101.3° (*c* 1, CH_3OH); R_f 0.37 (ethyl acetate).

2-Acetamido-4,6-O-benzylidene-2-deoxy-3-O-[2,3,4-tri-O-(4-methoxybenzyl)-α-L-fucopyranosyl]-β-D-glucopyranosyl azide (**7**). To a mixture of molecular sieves (3 Å, 10 g), ethyl 2,3,4-tri-*O*-(4-methoxybenzyl)-1-thio-β-L-fucopyranoside[20] (5.7 g, 10 mmol), *N*-acetylglucosaminylazide **6** (2.5 g, 7.5 mmol), and tetraethylammonium bromide is added 1 : 1 (v/v) dimethylformamide/dichloromethane (80 ml). After the suspension is stirred for 15 min at 20°, $CuBr_2$[35] (2.7 g, 12 mmol, dried under high vacuum) is added. The mixture is stirred at room temperature for 5 hr, then filtered through Celite, and the residue is washed with dichloromethane. The combined solutions are washed with saturated $NaHCO_3$ solution, then with water, dried with $MgSO_4$, and evaporated *in vacuo*. The residue is purified by flash chromatography in 3 : 1 (v/v) light petroleum ether/acetate: yield, 6.3 g (quantitative); mp 82°; $[\alpha]_D^{20}$ −132.6° (*c* 1, CH_3OH); R_f 0.66 (A).

Analysis: Calculated for $C_{45}H_{52}N_4O_{12}$ (840.9): C 64.27, H 6.23, N 6.66. Found: C 64.26, H 6.33, N 6.76.

[34] M. A. E. Shaban and R. W. Jeanloz, *Bull. Chem. Soc. Jpn.* **54,** 3570 (1981).
[35] S. Sato, M. Mori, Y. Ito, and T. Ogawa, *Carbohydr. Res.* **155,** C 6 (1986).

400 MHz ^1H NMR ($CDCl_3$): δ 0.87 (d, $J_{6,5}$ = 6.5 Hz, H-6, Fuc); 3.51–3.59 (m, 4H, H-2, H-5, GlcNAc, H-4, H-5, Fuc); 3.77, 3.78, 3.80 (3s, 9H, CH_3O—); 4.01 (dd, $J_{2,3}$ = 10.1 Hz, $J_{2,1}$ = 3.5 Hz, 1H, H-2, Fuc); 4.07 (t, $J_{3,4}$ = $J_{3,2}$ = 9.5 Hz, 1H, H-3, GlcNAc); 4.75 (d, $J_{1,2}$ = 9.1 Hz, 1H, H-1, GlcNAc); 5.03 (d, $J_{1,2}$ = 3.5 Hz, 1H, H-1, Fuc).

100.6 MHz ^{13}C NMR ($CDCl_3$): δ (ppm) 56.77 (C-2, GlcNAc); 67.27 (C-5, Fuc); 88.94 (C-1, GlcNAc); 98.93 (C-1, Fuc).

*2-Acetamido-4,6-O-benzylidene-2-deoxy-3-O-(2,3,4-tri-O-acetyl-α-L-fucopyranosyl)-β-D-glucopyranosylazide (**8**).* To **7** (4.2 g, 5 mmol) dissolved in 9 : 1 (v/v) aetonitrile/water (100 ml) is added dry ceric ammonium nitrate (16.5 g, 30 mmol). The orange-colored solution is stirred at room temperature until **7** is no longer detectable by TLC. After addition of saturated $NaHCO_3$ solution, the solvent is evaporated *in vacuo;* the remaining residue is dried by codistillation with toluene and finally dried under high vacuum. The residue is dissolved in pyridine (100 ml). Acetic acid anhydride (150 ml) is added dropwise, and the mixture is stirred at room temperature for 15 hr. After evaporation of reagents and solvent *in vacuo,* the crude **8** is dissolved in ethyl acetate (200 ml), washed with 0.5 *M* HCl, saturated $NaHCO_3$ solution, and water, dried with $MgSO_4$, and the solution again evaporated *in vacuo*. The residue is purified by flash chromatography in 3 : 1 (v/v) light petroleum ether/ethyl acetate. The yield is 2.5 g (82%); MP 195°; $[\alpha]_D^{20}$ −110.4° (*C* = 1, CH_3OH); R_f 0.20 (A).

Analysis: Calculated for $C_{27}H_{34}N_4O_{12}$ (606.6): C 53.46, H 5.65, N 9.24. Found: 53.32, H 5.70, N 9.04.

400 MHz ^1H NMR ($CDCl_3$): δ 0.42 (d, $J_{6,5}$ = 4.7 Hz, 3H, H-6, Fuc); 3.61 (q, $J_{2,3}$ = $J_{2,1}$ = $J_{2,NH}$ = 9.1 Hz, 1H, H-2, GlcNAc) 4.29 (q, $J_{5,6}$ = 6.2 Hz, 1H, H-5, Fuc); 4.75 (d, $J_{1,2}$ = 9.3 Hz, 1H, H-1, GlcNAc); 4.89 (dd, $J_{3,4}$ = 3.9 Hz, $J_{3,2}$ = 11.0 Hz, 1H, H-3, Fuc); 5.20 (d, $J_{1,2}$ = 2.9 Hz, 1H, H-1, Fuc); 5.22 (d, $J_{4,3}$ = 3.9 Hz, 1H, H-4, Fuc).

*2-Acetamido-2-deoxy-3-O-(2,3,4-tri-O-acetyl-α-L-fucopyranosyl)-β-D-glucopyranosylazide (**9**).* To a solution of **8** (3.6 g, 6 mmol) in acetonitrile (75 ml) is added aqueous tetrafluorohydroboric acid (48%, 2 ml). After the mixture is stirred at room temperature for 90 min, tri-*n*-butylamine (7.2 ml) is added and the solvent evaporated *in vacuo*. The remaining oil is purified by flash chromatography in 2 : 1 (v/v) ethyl acetate/light petroleum ether. The yield is 2.55 g (77%); mp 176°; $[\alpha]_D^{20}$ −130.8° (*c* 1, CH_3OH); R_f 0.44 (B).

Analysis: Calculated for $C_{20}H_{30}N_4O_{12} \cdot 2H_2O$ (554.5): C 43.32, H 6.18, N 10.10. Found: C 43.13, H 5.71, N 9.93.

400 MHz ^1H NMR (DMSO-d_6): δ 0.96 (d, $J_{6,5}$ = 6.4 Hz, 3H, H-6, Fuc); 1.75, 1.92, 1.98, 2.10 (4s, 12H, NAc, OAc), 3.30–3.41 (m, 2H,

H-2, H-4, GlcNAc); 4.73 (m, 1H, H-5, Fuc); 4.76 (d, $J_{1,2}$ = 8.7Hz, 1H, H-1, GlcNAc); 4.82 (dd, $J_{2,3}$ = 11.1 Hz, $J_{2,1}$ = 3.7 Hz, 1H,H-2, Fuc); 5.21 (d, $J_{1,2}$ = 4.1 Hz, 1H, H-1, Fuc).

50.3 MHz ^{13}C NMR ($CDCl_3$): δ 15.5 (C-6, Fuc); 56.14 (C-2, GlcNAc); 63.79 (C-5, Fuc); 88.18 (C-1, GlcNAc); 94.99(C-1, Fuc).

*2-Acetamido-4,6-di-O-acetyl-2-deoxy-3-O-(2,3,4-tri-O-acetyl-α-L-fucopyranosyl)-β-D-glucopyranosylazide (**10**).* To a solution of **9** (1.66 g, 3 mmol) in pyridine (50 ml) at 0° is added dropwise acetic acid anhydride (75 ml). The mixture is stirred at room temperature for 12 hr and then evaporated *in vacuo*. The residue is dissolved in ethyl acetate (100 ml), then washed with 1 *M* HCl, saturated $NaHCO_3$ solution, and water. After drying with $MgSO_4$, the solvent is evaporated *in vacuo,* and the product is purified by flash chromatography in 3 : 1 (v/v) light petroleum ether/ethyl acetate: yield, 1.73 g (96%); amorphous; $[\alpha]_D^{20}$ −110.4° (*c* 1, CH_3OH); R_f 0.20 (A).

Analysis: Calculated for $C_{24}H_{34}N_4O_{14}$ (602.6): C 47.84, H 5.69, N 9.30. Found: C 47.81, H 5.57, N 9.07.

400 MHz ^1H NMR ($CDCl_3$): δ 1.04 (d, $J_{6,5}$ = 6.5 Hz, 3H, H-6, Fuc); 1.97 (s, 3H, NAc); 1.93, 2.03, 2.05, 2.07, 2.10 (5s, 15H, OAc); 4.10 (dq, $J_{5,6}$ = 6.3 Hz, $J_{5,4}$ = 1.1 Hz, 1H, H-5, Fuc); 4.94 (dd, $J_{4,3}$ = 9.3 Hz, $J_{4,5}$ = 9.6 Hz, 1H, H-4, GlcNAc); 5.05 (d, $J_{1,2}$ = 9.5 Hz, 1H, H-1, GlcNAc); 5.19 (dd, $J_{4,3}$ = 3.3 Hz, $J_{4,5}$ = 1.2 Hz, 1H, H-4, Fuc); 5.21 (d, $J_{1,2}$ = 3.5 Hz, 1H, H-1, Fuc).

*2-Acetamido-3-O-acetyl-6-O-benzoyl-2-deoxy-β-D-glucopyranosylazide (**12**).* To a stirred solution of **6** (16.7 g, 50 mmol) in pyridine (100 ml) at 0° is added acetic acid anhydride (50 ml). The mixture freezes after some minutes and is kept for 16 hr at room temperature, after which the volatile components are evaporated *in vacuo*. Toluene (2 times 200 ml) is distilled from the residue, which then is dissolved in chloroform (100 ml), washed with 1 *M* HCl and saturated. $NaHCO_3$ solution, and dried with $MgSO_4$. After evaporation *in vacuo,* the residue is recrystallized from diethyl ether to give 2-acetamido-3-*O*-acetyl-4,6-benzylidene-2-deoxy-β-D-glucopyranosyl azide; yield, 17.0 g (90%); mp 202°–203°; $[\alpha]_D^{20}$ −116.8° (*c* 1; $CHCl_3$).

To a solution of this compound (15 g, 40 mmol) in acetonitrile (200 ml) is added aqueous tetrafluorohydroboric acid (48%, 6 ml). The clear mixture is stirred for 3 hr. Aqueous H_2O_2 (30%, 5 ml) is added, and stirring was continued for additional 90 min at room temperature. After addition of tri-*n*-butylamine (24 ml, 0.1 mol), the solvent is immediately evaporated *in vacuo*. The remaining reddish-brown oil is extensively extracted with diethyl ether (twice, 300 ml). The remainder is purified by flash chromatography in ethyl acetate to yield 2-acetamido-3-*O*-acetyl-2-deoxy-β-D-glucopyranosyl azide (10.5 g, 91%); mp 153°; $[\alpha]_D^{20}$ −67.9° (*c* 1, CH_3OH). (For comparison with the corresponding 3-*O*-monochloroacetyl analog, see Unverzagt and Kunz.[18]

The obtained compound (8.6 g, 30 mmol) is dissolved in acetonitrile (100 ml). At 0°, a freshly prepared solution of pyridine (2.4 ml, 30 mmol) and benzoyl chloride (4 ml, 32 mmol) in dry acetonitrile (25 ml) is added dropwise. After 30 min at 0°, the mixture is poured into cold water (600 ml). The precipitated product is isolated by filtration, washed with ice-cold water, and dissolved in ethyl acetate. Additionally, the filtered water solution is extracted with ethyl acetate (150 ml). The combined organic solutions are dried with $MgSO_4$. After evaporation *in vacuo,* the residue is purified by flash chromatography in 1 : 1 (v/v) light petroleum ether/ethyl acetate to give **12.** The overall yield is 7.5 g (52%); mp 185°; $[\alpha]_D^{20}$ −36.4° (*c* 1, CH_3OH); R_f 0.40 (ethyl acetate).

Analysis: Calculated for $C_{17}H_{20}N_4O_7$ (392.4): C 52.04, H 5.14, N 14.28. Found: C 52.02, H 5.15, N 14.19.

200 MHz ^1H NMR($CDCl_3$): δ 1.68 (s, 3H, NAc); 1.82 (s, 3H, OAc); 2.70 (1H, 4-OH); 4.50 (d, $J_{1,2}$ = 9.3 Hz, 1H, H-1); 4.81 (dd, $J_{3,4}$ = 8.7 Hz, $J_{3,2}$ = 10.2 Hz, 1H, H-3); 7.79 (dd, $J_{o,m}$ = 8.3 Hz, ^4J = 1.3 Hz, 2H, Ar-H_{ortho}).

*2-Acetamido-3-O-acetyl-6-O-benzoyl-2-deoxy-4-O-(2,3,4,6-tetra-O-acetyl-β-D-galactopyranosyl)-β-D-glucopyranosyl azide (**14**).* To a solution of **12** (3.9 g, 10 mmol) in dry dichloromethane (100 ml) are added dried molecular sieves (3 Å, 10 g) and 2,3,4,6-tetra-*O*-acetyl-*α*-D-galactopyranosyl bromide[23] (**13**) (5.1 g, 12.5 mmol). After being stirred for 15 min, the mixture is cooled to −10°. A solution of silver trifluoromethane sulfonate (3.0 g, 11.7 mmol) in dry toluene (40 ml) is added. After 5 hr the reaction is quenched by addition of saturated $NaHCO_3$ solution (100 ml). The mixture is filtered through Celite, and the organic layer is separated, diluted with dichloromethane (100 ml), washed with saturated $NaHCO_3$ solution and water, dried with $MgSO_4$, and evaporated *in vacuo*. The residue (*α*,*β* mixture, 6.4 g, 89%) is purified by flash chromatography in light petroleum ether/ethyl acetate to give **14,** which is recrystallized from dichloromethane/light petroleum ether. The yield is 5.1 g (71%); mp 107°–108°; $[\alpha]_D^{22}$ −6.1° (*c* 1, CH_3OH); R_f 0.22 (A).

Analysis: Calculated for $C_{31}H_{38}N_4O_{16}$ (722.7): C 51.52, H 5.30, N 7.75; Found: C 51.41, H 5.46, N 7.61.

400 MHz ^1H NMR($CDCl_3$): δ 3.77 (q, $J_{2,3}$ = $J_{2,1}$ = $J_{2,NH}$ = 8.0 Hz, 1H, H-2, GlcNAc); 4.51 (d, $J_{1,2}$ = 8.0 Hz, 1H, H-1, GlcNAc); 4.57 (d, $J_{1,2}$ = 9.0 Hz, 1H, H-1, Gal); 4.87 (dd, $J_{3,4}$ = 3.4 Hz, $J_{3,2}$ = 10.5 Hz, 1H, H-3, Gal); 5.05 (dd, $J_{3,4}$ = 10.1 Hz, 1H, H-3, GlcNAc); 5.29(d, $J_{4,3}$ = 3.3 Hz, 1H, H-4, Gal); 8.01 (d, $J_{o,m}$ = 8.1 Hz, 2H, Ar-H_{ortho}).

100.6 MHz ^{13}C NMR ($CDCl_3$): δ 53.12 (C-2, GlcNAc); 60.72 (C-6, GlcNAc); 62.32 (C-6, Gal); 88.56 (C-1, GlcNAc); 100.99 (C-1, Gal); 165.90, 169.97, 170.21, 170.36, 170.89 (CH_3*C*OO—, Ph*C*OO—).

Hydrogenolysis of Glycosylazides: General Procedure

To a solution of the glycosylazide **10** (10 mmol) in ethanol (100 ml), or **14** in 2-propanol (100 ml), is added Raney nickel (1 g) which has been extensively washed with water and, subsequently, with 2-propanol. The hydrogenation is carried under atmospheric pressure (TLC monitoring, reaction time 1–4 hr). After filtration, the solution is concentrated *in vacuo* (bath temperature maximum 40°). The remaining residue of 2-acetamido-4,6-di-*O*-acetyl-2-deoxy-3-*O*-(2,3,4-tri-*O*-acetyl-α-L-fucopyranosyl)β-D-gloucopyranosylamine (**11**) or 2-acetamido-3-*O*-acetyl-6-*O*-benzoyl-4-*O*(2,3,4,6-tetra-*O*-acetyl-β-D-galactopyranosyl)-β-D-glucopyranosylamine (**15**), respectively, is dissolved in dichloromethane (50 ml) and immediately used for further transformations.

4-*O*-Benzyl aspartate (**16**) is synthesized according to the procedure of Albert *et al.*[27] 4-*O*-Benzyl-1-*O*-*tert*-butyl aspartate (**17**) and 1-*O*-*tert*-butyl aspartate (**20**) are synthesized according to the procedure of Yang and Merrifield.[36]

N-(9-Fluorenylmethoxycarbonyl)-4-*O*-benzyl-l-*O*-*tert*-butyl aspartate (**18**) is synthesized from **17** according to the general procedure for Fmoc amino acid esters given by Carpino and Han[37]: yield, 89%; mp 73°–74°; $[\alpha]_D^{25}$ −5.4° (*c* 1, CH_3OH). *N*-9-Fluorenylmethoxycarbonylaspartic acid α-*tert*-butyl ester (**19**) is synthesized from **18** by hydrogenolysis analogously to described procedures[37,38]; yield, 84%; amorphous; $[\alpha]_D^{25}$ −15.5° (*c* 1, CH_3OH). *N*-2,2,2-Trichloroethoxycarbonylaspartic acid α-*tert*-butyl ester (**21**) is synthesized from **20** analogously to a general procedure described by Windholz and Johnston[28]: yield, 88%; mp 96°–98°; $[\alpha]_D^{20}$ −20.5° (*c* 1, CH_3OH).

Synthesis of Disaccharide Asparagine Conjugates: General Procedure

To a solution of the N-protected aspartic acid α-*tert*-butyl ester **19** or **21** (12 mmol), in dichloromethane (50 ml) are added 5 ml (17 mmol) of IIDQ[29]. After stirring for 5 min, the freshly prepared solution of the

[36] C. C. Yang and R. B. Merrifield, *J. Org. Chem.* **41,** 1032 (1976).

[37] L. A. Carpino and G. Y. Han, *J. Org. Chem.* **38,** 4218 (1973).

[38] C.-D. Chang, W. Waki, M. Ahmad, J. Meienhofer, E. O. Lundell, and J. D. Hang, *Int. J. Pept. Protein Res.* **15,** 59 (1980).

glycosylamine **11** or **15**, respectively (obtained from 10 mmol of the corresponding glycosylazide, see above), is added. The mixture is stirred until the amine is completely consumed (TLC monitoring, 8–24 hr). After evaporation of the solvent *in vacuo*, the crude product is purified by flash chromatography (150 g of silica gel) in 4 : 1 (v/v) light petroleum ether/ethyl acetate to give the conjugates **22, 23, 24,** or **25,** respectively.

For N^2-(9-fluoroenylmethoxycarbonyl)-N^4-[2-acetamido-4,6-di-*O*-acetyl-2-deoxy-3-*O*-(2,3,4-tri-*O*-acetyl-α-L-fucopyranosyl)-β-D-glucopyranosyl]asparagine *tert*-butyl ester (**22**), the reaction time is 24 hr: yield, 8.6 g (86%); mp 132°; $[\alpha]_D^{25}$ $-41.0°$ (c 1, CH_3OH); R_f 0.68 (B).

Analysis: Calculated for $C_{47}H_{59}N_3O_{19} \cdot 1.5H_2O$ (997.0): C 56.62, H 6.27, N 4.21. Found: C 56.54, H 6.60, N 4.54.

400 MHz ^1H NMR[^1H–^1H-correlated spectroscopy (COSY)] ($CDCl_3$): δ 1.04 (d, $J_{6,5}$ = 65 Hz, 3H, H-6, Fuc); 2.66 (dd, $J_{\beta,a}$ = 4.2 Hz, J_{gem} = 15.3 Hz, 1H, Asn-β-H_a); 2.74 (dd, $J_{\beta,\alpha}$ = 4.6 Hz, 1H, Asn-β-H_b); 4.13 (q, $J_{5,6}$ = 6.6 Hz, 1H, H-5, Fuc); 4.96 (t, $J_{1,2}$ = $J_{1,NH}$ = 9.0 Hz, 1H, H-1, GlcNAc); 5.02 (t, $J_{4,3}$ = $J_{4,5}$= 9.5 Hz, 1H, H-4, GlcNAc); 5.16 (d, $J_{1,2}$ = 3.6 Hz, 1H, H-1, Fuc); 5.18 (d, $J_{4,3}$ = 3.6 Hz, 1H, H4, Fuc) 7.58 (d, $J_{NH,\beta}$ = 11.4 Hz, 1H, Asn-NH_α); 7.70 (d, J_{vic} = 7.5 Hz, 2H, H-4, H-5, Fmoc).

100.6 MHz ^{13}C NMR ($CDCl_3$): δ 15.42 (C-6, Fuc); 37.95 (Asn-β-C); 47.08 (C-9, Fmoc); 51.05 (Asn-α-C); 54.15 (C-2, GlcNAc); 61.87 (C-6, GlcNAc); 65.96 (C-5, Fuc); 79.97 (C-1, GlcNAc); 82.09 [$C(CH_3)_3$]; 97.68 (C-1, Fuc).

For N^2-(2,2,2-trichloroethoxycarbonyl)-N^4-[2-acetamido-4,6-di-*O*-acetyl-2-deoxy-3-*O*-(2,3,4-tri-*O*-acetyl-α-L-fucopyranosyl)-β-D-glucopyranosyl]asparagine *tert*-butyl ester (**23**), the reaction time is 8 hr: yield, 8.1 g (88%); amorphous $[\alpha]_D^{25}$ $-53.1°$ (c 1, CH_3OH); R_f 0.75 (B).

Analysis: Calculated for $C_{35}H_{50}N_3O_{19}Cl_3$ (923.1): C 45.54, H 5.46, N 4.55, Cl 11.52. Found: C 45.44, H 5.49, N 4.46, Cl 11.25.

400 MHz ^1H NMR ($CDCl_3$): δ 1.04 (d, $J_{6,5}$ = 6.5 Hz, 3H, H-6, Fuc); 1.39 [s, 9H, $C(CH_3)_3$]; 2.68 (dd, $J_{\beta,\alpha}$ = 4.4 Hz, J_{gem} = 16.5 Hz, 1H, Asn-β-H_a); 2.82 (dd, $J_{\beta,\alpha}$ = 4.8 Hz, 1H, Asn-β-H_b); 3.97 (q, $J_{2,3}$ = $J_{2,1}$ = $J_{2,NH}$ = 9.9 Hz, 1H, H-2, GlcNAc); 4.13 (q, $J_{5,6}$ = 6.3 Hz, 1H, H-5, Fuc); 4.91 (dd, $J_{1,2}$ = 9.4 Hz, $J_{1,NH}$ = 8.1 Hz, 1H, H-1, GlcNAc); 5.12, (d, $J_{1,2}$ = 3.3 Hz, 1H, H-1, Fuc); 5.18 (dd, $J_{4,3}$ = 3.4 Hz, $J_{4,5}$ = 1.3 Hz, 1H, H-4, Fuc).

100.6 MHz ^{13}C NMR ($CDCl_3$): δ 15.39 (C-6, Fuc); 27.82 [$C(CH_3)_3$]; 37.60 (Asn-β-C); 51.17 (Asn-α-C); 54.14 (C-2, GlcNAc); 80.41 (C-1, GlcNAc); 82.28 [$C(CH_3)_3$]; 95.35 (Cl_3C_{Tcoc}); 98.26 (C-1, Fuc).

For N^2(9-fluorenylmethoxycarbonyl)-N^4-[2-acetamido-3-*O*-acetyl-6-*O*-benzoyl-2-deoxy-4-*O*-(2,3,4,6-tetra-*O*-acetyl-β-D-galactopyranosyl)-

β-D-glucopyranosyl]aspartic acid *tert*-butyl ester (*24*), the reaction time is 24 hr: yield, 10.0 g (92%); mp 137°; $[\alpha]_D^{22}$ −0.6° (*c* 1, CH_3OH); R_f 0.6 (B).

Analysis: Calculated for $C_{54}H_{63}N_3O_{21}$ (1090.1): C 59.50, H 5.82, N 3.85. Found: C 59.37, H 5.92, N 3.85.

400 MHz ^1H NMR ($CDCl_3$): δ 1.39 [s, 9H, $C(CH_3)_3$]; 2.63 (dd, $J_{\beta,\alpha}$ = 3.7 Hz, J_{gem} = 15.1 Hz, 1H, Asn-β-H_a); 2.81 (dd, $J_{\beta,\alpha}$ = 3.4 Hz, 1H, Asn-β-H_b); 3.91 (dd, $J_{4,5}$ = 9.3 Hz, $J_{4,3}$ = 8.9 Hz, 1H, H-4, GlcNAc); 4.48 (m, 1H, Asn-α-H); 4.51 (d, $J_{1,2}$ = 7.9 Hz, 1H, H-1, Gal); 4.85 (dd, $J_{3,4}$ = 3.0 Hz, $J_{3,3}$ = 10.4 Hz, 1H, H-3, Gal); 5.01–5.10 (m, 3H, H-3, H-1, GlcNAc, H-2, Gal); 5.29 (d, 1H, H-4,Gal).

100.6 MHz ^{13}C NMR ($CDCl_3$): δ 27.80 [$C(CH_3)_3$]; 37.92 (Asn-β-C); 47.06 (C-9, Fmoc); 50.96 (Asn-α-C); 53.55 (C-2, GlcNAc); 80.06 (C-1, GlcNAc); 82.11 [$C(CH_3)_3$]; 100.78 (C-1, Gal).

Isolation of N^2-(2,2,2-trichloroethoxycarbonyl)-N^4-[2-acetamido-3-*O*-benzoyl-2-deoxy-4-*O*-(2,3,4,6-tetra-*O*-acetyl-β-D-galactopyranosyl)-β-D-glucopyranosyl]aspartic acid *tert*-butyl ester (**25**) is carried out by dissolution of the crude product in ethyl acetate. This solution is washed with 0.5 *M* HCl, 0.5 *M* $NaHCO_3$ solution, and water. After drying with $MgSO_4$, the solvent is evaporated *in vacuo,* and the remaining residue is recrystallized from ethyl acctate/light petroleum ether. The reaction time is 8 hr: yield, 9.5 g (91%); mp 133°–134°; $[\alpha]_D^{21}$ −0.1° (*c* 1, CH_3OH).

Analysis: Calculated for $C_{42}H_{54}N_3O_{21}Cl_3$ (1043.3): C 48.35, H 5.22, N 4.03, Cl 10.19. Found: C 48.11, H 5.04, N 4.18, Cl 10.02.

400 MHz ^1H NMR ($CDCl_3$): δ 1.39 [s, 9H, $C(CH_3)_3$]; 2,64 (dd, $J_{\alpha,\beta}$ = 4.1 Hz, J_{gem} = 16.4 Hz, 1H, Asn-β-H_a); 2.81 (dd, $J_{\beta,a}$ = 4.4 Hz, Asn-β-H_b); 3.90 (dd, $J_{4,5}$ = 9.3 Hz, $J_{4,3}$ = 9.0 Hz, 1H, H-4, GlcNAc); 4.5 (d, $J_{1,2}$ = 7.8 Hz, 1H, H-1, Gal); 4.84 (dd, $J_{3,4}$ = 3.2 Hz, $J_{3,2}$ = 10.4 Hz, 1H, H-3, Gal); 4.98–5.09 (m, 3H, H-3, H-1, GlcNAc, H-2, Gal); 5.28 (d, $J_{4,3}$ = 2.7 Hz, 1H, H-4, Gal).

100.6 MHz ^{13}C NMR ($CDCl_3$): δ 20.38, 20.50, 20.74, 20.85 (CH_3COO); 23.03 (CH_3CO-NH); 27.81 [$C(CH_3)_3$]; 37.54 (Asn-β-C); 51.13 (Asn-α-C); 53.58 (C-2, GlcNAc); 60.79 (C-6, GlcNAc); 62.34 (C-6, Gal); 74.62 (CH_2O_{Tcoc}); 75.17 (C-3, GlcNAc); 80.10 (C-1, GlcNAc); 82.36 [$C(CH_3)_3$]; 95.32 (Cl_3C_{Tcoc}); 100.80 (C-1, Gal), 154.35 (urethane).

Reductive Removal of N-(2,2,2-Trichloroethoxycarbonyl) Group from Glycopeptides: General Procedure

To a solution of 3 mmol of the Tcoc-protected asparagine conjugate **23** or **25,** or the Tcoc–glycopeptide **30** or **31** (see below), in 20–25 ml of acetic acid is added zinc powder (3 g) which has been activated by a short treatment with dilute HCl. After 3 h, the zinc is filtered off. The acetic acid is evaporated *in vacuo*, and the remaining residue is distributed between ethyl acetate (70 ml) and an aqueous mixture obtained from 70 ml of 10% aqueous EDTA (disodium ethylenediaminetetraacetate) and saturated $NaHCO_3$ solution (35 ml). The aqueous layer should not drop below pH 7.5. The organic layer is separated, dried with $MgSO_4$, and evaporated *in vacuo*. Recrystallization from ethyl acetate/light petroleum ether yields the deblocked compounds **26, 27, 32,** or **33,** respectively, as colorless crystals, which are used for the condensation reactions.

For N^4-[2-acetamido-4,6-di-*O*-acetyl-2-deoxy-3-*O*-(2,3,4-tri-*O*-acetyl-α-L-fucopyranosyl)-β-D-glucopyranosyl]asparagine *tert*-butyl ester (**26**), the yield is 97%; mp 109°–111°; $[\alpha]_D^{25}$ −57.9° (*c* 1, CH_3OH) R_f 0.72 (CH_3OH). The yield of N^4-[2-acetamido-3-*O*-acetyl-6-*O*-benzoyl-2-deoxy-4-*O*-(2,3,4,6-tetra-*O*-acetyl-β-D-galactopyranosyl)-β-D-glucopyranosyl]asparagine *tert*-butyl ester (**27**) is 97%; mp 121°; $[\alpha]_D^{22}$ +6.9° (*c* 1.5, CH_3OH); R_f 0.73 (CH_3OH). For glycyl-N^4-[2-acetamido-4,6-di-*O*-acetyl-2-deoxy-3-*O*-(2,3,4-tri-*O*-acetyl-α-L-fucopyranosyl)-β-D-glucopyranosyl]asparagine *tert*-butyl ester (**32**) the yield is 87%; mp 77°–81°; $[\alpha]_D^{25}$ −47.4° (*c* 1, CH_3OH); R_f 0.49 (CH_3OH). Glycyl-N^4-[2-acetamido-3-*O*-acetyl-6-*O*-benzoyl-2-deoxy-4-*O*-(2,3,4,6-tetra-*O*-acetyl-β-D-galactopyranosyl)-β-D-glucopyranosyl]asparagine *tert*-butyl ester (**33**) is obtained in 98% yield; mp 131°–134°; $[\alpha]_D^{21}$ +8.9° (*c* 1, CH_3OH); R_f 0.43 (CH_3OH).

*Removal of 9-Fluorenylmethoxycarbonyl Group: General Procedure Exemplified for **24***

The Fmoc-(lactosamine)asparagine ester **24** (5.5 g, 5 mmol) is dissolved in 1 : 1 (v/v) morpholine/dichloromethane (100 ml) and stirred for 2 hr at room temperature. After evaporation of the reagent/solvent *in vacuo* (bath temperature 40°), the remaining residue is extensively washed with 1 : 1 (v/v) light petroleum ether/diethyl ether to give the N-deblocked conjugate **27** as colorless crystals. The yield is 4.3 g (quantitative). Data for this material are identical with those of **27** obtained from the Tcoc derivative (see above).

Removal of C-Terminal tert-Butyl Ester: General Procedure

The N-protected glycosyl asparagine *tert*-butyl ester **22, 23,** or **25** (5 mmol) is dissolved in dry trifluoroacetic acid (50 ml) and stirred for 30 min. The trifluoroacetic acid is evaporated *in vacuo* (bath temperature below 40°). From the remainder toluene (twice, 50 ml) is distilled off *in vacuo* (bath temperature below 40°). From the remainder toluene (twice, 50 ml) is distilled off *in vacuo*. The residue is washed with diethyl ether and recrystallized from ethyl acetate/light petroleum ether to give the carboxyl deblocked compounds.

N^2-(9-Fluorenylmethoxycarbonyl)-N^4-[2-acetamido-4,6-di-*O*-acetyl-2-deoxy-3-*O*-(2,3,4-tri-*O*-acetyl-α-L-fucopyranosyl)-β-D-glucopyranosyl]asparagine is obtained in quantitative yield (4.55 g); mp 156°–157°; $[\alpha]_D^{23}$ −47.6° (*c* 1, CH_3OH); R_f 0.26 (B).

Analysis: Calculated for $C_{43}H_{51}N_3O_{19}$ (913.9): C 56.51, H 5.62, N 4.60. Found: C 56.67, H 5.78, N 4.61.

400 MHz ^1H NMR ($CDCl_3$): δ 1.02 (d, $J_{6,5}$ = 6.2 Hz, 3H, H-6, Fuc); 2.75 (dd, $J_{\beta,\alpha}$ = 4.3 Hz, J_{gem} = 12.6 Hz, 1H, Asn-β-H_a); 2.89 (dd, $J_{\beta,\alpha}$ = 3.9 Hz, 1H, Asn-β-H_b); 3.89 (dd, $J_{3,4}$ = 8.9 Hz, $J_{3,2}$ = 9.3 Hz, 1H, H-3, GlcNAc); 5.00–5.07 (m, 3H, H-1, H-4, GlcNAc, H-2, Fuc); 5.20–5.28 (m, 2H, H-3, H-1, Fuc); 5.28 (d, $J_{4,3}$ = 2.7 Hz, 1H, H-4, Fuc); 7.67 (d, J_{vic} = 7.0 Hz, 2H, H-4, H-5, Fmoc).

100.6 MHz ^{13}C NMR ($CDCl_3$): δ15.48 (C-6, Fuc); 37.68 (Asn-β-C); 46.98 (C-9, Fmoc); 50.44 (Asn-α-C); 54.28 (C-2, GlcNAc); 79.45 (C-1, GlcNAc); 97.07 (C-1, Fuc); 119.93 (C-4, C-5, Fmoc).

For N^2(2,2,2-trichloroethoxycarbonyl)-N^4[2-acetamido-4,6-di-*O*-acetyl-2-deoxy-3-*O*-(2,3,4-tri-*O*-acetyl-α-L-fucopyranosyl)-β-D-glucopyranosyl]asparagine (**28**), the yield is 4.3 g (quantitative); mp 150°–153°; $[\alpha]_D^{23}$ −46.8° (*c* 1, CH_3OH); R_f 0.16 (B). For N^2-(2,2,2-trichloroethoxycarbonyl-N^4-[2-acetamido-3-*O*-acetyl-6-*O*-benzoyl-2-deoxy-3-*O*-(2,3,4,6-tetra-*O*-acetyl-β-D-galactopyranosyl)-β-D-glucopyranosyl]asparagine (**29**), the yield is 4.9 g (quantitative); mp 141°–143°; $[\alpha_D^{20}$ +5.8° (*c* 1, CH_3OH); R_f 0.26 (B).

Synthesis of Glycopeptides

To a solution of *N*-glycosylasparagine ester **26** or **27** (1.5 mmol) in dichloromethane(50 ml) are added 2,2,2-trichloroethoxycarbonylglycine[28] (0.5 g, 2 mmol) and IIDQ[29] (0.5 ml, 1.7 mmol). The mixture is stirred for 16 hr, washed twice each with 0.5 *M* HCl and 0.5 *M* $NaHCO_3$, and dried with $MgSO_4$, and the solvent is evaporated *in vacuo*. The residue is recrystallized from ethyl acetate/light petroleum ether to give **30** or **31,** respectively.

2,2,2-Trichloroethoxycarbonylglycyl-N^4-[2-acetamido-4,6-di-*O*-acetyl-2-deoxy-3-*O*-(2,3,4-tri-*O*-acetyl-α-L-fucopyranosyl)-β-D-glucopyranosyl]asparagine *tert*-butyl ester (**30**) is obtained in 88% yield (1.3 g); mp 135°–137°; $[\alpha]_D^{23}$ −44.0° (*c* 1, CH_3OH); R_f 0.60 (B).

Analysis: Calculated for $C_{37}H_{53}N_4O_{20}Cl_3$ (980.2): C 45.34, H 5.45, N 5.72, Cl 10.85. Found: C 45.29, H 5.42, N 5.42, Cl 10.56.

400 MHz ^1H NMR ($CDCl_3$): δ 1.05 (d, $J_{6,5}$ = 6.5 Hz, 3H, H-6, Fuc); 1.40 [s, 9H, $C(CH_3)_3$]; 2.70 (dd, $J_{\alpha,\beta}$ = 4.5 Hz, J_{gem} = 16.1 Hz, 1H, Asn-β-H_a); 2.81 (dd, $J_{\alpha,\beta}$ = 4.8 Hz, 1H, Asn-β-H_b); 4.12 (q, 1H, H-5, Fuc); 4.93 (t, $J_{1,2}$ = $J_{1,NH}$ = 8.9 Hz, 1H, H-1, GlcNAc); 5.03 (t, $J_{4,3}$ = $J_{4,5}$ = 12.1 Hz, 1H, H-4, GlcNAc); 5.17 (d, $J_{1,2}$ = 3.5 Hz, 1H, H-1, Fuc).

100.6 MHz ^{13}C NMR ($CDCl_3$): δ 15.47 (C-6, Fuc); 27.83 [$C(CH_3)_3$]; 37.56 (Asn-β-C); 44.29 (Gly-α-C); 49.49 (Asn-α-C); 54.17 (C-2, GlcNAc); 80.13 (C-1, GlcNAc); 82.49 [$C(CH_3)_3$]; 95.35 (Cl_3C_{Tcoc}); 97.88 (C-1, Fuc).

For 2,2,2-trichloroethoxycarbonylglycyl-N^4-[2-acetamido-3-*O*-acetyl-6-benzoyl-2-deoxy-3-*O*-(2,3,4,6-tetra-*O*-acetyl-β-D-galactopyranosyl)-β-D-glucopyranosyl]asparagine (**31**), the yield is 1.47 g (89%); mp 130°–132°; $[\alpha]_D^{23}$ +5.7° (*c* 1, CH_3OH); R_f 0.45 (B).

Analysis: Calculated for $C_{44}H_{57}N_4O_{22}Cl_3$ (1100.3): C 48.03, H 5.22, N 5.09, Cl 9.67. Found: C 47.81, H 5.22, N 5.22, Cl 9.13.

400 MHz ^1H NMR ($CDCl_3$): δ 1.35 [s, 9H, $C(CH_3)_3$]; 2.67 (dd, $J_{\alpha,\beta}$ = 4.3 Hz, J_{gem} = 16.3 Hz, 1H, Asn-β-H_a); 2.79 (dd, $J_{\alpha,\beta}$ = 4.0 Hz, 1H, Asn-β-H_b); 4.52 ($J_{1,2}$ = 7.8 Hz, 1H. H-1, Gal); 4.70 (s, 2H, CH_2O_{Tcoc}); 5.04–5.08 (m, 3H, H-1, H-3, GlcNAc, H-2, Gal); 5.28 (d, $J_{4,3}$ = 2.5 Hz, 1H, H-4, Gal).

100.6 MHz ^{13}C NMR ($CDCl_3$): δ 27.79 [$C(CH_3)_3$]; 37.51 (Asn-α-C); 44.24 (Gly-α-C); 49.39 (Asn-α-C); 53.35 (C-2, GlcNAc); 79.69 (C-1, GlcNAc); 82.52 [$C(CH_3)_3$]; 95.39 (Cl_3C_{Tcoc}); 100.81 (C-1, Gal).

Tripeptides with Two Disaccharide Side Chains

To a solution of the carboxyl deblocked asparagine conjugate **28** or **29** (1 mmol) is added IIDQ[29] (365 g, 1.2 mmol). After 10 min of stirring, the amino deblocked glycopeptide ester **32** or **33,** respectively (1.2 mmol), is added. The solution is stirred at room temperature for 5 days, washed twice each with 0.5 *M* HCl and saturated $NaHCO_3$ solution and with saturated NaCl solution, dried with $MgSO_4$, and concentrated *in vacuo*. The crude product **34** or **40,** respectively, is crystallized by addition of 1 : 1 (v/v) diethyl ether/light petroleum ether and purified by flash chromatography in ethyl acetate.

N^2-(2,2,2-Trichloroethoxycarbonyl)-N^4-[2-acetamido-4,6-di-*O*-acetyl-2-deoxy-3-*O*-(2,3,4-tri-*O*-acetyl-α-L-fucopyranosyl)-β-D-glucopyranosyl]asparaginylglycyl-N^4-[2-acetamido-4,6-di-*O*-acetyl-2-deoxy-3-*O*-(2,3,4-tri-*O*-acetyl-α-L-fucopyranosyl)-β-D-glucopyranosyl]asparagine *tert* butyl ester (**34**) is obtained in 51% yield (0.84 g); mp 186°–188°; $[\alpha]_D^{25}$ −56.8° (*c* 1, $CHCl_3$); R_f 0.28 (B).

Analysis: Calculated for $C_{65}H_{92}N_7O_{36}Cl_3$ (1653.8): C 47.21, H 5.61, N 5.93, Cl 6.43. Found: C 47.18, H 5.65, N 5.84, Cl 6.36.

400 MHz ^{1}H NMR ($CDCl_3$ plus 10% DMSO-d_6): δ 0.99 (d, $J_{6,5}$ = 6.3 Hz, 3H, H-6, Fuc); 1.00 (d, $J_{6,5}$ = 6.2 HZ, 3H, H-6, Fuc); 1.34 [s, 9H, $C(CH_3)_3$]; 2.55 (M, 2H, Asn-β-$H_{N\text{-term}}$); 2.70 (d, $J_{\alpha,\beta}$ = 4.8 Hz, 2H, Asn-β-$H_{C\text{-term}}$); 4.41 (m, 1H, Asn-α-$H_{N\text{-term}}$); 4.53 (m, 1H, Asn-α-$H_{C\text{-term}}$); 4.64 (s, 2H, CH_2O_{Tcoc}); 4.89 (t, $J_{1,NH}$ = 8.7 Hz, 1H, H-1, GlcNAc); 4.91 (t, $J_{1,2}$ = $J_{1,NH}$ = 9.7 Hz, 1H, H-1, GlcNAc); 5.11 (d, $J_{1,2}$ = 4.0 Hz, 2H, H-1, H-1, Fuc); 5.19 (d, $J_{4,3}$ = 3.5 Hz, 2H, H-4, H-4, Fuc).

100.6 MHz ^{13}C NMR ($CDCl_3$): δ 15.51, 15.58 (C-6, Fuc); 23.10, 23.19 (CH_3CONII); 27.87 [$C(CH_3)_3$]; 37.82 38.01 (Asn-β-C); 43.22 (Gly-α-C); 50.34, 51.55 (Asn-α-C); 54.30, 54.81 (C-2, GlcNAc); 79.52, 79.94 (C-1, GlcNAc);82.54 [$C(CH_3)_3$]; 95.19 (Cl_3C_{Tcoc}); 97.44, 97.65 (C-1, Fuc).

N[2]-(2,2,2-Trichloroethoxycarbonyl)-*N*[4]-[2-acetamido-3-*O*-acetyl-6-*O*-benzoyl-2-deoxy-4-*O*-(2,3,4,6-tetra-*O*-acetyl-β-D-galactopyranosyl)-β-D-glucopyranosyl]asparaginylglycyl-*N*[4]-[2-acetamido-3-*O*-acetyl-6-*O*-benzoyl-2-deoxy-4-*O*-(2,3,4,6-tetra-*O*-acetyl-β-D-galactopyranosyl)-β-D-glucopyranosyl]asparagine *tert*-butyl ester (**40**) is obtained in 51% yield (0.97 g); mp 185°–187°; $[\alpha]_D^{21}$ +3.8° (*c* 1; CH_3OH); R_f 0.29 (B).

Analysis: Calculated for $C_{79}H_{100}N_7O_{40}Cl_3$ (1894.0): C 50.10, H 5.32, N 5.18, Cl 5.62. Found: C 49.95, H 5.44, N 4.91, Cl 5.81.

400 MHz ^{1}H NMR (^{1}H–^{1}H-COSY) ($CDCl_3$ plus 10% DMSO-d_6): δ 1.29 [s, 9H, $C(CH_3)_3$]; 2.59 (d, $J_{\alpha,\beta}$ = 5.9 Hz, Asn-β-$H_{C\text{-term}}$); 2.66 (d, $J_{\alpha,\beta}$ = 5.2 Hz, 2H, Asn-β-$H_{N\text{-term}}$); 3.88 (t, $J_{4,5}$ = $J_{4,3}$ = 9.6 Hz, 2H, H-4, H-4; GlcNAc); 4.54 (d, $J_{1,2}$ = 7.9 Hz, 1H, H-1, Gal); 4.58 (d, $J_{1,2}$ = 8.0 Hz, 1H, H-1, Gal); 5.05–5.11 (m, 4H, H-1, H-1, H-3, H-3, GlcNAc); 5.18 (d, $J_{4,3}$ = 4.2, 1 Hz, 1H, H-4, Gal); 5.19 (d, $J_{4,3}$ = 3.8 Hz, 1H, H-4, Gal).

100.6 MHz ^{13}C NMR (^{1}H–^{13}C-COSY) ($CDCl_3$ plus 10% DMSO-d_d): δ 21.92, 21.96 (CH_3CONH); 26.95 [$C(CH_3)_3$]; 36.49 (Asn-β-$C_{C\text{-term}}$); 36.69 (Asn-β-$C_{N\text{-term}}$); 41.86 (Gly-α-C); 48.69 (Asn-α-$C_{C\text{-term}}$); 51.02 (Asn-α-$C_{N\text{-term}}$); 74.93, 74.99 (C-4, GlcNAc); 77.82, 77.94 (C-1, GlcNAc); 80.64 [$C(CH_3)_3$]; 94.68 (Cl_3C_{Tcoc}); 99.61 (C-1, Gal).

FAB-MS (3-NOBA, negative ionization): m/z (%) 1893.5 [2.2, $(M - H)^-$, calculated: 1893.0].

Removal of Tcoc Group and Introduction of N-Terminal Acetyl Group

To a solution of the glycopeptide **34** or **40** (50 μmol) in acetic acid (10 ml) is added zinc (15 mg) freshly activated with 0.5 *M* HCl. After 3 hr of stirring at room temperature, the zinc is filtered off through Celite and washed with acetic acid. The acetic acid is evaporated from the combined filtrates *in vacuo*. To the residue of **35** or **41,** respectively, is added 1 : 1 (v/v) pyridine/acetic acid anhydride (20 ml). After 2 hr at room temperature, the volatile components are evaporated *in vacuo*. The remainder of

36 or **42**, respectively, is purified by flash chromatography in a gradient from ethyl acetate to 1 : 1 (v/v) ethyl acetate/ethanol.

N^2-Acetyl-N^4-[2-acetamido-4,6-di-*O*-acetyl-2-deoxy-3-*O*-(2,3,4-tri-*O*-acetyl-α-L-fucopyranosyl)-β-D-glucopyranosyl]asparaginylglycyl-N^4-[2-acetamido-4,6-di-*O*-acetyl-2-deoxy-3-*O*-(2,3,4-tri-*O*-acetyl-α-L-fucopyrbanosyl)-β-D-glucopyranosyl]asparagine *tert*-butyl ester (**36**) is obtained in 84% yield (66 mg); mp 195°–197°; $[\alpha]_D^{20}$ −63.7° (*c* 1, dimethylformamide); R_f 0.80 (B).

Analysis: Calculated for $C_{64}H_{93}N_7O_{35} \cdot 3H_2O$ (1574.5): C 48.82, H 6.34, N 6.23. Found: C 48.56, H 6.55, N 6.32.

400 MHz ¹H NMR (DMSO-d₆): δ 1.03 (d, $J_{6,5}$ = 6.4 Hz, 6H, H-6, H-6, Fuc); 1.34 [s, 9H, $C(CH_3)_3$]; 2.33 (dd, $J_{\alpha,\beta}$ = 6.7 Hz, J_{gem} = 15.5 Hz, 2H, Asn-β-H_a); 2.59 (dd, $J_{\alpha,\beta}$ = 6.3 Hz, 2H, Asn-β-H_b, Asn-β-H_b); 3.71 (m, 2H, Gly-α-H); 4.47 (td, $J_{\alpha,\beta}$ = 6.5 Hz, $J_{\alpha,NH}$ = 7.3 Hz, 1H, Asn-α-$H_{N\text{-}term}$); 4.51 (td, $J_{\alpha,\beta}$ = 6.5 Hz, $J_{\alpha,NH}$ = 7.2 Hz, 1H, Asn-α-$H_{C\text{-}term}$); 4.79 (dd, $J_{1,2}$ = 9.1 Hz, $J_{1,NH}$ = 9.7 Hz, 1H, H-1, GlcNAc); 4.80 (dd, $J_{1,2}$ = 9.6 Hz, $J_{1,NH}$ = 9.1 Hz, 1H, H-1, GlcNAc); 5.00–5.03 (m, 2H, H-1, H-1, Fuc); 5.18 (d, $J_{4,3}$ = 2.8 Hz, 2H, H-4, H-4, Fuc).

100.6 MHz ¹³C NMR (DMSO-d₆): δ 15.38 (C-6, Fuc); 22.50, 22.59 (CH_3CONH); 27.48 [$C(CH_3)_3$]; 37.05 (Asn-β-C); 41.84 (Glyl-α-C); 48.98, 49.33 (Asn-α-C); 78.41 (C-1, GlcNAc); 80.64 [$C(CH_3)_3$]; 95.55 (C-1, Fuc).

The yield of N^2-acetyl-N^4-[2-acetamido-3-*O*-acetyl-6-*O*-benzoyl-2-deoxy-4-*O*-(2,3,4,6-tetra-*O*-acetyl-β-D-galactopyranosyl)-β-D-glucopyranosyl]asparaginylglycyl-N^4-[2-acetamido-3-*O*-acetyl-6-*O*-benzoyl-2-deoxy-4-*O*-(2,3,4,6-tetra-*O*-acetyl-β-D-galactopyranosyl)-β-D-glucopyr ano-syl]asparagine *tert*-butyl ester (**42**) is 76 mg (84%); mp 193°; $[\alpha]_D^{20}$ (*c* 1, dimethylformamide); R_f 0.75 (B).

Analysis: Calculated for $C_{78}H_{101}N_7O_{39} \cdot 3H_2O$ (1814.7): C 51.63, H 5.94, N 5.40. Found: C 51.54, H 5.96, N 5.51.

400 MHz ¹H NMR (DMSO-d₆): δ 1.25 [s, 9H, $C(CH_3)_3$]; 2.32 (dd, $J_{\beta,\alpha}$ = 5.8 Hz, J_{gem} = 15.9 Hz, 2H, Asn-β-H_a, Asn-β-H_a); 2.55 (dd, $J_{\alpha,\beta}$ = 5.9 Hz, 2H, Asn-β-H_b, Asn-β-H_b); 4.85 (d. $J_{1,2}$ = 7.9 Hz, 1H, H-1, Gal); 4.86 (d, $J_{1,2}$ = 7.9 Hz, 1H, H-1, Gal); 5.01 (t, $J_{1,2}$ = 1,NH = 8.7 Hz, 2H, H-1, H-1, GlcNAc); 5.18 (d, $J_{4,3}$ = 3.7 Hz, 2H, H-4, H-4, Gal).

100.6 MHz ¹³C NMR (DMSO-d₆): δ 22.47, 22,57 (CH_3CONH); 27.40 [$C(CH_3)_3$]; 37.04, 37.12 (Asn-β-C); 41.88 (Gly-α-C); 49.02, 49.41 (Asn-α-C); 77.82 (C-1, GlcNAc); 80.59 [$C(CH_3)_3$]; 99.69 (C-1, Gal).

Benzyloxycarbonylalanylserylalanine *tert*-butyl ester is synthesized according to conventional Z-*tert*-butyl strategy[39,40] and subsequently sub-

[39] E. Schröder and E. Klieger, *Liebigs Ann. Chem.* **673,** 208 (1964).
[40] W. Grassmann and E. Wünsch, *Chem. Ber.* **91,** 462 (1958).

jected to hydrogenolytic removal of the Z-group to give alanylserylalanine *tert*-butyl ester (**38**): mp 83°–85°; $[\alpha]_D^{21}$ −21.8° (*c* 1, $CHCl_3$); R_f 0.33 (CH_3OH).

200 MHz ¹H NMR (DMSO-d₆): δ 1.13 (d, $J_{\alpha,b}$ = 6.8 Hz, 3H, Ala-$CH_{3N\text{-term}}$); 1.24 (d, $J_{\alpha,\beta}$ = 7.2 Hz, 3H, Ala-$CH_{3C\text{-term}}$); 1.38 [s, 9H, $C(CH_3)_3$]; 3.57 (d, $J_{\beta,\alpha}$ = 4.9 Hz, 2H, Ser-β-H); 8.03 (m, 2H, -NH_2); 8.17 (d, $J_{NH,\alpha}$ = 7.0 Hz, 1H, Ser-NH, Ala-NH).

100.6 MHz ¹³C NMR (DMSO-d_d): δ 16.99 (Ala-$CH_{3C\text{-term}}$); 21.26 (Ala-$CH_{3N\text{-term}}$); 27.48 [$C(CH_3)_3$]; 48.22 (Ala-α-C); 50.10 (Ala-α-C); 54.23 (Ser-α-C); 61.86 (Ser-β-C); 80.34 [$C(CH_3)_3$]; 169.68, 171.48, 175.43 (amide, ester C=O).

Synthesis of Glycohexapeptides

The glycotripeptide *tert*-butyl ester **36** or **42** (50 μmol) is dissolved in formic acid (98%, 10 ml). After 5hr, the formic acid is evaporated *in vacuo*, and the remaining residue is washed with diethyl ether and dried under high vacuum. The crude **37** or **43,** respectively, and tripeptide ester **38** (60 mg, 200 μmol) are dissolved in dry 1 : 1 (v/v) acetonitrile/dimethylformamide (5 ml), and, subsequently, 1-hydroxybenzotriazole (40 mg, 300 μmol), *O*-(1*H*-benzotriazolyl)-*N,N,N'N'*-tetramethyluronium tetrafluoroborate[35] (TBTU) (32 mg, 100 μmol), and diisopropylethylamine (17 μl, 100 μmol), are added to the solution. After 24 hr, the mixture is concentrated *in vacuo* and the crude **39** or **44,** respectively, is purified by flash chromatography in a gradient from ethyl acetate to 1 : 1 (v/v) ethyl acetate/ethanol.

*N*²-Acetyl-*N*⁴-[2-acetamido-4,6-di-*O*-acetyl-2-deoxy-3-*O*-(2,3,4-tri-*O*-acetyl-α-L-fucopyranosyl)-β-D-glucopyranosyl]asparaginylglycyl-*N*⁴-[2-acetamido-4,6-di-*O*-acetyl-2-deoxy-3-*O*-(2,3,4-tri-*O*-acetyl-α-L-fucopyranosyl)-β-D-glucopyranosyl]asparaginylalanylserylalanine *tert*-butyl ester (**39**) is obtained in 71% yield (63 mg); mp 187°–189°; $[\alpha]_D^{20}$ −57.7° (*c* 1, CH_3OH); R_f 0.66 (B).

Analysis: Calculated for $C_{73}H_{108}N_{10}O_{39} \cdot H_2O$ (1767.7): C 49.60, H 6.27, N 7.92. Found: C 49.65, H 5.99, N 7.98.

400 MHz ¹H NMR (DMSO-d₆): δ *1.03 (d,* $J_{6,5}$ = 6.2 Hz, 6H, H-6, H-6, Fuc); 1.22 (d, $J_{\beta,\alpha}$ = 7.1 Hz, 6H, Ala-CH_3, Ala-CH_3); 1.36 [s, 9H, $C(CH_3)_3$]; 1.72 (s, 6H, NAc_{GlcNAc}); 1.82 (m, 3H, $NAc_{N\text{-term}}$); 2.34 (dd, $J_{\beta,\alpha}$ = 5.5 Hz, J_{gem} = 15.5 Hz, 1H, Asn-β-$H_{a,N\text{-term}}$); 2.38 (dd, $J_{\beta,\alpha}$ = 7.4 Hz, J_{gem} = 15.4 Hz, 1H, Asn-β-$H_{a,C\text{-term}}$); 2.61 (dd, $J_{\beta,\alpha}$ = 5.6 Hz, 1H, Asn-β-$H_{b,N\text{-term}}$); 2.66 (dd, $J_{\beta,\alpha}$ = 6.8 Hz, 1H, Asn-β-$H_{b,C\text{-term}}$); 4.81 (dd, $J_{1,2}$ = 9.0 Hz, $J_{1,NH}$ = 9.7 Hz, 2H, H-1, H-1, GlcNAc); 4.99 (d, $J_{1,2}$ = 3.3 Hz, 1H, H-1, Fuc); 5.01 (d, $J_{1,2}$ = 4.4 Hz, 1h, H-1, Fuc).

100.6 MHz ^{13}C NMR (DMSO-d_6): 15.34 (C-6, Fuc); 16.84, 17.46 (Ala-CH_3); 22.45, 22.56, (CH_3CO-NH); 27.48 [C(CH_3)$_3$]; 36.64, 36.93 (Asn-β-C); 42.24 (Gly-α-C); 48.69 (Asn-α-$C_{C\text{-term}}$); 61.49 (Ser-β-C); 78.30, 78.43 (C-1, GlcNAc); 80.25 [$C(CH_3)_3$]; 95.60 (C-1, Fuc).

The yield of N^2-acetyl-N^4-[2-acetamido-3-*O*-acetyl-6-*O*-benzoyl-2-*O*-deoxy-4-*O*-(2,3,4,6-tetra-*O*-acetyl-β-D-galactopyranosyl)-β-D-glucopyranosyl]asparaginylglycyl-N^4-[2-acetamido-3-*O*-acetyl-6-*O*-benzoyl-2-deoxy-4-*O*-(2,3,4,6-tetra-*O*-acetyl-β-D-galactopyranosyl)-β-D-glucopyranosyl]asparaginylalanylserylalanine *tert*-butyl ester (**44**) is 80 mg (80%); mp 234°–235°; $[\alpha]_D^{20}$ −13.3° (*c* 1, CH_3OH); R_f 0.77 (B).

Analysis: Calculated for $C_{87}H_{116}N_{10}O_{43}$ (1989.9): C 52.51, H 5.88, N 7.04. Found: C 52.41, H 5.94, N 7.21.

400 MHz ^{1}H NMR (DMSO-d_6): δ 1.16 (d, $J_{\beta,\alpha}$ = 7.0 Hz, 3H, Ala-CH_3); 1.17 (d, $J_{\beta,\alpha}$ = 7.0 Hz, 3H, Ala-CH_3); 1.34 [s, 9H, C(CH_3)]; 1.72, 1.73, (2s, 6H, NAc_{GlcNAc}); 1.77 (s, 3H $NAc_{N\text{-term}}$); 2.33 (dd, $J_{\beta,\alpha}$ = 6.2 Hz, J_{gem} = 15.3 Hz, 1H, Asn-β-$H_{a,N\text{-term}}$); 2.42 (m, 1H, Asn-β-$H_{a,C\text{-term}}$); 2.54–2.63 (m, 2H, Asn-β-$H_{b,N\text{-term}}$, Asn-β-$H_{b,C\text{-term}}$); 4.82–4.89 (m, 4H, H-3, H-3, GlcNAc, H-1, H-1, Gal); 5.02 (t, $J_{1,2}$ = $J_{1,NH}$ = 9.2 Hz, 2H, H-1, H-1, GlcNAc); 5.15 (m, 2H, H-3, H-3, Gal); 5.18 (m, 2H, H-4, H-4, Gal).

100.6 MHz ^{13}C NMR (DMSO-d_6): δ 16.84, 17.32 (Ala-CH_3); 22.43, 22.50, 22.54 (CH_3CO-NH); 27.49 [C(CH_3)$_3$]; 36.66 (Asn-β-$C_{C\text{-term}}$); 36.93 (Asn-β-$C_{N\text{-term}}$); 42.26 (Gly-α-C); 48.13 (Ala-α-$C_{N\text{-term}}$); 48.74 (Asn-α-$C_{C\text{-term}}$); 49.21 (Ala-αs-$C_{N\text{-term}}$); (Asn-α-$C_{N\text{-term}}$); 61.53 (C-6, GlcNAc, Ser-β-C) 77.69, 77.82 (C-1, GlcNAc); 80.25 [$C(CH_3)_3$]; 99.66 (C-1, Gal).

Deblocking of Glycosyl Hexapeptide Esters

The glycosyl hexapeptide ester **39** or **44** (220 μmol) is dissolved in formic acid (98%, 50 ml). After 4 hr at room temperature, the formic acid is evaporated *in vacuo*. The carboxyl deblocked compound **39a** or **44a,** respectively, is washed with diethyl ether and dried under high vacuum. For **39a,** the yield is 96%; mp 208°–210°; $[\alpha]_D^{20}$ −52.3° (*c* 1, CH_3OH); R_f 0.78 (CH_3OH). For **44a,** the yield is 91%; mp 205°–207°; $[\alpha]_D^{20}$ −0.1° (*c* 1, dimethylformamide); R_f 0.67 (CH_3OH).

Compound **39a** (203 mg, 120 μmol) or **44a** (390 mg, 202 μmol) is dissolved in 1 : 1 (v/v) methanol/water (80 ml). The solution is adjusted to pH 8.5 by dropwise addition of sodium methanolate (1%) in methanol. After 4 hr at room temperature, the pH was lowered to 6.5 by the addition of a solution of acetic acid (1% in methanol. The solvents are evaporated *in vacuo,* and the remaining crude **45** or **46,** respectively, is purified by gel-permeation chromatography (GPC) on Sephadex G-15 (350 g) in water.

For N^2-acetyl-N^4-[2-acetamido-2-deoxy-3-*O*-(α-L-fucopyranosyl)-β-D-glucopyranosyl]asparaginylglycyl-N^4-[acetamido-2-deoxy-3-*O*-(α-L-fucopyranosyl)-β-D-glucopyranosyl]asparaginylalanylserylalanine (**45**), the yield is 117 mg (77%); amorphous, mp 210° (dec.); $[\alpha]_D^{20} = -71.1°$ (*c* 1, water); R_f 0.43 [50 : 1 (v/v) CH_3OH, acetic acid].

400 MHz ^1H NMR (D_2O): δ 1.12 (d, $J_{6,5}$ = 6.5 Hz, 6H, H-6, H-6, Fuc); 1.34 (d, $J_{\beta,\alpha}$ = 7.2 Hz, 3H, Ala-CH_3); 1.38 (d, $J_{\beta,\alpha}$ = 7.2 Hz, 3H, Ala-CH_3); 1.95 (s, 6H, NAc_{GlcNAc}); 2.00 (s, 3H, $NAc_{N\text{-term}}$); 2.72 (dd, $J_{\beta,\alpha}$ = 7.3 Hz, J_{gem} = 16.2 Hz, 2H, Asn-β-H_a, Asn-β-H_a): 2.82 (dd, $J_{\beta,a}$ = 4.8 Hz, 1H, Asn-β-H_b); 2.83 (dd, $J_{\beta,\alpha}$ = 5.1 Hz, 1H, Asn-β-H_b); 3.69 (m, 1H, Ser-β-H_a); 3.76 (dd, $J_{\beta,\alpha}$ = 5.5 Hz, J_{gem} = 8.4 Hz, 1H, Ser-β-H_b); 3.79 (dd, $J_{3,4}$ = 3.2 Hz, $J_{3,2}$ = 10.5 Hz, 2H, H-3, H-3, Fuc); 3.90 (dd, $J_{2,3}$ = 10.2 Hz, $J_{2,1}$ = 9.6 Hz, 2H, H-2, H-2, GlcNAc); 4.32 (q, $J_{5,6}$ = 6.6 Hz, 2H, H-5, H-5, Fuc); 4.40 (t, $J_{\alpha,\beta}$ = 5.4 Hz, 1H, Ser-α-H); 4.96 (d, $J_{1,2}$ = 3.6 Hz, 2H, H-1, H-1, Fuc); 5.04 (d, $J_{1,2}$ = 9.8 Hz, 1H, H-1, GlcNAc); 5.05 (d, $J_{1,2}$ = 9.8 Hz, 1H, H-1, GlcNAc).

100.6 MHz ^{13}C NMR (D_2O): δ 15.18 (C-6, Fuc); 16.50, 17.03 (Ala-CH_3); 21.92 (CH_3CO-$NH_{N\text{-term}}$); 22.17 (CH_3CO-NH_{GlcNAc}); 36.55, 36.80 (Asn-β-C); 42.71 (Gly-α-C); 49.85 (Ala-α-C); 49.94, 50.23 (Asn-α-C); 53.93 (C-2, GlcNAc): 55.45 (Ser-α-C); 60.53 (C-6; GlcNAc); 61.11 (Ser-β-C); 66.94 (C-5, Fuc); 67.92 (C-2, GlcNAc); 68.22 (C-4, GlcNAc); 69.54 (C-3,Fuc); 71.83 (C-4, Fuc); 77.61 (C-5, GlcNAc); 78.12 (C-1, GlcNAc); 80.87 (C-3, GlcNAc); 99.89 (C-1, Fuc); 170.65, 171.18, 172.05, 172.45, 172.54, 173.21, 174.13, 174.60, 174.78, 178.35 (C=O).

FAB-MS (glycerol, negative ionization): *m/z* (%) 1271.0 [0.65, $(M - H)^-$, calculated: 1271.5].

N^2-Acetyl-N^4-[2-acetamido-2-deoxy-4-*O*-(β-D-galactopyranosyl)-β-D-glucopyranosyl]asparaginylglycyl-N^4-[2-acetamido-2-deoxy-4-*O*-(β-D-galactopyranosyl)-β-D-glucopyranosyl]asparaginylalanylserylalanine (**46**) is obtained in 79% yield (207 mg); amorphous, mp 215° (dec.); $[\alpha]_D^{20}$ −13.8° (*c* 1, water); R_f 0.37 [50 : 1 (v/v) CH_3OH/acetic acid].

400 MHz ^1H NMR (D_2O): δ 1.38 (d, $J_{\beta,\alpha}$ = 7.3 Hz, 6H, Ala-CH_3); 1.96 (s, 6H, NAc_{GlcNAc}, NAc_{GlcNAc}); 2.00 (s, 3H, $NAc_{N\text{-term}}$); 2.72 (dd, $J_{\beta,\alpha}$ = 6.7 Hz, J_{gem} = 16.1 Hz, 1H, Asn-β-H_a); 2.73 (dd, $J_{\beta,\alpha}$ = 7.3 Hz, J_{gem} = 16.1 Hz, 1H, Asn-β-H_a); 2.82 (dd, $J_{\beta,\alpha}$ = 5.6 Hz, 1H, Asn-β-H_b); 2.85 (dd, $J_{\beta,\alpha}$ = 6.3 Hz, 1H, Asn-β-H_b); 3.51 (dd, $J_{2,1}$ = 9.6 Hz, $J_{2,3}$ = 8.0 Hz, 2H, H-2, H-2, Gal); 3.60 (m, 2H, H-5, H-5, GlcNAc); 3.62 (dd, $J_{3,4}$ = 3.3 Hz, $J_{3,2}$ = 10.0 Hz, 2H, H-3, H-3, Gal); 3.68–3.75 (m, 10H. H-5, H-5, Gal, H-6, H-6, Gal, Ser-β-H, Gly-α-H); 3.77–3.90 (m, 10H, H-2, H-2, H-3, H-3, H-4, H-4, H-6, H-6, GlcNAc); 3.88 (d, $J_{4,3}$ = 3.5 Hz, 2H, H-4, H-4, Gal); 4.30 (q, $J_{\alpha,\beta}$ = 7.1 Hz, 1H, Ala-α-H); 4.31 (q, $J_{\alpha,\beta}$ = 7.3 Hz, 1H, Ala-α-H); 4.40 (dd, $J_{\alpha,\beta}$ = 4.9 Hz, $J_{\alpha,\beta}$ = 6.1 Hz, 1H, Ser-α-H); 4.44 (d, $J_{1,2}$ = 7.8 Hz, 2H, H-1, H-1, Gal); 4.70 (t, $J_{\alpha,\beta}$ = 6.4 Hz, 1H, Asn-α-H); 4.72 (t, $J_{\alpha,\beta}$ = 6.5 Hz, 1H, Asn-α-H); 5.03 (d, $J_{1,2}$ = 9.6 Hz, 1H, H-1, GlcNAc); 5.05 (d, $J_{1,2}$ = 9.6 Hz, 1H, H-1, GlcNAc).

100.6 MHz ^{13}C NMR (D_2O): δ 16.41 (Ala-CH_3); 21.92 (CH_3CO-$NH_{N\text{-term}}$); 22.09 (CH_3CO-NH_{GlcNAc}); 36.53, 36.83 (Asn-β-C); 42.74 (Gly-α-C); 49.05 (Ala-α-C); 49.96 (Asn-α-C); 53.83 (C-2, GlcNAc); 55.48 (Ser-α-C); 59.88 (C-6, GlcNAc); 61.01 (C-6, Gal, Ser-β-C); 68.56 (C-4, Gal); 70.96 (C-2, Gal); 72.53, 72.81 (C-3, Gal); 75.34 (C-5, GlcNAc); 76.42 (C-5, Gal); 78.00 (C-4, GlcNAc); 78.23 (C-1, GlcNAc); 102.86 (C-1, Gal); 171.01, 171.21, 172.16, 172.46, 172.56, 173.22, 174.14, 174.65, 174.85, 176.58 (C=O).

FAB-MS (glycerol, negative ionization): m/z (%) 1302.9 [0.19, $(M - H)^-$, calculated: 1303.5].

*Synthesis of Bovine Serum Albumin Conjugates of Glycopeptides **45** and **46***

To a solution of the glycopeptide **45** or **46** (30 μmol) in water (5 ml) are added *N*-hydroxysuccinimide[33] (17 mg, 0.15 mmol), EDC[12] (58 mg, 0.3 mmol), and a solution of BSA (Sigma, St. Louis, MO, 40 mg, 0.6 μmol) in 5 ml of water. The mixture is adjusted to pH 5.7–5.8. After 22 hr of shaking at room temperature, the solution is subjected to dialysis against water for 48 hr and finally lyophilized to give the BSA conjugate **47** of glycopeptide **45** (yield 73.6 mg) or the BSA conjugate **48** of glycopeptide **46** (yield 69.6 mg).

*Determination of Carbohydrate Content Neoglycoproteins **57** and **58***

The BSA conjugate **47** or **48** (1.5 mg) is dissolved in water (10 ml). To part (2 ml) of the solution are added an aqueous solution of phenol in water (80%, 50 μl) and, after shaking for a few seconds, slowly and dropwise concentrated sulfuric acid.[14] After 10 min the mixture is shaken. After an additional 30 min, photometeric measurement is carried out at a wavelength 490 nm in comparison to a solution of BSA (1.3 mg, determined by weight) in water (10 ml) which is subjected to identical treatment. As standards for the measurements are used the corresponding disaccharide azides obtained from the glycosyl azides **10** and **14** by transesterification with 0.01% sodium methanolate in methanol. Physical data for the compounds are as follows: for the unprotected azide obtained from **10,** mp 204°; $[\alpha]_D^{21}$ −128.4° (*c* 1, water); R_f 0.67 (methanol); for the unprotected lactosamine azide obtained from **14,** mp 205°–207°; $[\alpha]_D^{21}$ −19.0° (*c* 1, water); R_f 0.64 (CH_3OH).

Content of carbohydrate: for the fucosylglucosamine peptide neoglycoprotein **47,** 126 μg carbohydrate/1 mg neoglycoprotein, or 24 molecules glycopeptide **45** per molecule BSA; for the neolactolactosamine peptide neoglycoprotein **48,** 139 μg carbohydrate/1 mg neoglycoprotein, or 27 molecules glycopeptide **46** per molecule BSA. The measurements are performed on a weight basis.

[2] Preparation of Fluorescent Glycoconjugates for Energy Transfer Studies

By KEVIN G. RICE

Introduction

Oligosaccharide flexibility influences the avidity of carbohydrate–protein interactions by allowing repositioning of terminal sugar residues into a three-dimensional array which is complementary to the binding site topography on carbohydrate receptors. It is primarily the underlying oligosaccharide "scaffolding" which folds at one or more of the glycosidic linkages, resulting in multiple configurations of terminal sugar residues, only one of which is recognized by the receptor.

The primary tool for analyzing oligosaccharide conformational heterogeneity is ^{1}H nuclear magnetic resonance (NMR) spectroscopy, which utilizes nuclear Overhauser effect (NOE) measurements to generate interatomic distances used to predict a range of three-dimensional (3D) structures for oligosaccharides.[1,2] Although this approach has proved to be powerful at elucidating certain conformational features for underivatized oligosaccharides, owing to the poor time resolution it is incapable of directly detecting and quantifying the range of conformers present in N-linked oligosaccharides.

A more recent development in studying oligosaccharide conformation is the application of fluorescence energy transfer. When applied to N-linked oligosaccharides, this approach is able to measure experimentally the end-to-end distance across the entire length of the oligosaccharide and the flexibility of each of the antennas. As proposed by Förster and demonstrated by Stryer, the intramolecular distance separating two fluorophores which have a donor and acceptor relationship can be determined

[1] A. C. Bush, *Curr. Opin. Struct. Biol.* **2,** 655 (1992).
[2] S. W. Homans, R. A. Dwek, and T. W. Rademacher, *Biochemistry* **26,** 6553 (1987).

by steady-state energy transfer experiments.[3,4] The selection of a donor and acceptor fluorophore in energy transfer experiments is primarily dictated by the anticipated distance to be measured and, in proteins, is usually restricted by the presence of intrinsic fluorophores. In contrast, two extrinsic fluorophores are chosen for energy transfer experiments on oligosaccharides, allowing a tailor-made fit of the fluorescence measurement to the anticipated distance to be measured.

This is important because accurate distance determinations are only possible within 10 Å of R_0 (Förster's constant) for each donor–acceptor pair. An additional concern is the lifetime of the donor, which must possess certain fluorescent properties if one intends to utilize time-resolved fluorescence energy transfer to examine oligosaccharide flexibility. Under optimal conditions the donor fluorophore should have a maximal excitation wavelength in the upper UV range with a high quantum yield, should exhibit a monoexponential decay with a lifetime in the nanosecond range, and should be paired with a suitable acceptor fluorophore providing an R_0 of 20 Å or less.[5]

In addition to these considerations, each fluorophore must contain the proper functionality to allow it to be coupled selectively to the oligosaccharide. Tethering fluorophores to the oligosaccharide through an alkyl chain possessing two or more methylene groups is advantageous in energy transfer experiments. It allows free rotation and thereby reduces the probability of rotational correlation between fluorophores.

In our experiments, a triantennary glycopeptide is selected for study because of its high affinity for the asialoglycoprotein receptor found on rat hepatocytes. Importantly, each fluorescent triantennary oligosaccharide maintains its high affinity for the asialoglycoprotein receptor, providing strong evidence that the conformation deduced by fluorescence energy transfer does not deviate considerably from that recognized by the receptor.[6] A strategy to attach two fluorophores to the glycopeptide is outlined in Fig. 1. Partial oxidation of glycopeptide **I** with galactose oxidase introduces an aldehyde into the C-6 position of terminal galactose residues. Because the enzyme oxidizes each of the three terminal galactose residues of the oligosaccharide at an equal rate, early termination of the reaction produces three isomeric monoaldehyde glycopeptides.

The separation of each monoaldehyde glycopeptide is accomplished through derivatization with 2,4-dinitrophenylhydrazine (DNPH) in order

[3] T. Förster, *Ann. Phys. (Leipzig)* **2,** 55 (1948).

[4] L. Stryer, *Annu. Rev. Biochem.* **47,** 819 (1978).

[5] R. H. Fairclough and C. R. Cantor, this series, Vol. 48, p. 347.

[6] K. G. Rice, O. A. Weisz, T. Barthel, R. T. Lee, and Y. C. Lee, *J. Biol. Chem.* **265,** 18429 (1990).

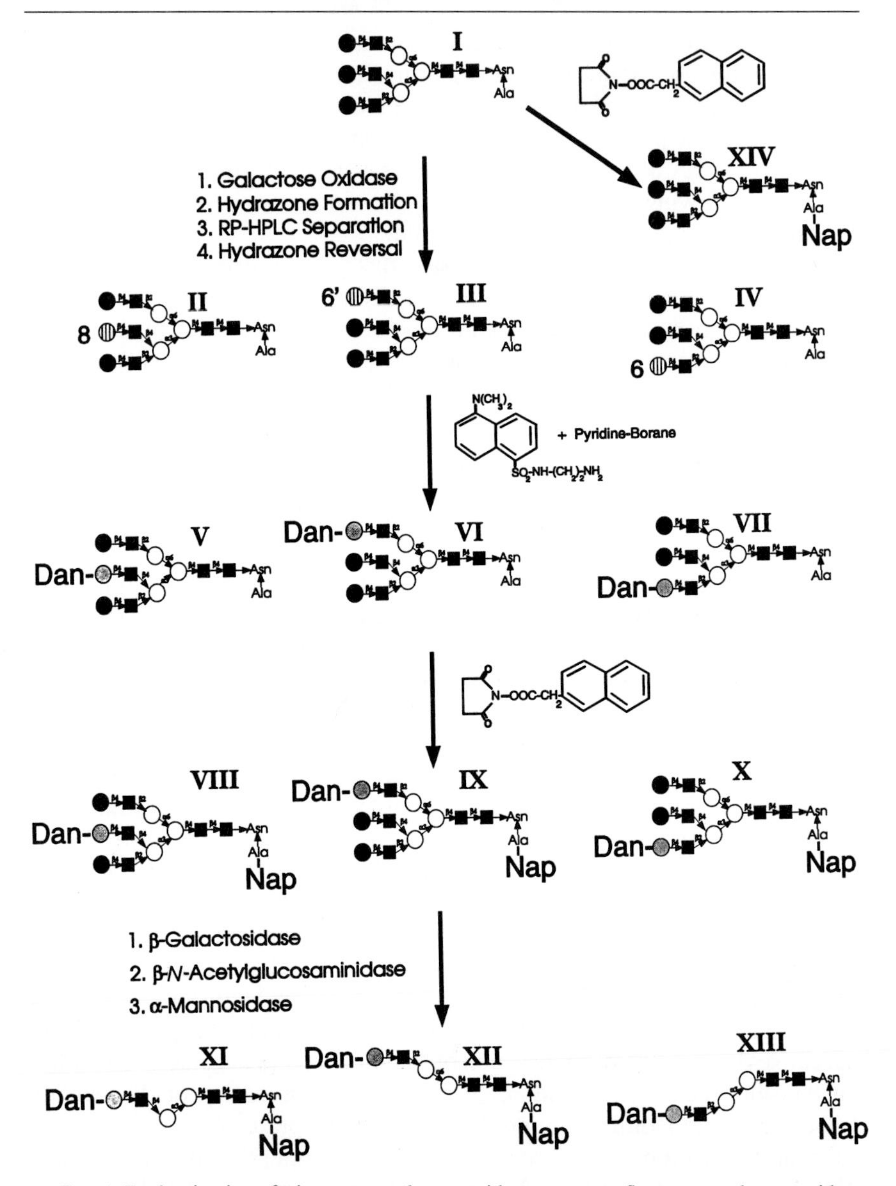

FIG. 1. Derivatization of triantennary glycopeptides to prepare fluorescent glycopeptides. The roman numerals are used in the text to discuss each glycopeptide. The galactose residues numbered 8, 6, and 6′ indicate the location of an oxidized residue according to nomenclature used in ^{1}H NMR.

to form glycopeptide hydrazones.[7] The isomeric glycopeptide hydrazones are resolved into single isomers using reversed-phase high-performance liquid chromatography (RP-HPLC). On regeneration, a single galactose C-6-aldehyde is formed from each isomeric glycopeptide hydrazone (Fig. 1, structures **II**, **III**, and **IV**). The three glycopeptide isomers differ only in the branch location of the single oxidized galactose residue. Most importantly, the aldehyde provides a signature in the ^{1}H NMR spectrum, allowing the assignment of the branch location of the galactose aldehyde in each isomer.[7]

2-Dansylaminoethylamine is chosen as the acceptor fluorophore, and naphthyl-2-acetic acid is selected as the donor fluorophore. This donor–acceptor pair has an R_0 of 20.8 Å and has been used previously to measure distances ranging from 10 to 30 Å in peptides using steady-state and time-resolved energy transfer techniques.[8,9] According to our scheme, 2-dansylaminoethylamine is introduced into the glycopeptide by reductively aminating the Schiff base formed between the primary amine of dansyl and the galactose C-6-aldehyde. This produces three isomeric glycopeptides each containing an acceptor fluorophore attached to the C-6 position of a single galactose residue (Fig. 1, structures **V**, **VI**, and **VII**). The dansylated glycopeptides are valuable intermediates and are necessary for steady-state energy transfer studies.

Naphthyl-2-acetic acid is then coupled to the N terminus of each dansylglycopeptide. This results in the formation of three dilabeled fluorescent glycopeptides that contain the donor fluorophore in a single environment and differ only in the branch location of the acceptor fluorophore (Fig. 1, structures **VIII**, **IX**, and **X**). The dilabeled fluorescent glycopeptides are used to measure three steady-state distances between the naphthyl group located at the base of the glycopeptide and the dansyl attached to the end of each antenna.

Once steady-state energy transfer experiments have determined the average distance separating the donor and acceptor, time-resolved energy transfer is used to analyze the flexibility of each antenna of the oligosaccharide.[10] This measurement differs from steady-state energy transfer in that the fluorescence lifetime of the donor fluorophore is monitored on the nanosecond time scale. The time resolution provided by this technique is faster than many of the intramolecular rotations and thereby reveals both extended and folded conformations for flexible antenna of the triantennary

[7] K. G. Rice and Y. C. Lee, *J. Biol. Chem.* **265,** 18423 (1990).

[8] L. Stryer and R. P. Haugland, *Proc. Natl. Acad. Sci. U.S.A.* **58,** 719 (1967).

[9] E. Haas, M. Wichek, E. Katchalski-Katzur, and I. Z. Steinberg, *Proc. Natl. Acad. Sci. U.S.A.* **72,** 1807 (1975).

[10] K. G. Rice, P. Wu, L. Brand, and Y. C. Lee, *Biochemistry* **30,** 6646 (1991).

glycopeptide.[10] The two conformers are spatially resolved because on folding the donor/acceptor distance changes from 20 to 10 Å. An intriguing observation is the modulation of the ratio of the extended and folded conformation for each flexible antenna when changing the temperature between 0° and 40°.[11] In this way, time-resolved fluorescence energy transfer provides data which can be used to estimate the value of thermodynamic constants for flexible glycosidic linkages in complex oligosaccharides.[11]

Fluorescent oligosaccharides are also easily modified by enzymatically trimming unlabeled branches of the oligosaccharide to prepare novel glycopeptides (Fig. 1, structures **XI**, **XII**, and **XIII**). This provides the opportunity to examine the influence of branching on the conformational flexibility of each antenna of the oligosaccharide.[12] However, the application of fluorescence energy transfer to examine complex oligosaccharide flexibility remains a challenge owing to the difficulty in introducing both a donor and an acceptor fluorophore at defined sites on the oligosaccharide. The following describes a chemoenzymatic approach used to synthesize triantennary N-linked glycopeptides containing two fluorophores.

Materials

Bovine fetuin is purchased from GIBCO Laboratories (Grand Island, NY) although the fetuin from Sigma (St. Louis, MO) may be substituted. Trypsin treated with L-1-*p*-tosylamino-2-phenylethyl chloromethyl ketone (TPCK) is purchased from Worthington Biochemical Corp. (Freehold, NJ). Pronase is obtained from CalBiochem (La Jolla, CA). *Arthrobacter ureafaciens* neuraminidase (EC 3.2.1.18), β-D-galactosidase (EC 3.2.1.23), and β-*N*-acetylglucosaminidase (EC 3.2.1.52) are obtained from Boehringer Mannheim (Indianapolis, IN). α-Mannosidase (EC 3.2.1.24) is purchased from V-Labs (Covington, LA). Triethylamine (sequenal grade), trifluoroacetic acid, and amino acid standard H are obtained from Pierce Chemicals (Rockford, IL). Boric acid is purchased from J. T. Baker Chemical Co. (Phillipsburg, NJ). Spectrapore dialysis tubing (10,000 molecular weight cutoff) is obtained from Spectrum Medical Industries (Los Angeles, CA). Sephadex G-50 and G15, dithiothreitol, iodoacetamide, galactose oxidase (EC 1.1.3.9), and catalase (EC 1.11.1.6), are from Sigma Chemical Co. 2-Naphthylacetic acid, *N*-hydroxysuccinimide, dicyclohexylcarbodiimide, and levulinic acid are purchased from Aldrich Chemical Co. (Milwaukee, WI). 2-Dansylaminoethylamine is obtained from Molecular

[11] P. Wu, K. G. Rice, L. Brand, and Y. C. Lee, *Proc. Natl. Acad. Sci. U.S.A.* **88,** 9355 (1991).
[12] K. G. Rice, P. Wu, L. Brand, and Y. C. Lee, *Biochemistry* **32,** 7264 (1993).

Probes Inc. (Eugene, OR). BioGel P-6 is obtained from Bio-Rad (Hercules, CA).

A dual-pump preparative/analytical HPLC system which is capable of eluting semipreparative columns at 10 ml/min is recommended (e.g., Gilson, Middleton, WI, and ISCO, Lincoln, NE, manufacture suitable systems). The system should include a column oven, a variable wavelength UV detector with peak separator, and an automated fraction collector.

The HPLC columns are purchased from Phase Separations (Norwalk CT). The manufacturer provides columns packed with 5 μm silica C_8 or C_{18} packing in both analytical (0.47 × 25 cm) and semipreparative (2 × 25 cm) scales. Small particle size semipreparative columns (5 μm) are recommended over the more common 10 μm semipreparative columns owing to the resolution required to separate structurally related glycopeptides. A guard column containing 10 μm C_{18} packing is recommended.

Preparation of Triantennary Glycopeptides from Bovine Fetuin

Step 1: Separation of Tryptic Glycopeptides

One gram of bovine fetuin is reduced and alkylated by dissolving the glycoprotein in 8 ml of 0.2 *M* Tris-HCl (pH 8.2). Then 8 ml of 8 *M* guanidine hydrochloride (in Tris buffer) is added, and the solution is stirred at room temperature until the glycoprotein dissolves. Dithiothreitol is added (8 ml of 0.18 *M* in 0.2 *M* Tris, 8 *M* guanidine hydrochloride, pH 8.2) and allowed to react for 1 hr while stirring. The reaction is quenched by adding 31 ml of 0.18 *M* iodoacetamide (prepared in 0.2 *M* Tris-HCl, 8 *M* guanidine hydrochloride, pH 8.2) followed by stirring at room temperature for 30 min. The reduced and alkylated glycoprotein is dialyzed overnight against running water in 10,000 molecular weight cutoff dialysis tubing. The retentate is freeze-dried.

The glycoprotein is dissolved in 100 ml of 50 m*M* sodium phosphate, pH 8.2, to which 10 mg of trypsin in 1 ml phosphate buffer is added. The tryptic digestion is allowed to proceed for 12 hr at 37° then terminated by freeze-drying. The dry glycopeptide is dissolved in 15 ml of pyridine–acetic acid–water [2.5 : 2.5 : 95 (v/v)], then applied to a Sephadex G-50 column (5 × 200 cm) eluted at 2 ml/min with the same buffer while collecting 25-ml fractions. Each fraction is assayed for carbohydrate using the phenol–sulfuric acid procedure.[13] The carbohydrate-positive fractions are pooled and freeze-dried.

[13] J. F. Mckelvy and Y. C. Lee, *Arch. Biochem. Biophys.* **132,** 99 (1969).

The glycopeptide is dissolved in 5 ml of water, and 1-ml portions are applied to a semipreparative C_8 RP-HPLC column (5 μm particle size, 2×25 cm) eluted at 10 ml/min with triethylamine–boric acid buffer (20 m*M* triethylamine, 300 m*M* boric acid, pH 7.0) and a gradient of acetonitrile (10 to 30% in 90 min). The eluent is monitored by $A_{220\,nm}$ [2 AUFS (absorbance units full scale)], and the eight major glycopeptide peaks are pooled from five consecutive chromatograms. The structures of the tryptic glycopeptides have been described previously.[14]

Each pooled fractions is adjusted to pH 5.0 with hydrochloric acid, and the samples are then dried by rotary evaporation. Boric acid is removed by repeatedly adding methanol and evaporating the methyl borate on a rotary evaporator. Residual salt is then removed from the glycopeptide by dissolving the sample in 2 ml of the pyridine–acetic acid buffer and applying it to a Sephadex G-50 column (2.5×50 cm) eluted with the same buffer while collecting 9-ml fractions. Fractions which are positive for carbohydrate by the phenol–sulfuric acid assay are pooled and freeze-dried.

Step 2: Preparation of Pronase Glycopeptides

Each tryptic glycopeptide is desialylated by preparing it in 50 m*M* sodium acetate, pH 5.0, at a concentration of 10 μmol/ml and adding 1 U of neuraminidase to 1 ml of the glycopeptide followed by incubation at 37° for 24 hr. At completion, 1 ml of 100 m*M* Tris buffer (pH 7.5) containing 4 mg of Pronase is added, and incubation is continued at 37° for 48 hr. Pronase glycopeptides are purified on a Sephadex G-50 column (2.5×100 cm) eluted with the pyridine–acetic acid buffer. The column fractions are analyzed for carbohydrate, and the glycopeptide peak is pooled and freeze-dried.

Step 3: Purification of Pronase Glycopeptides by Chromatography

Pronase glycopeptides are purified to homogeneity on a semipreparative C_8 RP-HPLC system eluted at 10 ml/min with 10 m*M* phosphoric acid. The elution time of each glycopeptide is dependent on the composition of the peptide sequence. The phosphoric acid concentration should be adjusted between 1 and 20 m*M* to optimize the separation for glycopeptides containing between 2 and 4 amino acids. Higher concentrations of acid reduces the retention times for more hydrophobic glycopeptides. The glycopeptide is prepared in water and applied in 2-μmol portions to the column. The major peak detected at $A_{220\,nm}$ is collected, neutralized with sodium hydroxide, and concentrated by rotary evaporation. The sodium

[14] K. G. Rice, N. B. N Rao, and Y. C. Lee, *Anal. Biochem.* **184,** 249 (1990).

phosphate is then removed by desalting the glycopeptide on a BioGel P-6 column (1 × 30 cm) eluted with water, detecting the glycopeptide by $A_{214\ nm}$.

Two triantennary glycopeptides (5 μmol each), which originate from different glycosylation sites and contain different peptide sequences (Ala-Asn and Asn-Asp-Ser), are the major products isolated from this purification. It is important to note that the remodeling scheme described below utilizes 75 μmol of a single triantennary glycopeptide as substrate to prepare dilabeled fluorescent glycopeptides. Therefore, it is necessary to process 15 g of bovine fetuin to prepare this quantity of triantennary glycopeptide substrate.

^{1}H NMR (400 MHz or higher) is used to deduce the glycopeptide structure based on characteristic signals for the anomeric and *N*-acetyl protons in the oligosaccharide. Each glycopeptide must be prepared in 0.5 ml of D_2O containing 0.1% acetone as an internal standard in order to compare the proton resonances to those in a structural reporter group library.[15] It is also necessary to add a small amount of sodium phosphate buffer (10 μl of 50 m*M*, pH 7.0) during D_2O exchange to buffer the pH of the glycopeptides.

The composition of the peptide portion is conveniently analyzed by Waters (Milford, MA) Picotag amino acid analysis following acid hydrolysis of the glycopeptide. In addition, monosaccharide compositional analysis should be conducted utilizing high-performance anion-exchange chromatography (HPAEC) as previously described by Hardy *et al.*[16]

Oxidation of Triantennary Glycopeptides and Separation of Positional Isomers

Step 1: Galactose Oxidase Reaction on Triantennary Glycopeptides

Triantennary glycopeptide (5 μmol containing the Ala-Asn peptide) is prepared in 1 ml of 100 m*M* sodium phosphate buffer, pH 7.0 (Fig. 1, structure **I**). Galactose oxidase (30 U) in 30 μl of phosphate buffer containing 30 μg of catalase is added, and the mixture is incubated at 37°. The reaction is monitored every 15 min by injecting 2 nmol onto an analytical C_8 RP-HPLC column eluted isocratically at 1 ml/min with 10 m*M* phosphoric acid detected by absorbance ($A_{210\ nm}$ 0.01 AUFS). Each oxidation product (mono-, di-, and trialdehyde) elutes successively earlier than the

[15] J. F. G. Vliegenthart, L. Dorland, and H. Van Halbeek, *Adv. Carbohydr. Chem. Biochem.* **41,** 209 (1983).

[16] M. R. Hardy, R. R. Townsend, and Y. C. Lee, *Anal. Biochem.* **170,** 54 (1988).

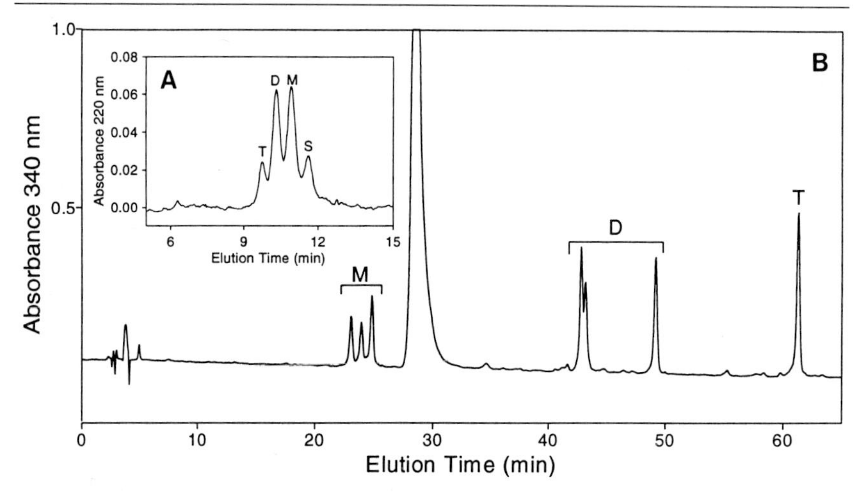

FIG. 2. Separation of oxidized triantennary glycopeptides. (A) Separation of partially (30 min) oxidized triantennary glycopeptides on RP-HPLC. Resolution of substrate (S), mono- (M), di- (D), and tri- (T) oxidized triantennary glycopeptides provides a rapid analysis of the progress of the reaction. (B) Separation of partially oxidized (30 min) triantennary glycopeptide after derivatization with DNPH. The peaks represent mono- (M), di- (D), and tri- (T) glycopeptide hydrazones with resolution between three isomeric mono- and dihydrazones of the triantennary glycopeptides.

unoxidized triantennary glycopeptide (Fig. 2A). The reaction is terminated by rapid freezing after 60 min when equal quantities of monoaldehyde and dialdehyde glycopeptides are produced.

Galactose oxidase acts rapidly on the monosaccharide galactose, whereas the reaction proceeds at a slower rate on triantennary or biantennary oligosaccharides containing either terminal Galβ1-4 or Galβ1-3 (and perhaps other) linkages.[17] It is important that the reaction pH not deviate from pH 7.0 in order to preserve the enzyme activity.

Step 2: Separation of Oxidized Glycopeptides as 2,4-Dinitrophenylhydrazine Derivatives

The partially oxidized glycopeptide (5 μmol in 1 ml of phosphate buffer) is reacted for 30 min at room temperature with 1 ml of DNPH (3 mg/ml in ethanol) containing 250 m*M* hydrochloric acid. The reaction is neutralized by adding 0.4 ml of 1 *M* sodium phosphate (pH 9.0), then dried by evaporation in a Speed-Vac concentrator (Savant Instruments, Farming-

[17] G. Avigad, *Arch. Biochem. Biophys.* **239,** 531 (1985).

dale, NY). The glycopeptide hydrazones are recovered in two extractions with 1 ml of water, whereas insoluble DNPH is removed by centrifugation.

Glycopeptide hydrazines are resolved by injecting 1 μmol onto semipreparative C_8 RP-HPLC columns eluted at 10 ml/min with 10 mM ammonium acetate, pH 5.0, and a stepwise increment of acetonitrile (0 to 36 min, from 4 to 6%; 39 min, 18%; and 55 min, 25%), resulting in the clution of three groups of peaks detected at $A_{340\ nm}$ (Fig. 2B). Note that the separation must be optimized for each glycopeptide by varying the acetonitrile concentration and pH of the buffer to achieve resolution between isomers. Three isomeric monohydrazones, differing only in the location of the oxidized galactose residue, elute at low acetonitrile concentrations (4–6%). A second cluster of three isomeric dihydrazones is eluted isocratically with 18% acetonitrile, and a single trihydrazone elutes with 25% acetonitrile. Note that residual DNPH elutes between mono- and dihydrazones and that unoxidized triantennary glycopeptide can be collected from the void volume of the column.

Each mono-, di-, and trihydrazone is isolated preparatively from successively 1-μmol injections in order to process 75 μmol of glycopeptide. The purification yields 5-μmol quantities of each of the seven glycopeptide hydrazones. This low yield is due in part to the typical recovery of 50% for glycopeptides from silica-based RP-HPLC columns and in part to the galactose oxidase reaction itself which generates seven isomeric glycopeptides and some residual starting material from a single substrate. Each purified glycopeptide hydrazone is freeze-dried to remove solvents and buffer.

Step 3: Regeneration and Characterization of Glycopeptide Aldehydes

Glycopeptide hydrazones are prepared in 0.5 ml water, and 90% levulinic acid is added and reacted for 4 hr at room temperature. The reversal is monitored by injecting 1 nmol onto an analytical C_8 RP-HPLC column eluted at 1 ml/min with 10 mM ammonium acetate and a gradent of acetonitrile (10 to 30% in 30 min). While monitoring $A_{340\ nm}$ to detect the loss of glycopeptide hydrazone and the formation of DNPH. The regenerated glycopeptide is recovered from a BioGel P-6 column (1 × 30 cm) eluted with 10 mM ammonium acetate (detecting by $A_{214\ nm}$). The glycopeptide aldehyde elutes at a higher molecular weight relative to DNPH and levulinic acid, which appear in the total volume of the column. Each regenerated glycopeptide is stored in 1 ml water and is stable for 1 year.

^{1}H NMR is used to analyze the galactose anomeric protons of each isomeric monoaldehyde glycopeptide.[7] A unique upfield shift (0.01 ppm at 400 MHz) is observed for the anomeric proton doublet of an oxidized galactose residue. In the case of triantennary glycopeptides, each isomer demonstrates this shift for only one of the galactose anomeric protons (Fig. 3). Proton NMR analysis of each dialdehyde glycopeptide also indicates this upfield shift for two of the galactose anomeric protons for each isomer. The trialdehyde glycopeptide contains all three anomeric protons shifted upfield by 0.01 ppm.[7]

Conjugation of Fluorophores to Triantennary Glycopeptides

Step 1: Reductive Amination with 2-Dansylaminoethylamine

Monoaldehyde glycopeptide (0.5 μmol) is prepared in 2 ml of 100 m*M* sodium phosphate buffer, pH 6.0, containing 10 mg of 2-dansylaminoethylamine. The reductive amination is performed at 37° by adding 5 μl of pyridine borane, which is reacted for 24 hr. The dansylated glycopeptide is purified on Sephadex G-15 (1 × 25 cm) eluted with 100 m*M* ammonium acetate, detecting the glycopeptide by $A_{254\ nm}$. The glycopeptide is purified to homogeneity on an analytical C_{18} RP-HPLC column eluted at 1 ml/min with 50 m*M* ammonium acetate (pH 6.5) and a gradient of acetonitrile (10–35% in 45 min). The major peak for each isomer elutes between 15 and 17 min and is collected and freeze-dried providing glycopeptides **V**, **VI**, and **VII** (Fig. 4).

Step 2: Coupling Succinimidyl Naphthyl 2-Acetate to Glycopeptides

Naphthyl-2-acetic acid (10 μmol), *N*-hydroxysuccinimide (10 μmol), and dicyclohexylcarbodiimide (10 μmol) are mixed in 1 ml of dimethylformamide and reacted at room temperature for 24 hr in order to form succinimidyl naphthyl 2-acetate. The reaction is filtered through Whatman (Clifton, NJ) No. 1 paper to remove crystalline dicyclohexylurea. Dansyl-glycopeptide (100 nmol) is prepared in 0.5 ml of 100 m*M* sodium bicarbonate (pH 8) to which 0.5 ml of dimethylformamide containing 5 μmol of succinimidyl naphthyl 2-acetate is added and reacted at room temperature for 4 hr. The glycopeptide product is purified from the reaction mixture first by chromatography on a Sephadex G-15 column and then by C_{18} RP-HPLC as described above for dansyl-glycopeptides to prepare products **VIII**, **IX**, and **X** (Fig. 1). Each isomeric dilabeled fluorescent glycopeptide has a unique elution position between 26 and 30 min on RP-HPLC (Fig. 4). Each glycopeptide contains both a dansyl (excitation 334 nm, emission 520 nm) and naphthyl (excitation 280 nm, emission 340 nm) fluorophore on analysis by scanning fluorescence.

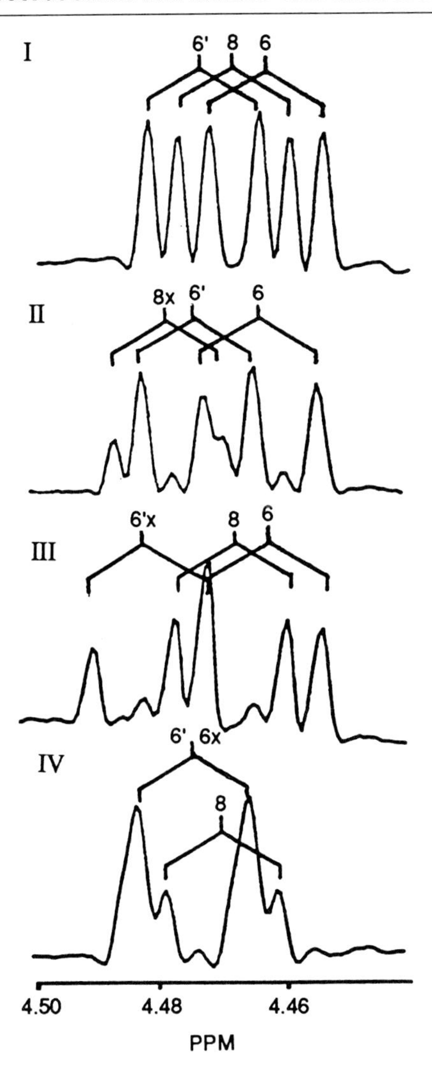

FIG. 3. ^{1}H NMR analysis of monooxidized triantennary glycopeptides. The galactose anomeric region of the 400 MHz ^{1}H NMR spectra for unoxidized triantennary glycopeptide (**I**) and three isomeric monooxidized triantennary glycopeptides (**II, III,** and **IV**) is shown (see Fig. 1 for structures). A chemical shift for only one of the anomeric protons (either 6x, 6′x, or 8x) is observed in each isomer, allowing identification of the location of each oxidized galactose residue.

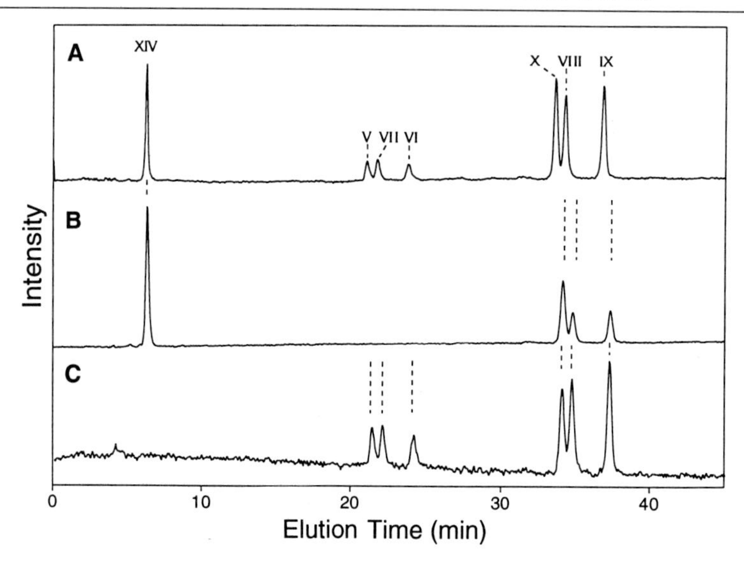

FIG. 4. Analysis of fluorescent glycopeptides by RP-HPLC. The separation of three isomeric triantennary glycopeptides containing either naphthyl (**XIV**), dansyl (**V, VI,** and **VII**), or both dansyl and naphthyl groups (**VIII, IX,** and **X**) is shown. An equimolar admixture containing each fluorescent glycopeptide was prepared and analyzed on RP-HPLC dectected by either $A_{220\ nm}$ (A), naphthyl fluorescence (excitation 280 nm, emission 340 nm) (B), or dansyl fluorescence (excitation 334 nm, emission 520 nm) (C).

In addition, naphthyl-2-acetic acid is coupled to the N terminus of glycopeptide **I** utilizing the procedure described above, resulting in the preparation of **XIV** (Fig. 1). The naphthyl-glycopeptide elutes earlier (10 min) than the dansylated glycopeptides on RP-HPLC eluted as described above.

Step 3: Trimming Fluorescent Glycopeptides with Exoglycosidases

Each fluorescent glycopeptide (600 nmol) is prepared in 200 μl of 0.1 *M* citrate–phosphate buffer, pH 4.3. β-Galactosidase (50 mU) is added, and the reaction is incubated at 37° for 24 hr. At completion of the reaction, 2.5 U of β-*N*-acetylglucosaminidase is added, and the mixture is incubated at 37° for an additional 24 hr. The product of the reaction is treated with 2 U of α-mannosidase and incubated at 37° for an additional 24 hr. Each reaction can be monitored by RP-HPLC, which results in a shift of the glycopeptide to longer retention times with each successive trimming of sugars.

The trimmed glycopeptide is purified by injecting 100-nmol portions onto a C_8 RP-HPLC column (0.47 × 25 cm) eluted at 1 ml/min with 10 m*M* ammonium acetate, pH 5.0, and a gradient of acetonitrile (0 time, 10%; 7 min, 10%; 15 min, 18/%; 40 min, 27%; 50 min, 27%) while detecting eluting peaks with $A_{220\ nm}$ in order to prepare glycopeptides **XI**, **XII**, and **XIII** (Fig. 1).

Conclusion

The procedure outlined here has been developed to attach a donor and an acceptor fluorophore to galactose-terminated triantennary glycopeptides for the purpose of performing fluorescence energy transfer distance measurements. In addition, the synthetic strategy provides glycopeptides containing dansyl or naphthyl groups attached individually. These products are indispensible in order to investigate fully donor–acceptor distances by energy transfer.

Steady-state energy transfer experiments resulted in a donor–acceptor distance of approximately 20 Å for each glycopeptide isomer.[10] Time-resolved energy transfer studies established that placement of the donor and acceptor at the ends of the glycopeptide provided the maximal distance resolution in order to reveal the existence of antenna folding.[10,11] An additional advantage of the fluorophore placement is the protection provided by the acceptor, which allows selective enzymatic trimming of unlabeled antennas.[12] Surprisingly, this modification results in diminished antenna flexibility.

These studies illustrate just some of the unique attributes of applying energy transfer to study complex oligosaccharide conformations. The approach should be easily adapted and elaborated using other complex oligosaccharides whose solution conformations are under investigation.

[3] Preparation of Tyrosinamide–Oligosaccharides as Iodinatable Glycoconjugates

By TOSHIAKI TAMURA, MANPREET S. WADHWA, MING H. CHIU, M. L. CORRADI DA SILVA, TAMARA MCBROOM, and KEVIN G. RICE

Introduction

Elucidating the function(s) of N-linked oligosaccharides on glycoproteins often necessitates their removal from the protein and purification into single structures. Once the compounds are purified, the biological activity of each N-linked oligosaccharide can be assessed individually.[1] In addition to simplifying structure–function assignments, purified oligosaccharides are much easier to transform using exoglycosidases and glycosyltransferases to obtain novel structures which otherwise occur only rarely on glycoproteins.[2] Purified oligosaccharides can also be remodeled by attaching fluorophores or photoaffinity labels, resulting in the preparation of probes that are used to study the solution conformation of oligosaccharides and the specificity by which they combine with receptors.[3–5]

Many schemes have been developed to separate oligosaccharides, most of which are directed at quantifying the distribution of N-linked oligosaccharides derived from analytical quantities of glycoproteins.[6,7] Comparatively few preparative purification methods have been developed for isolating micromole quantities of N-linked oligosaccharides. This is due in part to the high cost of enzymes that release oligosaccharides from glycoproteins and the technical difficulty of carrying out large-scale hydrazinolysis reactions.[8] Although glycopeptides are easier and less expensive to prepare, their purification is complicated by peptide heterogeneity which arises from incomplete or heterogeneous proteolysis.[9] Following the release of oligosaccharides from the glycoprotein, efficient chromatographic schemes are needed that separate oligosaccharides with both high resolution and sufficient capacity.

If purified oligosaccharides are to be used for biological studies they should also be capable of incorporating a radiolabel tracer. Reducing oligosaccharides are readily labeled with NaB^3H_4, resulting in the generation of an alditol.[10] However, oligosaccharide alditols may, in certain cases, compromise the binding affinity between the oligosaccharide and a lectin.[11] Furthermore, reductive labeling limits the range of oligosaccharide purification methods to only those which either fractionate reducing

[1] A. Varki, *Glycobiology* **3,** 97 (1993).

[2] O. Hindsgaul, K. J. Kaur, U. B. Gokhale, G. Srivasta, and M. M. Paulcic, *ACS Symp. Ser.* **466,** 38 (1991).

[3] K. G. Rice and Y. C. Lee, *J. Biol. Chem.* **265,** 18423 (1990).

[4] K. G. Rice, P. Wu, L. Brand, and Y. C. Lee, *Biochemistry* **30,** 6646 (1991).

[5] K. G. Rice, O. A. Weisz, T. Barthel, R. T. Lee, and Y. C. Lee, *J. Biol. Chem.* **265,** 18429 (1990).

[6] Y. C. Lee, *Anal. Biochem.* **189,** 151 (1990).

[7] M. W. Spellman, *Anal. Chem.* **62,** 1714 (1990).

[8] D. A. Cummings, C. G. Hellerqvist, M. Harris-Brandts, S. W. Michnick, J. P. Carver, and B. Bendiak, *Biochemistry* **28,** 6500 (1989).

[9] K. G. Rice, N. B. N. Rao, and Y. C. Lee, *Anal. Biochem.* **184,** 249 (1990).

[10] A. Kobata and K. Yamashita, this series, Vol. 179, p. 46.

[11] I. D. Manger, Y. C. Wong, T. W. Rademacher, and R. A. Dwek, *Biochemistry* **31,** 10733 (1992).

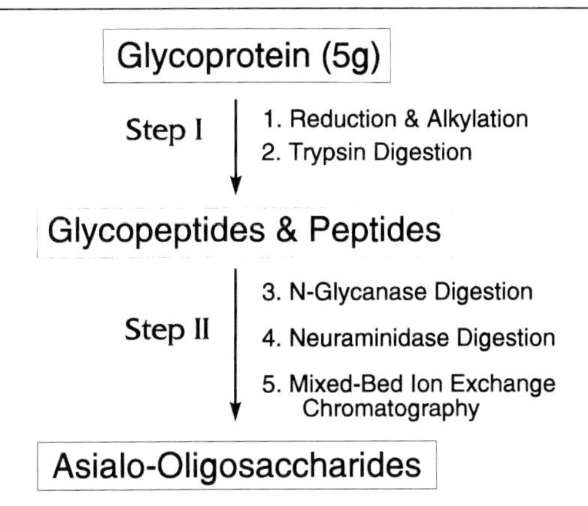

FIG. 1. Isolation of N-linked oligosaccharides from glycoproteins. A batch purification of N-linked oligosaccharides is described starting with 5 g of glycoprotein. The example of bovine fetuin is used in the text.

oligosaccharides or regenerates these from purified oligosaccharide derivatives.

Derivatization of oligosaccharides with tyrosine circumvents some of these complications by increasing the hydrophobic character of oligosaccharides. This not only facilitates their purification on large-capacity reverse-phase high-performance liquid chromatography (RP-HPLC) columns, but also introduces a phenolic group which can be radioiodinated for tracking during biological experiments.[12] Therefore, tyrosinamide–oligosaccharides are structurally defined neoglycoconjugates that are well-suited for studying the biological function(s) of N-linked oligosaccharides.

In the procedure developed, a batch process is used to reduce and alkylate 5 g of glycoprotein followed by trypsin digestion to prepare glycopeptides (Fig. 1, step I). N-Linked oligosaccharides are released enzymatically from glycopeptides utilizing *N*-glycosidase F which hydrolyzes the glycosylamide linkage on glycopeptides with improved efficiency compared with the enzyme activity on native glycoproteins. Following desialylation, ion-exchange chromatography is used to remove peptides and recover neutrally charged asialooligosaccharides (Fig. 1, step II).

Once purified from the protein, the reducing oligosaccharides are converted to tyrosinamide–oligosaccharides in two stages. First, an oligosaccharide–glycosylamine is generated by reaction of a reducing oligosaccha-

[12] T. Tamura, M. S. Wadhwa, and K. G. Rice, *Anal. Biochem.* **216,** 335–344 (1994).

ride with ammonium bicarbonate (Fig. 2, step III).[13–15] The resulting primary amine is then reacted with the succinimidyl ester of Boc-tryosine resulting in the formation of a β-glycosylamide linkage (Fig. 2, step IV). If necessary, the Boc (*N-tert*-butoxycarbonyl) group can be removed by treatment with trifluoroacetic acid (TFA) to expose a primary amine. This dramatically influences the resolution between closely related oligosaccharides in subsequent RP-HPLC separations (Fig. 2).

The tyrosinamide–oligosaccharides are first purified by gel-filtration chromatography and then by semipreparative RP-HPLC in order to isolate each oligosaccharide of the mixture. These are characterized by ^{1}H nuclear magnetic resonance (NMR) and fast atom bombardment-mass spectrometry (FAB-MS) prior to radioiodination (Fig. 2, step V).

The following protocol describes the batch preparation of N-linked oligosaccharides from bovine fetuin, derivatization with Boc-tyrosine, RP-HPLC purification, NMR and FAB-MS characterization, and radioiodination. With simple modifications, the approach is easily adapted to purify tyrosinamide N-linked oligosaccharides (sialyl or asialyl) from a variety of glycoproteins.

Materials and Methods

Bovine fetuin, iodoacetamide, dithiothreitol, Boc-tyrosine succinimidyl ester (Boc-Tyr-NHS), Sephadex G-25, and α-monothioglycerol are purchased from Sigma Chemical Co. (St. Louis, MO). Trypsin treated with L-1-*p*-tosylamino-2-phenylethyl chloromethyl ketone (TPCK) is purchased from Worthington Enzymes (Freehold, NJ). Anion-exchange resin Dowex AG50W-X2 and cation-exchange resin Dowex AG1-X2 are purchased from Bio-Rad (Richmond, CA). *N*-Glycosidase F (EC 3.5.1.52) (GPF) is purchased from New England BioLabs (Beverly, MA). Neuramidase from *Clostridium perfringens* (EC 3.2.1.18) and *Arthrobacter ureafaciens* (EC 3.2.1.18) are purchased from Boehringer Mannheim (Indianapolis, IN). Dialysis tubing is from Spectrum (Houston, TX). All other chemicals are of reagent grade or better.

High pH anion-exchange chromatography (HPAEC) is performed using a gradient pump, a PA1 anion-exchange column, and a pulsed amperometric detector (PAD), all of which are obtained from Dionex (Sunnyvale, CA). Preparative purification of tyrosinamide–oligosaccharides utilizes a

[13] I. D. Manger, T. W. Rademacher, and R. A. Dwek, *Biochemistry* **31,** 10724 (1992).

[14] E. Kallin, H. Loenn, T. Norberg, and M. Elofsson, *J. Carbohydr. Chem.* **8,** 597 (1989).

[15] L. M. Lihkosherstov, O. S. Novikova, V. A. Derevitskaja, and N. K. Kochetkov, *Carbohydr. Res.* **146,** C1 (1986).

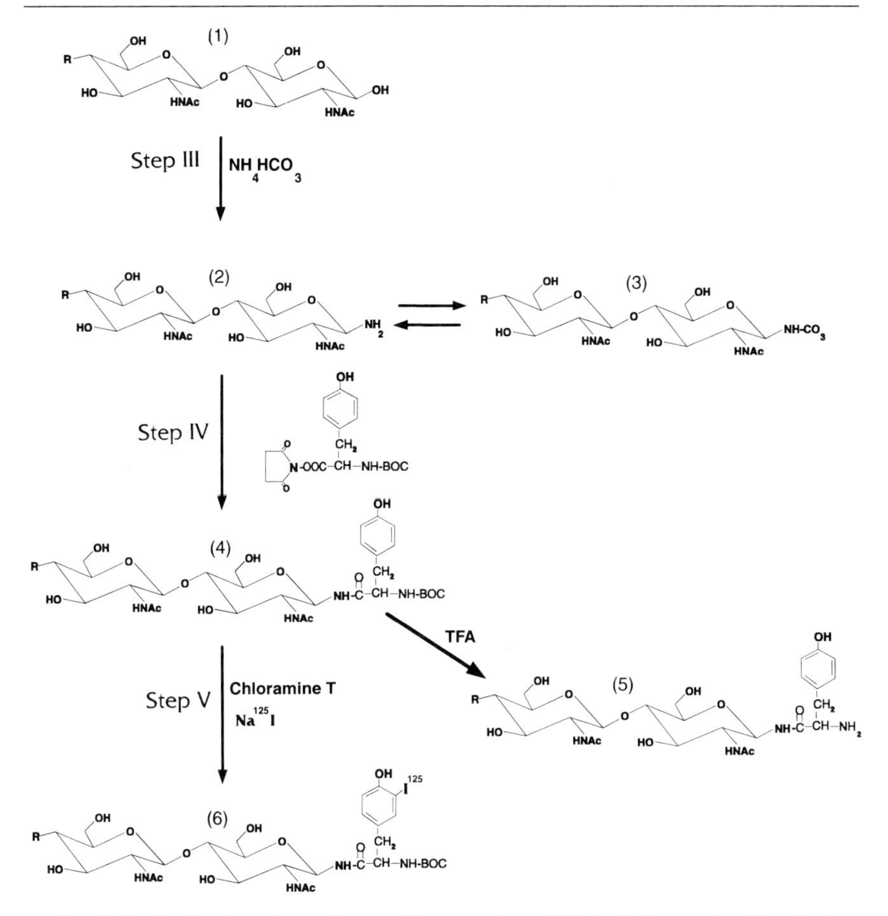

FIG. 2. Derivatization scheme for attaching tyrosine to N-linked oligosaccharides. Reducing oligosaccharides (**1**) are reacted with ammonium bicarbonate to form oligosaccharide–glycosylamines (**2**) and oligosaccharide–glycosylamine–carbonate (**3**). Following desalting, the oligosaccharide–glycosylamine is reacted with Boc-Tyr-NHS ester to prepare tyrosinamide–oligosaccharides (**4**). If necessary, the Boc group is removed by treatment with TFA in order to reveal a primary amine (**5**). Once purified and characterized, tyrosinamide–oligosaccharides are iodinated utilizing chloramine-T, resulting in iodinated glycoconjugates (**6**). R represents the remainder of the N-linked oligosaccharide structure.

dual pump preparative/analytical HPLC system with V4 variable wavelength detector and Foxy fraction collector from ISCO Inc. (Lincoln, NE).

Reverse-phase (RP) HPLC columns are purchased from Phase Separations (Norwalk, CT). The columns are packed with 5-μm silica C_8 packing in both analytical (0.47 × 25 cm) and semipreparative (2 × 25 cm) scales. Small particle size semipreparative columns (5-μm packing) are recommended over the more common 10-μm semipreparative columns owing to resolution demands needed to separate structurally related tyrosinamide–oligosaccharides. A guard column containing 10-μm C_8 packing material is utilized to protect the resolving column.

Step I: Reduction and Alkylation and Tryptic Digestion of Glycoproteins

Fetuin (5 g) is dissolved in 40 ml of 8 *M* guanidine hydrochloride, 0.2 *M* Tris, pH 8.2, while stirring at room temperature. Dithiothreitol (DTT) is added (1.1 g) and allowed to react at room temperature for 1 hr, then quenched by the addition of iodoacetamide (5.1 g), which is reacted for an additional 30 min.

The glycoprotein is transferred to 1000 molecular weight cutoff (MWCO) dialysis tubing and dialyzed against running water for 24 hr and then against 4 liters of 50 m*M* ammonium bicarbonate, pH 8.0, for an additional 12 hr. Trypsin (50 mg) is added to the retentate and allowed to react for 24 hr at 37°. The glycopeptides are then dialyzed for an additional 24 hr in 1000 MWCO tubing against running water in order to remove low molecular weight peptides and buffer salts, after which the retentate is freeze-dried.

Step II: Enzymatic Release and Desialylation of Oligosaccharides

The glycopeptide is dissolved in 100 ml of 50 m*M* ammonium acetate, pH 7.5, and the solution is carefully adjusted to pH 7.5 by adding 1 *M* ammonium hydroxide. The GPF enzyme is added (100 μl of 1,000,000 U/ml, 1 U is 2 μIU), and the reaction is allowed to proceed for 72 hr at 37°. The complete release of oligosaccharides from glycopeptides is ascertained utilizing HPAEC to analyze an aliquot (100 μl) of the reaction spiked with excess (2 μl) GPF.

The reaction mixture is adjusted to pH 5.0 by adding 450 μl of glacial acetic acid. Neuraminidase from *Clostridium perfringens* (500 μl of 1 mg/ml, 3.5 U/mg) is added followed by incubation at 37° for 24 hr. Alternatively, neuraminidase from *Athrobacter ureafaciens* (3.5 U) is used. The protein precipitate that forms during the reaction is removed by centrifugation for 5 min at 5000 *g*, and the supernatant is collected and freeze-dried.

The residue is dissolved in 40 ml of water and applied to a mixed-bed ion-exchange column (2.5 × 40 cm; top, Dowex AG50W-X2 acid form; bottom, Dowex AG1-X2 acetate form) eluted with water while monitoring $A_{214\,nm}$ [2 absorbance units full scale (AUFS)]. Asialooligosaccharides elut-

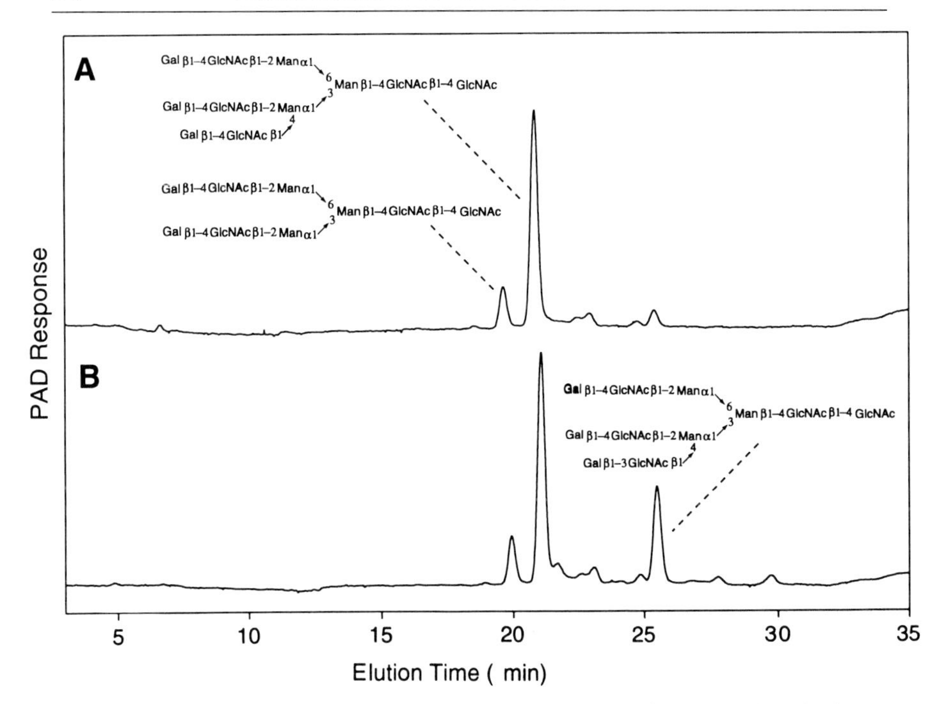

FIG. 3. Comparison of neuraminidase from *Clostridium perfringens* and *Arthrobacter ureafaciens*. The HPAEC technique was used to analyze the asialooligosaccharides recovered from mixed-bed ion-exchange chromatography when utilizing neuraminidase from either *Clostridium perfringens* (A) or *Arthrobacter ureafaciens* (B). A reduced recovery of the Galβ1-3 triantennary oligosaccharide occurs when using neuraminidase from *Clostridium perfringens*. This oligosaccharide may be recovered as a monosialyl structure by elution of the mixed-bed ion-exchange column with 1 *M* acetic acid.[12]

ing at the void of the column, as determined by the phenol–sulfuric acid assay, are pooled and freeze-dried.[16]

It should be noted that the choice of neuraminidase influences the distribution of asialooligosaccharides recovered from fetuin (Fig. 3). Neuraminidase from *Clostridium perfringens* selectively desialylates fetuin oligosaccharide, resulting in the formation of a monosialylated derivative of a Galβ1-3 triantennary oligosaccharide.[12] This oligosaccharide binds to the anion-exchange resin during ion-exchange chromatography and may be recovered by elution with 1 *M* acetic acid. This provides a facile route to prepare this unique oligosaccharide and simplifies the separation of triantennary asialooligosaccharides from fetuin.

[16] J. F. Mckelvy and Y. C. Lee, *Arch. Biochem. Biophys.* **132,** 99 (1969).

Step III: Oligosaccharide–Glycosylamine Formation

The asialooligosaccharides (10–100 μmol) are prepared in 1 ml of water in a 20-ml screw-top vial. To this is added, 1 g of ammonium bicarbonate, and the vial is sealed and heated at 50° in a water bath for 24 hr. Glycosylamine formation is determined by analyzing an aliquot of the reaction (1 nmol diluted 1 : 100 with water) on an HPAEC column eluted with 100 m*M* sodium hydroxide and a gradient of 0 to 250 m*M* sodium acetate in 30 min. Note that the dilution should be done immediately before injection because on standing the diluted glycosylamine will revert to the reducing oligosaccharide.

The reaction products are compared chromatographically with reducing oligosaccharide. In general, for each reducing oligosaccharide, two new products are observed in HPAEC. One elutes later than the reducing oligosaccharide and is assigned as oligosaccharide–glycosylamine–carbonate, whereas an earlier eluting peak is the oligosaccharide–glycosylamine. The ratio of products at this stage typically represents 40% oligosaccharide–glycosylamine, 50% oligosaccharide–glycosylamine–carbonate, and 10% reducing oligosaccharide (Fig. 4b).

Two approaches have been used to desalt oligosaccharides. Repeated freeze-drying (with the addition of 1 ml of water after each drying cycle) removes the ammonium bicarbonate and converts the oligosaccharide–glycosylamine–carbonate to the oligosaccharide–glycosylamine. However, this usually requires five or more drying cycles, and if the process is continued beyond the removal of the ammonium bicarbonate, the oligosaccharide–glycosylamine will begin to revert back to the reducing oligosaccharide.

An alternative approach is to desalt oligosaccharides by gel-filtration chromatography on a Sephadex G-25 column (2 × 50 cm) eluted with 10 m*M* ammonium bicarbonate (pH 8.0). The oligosaccharides are detected by $A_{214\,\mathrm{nm}}$, pooled, and freeze-dried. This approach is more controlled than repeated freeze-drying, although it can only be used to desalt oligosaccharides of sufficient size. Following reconstitution in water (pH 8.0), HPAEC analysis reveals that the oligosaccharides recovered are predominantly in the oligosaccharide–glycosylamine (90%) form, with the remainder being reducing oligosaccharides (Fig. 4C). The oligosaccharides are freeze-dried completely prior to coupling.

Step IV: Tyrosine Coupling and Preparative Purification of Tyrosinamide–Oligosaccharides by Chromatography

The oligosaccharide–glycosylamine (10 μmol) is derivatized with a 100 molar excess (278 mg) of Boc-Tyr-NHS ester prepared in 1.2 ml of

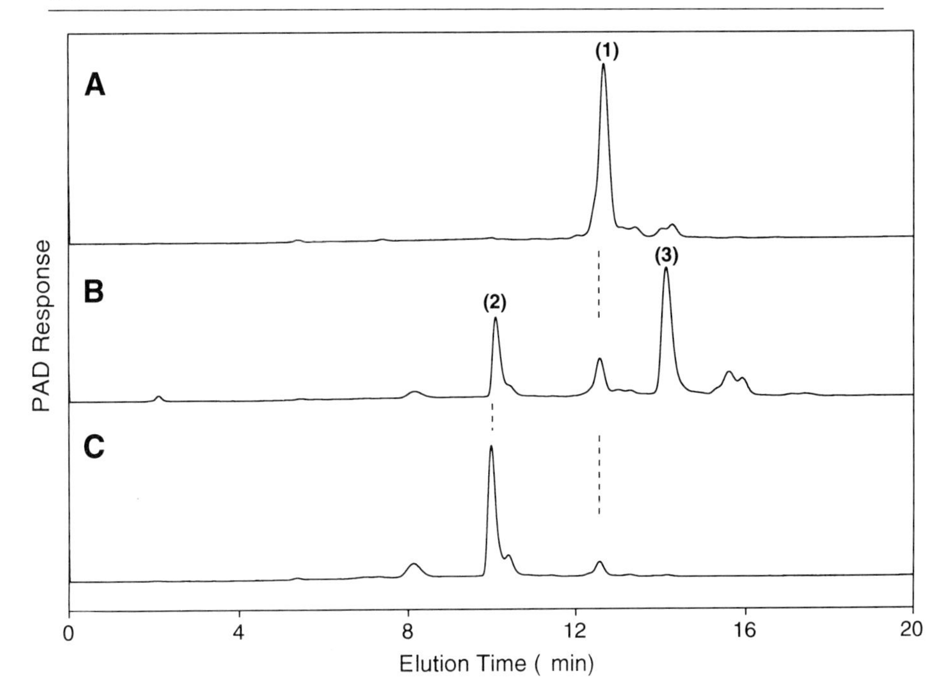

FIG. 4. Analysis of glycosylamine formation by HPAEC. The HPAEC analysis of triantennary reducing oligosaccharide (**1**) from fetuin is shown (A). After reaction of **1** with ammonium bicarbonate, HPAEC is used to quantify the ratio of oligosaccharide–glycosylamine (**2**) and oligosaccharide–glycosylamine–carbonate (**3**) that form (B). Following desalting, the oligosaccharide–glycosylamine–carbonate is converted to the oligosaccharide–glycosylamine (C). See Fig. 2 for the putative structure of each peak.

dimethylformamide (DMF), containing 5 μl of triethylamine as an organic base. The reaction is heated to 50° in a water bath for 3 hr, during which time the oligosaccharide is solubilized as the Boc-tyrosine couples to the glycosylamine as well as the hydroxyl groups on the oligosaccharide, resulting in the formation of tyrosine esters.

The tyrosine esters are hydrolyzed at adding 2 ml of 1 *M* sodium hydroxide. The precipitate that forms is centrifuged, and the supernatant containing the derivatized oligosaccharide is combined with two additional extracts (2 ml each) with 1 *M* sodium hydroxide. The combined extract is applied to a Sephadex G-25 column (2.5 × 30 cm) eluted with 1% (v/v) pyridine–acetic acid (1 : 1) buffer, detecting the oligosaccharides by $A_{280\,nm}$ while collecting fractions. The tyrosinamide–oligosaccharides elute between 75 and 125 ml, whereas the excess tyrosine reagent elutes between 150 and 250 ml. The oligosaccharide peak is pooled and freeze-dried. Note that the phenol–sulfuric acid analysis of column fractions indicates that reducing oligosaccharides elute coincidentally with tyrosinamide–oligosaccharides. The amount of tyrosinamide–oligosaccharide recovered from the reaction is determined by $A_{280\,nm}$ (ε 1330 M^{-1}). The reaction yield (typically 80%) is determined by HPAEC analysis of the ratio of tyrosinamide–oligosaccharide to reducing oligosaccharide.

Tyrosine derivatization of the monosialyloligosaccharide from fetuin (or other sialyl oligosaccharides) is achieved as described above. Note that the derivatization of sialylated oligosaccharides in pure DMF results in low product yields (20%). This may be caused by the formation of an intermolecular ion pair between the carboxyl group of sialic acid and an oligosaccharide–glycosylamine. One solution to this is to include triethylamine (as above), which deprotonates the amine and results in improved derivatization yields (70%) for sialylated oligosaccharides.

Tyrosinamide–oligosaccharides are purified by injecting 2-μmol portions onto a C_8 semipreparative HPLC column eluted at 10 ml/min with 35 mM acetic acid and 8% acetonitrile while detecting the oligosaccharides by $A_{280\,nm}$ 0.2 AUFS (Fig. 5). The isolated peaks from consecutive chromatograms are pooled, then concentrated by rotary evaporation and freeze-dried.

The separation between oligosaccharides can often be dramatically improved by removing the Boc group (Fig. 2). This is achieved by reacting 1 μmol of dry oligosaccharide with 100 μl of anhydrous TFA for 10 min at room temperature followed by freeze-drying to remove TFA. The oligosaccharides are chromatographed on C_{18} RP-HPLC column eluted with 0.1 (v/v)% TFA and acetonitrile (0 to 5% depending on the structure). Under these conditions, the tyrosinamide–oligosaccharides separate on RP-HPLC under a similar mode as separations of peptides from the tryptic digestion of a protein.

The purified oligosaccharide (1–2 nmol) is dissolved in 20 μl of water, 1 μl of α-thioglycerol is added, and the sample is dried on a Speed-Vac (Savant Instruments, Farmingdale, NY) evaporator, then applied to the probe of a Finnigan Matt 900 mass spectrometer operated in the positive ion mode. Tyrosinamide–biantennary and –triantennary oligosaccharides have m/z ions of 2290.9 and 1925.5, respectively, corresponding to the M + Na of the Boc-tyrosinamide oligosaccharides.

Oligosaccharides (1 μmol) are prepared for ^{1}H NMR by freeze-drying in D_2O twice. The deuterium-exchanged sample is prepared in 500 μl of D_2O containing 0.01% acetone as an internal standard. Tyrosinamide–biantennary and –triantennary oligosaccharides display anomeric and N-acetyl signals which are directly comparable with those from similar purified glycopeptides from Fetuin (Table I).[9] Notably, the glycosylamide

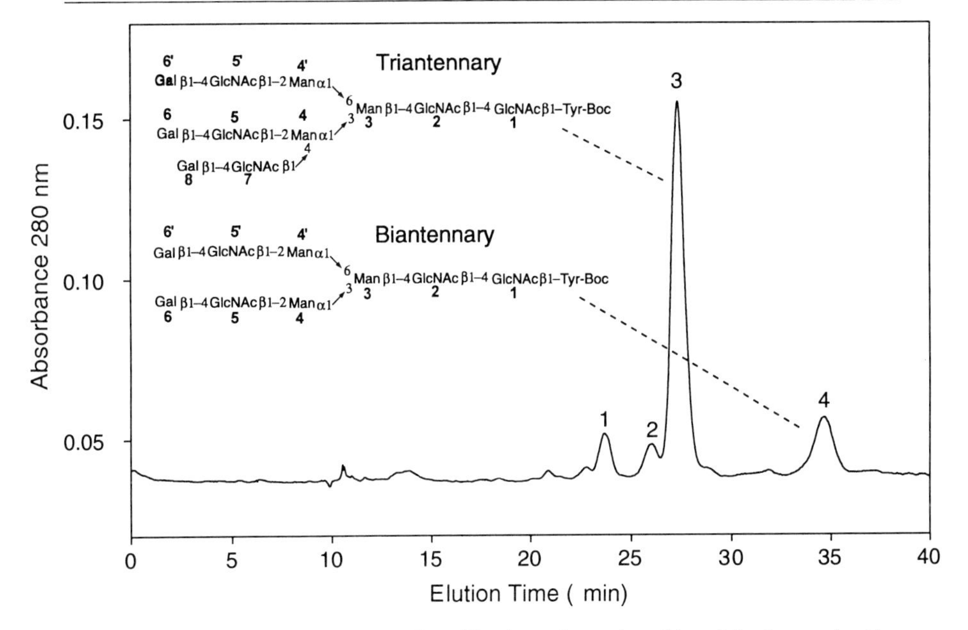

FIG. 5. Semipreparative RP-HPLC purification of tyrosinamide asialooligosaccharides from bovine fetuin. The chromatogram illustrates the resolution of 2 μmol of tyrosinamide–asialooligosaccharides on semipreparative RP-HPLC as described in the text. Triantennary and biantennary oligosaccharides (**3** and **4**) are the major products isolated. Peak **1** is a triantennary oligosaccharide containing an *N*-acetylmannosamine at the reducing end, whereas peak **2** contains a mixture of the α isomer of triantennary **3** and an isomeric Galβ1-3 triantennary structure.[12]

of each major tyrosinamide–oligosaccharide is exclusively in a β configuration as evidenced by the coupling constant ($J_{1,2}$) of 10 Hz for the anomeric proton of GlcNAc 1.

Step V: Radioiodination of Tyrosinamide–Oligosaccharides

The tyrosinamide–oligosaccharide (2 nmol) is prepared in 60 μl of 0.5 *M* sodium phosphate (pH 7.0). To this is added 0.5 mCi of sodium [^{125}I]iodide prepared in 50 μl of 0.1 *M* sodium hydroxide. The reaction is initiated by adding 20 μl of 10 m*M* chloramine-T prepared in 0.5 *M* phosphate buffer. After 3 min, 80 μl of 10 m*M* sodium metabisulfite in phosphate buffer is added to quench the reaction.

Radioiodinated oligosaccharides are chromatographed on a Sephadex G-10 column (0.8 × 25 cm) eluted with 0.15 *M* sodium chloride, pH 7.0, while collecting 0.5 ml fractions. The oligosaccharide elutes between 3

TABLE I
^{1}H NMR RESONANCES FOR TYROSINAMIDE–TRIANTENNARY AND –BIANTENNARY OLIGOSACCHARIDES

Proton[a]	Triantennary[b]	Biantennary
H-1 of		
1	5.017	5.015
2	4.617	4.616
4	5.119	5.122
4′	4.923	4.927
5	4.568	4.583
5′	4.583	4.583
6	4.463	4.473
6′	4.475	4.467
7	4.547	—
8	4.469	—
N-Ac of		
1	1.979	1.978
2	2.082	2.082
5	2.049	2.047
5′	2.046	2.050
7	2.076	—

[a] See Fig. 5 for residue nomenclature.
[b] Values are the chemical shift (ppm) recorded relative to an internal standard of acetone (2.225 ppm) at 23°.

and 4 ml as determined by gamma counting. The specific activity of the recovered oligosaccharide is typically 125 μCi/nmol.

Conclusions

The described protocol is an approach to isolate large quantities of reducing N-linked oligosaccharides from glycoproteins utilizing a batch procedure (Fig. 1). The oligosaccharides are recovered as a protein-free mixture, converted to oligosaccharide–glycosylamines, then conjugated with tyrosine (Fig. 2). The tyrosinamide–oligosaccharides are resolved into single structures on preparative RP-HPLC (Fig. 5). Each major oligosaccharide contains a β-glycosylamide linkage between GlcNAc and the carboxyl group of tyrosine with only minor amounts ($<5\%$) of the α isomer recovered. The approach is versatile enough to allow the derivatization of sialyloligosaccharides and is capable of resolving complicated oligosaccharide mixtures after removing Boc. This modification also exposes a primary amine which can be further derivatized if desired.

The resulting oligosaccharides are chemically and enzymatically stable and are conveniently quantified by the absorbance of tyrosine. Most importantly, tyrosinamide–oligosaccharides are iodinatable, providing glycoconjugates with a high specific activity that may be used to investigate the biological function of individual N-linked oligosaccharides.

Acknowledgments

The authors acknowledge the technical assistance of Dr. Charles Cottrell and David Chang for performing NMR and FAB-MS services. Support for this work provided by National Institutes of Health Grants DK45742 and GM48048 (K.G.R) and by the Ohio State University Office of Research is gratefully acknowledged.

[4] Glycamine Formation via Reductive Amination of Oligosaccharides with Benzylamine

By TOMOAKI YOSHIDA

Introduction

Lectins or lectinlike molecules, which recognize certain carbohydrate structures, are widely distributed throughout the plant and animal kingdoms. Although the biological functions of lectins are not always delineated unambiguously, some are known to be involved in metastasis,[1] fertilization,[2,3] protein trafficking,[4] microbial infection,[5] and cell adhesion.[6–8] One may gain insights into the specificity and the binding mode of lectins by utilizing suitable neoglycoconjugates. A convenient and useful method to obtain such conjugates is to attach the carbohydrates of interest to carrier proteins. This allows us to elucidate the effect of both carbohydrate

[1] J. Finne, T.-W. Tao, and M. M. Burger, *Cancer Res.* **40,** 2580 (1980).

[2] C. G. Glabe, L. B. Grabel, V. D. Vacquier, and S. D. Rosen, *J. Cell Biol.* **94,** 123 (1982).

[3] B. D. Shur and N. G. Hall, *J. Cell Biol.* **95,** 574 (1982).

[4] N. M. Dahms, P. Lobel, and S. Kornfeld, *J. Biol. Chem.* **264,** 12115 (1989).

[5] G. D. Glick, P. L. Toogood, D. C. Wiley, J. J. Skehel, and J. R. Knowles, *J. Biol. Chem.* **266,** 23660 (1991).

[6] Q. Zhou, K. L. Moore, D. F. Smith, A. Varki, R. P. McEver, and R. D. Cummings, *J. Cell Biol.* **115,** 557 (1991).

[7] M. L. Philips, E. Nudelman, F. C. A. Gaeta, M. Perez, A. K. Singhal, S. Hakomori, and J. C. Paulson, *Science* **250,** 1130 (1990).

[8] Y. Imai, D. D. True, M. S. Singer, and S. D. Rosen, *J. Cell Biol.* **111,** 1225 (1990).

structure and clustering.[9,10] The advantages of such neoglycoproteins lie in the relative ease of production as well as the definitive carbohydrate structures they contain. In many cases, clustering of carbohydrate ligands and lectins leads to a higher binding affinity.[9]

When a glycopeptide is to be used as a carbohydrate donor, its amino group can be used for conjugation via homobifunctional[11] or heterobifunctional coupling reagents[12] under mild conditions. In conjugating oligo- or polysaccharides that have no amino group, however, the only readily available functional group for conjugation is the carbonyl group of the reducing sugar unit. The carbonyl group of sugars, normally being in hemiacetal form in aqueous solution, is not very reactive. Conjugating reducing oligosaccharides to a protein by direct reductive amination typically requires many days, even when a large excess of the reducing oligosaccharide is used.[13]

In our attempts to prepare neoglycoproteins from ribonuclease A and pentamannose phosphate (PMP) from yeast mannan,[14] neither direct reductive amination nor reductive amination after selective periodate oxidation of the reduced PMP gave satisfactory results. Therefore, we have adopted a strategy of forming a glycamine to which a long heterobifunctional spacer[12] is attached (Scheme 1). The unmasked aldehydo group of such a product is more reactive, and the long spacer can alleviate the problem of steric hindrance.

For converting reducing oligosaccharides to glycamines, ammonia or ammonium salts have been used conventionally as the source of the amino group. However, the yield is rather poor even under harsh conditions. Here we describe a more effiient formation of glycamines through the use of benzylamine as the amino donor. The method is applicable even to unstable oligosaccharides in minute quantities.[15]

Methods

Chemicals

Borane–pyridine complex, palladium (10% (w/w) on activated carbon), trifluoroacetic acid (TFA) (98%), and *tert*-butyl nitrite are from Aldrich

[9] R. T. Lee and Y. C. Lee, *Glycoconjugate J.* **4,** 317 (1987).

[10] Y. C. Lee and R. T. Lee *in* "The Glycoconjugates" (M. I. Horowitz, ed.), Vol. 4, p. 57. Academic Press, New York, 1982.

[11] J. C. Rogers and S. Kornfeld, *Biochem. Biophys. Res. Commun.* **45,** 622 (1971).

[12] R. T. Lee, T.-C. Wong, R. Lee, L. Yue, and Y. C. Lee, *Biochemistry* **28,** 1856 (1989).

[13] R. G. Gray, *Arch. Biochem. Biophys.* **163,** 426 (1974).

[14] M. E. Slodki, R. M. Ward, and J. A. Boundy, *Biochim. Biophys. Acta* **304,** 449 (1973).

[15] T. Yoshida and Y. C. Lee, *Carbohydr. Res.* **251,** 175 (1994).

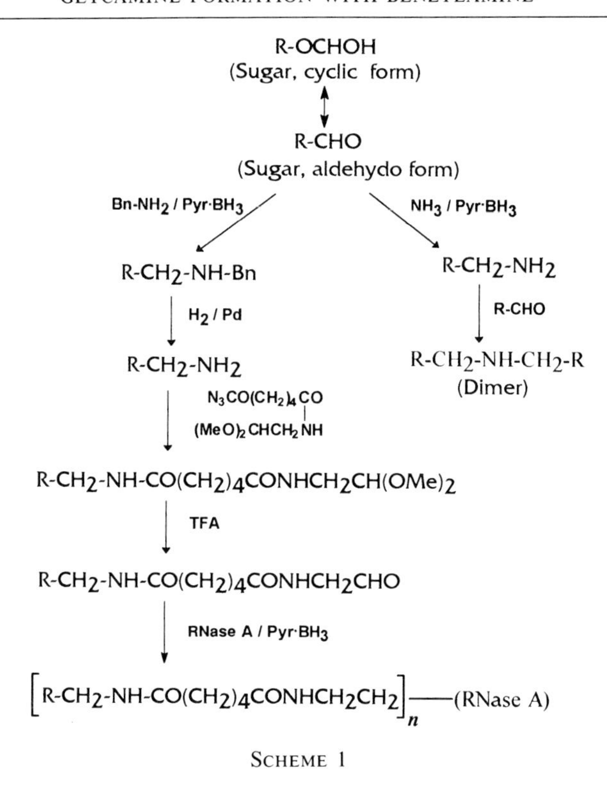

SCHEME 1

Chemical Co. (Milwaukee, WI). Fluorescamine and ribonuclease A (RNase A) are from Sigma Chemical Co. (St. Louis, MO). Benzylamine (Bn-NH_2) is from J. T. Baker Chemicals Co (Phillipsburg, NJ). Laminarihexaose (fine) is purchased from Seikagaku America Inc. (Rockville, MD). Pentamannose phosphate (PMP) is prepared by the established method[16] from phosphomannan from *Hansenula holstii* (Y-2448) donated by Dr. Morey Slodki of the Northern Regional Research Center, U.S. Department of Agriculture (Peoria, IL).

Assay Methods

The yields of reductive amination of laminarihexaose and PMP are examined by a combination of high-performance anion-exchange chromatography (HPAEC) and detection with a pulsed amperometric detector (PAD),[17] using a BioLc system and PAD-II (Dionex, Sunnyvale, CA). A

[16] R. K. Bretthauer, G. J. Kaczorowski, and M. J. Weise, *Biochemistry* **12,** 1251 (1973).
[17] M. R. Hardy, R. R. Townsend, and Y. C. Lee, *Anal. Biochem.* **170,** 54 (1988).

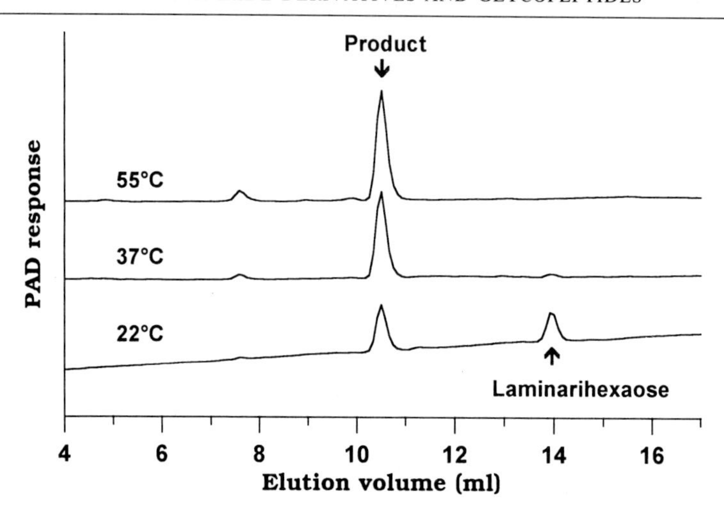

FIG. 1. Analysis by HPAEC of reductive amination of laminarihexaose with benzylamine. Laminarihexaose (20 m*M*) was reacted with pyridine–borane (100 m*M*) and the benzylamine reagent (500 m*M*) at different temperatures for 18 hr.

Carbopac PA-1 column (4.6 × 250 mm) is eluted with 100 m*M* NaOH with a linear sodium acetate gradient increasing from 100 to 300 m*M* in 20 min. The relative response factor of reductively aminated laminarihexaose is virtually identical to that of laminarihexaose. Examples of elution profiles on HPAEC are shown in Fig. 1. For simultaneous analysis of mannose and mannose 6-phosphate, the Carbopac PA-1 column is eluted first isocratically with 14 m*M* NaOH containing 10 m*M* sodium acetate for 5 min, then with a linear increases of sodium acetate to 500 m*M* in the subsequent 25 min.

Carbohydrates are quantified by the phenol–sulfuric acid reaction[18] using glucose (for laminarihexaose) and mannose (for pentamannose phosphate) as the standards. The color yield of mannose 6-phosphate is 95% of that of mannose. Protein is determined by the fluorescamine assay after 11 hr of hydrolysis in 2*M* NaOH at 100°,[19] using ribonuclease A as standard.

Efficiency of Benzylamine as Amino Donor to Laminarihexaose

To determine the optimal conditions for reductive amination with benzylamine, we employ laminarihexaose as a model oligosaccharide, because

[18] J. F. Mckelvy and Y. C. Lee, *Arch. Biochem. Biophys.* **132,** 99 (1969).

[19] C. P. Stowell, T. B. Kuhlenschumidt, and C. A. Hoppe, *Anal. Biochem.* **85,** 572 (1978).

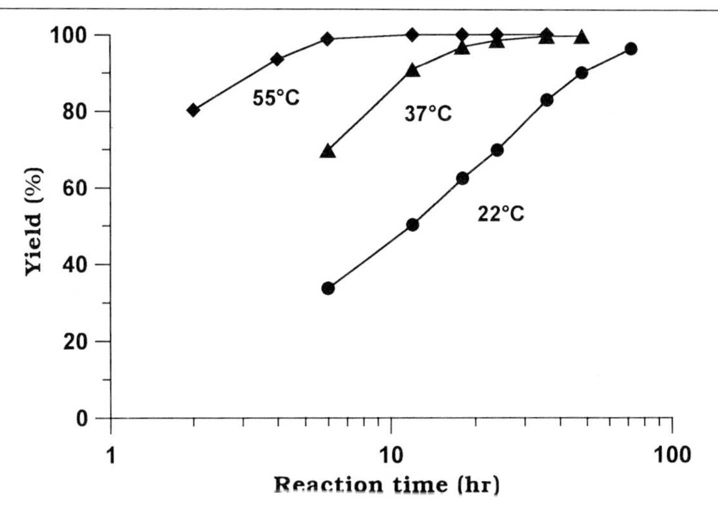

FIG. 2. Time course of reductive amination of laminarihexaose with benzylamine at various temperatures. Laminarihexaose (20 m*M*) was reacted with pyridine–borane (100 m*M*) and the benzylamine reagent (500 m*M*) at different temperatures as described in the text. The yields were determined by HPAEC with a Carbopac PA-1 column attached to a PAD-II detector.

it is about the same size as PMP and also contains an alkali-sensitive β1,3-linkage. First, the benzylamine reagent for reductive amination is prepared by mixing equimolar glacial acetic acid and benzylamine (e.g., mix 1 ml glacial acetic acid and 1.9 ml benzylamine). The final concentrations of the reactants in a reaction medium of 25% (v/v) methanol in water are as follows: laminarihexaose, 20 m*M*; the benzylamine reagent, 500 m*M*; and pyridine–borane, 100 m*M*. The reaction yields the product in 83, 86, 83, and 73% at pH 6, 7, 8, and 9, respectively, after 48 hr at room temperature. The optimal pH range found agrees with that for the reaction between the unprotected aldehyde and the ε-NH_2 groups.[20] Preliminary results indicate that 500 m*M* of the benzylamine reagent results in a good yield within a reasonable reaction time, and thus is chosen for further experiments below. As expected, the rate of reaction is considerably accelerated at 37° and 55° (Fig. 2). At 55°, more than 95% of the oligosaccharide is aminated in 6 hr as compared with 33% at room temperature. Furthermore, even when lower concentrations of the oligosaccharide are used (such as 10 or 5 m*M*), a 94% yield is attained by 12 hr of reaction at 55° (Table I).

[20] R. T. Lee and Y. C. Lee, *Biochemistry* **19,** 156 (1980).

TABLE I
EFFICIENCY OF BENZYLAMINATION OF LAMINARIHEXAOSE AT 55°

Reaction time[a] (hr)	Yields (%)[b] using laminarihexaose (m*M*)				
	20	10	5	2.5	1.25
6	98.9	90.8	83.0	64.0	43.2
12	ND[c]	99.0	94.9	85.3	65.8

[a] Laminarihexaose at each concentration was reacted with 500 m*M* benzylamine and 100 m*M* pyridine–borane.
[b] The yields were determined by HPAEC analysis.
[c] Not determined.

Comparison with Reductive Amination with Ammonium Acetate

The comparison of benzylamine versus ammonium acetate as the amino donor reveals that the former reacts considerably faster than the latter at all the concentrations of laminarihexaose tested (Table II). The advantage of benzylamine is most evident at room temperature. This makes the benzylamine method particularly useful for some heat-sensitive oligosaccharides such as phosphorylated or sialylated oligosaccharides. Moreover, formation of dimeric by-products is negligible (<2%) at all temperatures when benzylamine is used, whereas ammonium acetate

TABLE II
COMPARISON OF AMMONIUM ACETATE AND BENZYLAMINE FOR REDUCTIVE AMINATION OF LAMINARIHEXAOSE

Reagents	Incubation time[a] (hr)	Yields[b] (%) using laminarihexaose (m*M*)		
		20	100	400
1 *M* Ammonium acetate	18	5.2	8.3	2.1
	36	8.6	12.3[c]	6.9
500 m*M* Benzylamine	18	45.4	69.0	74.1
	36	74.9	89.7	88.5

[a] Laminarihexaose was mixed with either ammonium acetate or benzylamine and 5-fold excess of pyridine–borane. The reaction was allowed to proceed at room temperature for the indicated duration.
[b] The yield was determined by HPAEC analysis as described in the text.
[c] The dimeric by-product (2.1%) was detected under these conditions, whereas less than 2% by-product was detected for others.

yields a considerable amount (11.4%) of dimeric by-product when 20 m*M* laminarihexaose is reacted at 55° to give 74% product (data not shown). The difference between the two methods becomes more pronounced when lactose is reductively aminated; lactose reacts noticeably slower than laminarihexaose, and the dimeric by-product (Scheme 1) increases significantly.

Reductive Amination of Pentamannose Phosphate wth Benzylamine and Formation of Pentamannose Phosphate-Glycamine

Pentamannose phosphate is reductively aminated with benzylamine based on the optimal conditions determined with laminarihexaose. A mixture of 400 m*M* PMP, 500 m*M* benzylamine, and 1 *M* pyridine–borane in 34% (v/v) methanol/water is incubated at room temperature (chosen to avoid dephosphorylation). A higher concentration of oligosaccharide and pyridine–borane is employed to accelerate the reaction. After 48 hr, when PMP has completely disappeared by HPAEC analysis, a by-product, most likely a product of PMP dimerized through benzylamine, is detected to the extent of 2.0–6.5%. The reaction mixture is dried under vacuum with a Speed-Vac concentrator (Savant Instruments, Farmingdale, NY), dissolved in 50% methanol, and subjected to hydrogenolysis in a Brown hydrogenator[21] using palladium on activated carbon as a catalyst. After evaporating methanol, PMP-glycamine is purified on a BioGel (Bio-Rad, Richmond, CA) P-4 column (1.5 × 92 cm) in water to remove remaining excess reagents and dimerized PMP (the first peak in Fig. 3). The fractions collected are positive to both the phenol–sulfuric acid and the trinitrobenzene sulfonate reactions,[22] which indicate the presence of a primary amino group and carbohydrates. The overall yield is 74% based on the phenol–sulfuric acid reaction. In contrast, 1.9 *M* of ammonium acetate and 100 m*M* of PMP result in less than 50% yield even with incubation at 37°, a condition which causes some loss of phosphate groups (data not shown).

Coupling of Pentamannose Phosphate-Glycamine to Heterobifunctional Spacer

A heterobifunctional reagent containing an acyl hydrazide and masked aldehyde group at opposite ends[12] is converted to the acyl azide by reacting with *tert*-butyl nitrite and trifluoroacetic acid.[23] Briefly, the acyl hydrazide (9.1 mg, 37 μmol) is dissolved in 500 μl of dimethylformamide (DMF) and

[21] C. A. Brown and H. C. Brown, *J. Am. Chem. Soc.* **31,** 3989 (1966).
[22] Y. C. Lee, *Carbohydr. Res.* **67,** 509 (1978).
[23] R. U. Lemieux, D. R. Bundle, and D. A. Baker, *J. Am. Chem. Soc.* **97,** 4076 (1975).

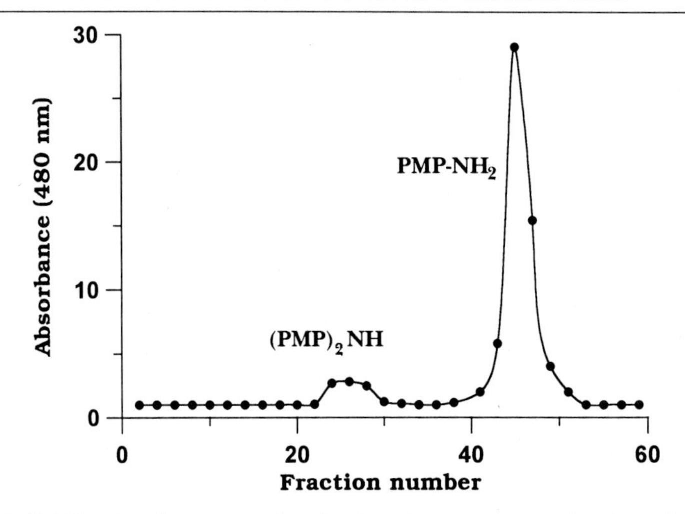

FIG. 3. Gel-filtration chromatography of aminated pentamannose phosphate. Pentamannose phosphate (PMP) was reductively aminated with the benzylamine reagent and hydrogenolyzed to give a primary amine. The reaction mixture was fractionated with a BioGel P-4 column (1.5 × 92 cm) in water. Each fraction was analyzed for carbohydrate content by the phenol–sulfuric acid method. The peak at fractions 21–30 contained the dimeric by-product.

chilled to −25°. Trifluoroacetic acid (23 μl, 300 μmol) and 5 mg (44 μmol) of *tert*-butyl nitrite in 50 μl of DMF are added to the hydrazide solution, and the reaction mixture is stirred in an ice bath for 30 min. Assay with trinitrobenzene sulfonate[24] confirms the total disappearance of the hydrazide group after the reaction. Triethylamine (80 μl) is added to neutralize the reaction mixture, and the acyl azide thus formed is mixed immediately with PMP-glycamine in water using a 2.4- to 11.7-fold excess of the acyl azide, after which the reaction mixture is left for 14 hr at 4°. The spacer-extended PMP-amine is purified on a Sephadex G-10 column (2.5 × 92 cm) in 0.1 *M* acetic acid, the effluent being monitored at $A_{214\ nm}$. The first peak is positive in the phenol–sulfuric acid assay (i.e., contains carbohydrate) and is pooled and evaporated to dryness. The pooled intermediate is treated with 50% trifluoroacetic acid for 5 hr at room temperature to unmask the aldehyde group,[12] and the reaction mixture is evaporated on a Speed-Vac.

[24] X.-Y. Qi, N. O. Keyhani, and Y. C. Lee, *Anal. Biochem.* **175,** 139 (1988).

Conjugation of Pentamannose Phosphate to Ribonuclease A

To a solution of 3.3 μmol of the spacer-extended PMP-glycamine in water are added 75 nmol of ribonuclease A and 3.3 μmol of pyridine–borane, and the reaction mixture is adjusted to 50 m*M* phosphate buffer, pH 7.5 (300 μl). Because ribonuclease A possesses 11 primary amino groups including the N terminal, there is a 4-fold excess of aldehyde per amino group. The reaction is allowed to proceed for 2 days at room temperature. Doubling the amount of aldehydic reagent does not improve the extent of coupling (Table III). The reaction mixture is fractionated on a Sephadex G-50 column (2.5 × 90 cm) eluted with 0.5 m*M* phosphate buffer, pH 7.0, containing 50 m*M* Na_2SO_4 into two broad peaks (monitored at $A_{280\,nm}$ as well as a peak attributable to pyridine. The fluorescamine assay and phenol–sulfuric acid assay reveal that the earlier peak contains 15.1 mol of PMP per mole of ribonuclease. Interestingly, the later peak shows 19.1 mol of PMP per mole of ribonuclease, nearly the maximal level of coupling where all amino groups are converted to tertiary amines, with each amino group accepting two aldehyde groups.[20] The late elution of this component may be rationalized by the hydrophobic interaction between the spacer groups and the column packing.

TABLE III
CONJUGATING PENTAMANNOSE PHOSPHATE TO RIBONUCLEASE A

PMP/RNase A[a] (molar ratio)	Eluted peaks from G-50 column[b] for conjugated PMP/RNase A[c]	
	Early	Late
44	15.1	19.9
88	13.3	19.1

[a] PMP, reductively aminated and modified with a heterobifunctional spacer, was reacted with RNase A as described in the text.
[b] The reaction mixture was fractionated with a Sephadex G-50 column in 50 m*M* Na_2SO_4.
[c] The contents of carbohydrate and protein were determined by the phenol–sulfuric acid and fluorescamine methods, respectively.

The presence of phosphate groups on the conjugated PMP is confirmed by monosaccharide analysis by HPAEC after hydrolysis in 2 *M* trifluoroacetic acid at 100° for 4 hr.[17] Standard mannose and mannose 6-phosphate are also treated under the same conditions for comparison and to obtain correction factors. Hydrolysis under the employed conditions converted 8% of mannose 6-phosphate to mannose. Taking into account the extent of dephosphorylation (8%) and degradation of mannose (9%) during the hydrolysis, the ratio of mannose to mannose 6-phosphate is calculated to be 1 : 3.2, which is consistent with the expected value (1 : 3) and suggests negligible (<7%) dephosphorylation during the reaction sequences.

Concluding Remarks

Here we present a more efficient method for producing glycamine by employing benzylamine as an amino donor. The higher efficiency of this method is due to the ease of Schiff base formation attributable to the lower basicity of benzylamine as well as to the lesser amount of dimeric by-product formed owing to the difficulty of tertiary amine formation. Although this approach requires an extra step of hydrogenolysis to unmask the amino group, the advantages outweigh the minor inconvenience. Although we have described only two examples of oligosaccharide–glycamine formation, and coupling of one of the glycamines to ribonuclease A, the methodology established here should be applicable to many other oligosaccharide–protein combinations, including sialylated oligosaccharides.

[5] Preparation, Isolation, and Analysis of Heterogeneous Branched Cyclodextrins

By KYOKO KOIZUMI, SUMIO KITAHATA, and HITOSHI HASHIMOTO

Introduction

Cyclodextrins (CDs) are the cyclic, nonreducing maltooligosaccharides produced by the action of an enzyme (cyclomaltodextrin glucanotransferase, EC 2.4.1.19, CGTase) on starch. Among them, the α, β, and γCDs, which are composed of 6, 7, and 8 $\alpha(1 \rightarrow 4)$-linked D-glucosyl residues, respectively, are well known and referred to as the first generation or parent CDs.

The internal cavity of the doughnut-shaped CD molecule is hydrophobic, whereas the external surface is hydrophilic. The most characteristic and unique function of CDs is the formation of complexes by the inclusion of various kinds of molecules in the hydrophobic cavity. This unique function of CDs has been utilized in such diverse areas as the pharmaceutical, food and cosmetic industries.[1,2]

As the applications of the parent CDs expanded, CDs with specific properties and capable of fulfilling more specific demands have been developed by conjugating different substances with the parent CDs via chemical or enzymatic reactions. They are called second generation or modified CDs. Branched CDs are a group of modified CDs and may be classified into two categories: (1) homogeneous branched CDs, which have side chains composed only of glucose (G1) or such maltooligosaccharides as maltose (G2), maltotriose (G3), and more generally maltooligosaccharide (G*n*) bound to the parent CDs,[3-7] and (2) heterogeneous branched CDs, which carry one or more galactose or mannose residues in the side chain of homogeneous branched CDs or directly on the parent CDs.[8-11]

Homogeneous branched CDs have the same ability to form inclusion complexes as the parent CDs, and they possess higher aqueous solubility and lower hemolytic activity against human erythrocytes compared to the parent CDs.[12] Consequently branched CDs are finding many uses beyond the conventional applications. Heterogeneous branched CDs were pre-

[1] D. Duchene (ed.), "Cyclodextrins and Their Industrial Uses." Editions de Sante, Paris, 1987.

[2] D. Duchene (ed.), "New Trends in Cyclodextrins and Derivatives." Editions de Sante, Paris, 1991.

[3] J. Abe, S. Hizukuri, K. Koizumi, Y. Kubota, and T. Utamura, *Carbohydr. Res.* **154,** 81 (1986).

[4] J. Abe, S. Hizukuri, K. Koizumi, Y. Kubota, and T. Utamura, *Carbohydr. Res.* **176,** 87 (1988).

[5] S. Hizukuri, J. Abe, K. Koizumi, Y. Okada, Y. Kubota, S. Sakai, and T. Mandai, *Carbohydr. Res.* **185,** 191 (1989).

[6] S. Hizukuri, S. Kawano, J. Abe, K. Koizumi, and T. Tanimoto, *Biotechnol. Appl. Biochem.* **11,** 60 (1989).

[7] T. Shiraishi, S. Kusano, Y. Tsumuraya, and Y. Sakano, *Agric. Biol. Chem.* **53,** 2181 (1989).

[8] S. Kitahata, K. Fujita, Y. Takagi, K. Hara, H. Hashimoto, T. Tanimoto, and K. Koizumi, *Biosci. Biotech. Biochem.* **56,** 242 (1992).

[9] S. Kitahata, K. Hara, K. Fujita, N. Kuwahara, and K. Koizumi, *Biosci. Biotech. Biochem.* **56,** 1518 (1992).

[10] K. Hara, K. Fujita, H. Nakano, N. Kuwahara, T. Tanimoto, H. Hashimoto, K. Koizumi, and S. Kitahata, *Biosci. Biotech. Biochem.* **58,** 60 (1994).

[11] K. Hara, K. Fujita, N. Kuwahara, T. Tanimoto, H. Hashimoto, K. Koizumi, and S. Kitahata, *Biosci. Biotech. Biochem.* **58,** 652 (1994).

[12] Y. Okada, Y. Kubota, K. Koizumi, S. Hizukuri, T. Ohfuji, and K. Ogata, *Chem. Pharm. Bull.* **36,** 2176 (1988).

pared originally for the purpose of obtaining drug carriers in drug delivery systems. The ultimate purpose of drug delivery systems is the properly timed delivery to a specific lesion of the required dose of a specific drug. As medical studies have succeeded in characterizing the pathogeneses of numerous diseases, the pathogenic substances and cellular nature of lesions have now been characterized in many cases, and the target tissues for medical treatment are now often clearly established.

The modern era has also witnessed rapid progress in carbohydrate chemistry and glycobiology. The understanding of the functions of sugar chains *in vivo* has been substantially advanced. Carbohydrates can be utilized either as carriers of drugs or as ligands, or as regulators of the physical properties of drugs. The utilization of carbohydrates as ligands for targeting of drugs is especially noteworthy in this respect. Liver lectin, for example, is a sugar-recognizing protein present on the surface of the parenchymal cells of the liver, which recognizes specifically sugar chains that have D-galactose and *N*-acetyl-D-galactosamine at the nonreducing terminal.[13] The requirements of specific sugar chain structures for recognition by liver lectin has been studied in detail by Lee and co-workers.[14] Heterogeneous branched CDs with a galactosyl residue(s) at the nonreducing terminal may be used as carriers of drugs to liver parenchymal cells. In addition, the existence of lectins which specifically recognize sugar chains with D-mannose is also well established.[15]

We describe here the preparation, isolation, and structural analysis of heterogeneous branched CDs, which are expected to be useful as possible drug carriers for targeting.

Preparation

Transgalactosylation by β-Galactosidase

β-Galactosidase (β-D-galactopyranoside galactohydrolase, EC 3.2.1.23) is an exoglycosidase which normally hydrolyzes D-galactopyranosyl linkages. The enzyme also catalyzes transgalactosylation when high substrate concentrations are used. The ratio of transgalactosylation to hydrolysis, the acceptor specificity, and the structures of transgalactosylated products depend on the source of the enzyme and the reaction

[13] G. Ashwell and A. G. Morell, *Adv. Enzymol.* **41,** 99 (1974).

[14] R. R. Townsend, M. R. Hardy, T. C. Wong, and Y. C. Lee, *Biochemistry* **25,** 5716 (1986).

[15] P. Chakraborty, A. N. Bhaduri, and P. K. Das, *Biochem. Biophys. Res. Commun.* **166,** 404 (1990).

conditions. For example, *Kluyveromyces lactis* β-galactosidase catalyzes mostly hydrolytic reactions even under conditions of high concentrations of lactose. *Escherichia coli* β-galactosidase catalyzes transgalactosylation to lactose, but not to parent CDs or branched CDs. On the other hand, *Bacillus circulans* and *Penicillium multicolor* β-galactosidases synthesize three types of transgalactosylated products (**1**, **2**, and **3**) from each branched CD (Table I and Fig. 1). None of the β-galactosidases examined in this study synthesize a transgalactosylation product of parent CDs, but they will use the branching glucose as an acceptor.[8] As a rule, in the synthesis of heterobranched CDs, the yield of transglycosylated products of CDs or branched CDs increases with increasing donor substrate concentration, as the efficiency of transglycosylation is enhanced, but is little affected by acceptor concentration, although the amount of transfer product increases with increasing acceptor substrate concentration.

TABLE I
TRANSGALACTOSYLATED PRODUCTS BY β-GALACTOSIDASES, USING LACTOSE AS DONOR

Source of β-galactosidase	Acceptor	Product	Yield (%)
Bacillus circulans	G1-αCD	Gal[β]-G1-α**1**	20
		Gal[β]-G1-α**2**	6
		Gal[β]-G1-α**3**	1
	G2-αCD	Gal[β]-G2-α**1**	22
		Gal[β]-G2-α**2**	5
		Gal[β]-G2-α**3**	3
	G1-βCD	Gal[β]-G1-β**1**	20
		Gal[β]-G1-β**2**	4
		Gal[β]-G1-β**3**	1
	G2-βCD	Gal[β]-G2-β**1**	22
		Gal[β]-G2-β**2**	5
		Gal[β]-G2-β**3**	2
	G1-γCD	Gal[β]-G1-γ**1**	17
		Gal[β]-G1-γ**2**	3
		Gal[β]-G1-γ**3**	0
	G2-γCD	Gal[β]-G2-γ**1**	21
		Gal[β]-G2-γ**2**	8
		Gal[β]-G2-γ**3**	2
Penicillium multicolor	G1-αCD	Gal[β]-G1-α**3**	7
	G2-αCD	Gal[β]-G2-α**3**	10
	G1-βCD	Gal[β]-G1-β**3**	14
	G2-βCD	Gal[β]-G2-β**3**	13
	G1-γCD	Gal[β]-G1-γ**3**	10
	G2-γCD	Gal[β]-G2-γ**3**	8

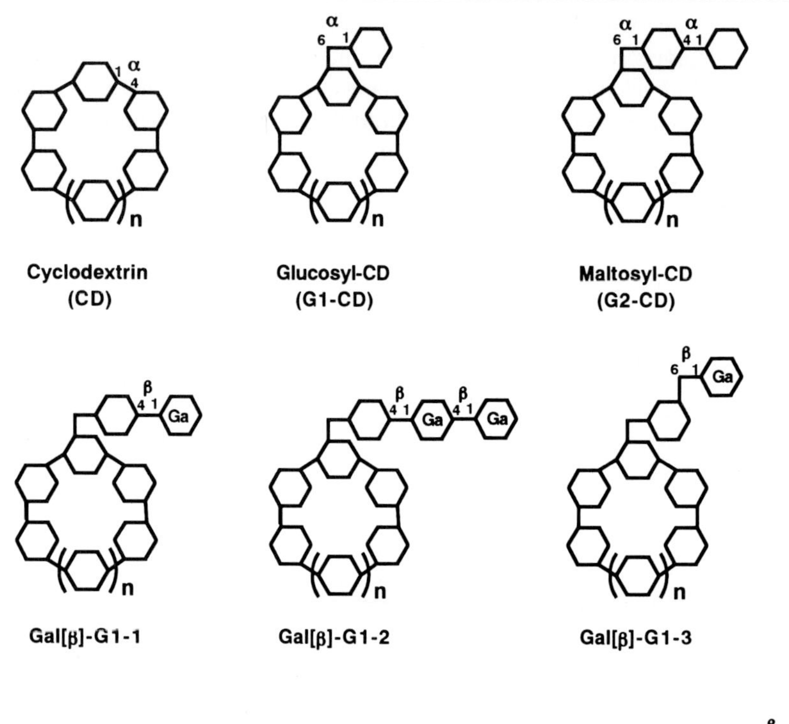

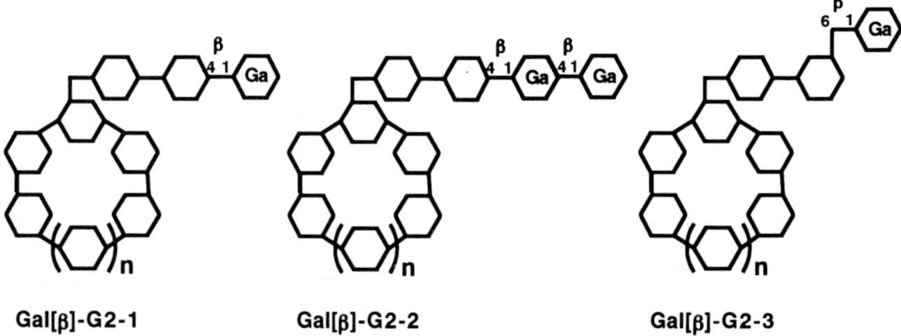

FIG. 1. Abbreviated structural diagrams for CDs and trans-β-D-galactosylated derivatives. n = 1, αCD; n = 2, βCD; n = 3, γCD; ⬡, glucosyl residue; (Ga), galactosyl residue.

Bacillus circulans β-galactosidase produces **1**, **2**, and **3** from each branched CD. Their yields, which are calculated as [transfer product (mol)/acceptor (mol)] × 100 (%), are, respectively, approximately 20, 5, and 2% when the enzyme is incubated with a mixture of 20% (w/v) lactose and 40% (w/v) branched CDs. Over a long period of incubation, the yield of **3** increases as the amount of product **1** decreases owing to hydrolysis.

On the other hand, *P. multicolor* β-galactosidase produces from each branched CD mainly product **3**, in yields of approximately 10%, and **1** and **2** in very small yields (<1%). *Aspergillus oryzae* β-galactosidase also produces the same three transgalactosylated products from branched CDs, but the amounts are less than those produced by *P. multicolor* β-galactosidase.

Materials

- Lactose, reagent grade
- G1-α, -β, and -γ cyclodextrins and G2-α, -β, and -γ cyclodextrins, highest grade
- *Bacillus circulans* β-galactosidase (crude preparation, 7 units/mg; Daiwa Kasei Co. Ltd., Osaka, Japan)
- *Penicillium multicolor* β-galactosidase (crude preparation, 60 units/mg; K · I Chemical Industry Co., Ltd., Shizuoka, Japan)

Assay of β-Galactosidase Activity. Fifty microliters of an enzyme solution is incubated with 400 μl of 5 m*M* *p*-nitrophenyl β-galactopyranoside in 25 m*M* acetate buffer (at the optimum pH for each enzyme) at 40° for 10 min. The reaction is stopped by the addition of 0.5 ml of 0.2 *M* Na_2CO_3, and the release of *p*-nitrophenol from the substrate is measured spectrophotometrically at 400 nm. One unit of the enzyme activity is defined as the amount liberating 1 μmol of *p*-nitrophenol per minute.

Procedure. Five grams of lactose and 5 g of branched CDs (G1-α, G1-β, G1-γ, G2-α, G2-β, or G2-γ cyclodextrins are dissolved in 30.6 ml of 50 m*M* hot acetate buffer (pH 6.0 for *B. circulans* and pH 4.5 for *P. multicolor*). After cooling to 40°, β-galactosidases from *B. circulans* or *P. multicolor* (70 units, 2.8 ml) are added to separate solutions and incubated at 40° for 1–2 hr. The reaction mixtures are heated at 100° for 15 min to stop the enzyme action, then centrifuged to remove insoluble materials. After removal of monosaccharides and acyclic oligosaccharides from the reaction mixtures by preparative high-performance liquid chromatography (HPLC) on an octadecylsilyl (ODS) column, galactosylated branched CDs, are isolated by HPLC.

Transgalactosylation by α-Galactosidase

α-Galactosidase (α-D-galactopyranoside galactohydrolase, EC 3.2.1.22) catalyzes the hydrolysis of α-galactopyranosidic linkages, such as those of melibiose and synthetic α-galactopyranosides, and also transgalactosylation, particularly at high substrate concentrations. α-Galactosidases are distributed in various plants, animal tissues, and microorganisms. They differ in the ratio of hydrolysis to transgalactosylation catalyzed and in acceptor specificity. For example, microbial enzymes such as *Mortierella vinacea, Candida guilliermondii* H-404, and *Absidia reflexa* α-galactosidases possess transfer activity to acceptors such as melibiose, glucose, and G2-CDs, but not to CDs or G1-CDs. On the other hand, the enzyme from coffee bean has a stronger transfer activity and wider acceptor specificity than do the microbial enzymes, and it produces transgalactosylated derivatives of CDs, G1-CDs, and G2-CDs.[11]

Coffee bean α-galactosidase synthesizes two kinds of transgalactosylated products (**I** and **II**) from each parent CD, and three kinds of transfer products (**I**, **II**, and **III**) from each branched CD (Table II and Fig. 2). Gal[α]-α**II** and Gal[α]-β**II** comprise about 5% of the total Gal[α]-CD product, whereas the yield of Gal[α]-γ**II** is somewhat low (2–3%). The CDs are more effective acceptors than are branched CDs for the coffee bean enzyme. Therefore, the coffee bean enzyme synthesizes mostly monogalactosylated products from them and only very small amounts of doubly branched CDs. It synthesizes transgalactosylated products of CDs and branched CDs in yields of 38% (**I** and **II**) from CDs and 15% (**I**, **II**, and

TABLE II
TRANSGALACTOSYLATED PRODUCTS BY α-GALACTOSIDASES, USING MELIBIOSE AS DONOR

Source of α-galactosidase	Acceptor	Product	Yield (%)
Coffee bean	αCD	Gal[α]-α**I**	36
		Gal[α]-α**II**	2
	βCD	Gal[α]-β**I**	36
		Gal[α]-β**II**	2
	γCD	Gal[α]-γ**I**	37
		Gal[α]-γ**II**	1
	G1-αCD	Gal[α]-G1-α**I**	5
		Gal[α]-G1-α**II**	7
		Gal[α]-G1-α**III**	3
	G1-βCD	Gal[α]-G1-β**I**	10
		Gal[α]-G1-β**II**	2
		Gal[α]-G1-β**III**	3
	G2-αCD	Gal[α]-G2-α**I**	6
		Gal[α]-G2-α**II**	5
		Gal[α]-G2-α**III**	4
	G2-βCD	Gal[α]-G2-β**I**	8
		Gal[α]-G2-β**II**	2
		Gal[α]-G2-β**III**	5
Mortierella vinacea	G2-αCD	Gal[α]-G2-α**III**	6
	G2-βCD	Gal[α]-G2-β**III**	6

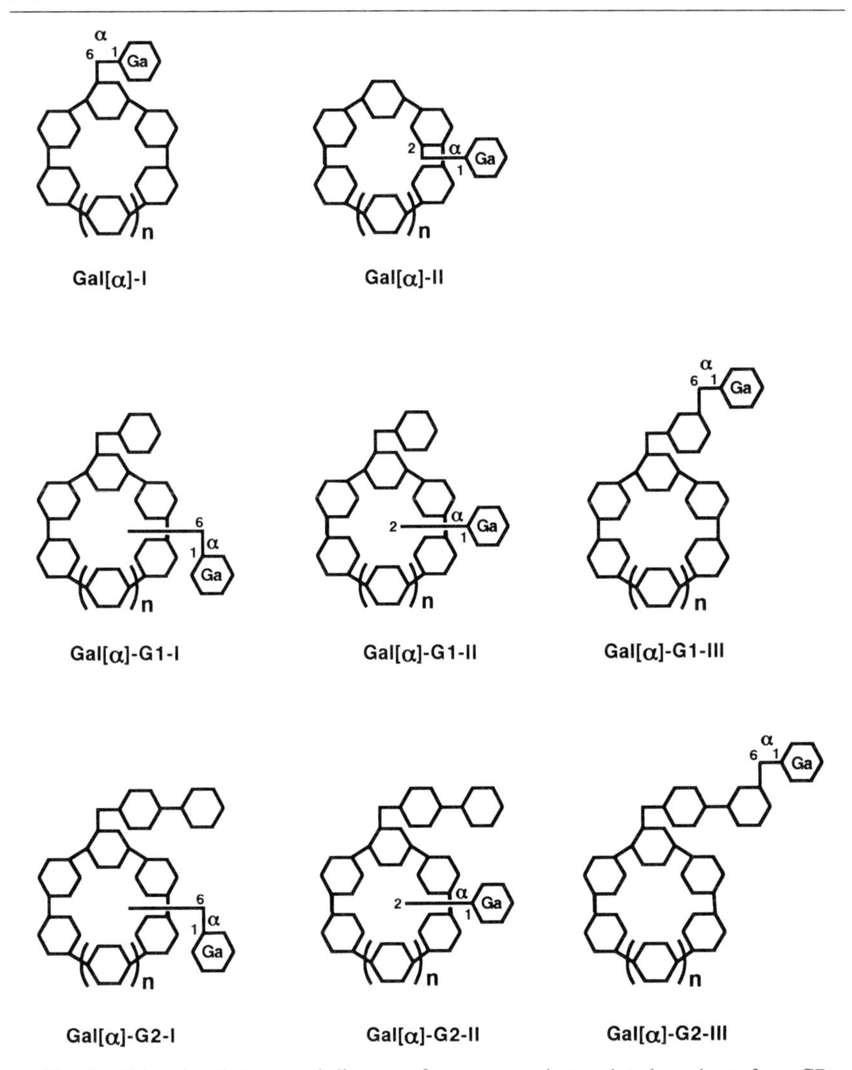

FIG. 2. Abbreviated structural diagrams for trans-α-galactosylated products from CDs, G1-CDs, and G2-CDs. The abbreviations are the same as those in Fig. 1.

III) from each branched CD when coffee bean α-galactosidase is incubated with 27% (w/v) melibiose as donor substrate and 13% (w/v) CDs or branched CDs as acceptor. Furthermore, the amount of transgalactosylated products on the side chains of branched CDs (**III**) is less than that formed directly on the CD rings (**I** and **II**).[11]

The *M. vinacea* α-galactosidase synthesizes only one kind of transgalactosylated product from G2-CDs in yields of 6% under the same reaction conditions as for the coffee bean enzyme.

Materials

Melibiose, reagent grade
α, β, and γ cyclodextrins, G1-α and -β cyclodextrins, and G2-α and -β cyclodextrins, highest grade
Coffee bean α-galactosidase [suspension in 3.2 M $(NH_4)_2SO_4$ solution, pH 6.0, containing bovine serum albumin (BSA);
specific activity, 10 units/mg; Sigma Chemical Co., St. Louis, MO], dialyzed against 50 mM acetate buffer, pH 6.0, prior to use
Mortierella vinacea α-galactosidase (lyophilized powder; specific activity, 200 units/mg; Seikagaku Kogyo Co., Ltd., Tokyo, Japan)

Assay of α-Galactosidase Activity. The method of assay and the definition of enzyme activity are the same as for studies of β-galactosidase, except for the use of *p*-nitrophenyl α-galactopyranoside as substrate.

Preparation of Galactosylated α, β, and γCDs. One gram of melibiose and 0.5 g of CD (α, β, or γCD) are dissolved or suspended in 3 ml of hot phosphate buffer (50 mM, pH 6.5). After cooling to 40°, coffee bean α-galactosidase (16 units, 0.5 ml) is added to the solution, and the mixture is incubated at 40° for 48 hr. The reaction mixtures are then heated at 100° for 15 min to stop the reaction, then centrifuged to remove insoluble materials. The Gal[α]-CDs are isolated by HPLC on an amino column and lyophilized. Each of the products is separated into one major (Gal[α]-α**I**, Gal[α]-β**I**, and Gal[α]-γ**I**) and one minor (Gal[α]-α**II**, Gal[α]-β**II**, and Gal[α]-γ**II**) peak by HPLC on an ODS column.

Preparation of Transgalactosylated Products of G1-αCD and G1-βCD. One-half gram of melibiose and 0.5 g of G1-CD (G1-αCD or G1-βCD) are dissolved in 2 ml of hot phosphate buffer (50 mM, pH 6.5). After cooling to 40°, coffee bean α-galactosidase (8 units, 0.5 ml) is added to the solution, and the mixture is incubated at 40° for 48 hr. The reaction mixture is then heated at 100° for 15 min to stop the reaction and subsequently centrifuged to remove insoluble materials. Galactosylated branched CDs, Gal[α]-G1-α**I**, **II**, **III** and Gal[α]-G1-β**I**, **II**, **III**, are isolated by HPLC.

Preparation of Transgalactosylated Products of Gal-G2-αCD and Gal-G2-βCD. Three grams of melibiose and 3 g of G2-αCD or G2-βCD are dissolved in 10 ml of hot acetate buffer (50 mM, pH 6.5). After cooling to 40°, coffee bean α-galactosidase (48 units) is added to the solution, and the mixture is incubated at 40° for 48 hr. The reaction mixture is then heated at 100° for 15 min to stop the reaction and subsequently centrifuged

TABLE III
TRANSMANNOSYLATED PRODUCTS BY α-MANNOSIDASE, USING METHYL α-MANNOSIDE AS DONOR

Source of α-mannosidase	Acceptor	Product	Yield (%)
Jack bean	G1-αCD	Man[α]-G1-αCD	6
	G1-βCD	Man[α]-G1-βCD	5
	G2-αCD	Man[α]-G2-αCD	6
	G2-βCD	Man[α]-G2-βCD	6

to remove insoluble materials. Gal[α]-G2-α**I**, **II**, **III** and Gal[α]-G2-β**I**, **II**, **III** are isolated by HPLC and lyophilized.

Transmannosylation by α-Mannosidase

α-Mannosidase (α-D-mannopyranoside mannohydrolase, EC 3.2.1.24) is a carbohydrolase that catalyzes the hydrolysis of substrates such as synthetic α-mannopyranosides. It catalyzes transmannosylation as well as hydrolysis, particularly under the conditions of high substrate concentrations. Jack bean α-mannosidase produces transmannosylated products from branched CDs, but not from parent CDs, when incubated with 7% (w/v) methyl α-mannopyranoside as donor substrate and 14% (w/v) branched CDs or CDs as acceptor (Table III and Fig. 3). The yield of transmannosylated products increases with increasing methyl α-mannopyranoside concentration, as the efficiency of transmannosylation is en-

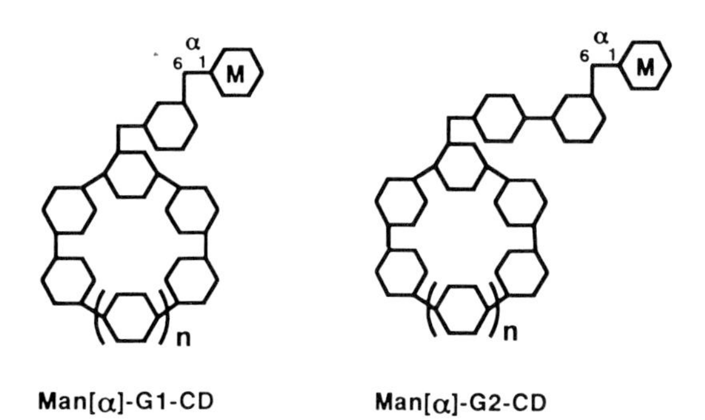

FIG. 3. Abbreviated structural diagrams for trans-α-mannosylated products from G1-CDs and G2-CDs. ⟨M⟩, Mannosyl residue; other abbreviations are the same as those in Fig. 1.

hanced. Almond α-mannosidase produces the same transmannosylated products from branched CDs, but the amounts of products are lower than those obtained with the jack bean enzyme.[10]

Materials

- Methyl α-mannopyranoside, reagent grade
- G1-α and -β cyclodextrins, and G2-α and -β cyclodextrins, highest grade
- Jack bean α-mannosidase [suspension in 3.0 *M* $(NH_4)_2SO_4$ and 0.1 m*M* zinc acetate solution, pH 7.5; specific activity, 20 units/mg; Sigma], dialyzed against 50 m*M* phosphate buffer, pH 7, before use
- Almond α-mannosidase (solution in 10 m*M* potassium phosphate buffer, pH 6.0, containing 0.1% sodium azide; specific activity, 15–30 units/mg; Sigma), dialyzed against 50 m*M* phosphate buffer, pH 7, before use

Assay of α-Mannosidase Activity. The method of assay and the definition of enzyme activity are the same as those used for studies of β-galactosidase, except for the use of *p*-nitrophenyl α-mannopyranoside as substrate.

Procedure. Three grams of methyl α-mannopyranoside and 3 g of branched CDs (G1-α, G1-β, G2-α, or G2-βCD) are dissolved in 9 ml of 50 m*M* hot acetate buffer (pH 4.5). After cooling to 40°, jack bean α-mannosidase (25 units, 1.0 ml) is added to the solution, and the mixture is incubated at 40° for 48 hr. The reaction mixture is then heated at 100° for 15 min to stop the reaction and subsequently centrifuged to remove insoluble materials. Each of the transfer products is isolated by HPLC.

Isolation and Analysis

High-Performance Liquid Chromatography

High-performance liquid chromatography (HPLC) is useful for the isolation and characterization of heterogeneous branched CDs. In general, so-called amino columns and reversed-phase columns can be employed for efficient isolation of individual heterogeneous branched CDs from a mixture of transglycosylation products.

The aminopropyl-bonded silica column is a typical amino column. The sequence of elution of carbohydrates with this column and acetonitrile–water system follows the order of molecular size. The elution time therefore gives qualitative information about the molecular size of the carbohydrate, and components of a reaction mixture are initially divided into molecular size groups on this type of column. However, aminopropyl-bonded silica columns have limited lifetimes, owing to hydrolysis of the bonded phase. Consequently, for this separation step we have usually used two new HPLC media, TSK-Gel Amide-80 (Tosoh Co. Ltd., Tokyo, Japan), containing carbamoyl instead of aminopropyl groups, and Asahipak NH2P-50 (Asahi Kasei), a polyamine-bonded vinyl alcohol copolymer gel. The packing stability of these new media is much better than that of aminopropyl-bonded silica.

In contrast, octadecylsilyl (ODS) silica is used in reversed-phase columns, and its mechanism of separation is therefore probably an example of hydrophobic chromatography, that is, increased retention with decreasing solubility in water. Isomers of the same molecular size in a fraction obtained by HPLC on an amino column can be isolated by repeat chromatography on an ODS column.

*Isolation of Gal[α]-G1-α**I**, -α**II**, and -α**III**.* The by-products (monosaccharides and acyclic oligosaccharides) from the reaction mixture prepared from G1-αCD by transgalactosylation with coffee bean α-galactosidase are removed during preparative HPLC on an ODS column (100 × 26.4 cm i.d.) by elution with water, and then the branched CDs are eluted from the column with 10% ethanol. The fraction containing the desired monogalactosylated G1-αCDs [degree of polymerization (DP) 8] is separated from the starting material G1-αCD (DP 7) and overgalactosylated CDs (DP >8) on a TSK-Gel Amide-80 column (250 × 4.6 mm i.d.) with acetonitrile–water (63 : 37, v/v) at 35° and flow rate of 1 ml/min.[11]

Figure 4 shows a chromatogram of the DP 8 fraction obtained with a YMC-Pack A-312-3 ODS column (3 μm, 150 × 6 mm i.d.) (YMC, Kyoto, Japan). The main products are three isomers of Gal[α]-G1-αCD (**I**, **II**, and **III**); they are isolated by repetition of the chromatography on a YMC-Pack SH-345-5 semipreparative ODS column (5 μm, 500 × 20 mm i.d.) with 5–8% methanol at 30° and a flow rate of 3–5 ml/min. Fast atom bombardment-mass spectrometry (FAB-MS) and nuclear magnetic resonance (NMR) spectroscopies demonstrate that Gal[α]-G1-α**III** is a 6-*O*-α-D-galactopyranosylated derivative on the glucose unit in the side chain of G1-αCD and a single compound, and that Gal[α]-G1-α**I** and -α**II** are, respectively, 6-*O*- and 2-*O*-α-D-galactopyranosylated derivatives on a glucose unit in the αCD ring, each a mixture containing positional isomers. On HPLC with a graphitized carbon column, Carbonex (Tonen, Tokyo, Japan) or Hypercarb (Shandon, Cheshire, England), **I** and **II** manifest two and four peaks, respectively (K. Koizumi and Y. Okada, 1993, unpublished data). The retention characteristics of carbohydrates on the graphitized

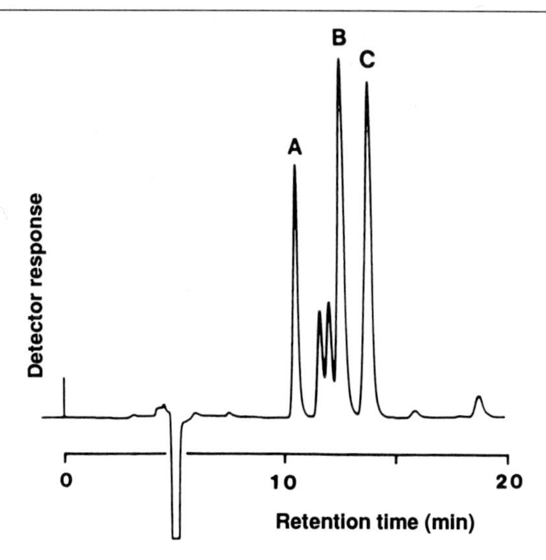

FIG. 4. Chromatogram of a mixture containing three isomers of Gal[α]-G1-αCD (**I, II, III**). Peaks A, B, and C are **III, II,** and **I,** respectively. Chromatographic conditions: column, YMC-Pack A-312-3; eluent, CH_3OH–water (5 : 95, v/v); flow rate, 0.7 ml/min; temperature, 30°; detector, Shodex RI-71.

carbon column are principally the manifestation of an adsorption mechanism, and it has been shown that the unique power of resolution of this column enables an excellent separation of positional isomers.[16]

Information regarding the characteristic chromatographic behavior of heterogeneous CDs on HPLC columns of differing separation modes is useful for predicting their structures.

Elution Profiles of β-D-Galactosylated G1- and G2-βCDs on Reversed-Phase Column and Amino Column. Figure 5 shows the elution profiles of Gal[β]-G1 and G2-β**1**, **2**, **3** on a YMC-Pack A-312-3 column. The 6-*O*-β-D-galactosylated isomer (**3**) moves faster than the corresponding 4-*O*-β-D-galactosylated isomer (**1**). On the other hand, the order of elution of the two isomers on an Asahipak NH2P-50 column is reversed (Fig. 6). In the αCD and γCD series, the retention time (t_R) of each corresponding derivative differs; nevertheless, the order of elution of isomers is the same as that in the βCD series.[17]

The order of elution on a reversed-phase column may suggest the relative solubility of CDs in water.

[16] K. Koizumi, Y. Okada, and M. Fukuda, *Carbohydr. Res.* **215,** 67 (1991).

[17] K. Koizumi, T. Tanimoto, K. Fujita, K. Hara, N. Kuwahara, and S. Kitahata, *Carbohydr. Res.* **238,** 75 (1993).

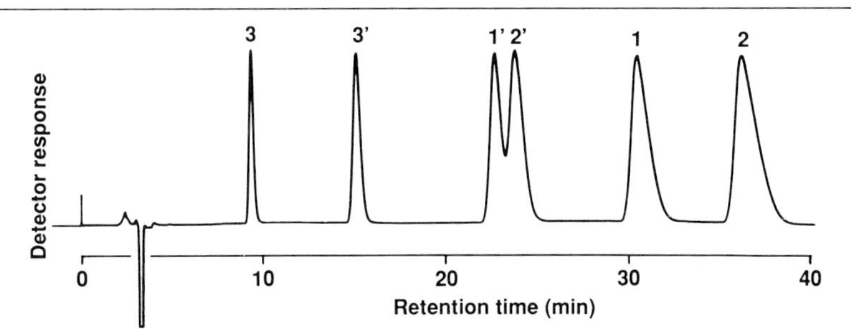

FIG. 5. Elution profiles of Gal[β]-G1 and G2-β**1**, **2**, **3** on an ODS column. Peaks 1, 2, and 3 are Gal[β]-G1-β**1**, **2**, and **3**; peaks 1', 2', and 3' are Gal[β]-G2-β**1**, **2**, and **3**, respectively. Chromatographic conditions: column, YMC-Pack A-312-3; eluent, CH_3OH–water (7 : 93, v/v); flow rate, 1 ml/min; temperature, 25°. (Adapted from Koizumi *et al.*,[17] with permission of Elsevier Science Publishers BV.)

Order of Elution of α-D-Glucosyl-, α-D-Galactosyl-, α-D-Mannosyl-αCDs, and Parent αCD. *O*-α-D-Glucosyl-(1 → 6)-αCD (Glc-αCD) is more soluble in water than its parent αCD, and the solubility of *O*-α-D-galactosyl-(1 → 6)-αCD (Gal-αCD) in water is higher than that of Glc-αCD. The solubilities in water (millimoles/100 ml) at 25° of αCD, Glc-αCD, and Gal-αCD are 18, 80, and 95, respectively. It can be predicted from the t_R values shown in Fig. 7 that *O*-α-D-mannosyl-(1 → 6)-αCD (Man-αCD, chemically synthesized) should be the most soluble in water.

*Order of Elution of 6-O-α-D-Galactosylated CD (Gal[α]-**I**) and 2-O-α-D-Galactosylated CD (Gal[α]-**II**).* On a YMC-Pack A-312-3 reversed-phase column, with methanol–water as eluent, the minor product (**II**) of transga-

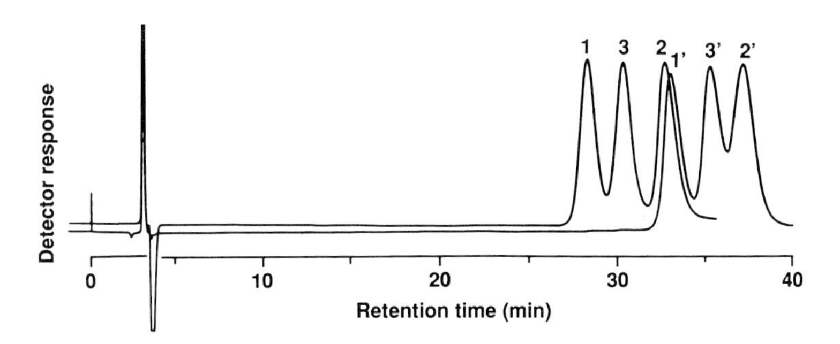

FIG. 6. Elution profiles of Gal[β]-G1 and G2-β**1**, **2**, **3** on an amino column. Peaks 1, 2, and 3 are Gal[β]-G1-β**1**, **2**, and **3**; peaks 1', 2', and 3' are Gal[β]-G2-β**1**, **2**, and **3**, respectively. Chromatographic conditions: column, Asahipak NH2P-50; eluent, CH_3CN–water (64 : 36, v/v); flow rate, 0.8 ml/min; temperature, 40°. (Adapted from Koizumi *et al.*,[17] with permission of Elsevier Science Publishers BV.)

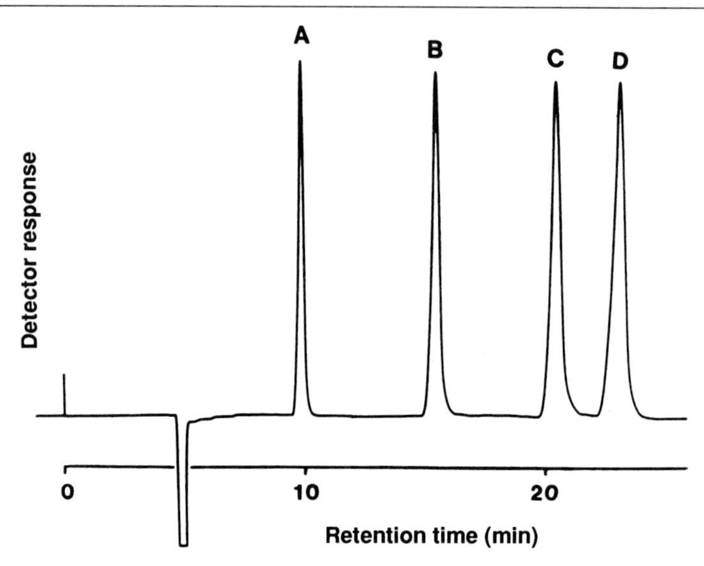

FIG. 7. Chromatogram of (A) α-D-mannosyl-, (B) α-D-galactosyl-, (C) α-D-glucosyl-αCDs, and (D) parent αCD on a reversed-phase column. Chromatographic conditions were as in Fig. 4.

lactosylation to CDs by coffee bean α-galactosidase is in the case of αCD derivatives eluted prior to the major product (**I**), whereas **II** elutes after **I** in the case of βCD and γCD derivatives.[11] The t_R values (minutes) of γCD derivatives, which move most rapidly, are 13.2 for Gal[α]-γ**I** and 13.8 for **II** with methanol–water (5 : 95, v/v) at 30° and a flow rate of 0.7 ml/min; the t_R values of Gal[α]-α**I** and **II** under the same conditions are 16.5 and 15.4, respectively. The βCD derivatives elute too slowly under these conditions, and the rate of flow of eluent is therefore increased to 1 ml/min. At this flow rate, the t_R values of Gal[α]-β**I** and **II** are 24.2 and 27.7, respectively.

Enzymatic Analysis

Heterogeneous branched CDs are hydrolyzed by the enzyme used for their syntheses to the parent CD or branched CD and D-galactose or D-mannose. The molar ratio calculated for this reaction can be used to help determine the composition of a heterogeneous branched CD.

*Hydrolysis of Gal[α]-α**I***. Gal[α]-α**I** (1 mg) in 50 m*M* acetate buffer (pH 6.5, 0.1 ml) is incubated with coffee bean α-galactosidase (0.16 units) at 40° for 30 min, and the reaction product is analyzed by HPLC on a TSK-Gel Amide-80 column (250 × 4.6 mm i.d.) with acetonitrile–water

(63 : 37, v/v) at 35° and a flow rate of 1 ml/min.[9] Gal[α]-α**I** is completely hydrolyzed to galactose and αCD in a molar ratio of 1 : 1. This finding suggests that Gal[α]-α**I** is mono-*O*-α-D-galactosyl-αCD.

Fast Atom Bombardment-Mass Spectrometry

The FAB-MS study of heterogeneous branched CDs can be used to determine not only estimates of the molecular weights but also whether a galactose or a mannose residue is linked to the side chain of the homogeneous branched CD or directly to the parent CD ring.

*FAB-MS of Gal[α]-G1-α**I**, -α**II**, and -α**III**.* Figure 8 shows FAB-MS spectra for Gal[α]-G1-α**I**, -α**II**, and -α**III**.[11] The spectra are obtained in the negative ion mode with a JEOL JMS-DX 303 mass spectrometer (JEOL

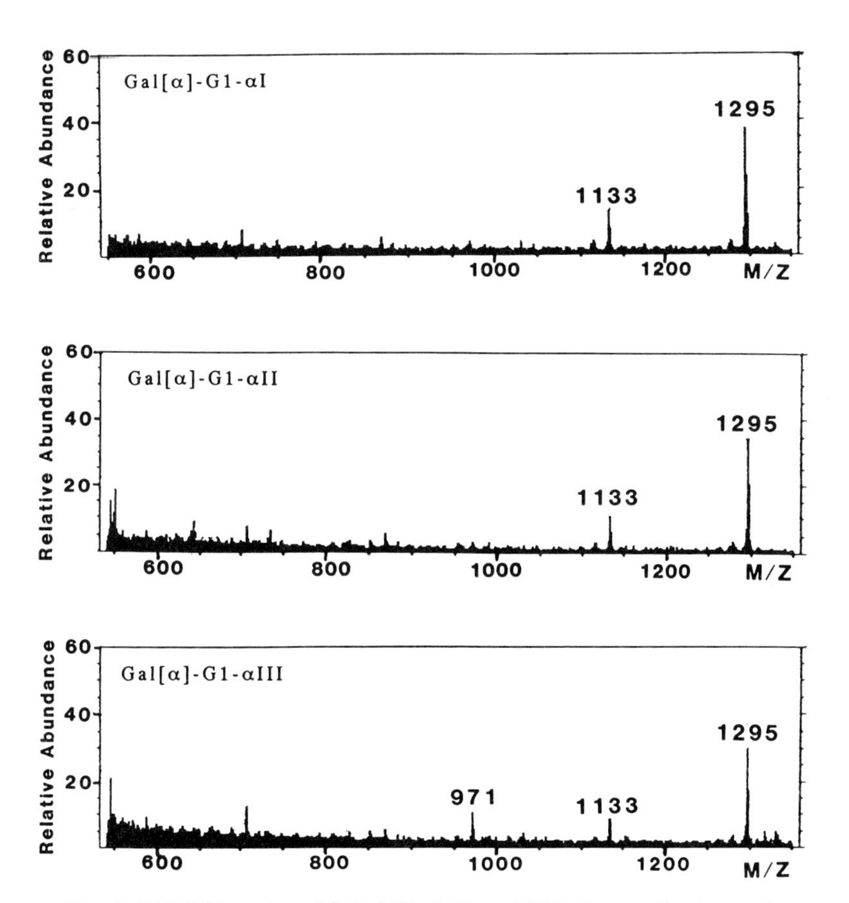

FIG. 8. FAB-MS spectra of Gal[α]-Gl-α**I**, **II**, and **III** in the negative ion mode.

Ltd. Tokyo, Japan) using Xe atoms having a kinetic energy equivalent to 6 kV at an accelerating voltage of 3 kV. The mass marker is calibrated with perfluoroalkylphosphazine (Ultra Mark, JEOL), and glycerol is used as the matrix.

The $[M - H]^-$ peak is clearly observed at *m/z* 1295 in each spectrum (Fig. 8), indicating that these compounds are all monogalactosylated glucosyl-αCDs composed of 8 hexose units. A peak for the fragment ion $[M - G - H]^-$ (*m/z* 1133) is present in all spectra, but a $[M - 2G - H]^-$ peak (*m/z* 971) is observed only in the Gal[α]-G1-α**III** spectrum. Because all the fragment ions must have been formed through one cleavage of the side chain (primary fragments), the side chain of Gal[α]-G1-α**III** must be a disaccharide, whereas those of Gal[α]-G1-α**I** and -α**II** must be two monosaccharides; more specifically, a galactosyl residue is on the side-chain glucose in the Gal[α]-G1-α**III** molecule but on the ring glucose in the case of Gal[α]-G1-α**I** and -α**II**.

Nuclear Magnetic Resonance Spectroscopy

^{13}C NMR spectroscopy is a very powerful tool in the structural analysis of heterogeneous branched CDs. The ^{13}C resonances of all carbons in the spectra of certain typical heterogeneous branched CDs can be assigned using ^{1}H–^{1}H correlation spectroscopy (COSY) and ^{1}H–^{13}C COSY methods. Spectra of members of a homologous series are similar to one another, and hence assignments of signals can be made by analogy.

The position at which substitution with a galactosyl or mannosyl residue occurs on the acceptor CD molecule is readily determined, since a substituent on the oxygen atom attached to any carbon of the carbohydrate affects the chemical shift of that carbon atom, moving it downfield by 8–11 ppm. Fortunately, assignments of the signals of C-6, which is the most favorable position for transgalactosylation and transmannosylation, can be confirmed by the INEPT (insensitive nucleus enhancement by polarization transfer) method,[18] using $\Delta = 3/4J$, or the DEPT (direct enhancement by polarization transfer) method,[19] since the CH_2 signal can easily be recognized as the negative peak with either of the methods.

*Assignments of Signals in ^{13}C spectrum of Gal[β]-G1-β**1***. Figure 9 shows the ^{1}H–^{13}C COSY spectrum of Gal[β]-G1-β**1** in D_2O at 50°.[17] All proton signals in the ^{1}H spectrum of the compound are assigned using the ^{1}H–^{1}H COSY method prior to assignments of ^{13}C resonances. The NMR spectra are obtained with a JEOL GSX-500 spectrometer, and chemical shifts are expressed as parts per million (ppm) downfield from the signal

[18] G. A. Morris and R. Freeman, *J. Am. Chem. Soc.* **101,** 760 (1979).
[19] D. M. Doddrell and D. T. Pegg, *J. Am. Chem. Soc.* **102,** 6388 (1980).

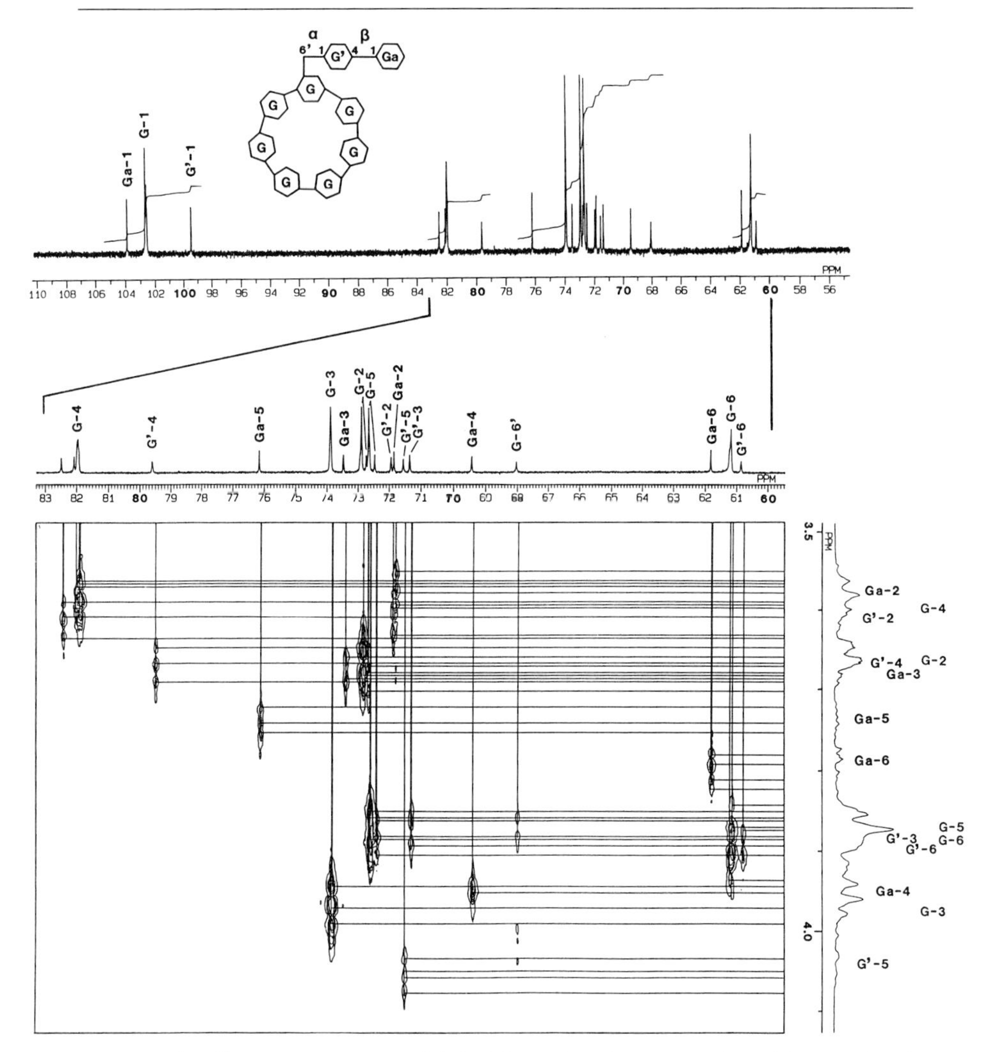

FIG. 9. $^1H-^{13}C$ COSY spectrum of Gal[β]-G1-β**1** in D_2O at 50°. (Adapted from Koizumi *et al.*,[17] with permission of Elsevier Science Publishers BV.)

of $(CH_3)_4Si$ referred to external 1,4-dioxane (67.40 ppm). The conditions used for NMR spectroscopy are listed in Table IV.

The C-6 signals of hexopyranose residues normally appear at the highest field (δ 61–62 ppm). The resonance of the branch-point C-6 (G-6′) of Gal[β]-G1-β**I** is shifted downfield by approximately 7 ppm from the usual range for the pyranose C-6. The C-1 resonances of the glucose residues

TABLE IV
CONDITION FOR NMR MEASUREMENTS

Parameter	^{13}C NMR	^{1}H–^{1}H COSY	^{1}H–^{13}C COSY
Frequency range (Hz)	7002.8	1000	7002.8
Acquisition time (sec)	4.679	1.024	0.292
Data points	65,536	2048	4096
Column points		1024	512
Column frequency		1000	1000

of the βCD ring usually appear at approximately 102.6 ppm, and those of side-chain glucose α(1 → 6)-linked to the CD ring and of galactose β(1 → 4)-linked to the side-chain glucose are observed at δ 99.5 and 103.9 ppm, respectively. Although the C-4 resonances of glucose residues composing the CD ring occur at δ approximately 82 ppm, having undergone a glycosylation shift of approximately 12 ppm downfield from the usual range for C-4 with free OH (δ approximately 70 ppm), the C-4 signal of side-chain glucose (G′-4) is also shifted downfield by about 9.6 ppm. These findings suggest that galactose is probably linked to C-4 of the side-chain glucose.

*Comparison of Chemical Shifts of C-1, C-4, and C-6 in ^{13}C NMR Spectra of Gal[β]-G1-β**1**, **2**, and **3**.* The ^{13}C chemical shifts of C-1, C-4, and C-6 of Gal[β]-G1-βCDs (**1**, **2**, and **3**), information which can be used for structural analysis, are summarized in Table V.[17] Large downfield shifts of one and two C-4 signal(s) are observed in the spectra of **1** and **2**, respectively, whereas one C-6 signal (G′-6) other than that of the branch-point C-6 (G-6′) shifts downfield in the spectrum of **3**. In all spectra of **1**, **2**, and **3**, the C-1 resonances of the side-chain glucose residues, which are α(1 → 6)-linked to the CD ring, (G′-1) appear at higher fields than did the C-1 signal of the CD ring (G-1), whereas β(1 → 4)- or β(1 → 6)-linked C-1 signals of the galactose residues (Ga-1 and Ga′-1) are observed at lower fields than G-1. For **1** and **3**, the ratio of signal intensities arising from C-1 atoms of the CD ring (G-1), the side-chain glucose (G′-1), and galactose (Ga-1) is 7 : 1 : 1, whereas in the spectrum of **2** there is one additional C-1 signal arising from galactose (Ga′-1).

These NMR findings confirm that Gal[β]-G1-β**1**, **2**, and **3** are *O*-β-D-galactopyranosyl-(1 → 4)-*O*-α-D-glucopyranosyl-(1 → 6)-βCD, *O*-β-D-galactopyranosyl-(1 → 4)-*O*-β-D-galactopyranosyl-(1 → 4)-*O*-α-D-glucopyranosyl-(1 → 6)-βCD, and *O*-β-D-galactopyranosyl-(1 → 6)-*O*-α-D-glucopyranosyl-(1 → 6)-βCD, respectively.

*Comparison of ^{13}C NMR Spectra of Gal[α]-G1-α**III** and Gal[β]-G1-α**3**.* Although Gal[α]-G1-α**III**[11] (**A**) and Gal[β]-G1-α**3**[17] (**B**) are both

TABLE V
CHEMICAL SHIFTS OF C-1, C-4, AND C-6 IN ^{13}C NMR SPECTRA[a] OF Gal[β]-G1-β-CYCLODEXTRINS (**1**, **2**, AND **3**)[b]

	1		**2**		**3**	
Atom[c]	δ (ppm)	Integral	δ (ppm)	Integral	δ (ppm)	Integral
Ga′-1			105.07	1		
Ga-1	103.90	1	103.82	1	104.14	1
G-1	102.68	7	102.66	7	102.68	7
	102.66		102.63		102.66	
	102.64		102.58		102.64	
	102.61		102.55		102.62	
	102.60				102.60	
	102.58					
	102.57					
G′-1	99.47	1	99.42	1	99.95	1
G-4[d]	82.53	7	82.52	7	82.45	7
	82.12		82.09		82.28	
	82.02		81.99		82.06	
	82.01		81.96		82.02	
	82.00		81.95		82.00	
	81.98				81.99	
	81.96				81.97	
G′-4	79.64[d]	1	79.60[d]	1	70.27	1
Ga-4	69.48	1	77.97[d]	1	69.58	1
Ga′-4			69.51	1		
G-6′[d]	68.07	1	68.06	1	68.12	1
Ga′-6			61.84	1		
Ga-6	61.87	1	61.59	1	61.91	1
G-6	61.30	6	61.20	6	61.32	6
	61.27				61.25	
	61.23					
G′-6	60.89	1	60.81	1	69.00[d]	1

[a] Spectra were acquired at 125.65 MHz, in D_2O at 50°.
[b] Adapted from Koizumi *et al.*,[17] with permission of Elsevier Science Publishers BV.
[c] G-1, -4, and -6 are the C-1, -4, and -6 atoms of the ring D-glucopyranose units. G-6′, is the C-6 atom of the ring D-glucopyranose unit involved in branching. G′-1, -4, and -6 are the carbon atoms of the side-chain D-glucopyranose unit. Ga and Ga′-1, -4, and -6 are the carbon atoms of D-galactopyranose units.
[d] C-4 and C-6 atoms involved in linkages.

D-galactopyranosyl-(1 → 6)-G1-αCD, the galactosyl residues of **A** and **B** are α- and β-linked to C-6 of the side-chain glucose, respectively. The structures are confirmed by the differences in chemical shifts of C-6 signals of the side-chain glucose (G′-6) and of C-1 signals of galactosyl residues (Ga-1) in the ^{13}C NMR spectra (Fig. 10).

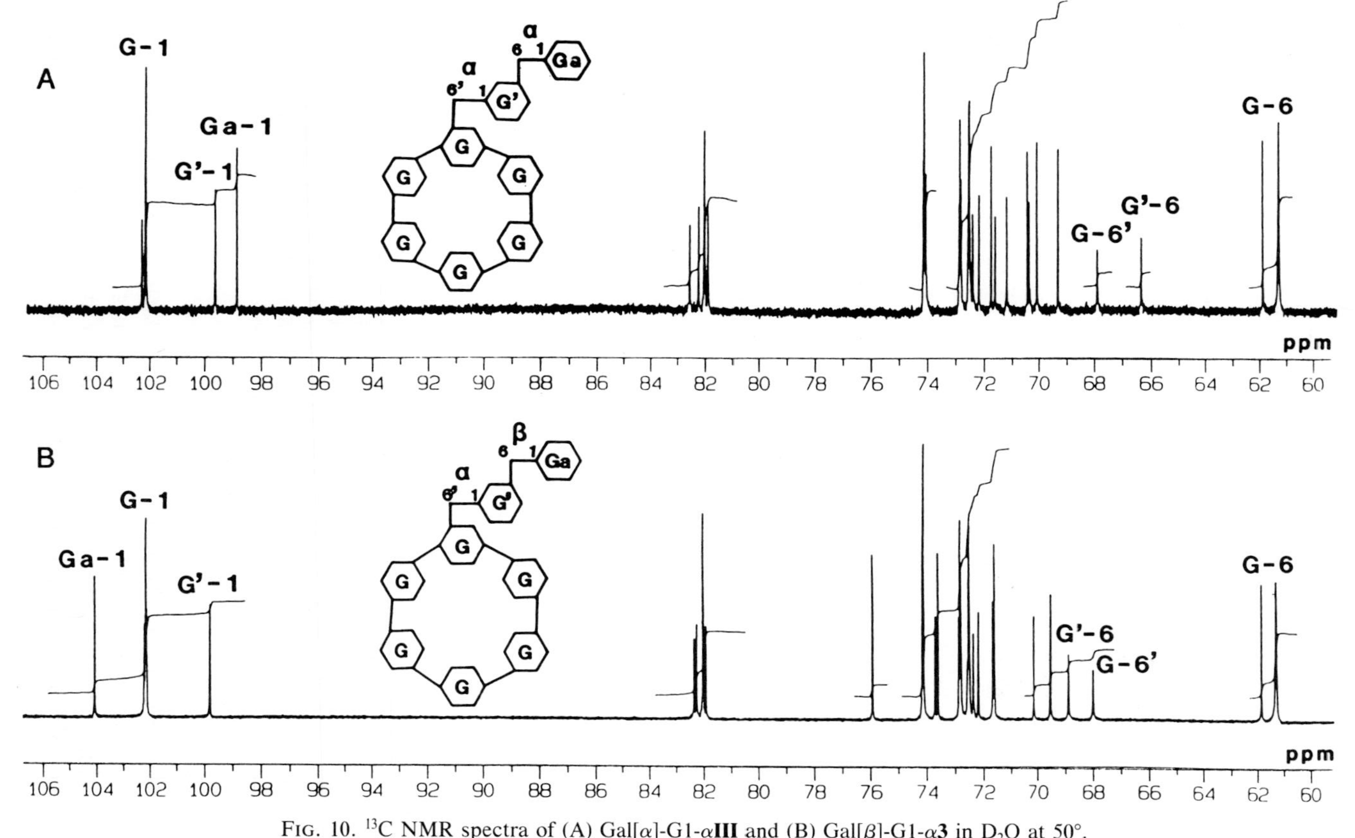

FIG. 10. ^{13}C NMR spectra of (A) Gal[α]-G1-α**III** and (B) Gal[β]-G1-α**3** in D_2O at 50°.

Comparison of ^{13}C NMR Spectra of Gal[α]-βI and -βII. One of the C-6 resonances of Gal[α]-β**I** is shifted to δ 68.2 ppm, away from the other C-6 resonances (δ 61.2–61.3 ppm), whereas in the ^{13}C NMR spectrum of Gal[α]-β**II**, one C-2 resonance appears at δ 78.3 ppm, having undergone a glycosylation shift of approximately 6 ppm downfield from the other C-2 resonances. These NMR findings indicate that the galactosyl residue of Gal[α]-β**I** is linked to the C-6 of glucose in the βCD ring, whereas that of Gal[α]-β**II** is bonded to the C-2 of a ring glucose residue.

Structural Analysis of Transmannosylated Product of G1-CD by ^{13}C NMR Spectroscopy. Figure 11 shows the ^{13}C NMR spectra of the trans-mannosylated product of G1-βCD obtained using the DEPT method in D_2O at 50° and 80°.[10] One of three signals, which appears at approximately 68 ppm, is a positive peak, and the other two are negative peaks. The positive peak is assigned to the C-4 of the mannosyl residue (M-4), whereas the negative peaks are the signals of glycosylated C-6, one of which is assigned to the carbon atom of the branch point on the CD ring (G-6′) and the other to C-6 on the side-chain glucose (G′-6). In the spectrum measured at 50°, the signal of G-6′ is considerably broadened, and this broadening is improved somewhat by the measurement of the spectrum at 80°. These findings suggest that free rotation of the linkage at the branch point is hindered. The C-1 resonance of the mannosyl residue (M-1) appears at δ 100.3 ppm, suggesting that the mannosyl residue is linked to C-6 of the side-chain glucose of G1-βCD in the α configuration.

The ^{13}C NMR spectrum of the transmannosylated product of G1-αCD obtained in D_2O at 50° is similar to that of the G1-βCD derivative described above, except that the signal of the C-6 atom of the branch-point ring glucose unit (G-6′) is normal.

Methylation Analysis

Methylation analysis is carried out in order to ensure that in 6-*O*-galactosylated or 6-*O*-mannosylated G2-CDs the galactosyl or mannosyl residue is substituted on the oxygen at C-6 of the nonreducing end of the side-chain maltose (G2). For example, a sample of Gal[α]-G2-α**III** (7 mg) is dried over molecular sieves (0.4 nm) for 29 hr at 100° and stirred with trimethyl phosphate (1 ml) at 45° until the powder completely dissolves. To the clear solution are added 2,6-di-(*tert*-butyl)pyridine (150 μl) and methyl trifluoromethane sulfonate (100 μl), and the mixture is allowed to stand for 3 hr at 50°. The mixture is then applied to a Sep-Pak C_{18} cartridge or Sep-Pak Plus C_{18} environmental cartridge (Waters, Milford, MA). After removing the remaining reagents by washing with water, permethylated Gal[α]-G2-α**III** is recovered with acetone, concentrated, and hydrolyzed

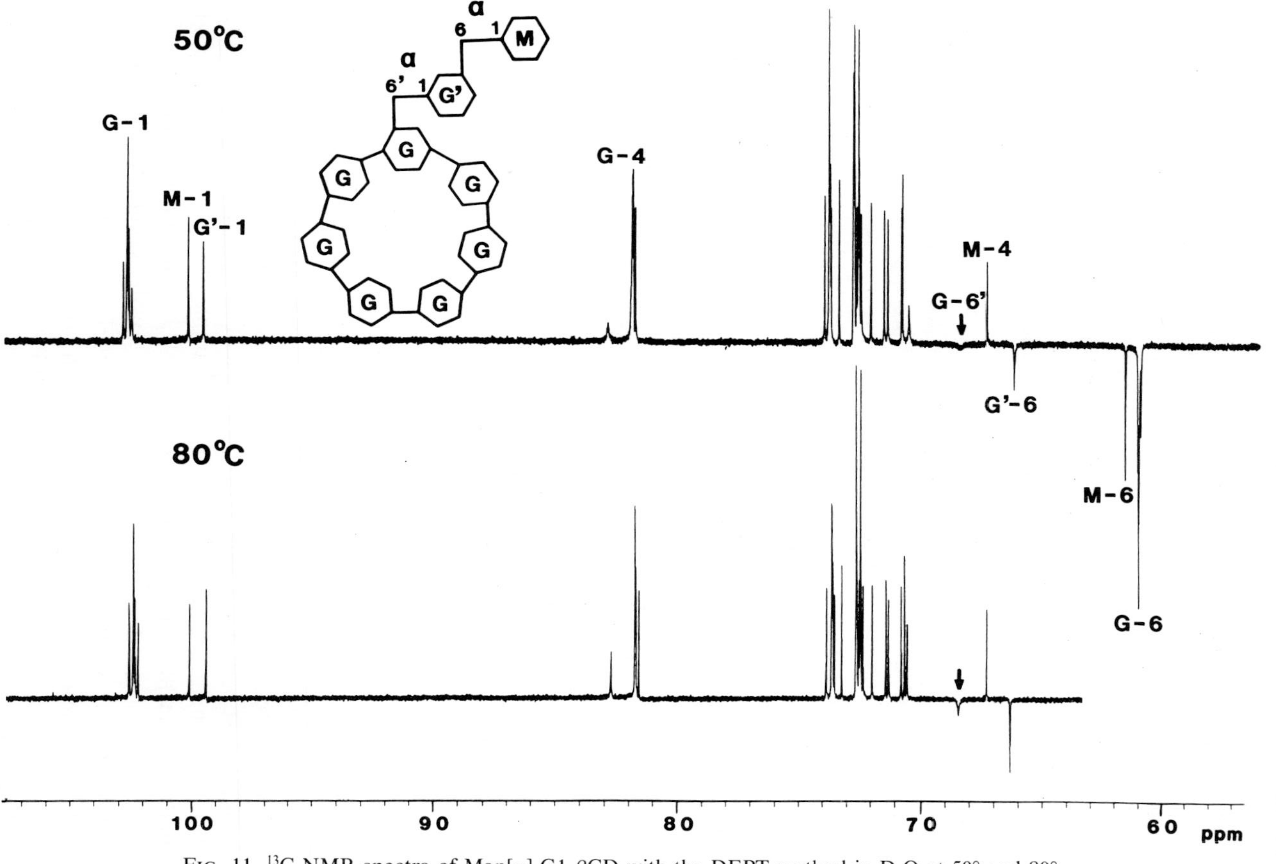

FIG. 11. ^{13}C NMR spectra of Man[α]-G1-βCD with the DEPT method in D_2O at 50° and 80°.

with 1.5 *M* trifluoroacetic acid (1 ml) at 100° overnight. The partially methylated hydrolyzates are reduced with sodium borohydrate (100 mg) in water (2 ml), and the solution is neutralized with acetic acid, treated with Amberlite IR-120B, filtered, and evaporated to dryness. The residue is acetylated with acetic anhydride (1 ml) in pyridine (2 ml) overnight. The partially methylated alditol acetates obtained are analyzed with a Shimadzu GCMS-QP2000 gas chromatograph–mass spectrometer on a ULBON HR-SS-10 column (50 m × 0.25 mm i.d.) (Shinwa Kako, Kyoto, Japan) and identified by the retention times and mass patterns. Results indicate the production of 2,3,4-tri-*O*-methyl-D-glucitol peracetate, confirming that Gal[α]-G2-α**III** is *O*-α-D-galactopyranosyl-(1 → 6)-*O*-α-D-glucopyranosyl-(1 → 4)-*O*-α-D-glucopyranosyl-(1 → 6)-cyclomaltohexaose.

[6] Solid-Phase Synthesis of *O*-Glycopeptides

By THOMAS NORBERG, BJÖRN LÜNING, and JAN TEJBRANT

Introduction

One of the most significant discoveries in protein chemistry is that proteins of living organisms have a number of different groups (e.g., sulfates, phosphates, and carbohydrates) covalently attached to the polypeptide chain. The most complex of these groups are the carbohydrates, the protein–carbohyrate combinations being known as glycoproteins (or proteoglycans, if the carbohydrate is a polysaccharide). In fact, a majority of proteins known today have been found, when studied more closely, to be glycoproteins. Biosynthetically, the carbohydrates of the glycoproteins are attached (co- or posttranslationally) by glycosyltransferase enzymes, each enzyme being specific for a particular monosaccharide unit and linkage type. Both stepwise and blockwise attachment of the carbohydrate residues can occur, depending on the type of linkage and carbohydrate. After attachment, the carbohydrate chains are sometimes modified ("trimmed") by glycosidase enzymes. The extent of glycosidation and trimming of a specific polypeptide chain is a complex function of factors such as the availability and reactivity of nucleotide sugars, enzymes, and cofactors. These factors vary according to the status and type of cell where biosynthesis takes place. Therefore, different glycoforms of the same protein are regularly found, differing in attached carbohydrate structure. Different glycoforms can have different net charges, electrophoretic mobilities, and molecular weights. The glycoforms can also differ in other properties, such as protease stability, affinity to receptors, and conformational rigidity. Perhaps the most well-known example of difference between glycoforms is the large change in serum clearance rate on desialylation of glycoproteins. The desialylated glycoforms expose a previously "hidden" terminal galactosyl unit, which results in their rapid clearance (by liver uptake) from serum.[1] Similarly, several glycoproteins considered for use as therapeutic agents [e.g., EPO, tissue plasminogen activator (t-PA)] display different activity and pharmacokinetic profiles, depending on the glycoform.[2–11]

To study the influence of carbohydrate variation on the properties of a peptide or a protein, it has become important to have access to several different glycoforms. This can be achieved by separating a mixture of glycoforms into components[12] or by using different biological sources that produce different glycoforms. Furthermore, enzymatic glycosylation or deglycosylation can convert one glycoform to another. However, the most generally applicable approach for obtaining defined glycoprotein fragments is chemical synthesis.[13] Technically, this can be carried out either in solution or, preferably for larger fragments, on a solid phase. The 9-fluorenylmethoxycarbonyl (Fmoc) modification of the orginal Merrifield scheme for solid-phase peptide synthesis is well suited for this purpose. The literature (1988 and onward) describes numerous successful syntheses of glycopeptides using this technique. This chapter reviews work published until mid-1993 on the synthesis of *O*-glycopeptides, with special emphasis on solid-phase syntheses. The experimental section then de-

[1] G. Ashwell and J. Hartford, *Annu.Rev. Biochem.* **51,** 531 (1982).

[2] S. Dubé, W. Fisher, and J. S. Powell, *J. Biol. Chem.* **263,** 17516 (1988).

[3] E. Delorme, T. Lorenzini, J. Giffin, F. Martin, F. Jacobsen, T. Boone, and S. Elliott, *Biochemistry* **31,** 9871 (1992).

[4] M. Higuchi, M. Oheda, H. Kuboniwa, K. Tomonoh, Y. Shimonaka, and N. Ochi, *J. Biol. Chem.* **267,** 7703 (1992).

[5] L. Narhi, T. Arakawa, K. Aoki, R. Elmore, M. Rodhe, T. Boone, and T. Strickland, *J. Biol. Chem.* **266,** 23022 (1991).

[6] M. Takeuchi, *et al., Proc. Natl. Acad. Sci. U.S.A.* **86,** 7819 (1989).

[7] M. Takeuchi, S. Takasaki, M. Shimada, and A. Kobata, *J. Biol. Chem.* **265,** 12127 (1990).

[8] E. Tsuda, G. Kawanishi, M. Ueda, S. Masuda, and R. Sasaki, *Eur. J. Biochem.* **188,** 405 (1990).

[9] K. Yamaguchi, K. Akai, G. Kawanishi, M. Ueda, S. Masuda, and R. Sasaki, *J. Biol. Chem.* **266,** 20434 (1991).

[10] T. Marti, J. Schaller, E. E. Rickli, K. Schmid, J. P. Kamerling, G. J. Gerwig, H. Van Halbeek, and J. F. G. Vliegenthart, *Eur. J. Biochem.* **173,** 57 (1988).

[11] A. J. Wittwer and S. C. Howard, *Biochemistry* **29,** 4175 (1990).

[12] P. Rudd, I. Scragg, E. Coghill, and R. Dwek, *Glycoconjugate J.* **9,** 89 (1992).

[13] H. Kunz, *Angew. Chem., Int. Ed. Engl.* **26,** 294 (1987).

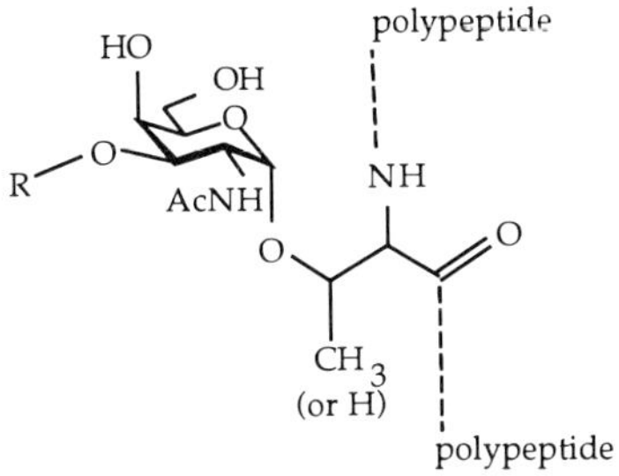

FIG. 1. The 2-acetamido-2-deoxy-α-D-galactopyranosyl-serine/threonine linkage. R is H or other sugar residue(s).

scribes details of the synthesis of a representative glycopeptide from our laboratory.

Strategies for Synthesis of *O*-Glycopeptides

An *O*-glycopeptide is a peptide where one or more of the amino acid side-chain hydroxyl groups are O-glycosylated with a mono- or oligosaccharide. The most common carbohydrate–peptide linkage type in *O*-glycopeptides is that involving 2-acetamido-2-deoxy-α-D-galactopyranosyl to serine or threonine (Fig. 1), but other linkages have also been found, such as β-D-xylopyranosyl-serine/threonine, α-D-mannopyranosyl-serine/threonine,[14] α-L-fucopyranosyl-serine/threonine,[15,16] β-D-glucopyranosyl-serine,[17,18] glucopyranosyl-tyrosine,[19,20] and β-D-galactopyranosyl-hydroxylysine.[21,22] In *O*-glycopeptide synthesis, a crucial step is to create the desired O-glycosidic linkage between a carbohydrate donor and either a peptide or an amino acid. To accomplish this, two different approaches can be used. The first is to synthesize an appropriately protected peptide

[14] P. Gellerfors, K. Axelsson, A. Helander, S. Johansson, L. Kenne, S. Lindqvist, B. Pavlu, A. Skottner, and L. Fryklund, *J. Biol. Chem.* **264,** 11444 (1989).
[15] H. Nishimura, T. Takao, S. Hase, Y. Shimonishi, and S. Iwanaga, *J. Biol. Chem.* **267,** 17520 (1992).
[16] P. Hallgren, A. Lundblad, and S. Svensson, *J. Biol. Chem.* **250,** 5312 (1975).
[17] H. Nishimura, S.-I. Kawabata, W. Kisiel, S. Hase, T. Ikenaka, T. Takao, Y. Shimonishi, and S. Iwanaga, *J. Biol. Chem.* **264,** 20320 (1989).
[18] S. Hase, *et al., J. Biochem. (Tokyo)* **104,** 867 (1988).
[19] I. Rodriguez and W. Whelan, *Biochem. Biophys. Res. Commun.* **132,** 829 (1985).
[20] C. Smythe, F. Caudwell, M. Ferguson, and P. Cohen, *EMBO J.* **7,** 2681 (1988).
[21] W. Butler and L. Cunningham, *J. Biol. Chem.* **241,** 3882 (1966).
[22] R. Spiro, *J. Biol. Chem.* **242,** 4813 (1967).

and then O-glycosylate.[23–32] This approach is seldom used nowadays, however, since larger glycosyl acceptors such as protected peptides display somewhat unpredictable reactivity and/or solubility under the glycosylation conditions. Once accomplished, the products of such a synthesis (e.g., α/β mixtures) have to be separated, also a difficult task.

The second approach is to use preglycosylated amino acids in a peptide synthesis. This approach is more common, utilizing either solution or solid-phase synthesis techniques. The presence of the glycosyl moiety does not seriously affect the coupling reactivity of the amino acid, and, with the proper glycosylated amino acid at hand, the peptide synthesis can therefore be carried out more or less as usual. However, extra deprotection steps have to be carried out (for the carbohydrate protective groups) after chain assembly, and repetitive use of strongly acidic conditions must be avoided, since the glycosidic linkages are acid-sensitive. The latter requirement limits the usefulness of the *tert*-butyloxycarbonyl (*t*-Boc) synthesis strategy for solid-phase glycopeptide synthesis, since it typically employs 95% (v/v) trifluoroacetic acid for each N-deprotection step, and liquid hydrogen fluoride for resin cleavage. The *t*-Boc strategy has, however, been used in some early, pioneering work utilizing simple, less acid-sensitive carbohydrate amino acid derivatives.[33,34] Most commonly nowadays, however, the Fmoc strategy is chosen for solid-phase glycopeptide synthesis, since the repeated use of acidic conditions is not an intrinsic part of this strategy, although strong acid (95% trifluoroacetic acid) is often used in the cleavage step at the end of the synthesis.

In the following, the various steps (synthesis of glycosylated amino acids, solid-phase synthesis, final deprotection, purification, and analysis) in Fmoc-type solid-phase synthesis of glycopeptides are considered separately.

[23] A. K. M. Anisuzzaman, L. Anderson, and J. L. Navia, *Carbohydr. Res.* **174,** 265 (1988).
[24] H. Garg and R. Jeanloz, *Carbohydr. Res.* **76,** 85 (1979).
[25] H. Garg, T. Hasenkamp, and H. Paulsen, *Carbohydr. Res.* **151,** 225 (1986).
[26] M. Hollosi, E. Kollat, I. Laczko, K. Medzihradszky, J. Thurin, and L. Otvos, *Tetrahedron Lett.* **32,** 1531 (1991).
[27] H. Koeners, C. Schattenkerk, J. Verhoeven, and J. van Boom, *Tetrahedron* **37,** 1763 (1981).
[28] M. Kottenhahn, A. Kling, and C. Kolar, *Angew. Chem., Int. Ed. Engl.* **26,** 888 (1987).
[29] H. Paulsen and M. Brenken, *Liebigs Ann. Chem.,* 649 (1988).
[30] S. Rio, J. Beau, and J. Jacquinet, *Carbohydr. Res.* **219,** 71 (1991).
[31] M. Schultz and H. Kunz, *Tetrahedron Lett.* **33,** 5319 (1992).
[32] N. Maeji, Y. Inoue, and R. Chujo, *Carbohydr. Res.* **146,** 174 (1986).
[33] S. Lavielle, N. Ling, and R. Guillemin, *Carbohydr. Res.* **89,** 221 (1981).
[34] S. Lavielle, N. Ling, R. Saltman, and R. Guillemin, *Carbohydr. Res.* **89,** 229 (1981).

Synthesis of O-Glycosylated Amino Acids

General

Crucial to the success of the Fmoc strategy for solid-phase glycopeptide synthesis is the preparation of O-glycosylated Fmoc-amino acids in a fair yield. This is often the most time-consuming activity (fortunately, the most commonly used glycosylated Fmoc-amino acid derivatives have become commercially available). The preparation requires a substantial amount of chemical steps, especially when the carbohydrate is complicated. However, the same general principles for synthesis of oligosaccharides apply, and it is normally possible, after some experimenting in each case, to achieve fair to good yields of the desired products.

The strategy most often used is to couple a properly protected monosaccharide or oligosaccharide derivative (prepared by some of the many methods now available for oligosaccharide synthesis) with a carboxyl-protected Fmoc-amino acid carrying a free hydroxyl group (for an example, see the synthesis in Fig. 2). The carboxyl protecting groups that have been used on the amino acid include *tert*-butyl,[35–37] *p*-nitrophenyl, [38–40] *N*-hydroxysuccinimidoyl,[39–41] phenacyl,[42–45] benzyl,[25,36,46–50] allyl,[51–53] 2-

[35] H. Paulsen and K. Adermann, *Liebigs Ann. Chem.* 751 (1989).
[36] H. Paulsen, W. Rauwald, and U. Weichert, *Liebigs Ann. Chem.*, 75 (1988).
[37] H. Paulsen, G. Merz, and U. Weichert *Angew. Chem. Int. Ed. Engl*, **27,** 1365 (1988).
[38] P. J. Garegg, S. Oscarson, I. Kvarnström, A. Niklasson, G. Niklasson, S. C. T. Svensson, and J. V. Edwards, *Acta Chem. Scand.* **44,** 625 (1990).
[39] B. Ferrari and A. Pavia, *Int. J. Pept. Protein Res.* **22,** 549 (1983).
[40] B. Ferrari and A. Pavia, *Tetrahedron* **41,** 1939 (1985).
[41] M. Gobbo, L. Biondi, F. Filira, R. Rocchi, and V. Lucchini, *Tetrahedron* **44,** 887 (1988).
[42] B. Lüning, T. Norberg, and J. Tejbrant, *J. Carbohydr. Chem.* **11,** 933 (1992).
[43] B. Lüning, T. Norberg, and J. Tejbrant, *Glycoconjugate J.* **6,** 5 (1989).
[44] Y. Nakahara, H. Iijima, S. Sibayama, and T. Ogawa, *Tetrahedron Lett.* **31,** 6897 (1990).
[45] Y. Nakahara, H. Iijima, S. Shibayama, and T. Ogawa, *Carbohydr. Res.* **216,** 211 (1991).
[46] L. Biondi, F. Filira, M. Gobbo, B. Scolaro, and R. Rocchi, *Int. J. Pept. Protein Res.* **37,** 112 (1991).
[47] H. Kessler, A. Kling, and M. Kottenhahn, *Angew. Chem., Int. Ed. Engl.* **29,** 425 (1990).
[48] H. Kunz and S. Birnbach, *Angew. Chem., Int. Ed. Engl.* **25,** 360 (1986).
[49] H. Kunz, S. Birnbach, and P. Wernig, *Carbohydr. Res.* **202,** 207 (1990).
[50] H. Paulsen and M. Schultz, *Carbohydr. Res.* **159,** 37 (1987).
[51] B. G. Delatorre, J. L. Torres, E. Bardaji, P. Clapes, N. Xaus, X. Jorba, S. Calvet, F. Albericio, and G. Valencia, *J. Chem. Soc., Chem. Commun.*, 965 (1990).
[52] H. Iijima, Y. Nakahara, and T. Ogawa, *Tetrahedron Lett.* **33,** 7907 (1992).
[53] S. Friedrich-Bochnitschek, H. Waldmann, and H. Kunz, *J. Org. Chem.* **54,** 751 (1989).

(hydroxymethyl)-9,10-anthraquinonyl[54] or pentafluorophenyl (Pfp) esters.[55–62] Coupling of carbohydrates to Fmoc amino acids without carboxyl protection is also possible in some cases.[63] The Pfp carboxyl protecting group provides a short route to useful amino acid derivatives, since carboxyl deprotection is not necessary after glycosidation. These esters are active enough to be used directly in peptide synthesis. They are, nevertheless, stable enough to survive the necessary column chromatographic purification after the glycosylation reaction, and they can be stored without decomposition at −20°. Glycosylated Fmoc-amino acid pentafluorophenyl esters are therefore an attractive alternative to free glycosylated Fmoc-amino acids in glycopeptide synthesis.

An alternative strategy for preparation of glycosylated Fmoc-amino acids is to attach the carbohydrate moiety to an N- and C-protected amino acid, then remove the C- and N-protecting groups, and finally attach the *N*-Fmoc group. Although considerably longer, this strategy has been used in some instances.[61,64–66]

In the following, syntheses of glycosylated amino acids are presented in groups according to linkage type. For a summary of prepared glycosylated Fmoc amino acid derivatives (free acids or activated ester derivatives), see Table I.

[54] P. Hoogerhout, C. Guis, C. Erkelens, W. Bloemhoff, K. Kerling, and J. VanBoom, *Recl. Trav. Chim. Pays-Bas* **104,** 54 (1985).

[55] T. Bielfeldt, S. Peters, M. Meldal, K. Bock, and H. Paulsen, *Angew. Chem., Int. Ed. Engl.* **31,** 857 (1992).

[56] A. Jansson, M. Meldal, and K. Bock, *Tetrahedron Lett.* **31,** 6991 (1990).

[57] M. Meldal and K. J. Jensen, *J. Chem. Soc., Chem. Commun.,* 483 (1990).

[58] S. Peters, T. Bielfeldt, M. Meldal, K. Bock, and H. Paulsen, *Tetrahedron Lett.* **32,** 5067 (1991).

[59] S. Peters, T. Bielfeldt, M. Meldal, K. Bock, and H. Paulsen, *J. Chem. Soc., Perkin Trans. 1,* 1163 (1992).

[60] S. Peters, T. Bielfeldt, M. Meldal, K. Bock, and H. Paulsen, *Tetrahedron Lett.* **33,** 6445 (1992).

[61] K. B. Reimer, M. Meldal, S. Kusumoto, K. Fukase, and K. Bock, *J. Chem. Soc., Perkin Trans. 1,* 925 (1993).

[62] L. Urge, L. Gorbics, and L. Otvos, *Biochem. Biophys. Res. Commun.* **184,** 1125 (1992).

[63] M. Elofsson, B. Walse, and J. Kihlberg, *Tetrahedron Lett.* **32,** 7613 (1991).

[64] E. Bardaji, J. Torres, J. Clapes, F. Albericio, G. Barany, R. Rodriguez, M. Sacristian, and G. Valencia, *J. Chem. Soc., Perkin Trans. 1,* 1755 (1991).

[65] F. Filira, L. Biondi, B. Scolaro, M. T. Foffani, S. Mammi, E. Peggion, and R. Rocchi, *Int. J. Biol. Macromol.* **12,** 41 (1990).

[66] R. Polt, L. Szabo, J. Treiberg, Y. S. Li, and V. J. Hruby, *J. Am. Chem. Soc.* **114,** 10249 (1992).

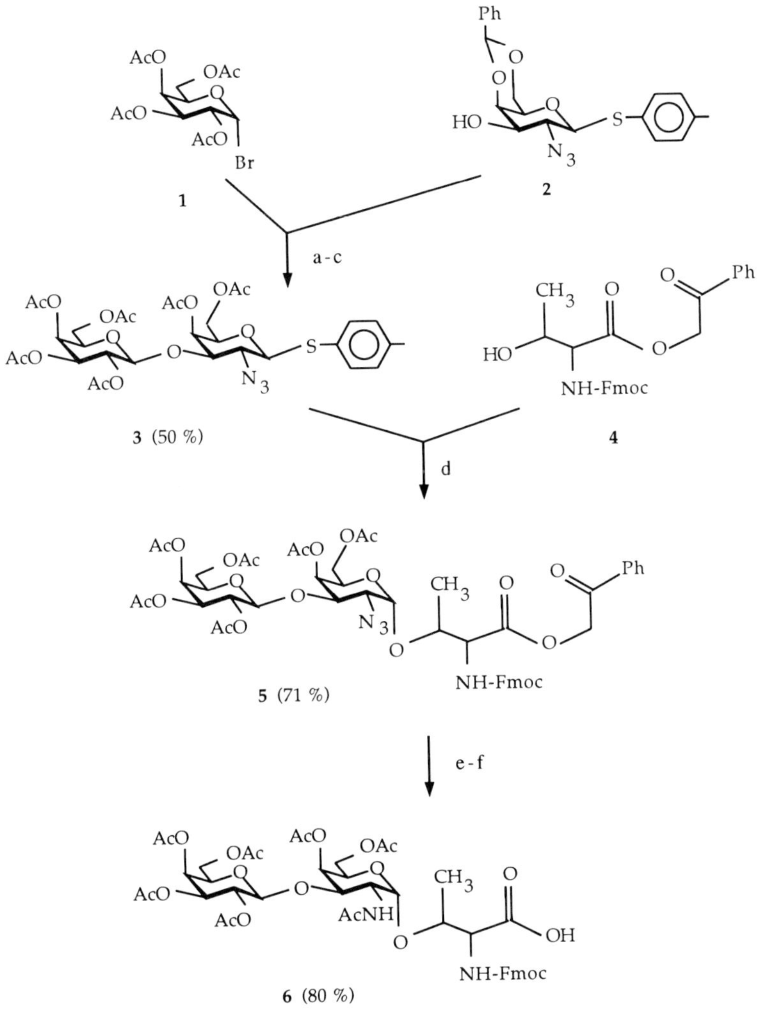

FIG. 2. Synthesis of a per-OAc-Galβ3GalNAcα(Fmoc)-Thr derivative. Reagents for step a, silver trifluoromethane sulfonate; b, acetic acid/water; c, acetic anhydride/pyridine; d, dimethyl(methylthio)sulfonium triflate; e, thioacetic acid; f, zinc/acetic acid.

GalNAc-Ser/Thr and GalGalNAc-Ser/Thr Linkage

The GalNAc-Ser/Thr linkage is the most common of the O-glycosidic carbohydrate–protein linkages found; it occurs in a wide variety of glycoproteins, particularly often in the mucins. Unprotected GalNAc-

TABLE I
GLYCOSYLATED Fmoc-AMINO ACID DERIVATIVES

Amino acid	Sugar	Derivative	O-Protecting group	Ref.
Thr	GalNAc	COOH	OAc	35, 43, 46
Thr	GalNAc	COOPfp	OAc	58, 59
Thr	GalN3	COOPfp	OAc	55
Thr	GalN3	COOSu	OBn	39
Ser	GalNAc	COOH	OAc	35, 43
Ser	GalNAc	COOPfp	OAc	58, 59
Ser	GalN3	COOPfp	OAc	55
Ser	GalN3	COOPnp	OBn	39
Thr	GalGalNAc	COOH	OAc/OBz	68
Thr	GalGalNAc	COOH	OAc	43
Thr	GalGalN3	COOPfp	OAc	60
Ser	GalGalNAc	COOH	OAc/OBz	50
Ser	GalGalNAc	COOH	OAc	43
Ser	NeuAcGalGalNAc	COOH	OBn	44, 45, 52
Thr	NeuAcGalGalNAc	COOH	OBn	52
Thr	ManMan	COOPfp	OAc	56
Thr	Glc	COOH	OAc	74
Ser	Glc	COOH	OAc	61, 66, 74
Ser	GlcGlcGlc	COOPnp	OAc	38
Ser	XylGlc	COOH	OAc	42
Ser	XylXylGlc	COOPfp	OAc	61
Ser	Xyl	COOH	OAc	25
Thr	Gal	COOH	OAc	51, 65, 74
Ser	Gal	COOH	OAc	51, 54
Ser	Gal	COOPfp	OBz	57
Hyp	Gal	COOH	OAc	64, 83
Hyp	Glc	COOH	OAc	64, 83

[a] The list includes free acids or activated esters useful in solid-phase peptide synthesis. Pfp, Pentafluorophenyl; Pnp, *p*-nitrophenyl; Su, *N*-hydroxysuccinimidoyl; Bn, benzyl; Ac, acetyl; Bz, benzoyl.

Ser and GalGalNAc-Ser structures were synthesized before 1985 (for a review, see Garg and Jeanloz[67]). Sugar-protected, free carboxylic acid Fmoc derivatives of these structures useful for glycopeptide synthesis were first synthesized by Paulsen and co-workers[35,50,68] Using alter-

[67] H. Garg and R. Jeanloz, *Adv. Carbohydr. Chem. Biochem.* **43,** 135 (1985).

native routes, the same structures were later synthesized by other groups.[43,46]

As an example, a synthesis of the GalGalNAc-threonine structure **6** from our laboratory is shown in Fig. 2.[43,69] A disaccharide building block is first constructed by glycosylating the monosaccharide donor **1** with the acceptor **2**. A participating acetyl group is used in the 2 position of the donor **1** to achieve stereoselective β-glycosylation. Protecting group manipulations on the product disaccharide then give the disaccharide donor **3**, carrying the nonparticipating azido group at the 2 position. Activation of **3** and glycosylation of the threonine Fmoc derivative **4** are then carried out. The desired α derivative **5** is obtained (71%), in admixture with the corresponding β derivative (18%), and the products are separated by column chromatography. Conversion of the azido group to an acetamido by thioacetic acid treatment and removal of the threonine phenacyl group with zinc in acetic acid finally give the target derivative **6** (80%), which is ready for use in solid-phase synthesis.

The alternative use of Pfp esters for carboxyl protection on the amino acid deserves special mention in the case of GalNAc-Ser/Thr derivatives. Glycosidation with 2-azido-2-deoxygalactopyranosyl derivatives like **7** (Fig. 3) of an Fmoc-threonine Pfp ester **8** to give **9** is possible, but the conversion from the 2-azido to a 2-acetamido functionality on the sugar cannot be carried out at this stage without damage to the sensitive Pfp ester. However, direct use of a 2-azido-2-deoxygalactosyl-Fmoc-serine Pfp ester like **9** (Fig. 3) in peptide synthesis is possible, and the necessary conversion of the azido function to an acetamido can be effected, with thioacetic acid, on the completed glycopeptide, while it is still attached to the resin.[55,60] Alternatively, a 2-acetamido-2-deoxygalactosyl-Fmoc-serine or -threonine derivative (**10**) is first synthesized using another carboxyl protecting group and is then esterified with pentafluorophenol[58,59] to give a derivative like **12**. This route is considerably longer, though, but does not, on the other hand, involve extra resin manipulations. The two routes are summarized in Fig. 3.

The core GalNAc-Ser/Thr or GalGalNAc-Ser/Thr structures of glycoproteins are often further glycosylated with other sugars, such as 2,6-linked *N*-acetylneuraminic acid. Fmoc-amino acid derivatives carrying such extended structures have been synthesized.[44,45,52]

[68] H. Paulsen and M. Schultz, *Liebigs Ann. Chem.*, 1435 (1986).

[69] J. Tejbrant, B. Lüning, and T. Norberg, unpublished results (1992).

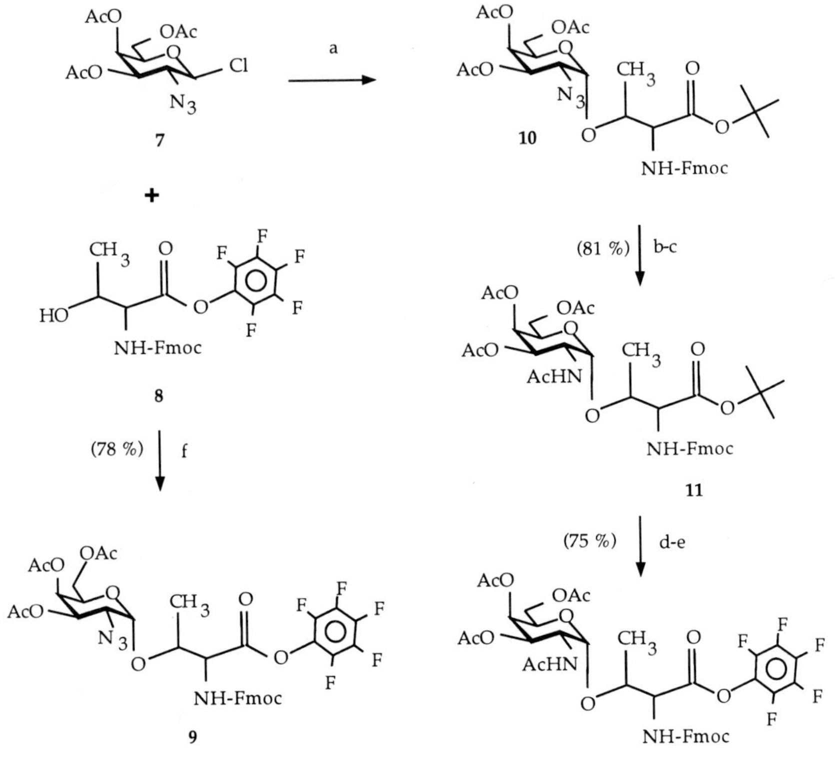

FIG. 3. Synthesis of two Fmoc-Thr Pfp derivatives of galactosamine. Reagents for step a, Fmoc-L-threonine *tert*-butyl ester/Ag_2CO_3/$AgClO_4$; b, H_2S; c, acetic anhydride/pyridine; d, HCOOH; e, pentafluorophenol/1,3-dicyclohexylcarbodiimide; f, Ag_2CO_3/$AgClO_4$.

Man-Ser/Thr Linkage

The Man-Ser/Thr linkage is found in glycoproteins from yeast and molds.[14,70] The per-*O*-acetyl-α-D-mannopyranosyl-(1-2)-α-D-mannopyranosyl-Fmoc-L-threonine Pfp derivative **13** (Fig. 4) has been synthesized and was used in one of the first examples of a solid-phase synthesis of a glycopeptide.[56] Peracetyl-α-D-mannopyranosyl and α-D-mannopyranosyl-(1-2)-α-D-mannopyranosyl derivatives of Fmoc-L-serine benzylester have also been synthesized for nuclear magnetic resonance (NMR) and conformational studies.[71]

[70] J. Montreuil, *Adv. Carbohydr. Chem. Biochem.* **37,** 157 (1980).

[71] A. Helander, L. Kenne, S. Oscarson, T. Peters, and J. Brisson, *Carbohydr. Res.* **230,** 299 (1992).

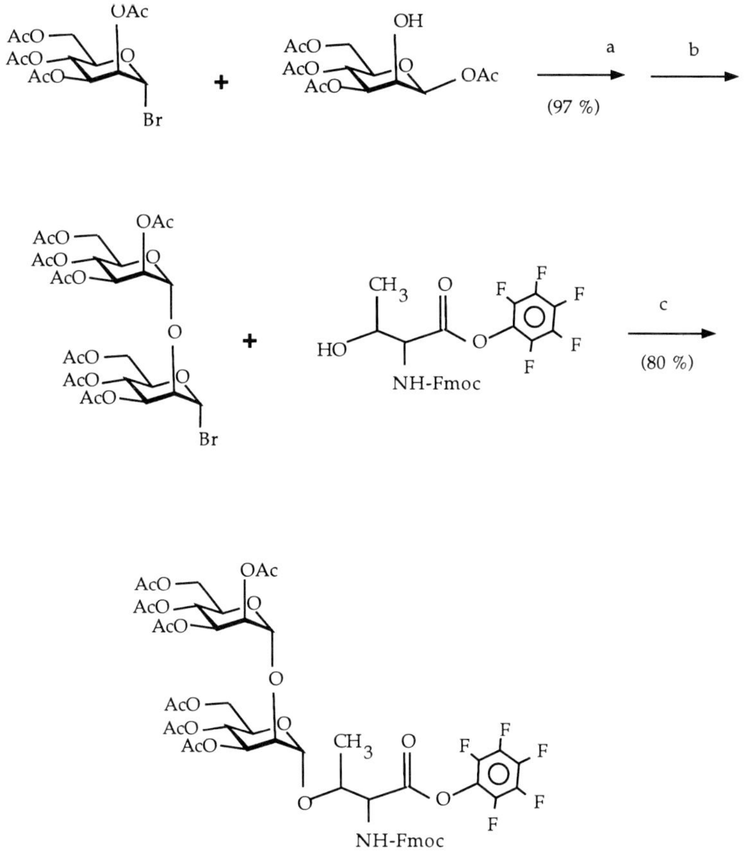

FIG. 4. Synthesis of a dimannosyl Fmoc-Thr Pfp derivative. Reagents for step a, silver trifluoromethane sulfonate; b, HBr–acetic acid; c, silver trifluoromethane sulfonate.

Glc-Ser/Thr Linkage

The α-D-xylopyranosyl-(1-3)-α-D-xylopyranosyl-(1-3)-β-D-glucopyranosyl-serine linkage has been demonstrated in factor IX glycoproteins.[17,18] Synthesis of the unprotected trisaccharide serine derivative has been accomplished,[72] and this derivative has also been acetylated and converted to the Fmoc-protected Pfp ester,[61] for use in solid-phase peptide synthesis. Similarly, peracetylated α-D-xylopyranosyl-(1-3)-β-D-glucopyranosyl-

[72] K. Fukase, S. Hase, T. Ikenaka, and S. Kusumoto, *Bull. Chem. Soc. Jpn.* **65,** 436 (1992).

Fmoc-serine has been synthesized[42] by a shorter route and used in peptide synthesis.[73]

The β-D-glucopyranosyl-L-serine linkage has also been "unnaturally" incorporated into peptides to investigate structure–activity relationships. For example, Fmoc-serine *p*-nitrophenyl esters containing 1,3-linked glucosyl mono-, di-, and trisaccharides have been synthesized[38] for use in peptide synthesis, the purpose being to study the phytoalexin elicitor activity of the glycopeptides.

Several authors[74,75] have prepared O-acetylated glucopyranosyl-Fmoc-threonine derivatives and used them in synthesis of glycosylated analogs of biologically active peptides, the purpose being to study changes in properties on addition of this simple glucosyl unit to the peptides.

Xyl-Ser Linkage

In proteoglycans (e.g., heparin) the linkage region between protein and carbohydrate most often has the following structure: polysaccharide-β-D-GlcA*p*-(1-3)-β-D-Gal*p*-(1-3)-β-D-Gal*p*-(1-4)-β-D-Xylp-serine. Unprotected mono- or oligosaccharide–serine[76–79] or oligosaccharide–peptide fragments[25,29,30]) of this structure have been synthesized for biochemical or structural studies, as well as protected xylopyranosylserines and xylopyranosylpeptides.[53,80] An Fmoc derivative of tri-*O*-acetyl-β-D-xylopyranosyl-L-serine, suitable for solid-phase peptide synthesis, has been prepared.[25]

Glc-Tyr Linkage

The Glc-Tyr linkage has been found to occur in glycogenin.[19,20] To study the biosynthesis of glycogen with the help of defined glycopeptide structures, peracylated α- and β-O-glucosylated Fmoc-tyrosine Pfp esters have been prepared[81] and used in solid-phase peptide synthesis.

[73] J. Tejbrant, *Chem. Commun. Stockholm Univ.* (Doctoral dissertation) (1992).

[74] F. Filira, L. Biondi, F. Cavaggion, B. Scolaro, and R. Rocchi, *Int. J. Pept. Protein Res.* **36,** 86 (1990).

[75] R. Rocchi, L. Biondi, F. Filira, M. Gobbo, S. Dagan, and M. Fridkin, *Int. J. Pept. Protein Res.,* **29,** 250 (1987).

[76] B. Erbing, B. Lindberg, and T. Norberg, *Acta. Chem. Scand. Ser. B* **32,** 308 (1978).

[77] P. J. Garegg, B. Lindberg, and T. Norberg, *Acta Chem. Scand. Ser. B* **33,** 449 (1979).

[78] F. Goto and T. Ogawa, *Tetrahedron Lett.* **33,** 5099 (1992).

[79] G. Ekborg, T. Curenton, N. R. Krishna, and L. Rodén, *J. Carbohydr. Chem.* **9,** 15 (1990).

[80] H. Kunz and H. Waldmann, *Angew. Chem., Int. Ed. Engl.* **23,** 71 (1984).

[81] K. Jensen, M. Meldal, and K. Bock, *J. Chem. Soc., Perkin Trans. 1,* 2119 (1993).

Other Linkages

To study the influence of glycosylation on peptides, tetra-*O*-acetyl-β-D-glucopyranosyl and -galactopyranosyl derivatives of Fmoc-L-hydroxyproline have been prepared and used to make "unnaturally" glycosylated variants of tuftsins or opioid peptides. [64,82,83] Similarly, tetra-*O*-acetyl-β-D-galactopyranosyl derivatives of Fmoc-serine[51,54] or Fmoc-threonine[51,65] have been prepared and used in peptide synthesis. Tetra-*O*-benzoyl-β-D-galactopyranosyl-L-serine Pfp ester has also been prepared,[57] the purpose being to demonstrate the usefulness of the Pfp ester as carboxyl protecting group under glycosylation conditions and to investigate the reactivity of glycosylated amino acid Pfp esters in peptide coupling reactions.

Solid-Phase Synthesis

General

Numerous reports have appeared on use of glycosylated amino acids in Fmoc-type solid-phase peptide synthesis.[26,37,46,55–62,64,65,83–88] The general picture emerging from these published syntheses is that glycosylated amino acids, in spite of their rather large carbohydrate substituents, maintain enough reactivity in the carboxyl-activated form to perform well in coupling reactions, and that, therefore, solid-phase synthesis of glycopeptides is generally possible (the methods section below describes the synthesis of the glycopeptide in Fig. 6).[86] Because of the special properties of the sugar moiety, precautions have to be taken, for example, in the choice of resin, linkage type, N-deprotection, coupling conditions, final deprotections, purification, and analysis. These issues are considered separately below.

Resin and Linkage Type

The solid-phase resins most often used for glycopeptide synthesis are polystyrene-based[37,83,86,88] or kiselguhr-supported, polydimethylacrylamide-based[46,55–57,59,60,65] resins. A few other, graft-type resins have also been

[82] J. L. Torres, I. Haro, E. Bardaji, G. Valencia, J. M. Garcia-Anton, and F. Reig, *Tetrahedron* **44,** 6131 (1988).

[83] E. Bardaji, J. L. Torres, P. Clapes, F. Albericio, G. Barany, and G. Valencia, *Angew. Chem., Int. Ed. Engl.* **29,** 291 (1990).

[84] H. Kunz and B. Dombo, *Angew. Chem., Int. Ed. Engl.* **27,** 711 (1988).

[85] B. Lüning, T. Norberg, and J. Tejbrant, *J. Chem. Soc., Chem. Commun.*, 1267 (1989).

[86] B. Lüning, T. Norberg, C. Rivera-Baeza, and J. Tejbrant, *Glycoconjugate J.*, **8,** 450–55 (1991).

[87] H. Paulsen, K. Adermann, G. Merz, M. Schultz, and U. Weichert, *Stärke*, 465 (1988).

[88] H. Paulsen, G. Merz, S. Peters, and U. Weichert, *Liebigs Ann. Chem.*, 1165 (1990).

tried.[59,62] The syntheses have been carried out manually,[86,88] or automatically in a continuous-flow apparatus.[56,60,62,65] The fact that carbohydrates are acid-sensitive has restricted the choice of resin linkage to groups that do not require hydrogen fluoride for cleavage. The PAL 5-(4-aminomethyl-3,5-dimethoxyphenoxy)-valeric acid] and Rink [*p*-(α-amino-2,4-dimethoxybenzyl)-phenoxyacetic acid] linkers require 95% aqueous trifluoroacetic acid for cleavage, and give peptide amides as products.[55,59,60,83] The 4-alkoxybenzyl alcohol linker,[37,46,65,88] which cleaves with 95% aqueous trifluoroacetic acid or with trifluoroacetic acid–anisole mixtures, and the 3-methoxy-4-alkoxybenzyl alcohol linker,[86] which cleaves with 1% trifluoroacetic acid, give free peptides on cleavage. The above acidic conditions, which are only employed once and for a short time (<2 hr) do not seem to seriously affect the carbohydrate chains, especially when they are stabilized against acids by *O*-acetyl protecting groups. Peptides with sialylated oligosaccharide chains should require special attention, however, since sialic acid, even in its O-acetylated form, is far more acid-sensitive than normal aldohexopyranoses. To date, only solution-phase conditions have been reported for preparation of such glycopeptide derivatives.[45] Other examples of resin linkage chemistry include the [4-(aminoacyloxy)crotonyl]aminomethyl (HYCRAM) moiety, which can be cleaved under neutral conditions using Pd(0) reagents.[84]

N-Deprotection

N-Deprotection is the first of the two repetitive chemical reactions carried out on the resin during a typical solid-phase synthesis. It has received less attention, as far as yields and monitoring techniques are concerned, than the subsequent step, coupling. However, quantitative deprotection of the Fmoc group, under conditions which do not affect other parts of the glycopeptide, is equally important for the purity and yield of the final product. In an investigation[89] of different deprotection agents, it has been shown that rapid, quantitative Fmoc deprotection is most consistently obtained with 2% DBU (1,8-diazabicyclo[5.4.0]undec-7-ene)[90] in dimethylformamide (DMF) containing 2% piperidine. Special attention was paid to the detection of base-catalyzed carbohydrate elimination products, but these were not detected with any of the bases investigated (piperidine, morpholine, and DBU). It was therefore concluded, especially for synthesis of glycopeptides with longer peptide chains, that the combination 2% DBU/2% piperidine in DMF is the Fmoc-deprotection reagent of choice.

[89] K. Jensen, M. Meldal, S. Bielfeldt, S. Peters, K. Bock, and H. Paulsen, *J. Pept. Protein Res.* **43,** 529 (1994).

[90] J. Wade, J. Bedford, R. Sheppard, and G. Tregear, *Pept. Res.* **4,** 194 (1991).

Carboxyl Activation and Coupling

As found by several investigators, activated glycosylated amino acids are reactive enough to couple well in solid-phase peptide synthesis. Molar excesses of activated species have been in the normal (1.5–4 times excess) range (fortunately, since the derivatives often have to be prepared by multistep synthesis and therefore command a high per gram price). The activation method most frequently used with free, glycosylated Fmoc-amino acids is the carbodiimide-mediated, *in situ* conversion to the reactive hydroxybenzotriazole (HOBt) ester. The Pfp esters of glycosylated Fmoc-amino acids in combination with 3,4-dihydro-3-hydroxy-4-oxo-1,2,3-benzotriazine (Dhbt-OH) have also been used extensively. In continuous-flow reactors, the progress of Pfp/Dhbt-OH coupling reactions has been monitored with a spectrophotometric device, measuring the concentration of the Dhbt-OH anion, which functions as an indicator of free amino groups.[59]

Monitoring of coupling reactions is important to avoid deletion sequences. In addition to the above-mentioned indicator monitoring, the original Kaiser ninhydrin test has been used in glycopeptide synthesis. Monitoring by sample resin cleavage and subsequent fast atom bombardment-mass spectrometry (FAB-MS) of the released peptides has also been carried out.[86] This method, although in its present form rather laborious, gives excellent structural information but cannot presently be used quantitatively. Another monitoring possibility is the trinitrobenzenesulfonic acid test.[65]

In a striking example[60] of the coupling effectiveness with glycosylated amino acids, the multiple substituted derivative **14** (Fig. 5) has been prepared, and the coupling efficiencies are not seriously affected by the steric hinderance from the bulky carbohydrate groups. The acylating amino acids are used only in 1.5-fold excess, and coupling times are 3–10 hr,

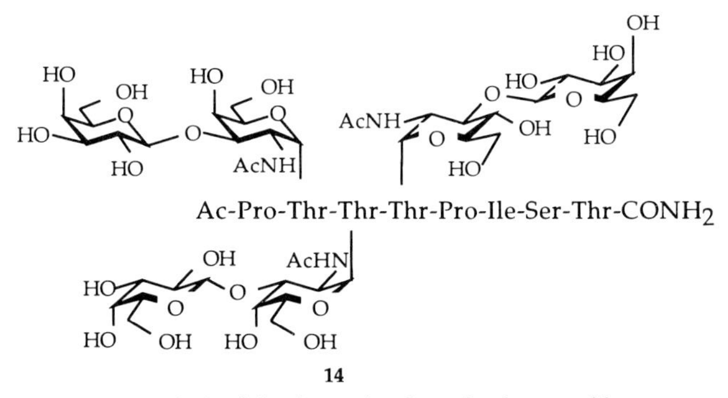

FIG. 5. A triple glycosylated mucin glycopeptide.

except for the first glycosylated serine to proline coupling, which requires 22 hr.

Final Deprotection

As noted above, the glycopeptide chain is usually detached from the resin using acidic conditions, typically 1–95% trifluoroacetic acid. In principle, the other protecting groups present on the glycopeptide can be removed either before, simultaneously with, or after resin cleavage. Conversion of 2-azido groups to acetamido groups (with thioacetic acid) have been done successfully before resin cleavage (see above). Removal of *tert*-butyl side-chain protecting groups occurs simultaneously with resin cleavage, if 95% trifluoroacetic acid is used. Removal of carbohydrate *O*-acetyl groups with basic reagents has been carried out before resin cleavage,[46,83] although it has been more common to carry out this transformation after the glycopeptide has been cleaved and brought into solution, where the reaction can be more easily monitored. Undesired base-catalyzed elimination of substituted serine and threonine derivatives during O-deacetylation can be avoided by choosing mildly alkaline reaction conditions (catalytic amount of sodium methoxide in methanol,[37,56,86] hydrazine in methanol,[46] or dilute sodium hydroxide.[62] Other protecting groups, such as *O*- or *N*-benzyl groups, have been removed after resin cleavage.[86]

Generally, it is desirable to have as few protecting groups as possible on the glycopeptide, to minimize the amount of deprotection steps to be carried out on the end product of the synthesis. This means that carbohydrate chains should preferably carry one type of O-protection, although their assembly might require several different types of protection. The hitherto most commonly used O-protecting group is the *O*-acetyl group. This group is easily introduced, enhances the acid stability and solubility of the sugar derivatives in organic solvents, and cleaves under mildly alkaline conditions.

Purification

As in the case in normal solid-phase peptide synthesis, the crude glycopeptide has to be purified, usually by one or more chromatographic steps. The most commonly used methods are gel filtration and reversed-phase high-performance liquid chromatography (HPLC) (for details, see Refs. 37, 46, 55–57, 59–61, 64, 65, 84, 86, and 88). Good separation from deletion and unglycosylated sequences is usually possible. Mobilities on chromatographic materials may, however, vary considerably between gly-

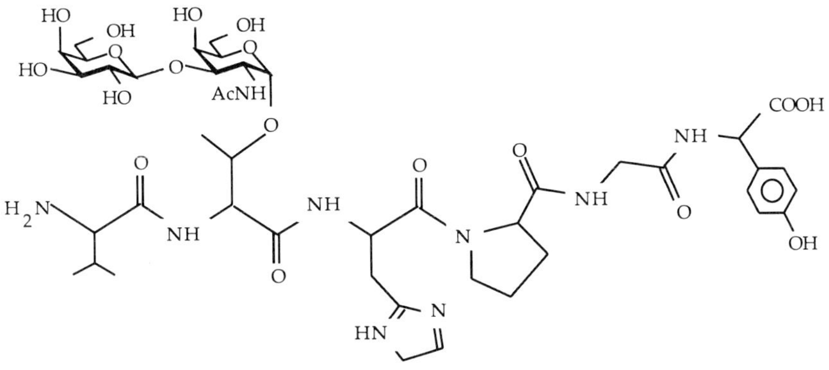

15

FIG. 6. An oncofetal fibronectin glycopeptide fragment (see text).

copeptides and unsubstituted counterparts, because of the increased hydrophilicity conferred to the peptide by the carbohydrate chain.[91]

Analysis

The two most powerful methods for assessing the purity and structure of synthesized glycopeptides is high-field NMR (^{13}C and ^{1}H) spectroscopy and FAB-MS. It is beyond the scope of this chapter to review the work published on this subject, and the interested reader is referred to articles that exemplify these applications.[59,86,92]

Methods

General

As mentioned above, many successful syntheses of glycopeptides have been carried out. The following preparation from our laboratory of a fibronectin fragment (Fig. 6) serves as a simple example. The resin used and all amino acid derivatives are commercially availale, and the details of the synthesis of these derivatives are therefore not included (the peracetylated Galβ3GalNAcα-Fmoc-L-threonine derivative is from Bachem California, Torrance, CA, and the other amino acid derivatives are either from this supplier or from Bachem Feinchemikalien AG, Bubendorf, Switzer-

[91] L. Otvos, L. Urge, and J. Thurin, *J. Chromatogr.* **599,** 43 (1992).

[92] H. Kessler, H. Matter, G. Gemmecker, M. Kottenhahn, and J. W. Bats, *J. Am. Chem. Soc.* **114,** 4805 (1992).

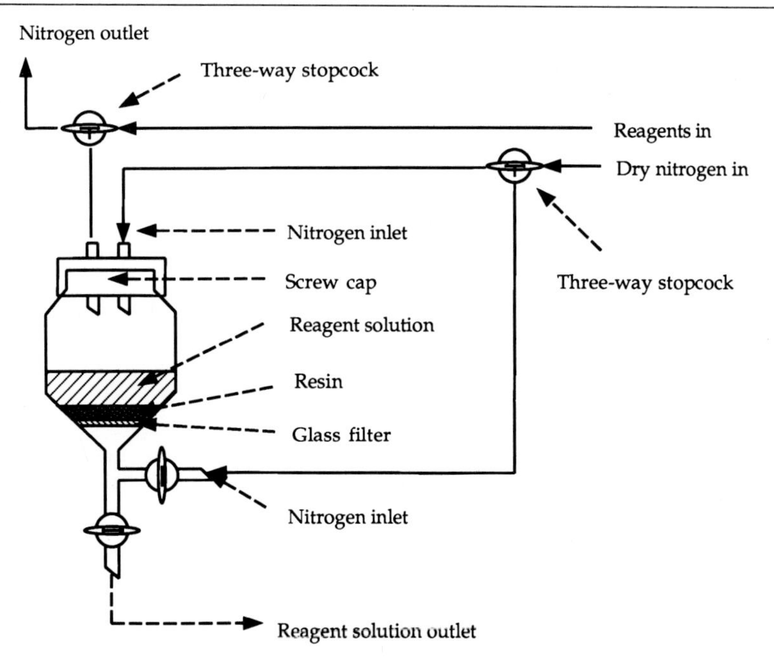

FIG. 7. Lüning apparatus for solid-phase synthesis.

land). The reaction vessel used is of the stationary, fritted glass filter bottom type, where nitrogen is pressed up through the filter to provide agitation (see Fig. 7). The Sasrin resin used is polystyrene-based and contains the 3-methoxy-4-alkoxybenzyl alcohol linker, which can be cleaved with 1% trifluoroacetic acid in dichloromethane. *In situ* generated HOBt amino acid esters or symmetrical anhydrides are used as coupling reagents. Piperidine or morpholine (after incorporation of sugar) is used for N-deprotection. The protecting groups used on the sugar are *O*-acetyl, and on the amino acids tyrosine is O-benzylated and histidine is N-benzylated.

More sophisticated procedures than the one used here, such as the use of automated, continuous-flow protocols and amino acid Pfp esters as coupling reagents, would probably have given similar or better yields and product purity. The interested reader is referred to the original articles on these esters[59] for more details. The present solid-phase protocol is, however, simple to perform and has been successfully repeated independently in several laboratories.[93]

[93] T. Norberg, unpublished results (1992).

TABLE II
AMOUNTS AND TYPES OF REAGENTS USED FOR DEPROTECTION AND COUPLING

Cycle	Deprotection	Amino acid	Quantity	HOBt	DIPCDI
1	20% piperidine in DMF	Fmoc-Gly-OH	0.30 g, 1.01 mmol	0.14 g, 1.04 mmol	158 μl, 1.02 mmol
2	20% piperidine in DMF	Fmoc-Pro-OH	0.69 g, 2.05 mmol	—	158 μl, 1.02 mmol
3	20% piperidine in DMF	Fmoc-His(Bn)-OH	0.48 g, 1.03 mmol	0.14 g, 1.04 mmol	158 μl 1.02 mmol
4	20% piperidine in DMF	Fmoc-Thr-perAc-(GalGalNAc)-OH	0.51 g, 0.53 mmol	0.073 g, 0.54 mmol	84 μl, 0.54 mmol
5	50% morpholine in DMF	Fmoc-Val-OH	0.35 g, 1.03 mmol	0.14 g, 1.04 mmol	158 μl, 1.02 mmol

NH_2-Val-(Galβ3GalNAcα)Thr-His-Pro-Gly-Tyr-OH

The Fmoc-Tyr(Bn)-Sasrin resin (0.64 g, degree of substitution 0.5 mmol/g, 0.32 mmol, from Bachem Feinchemikalien) is allowed to swell in DMF for 10 min in a synthesis vessel with a fritted glass bottom (see Fig. 7). The DMF solution is then removed by nitrogen pressure-assisted filtration through the glass filter in the bottom of the apparatus. All the protection–coupling cycles described below are then carried out in the same reaction vessel. Addition of solutions is done manually from sealed reagent bottles with the aid of Teflon tubing and a slight excess of nitrogen pressure. Agitation in the vessel is obtained by passing a slow stream of nitrogen gas from the bottom of the vessel through the fritted glass filter. Removal of solutions from the vessel is done by nitrogen pressure-assisted filtration through the filter. The following deprotection–coupling cycles are performed:

Deprotection: The resin is treated with deprotection solution (4 ml, treatment with agitation for 2 min), washed with DMF (2–4 ml) three times, treated again (10 min agitation) with deprotection solution, and then washed 10 times with DMF.

Coupling: A solution of the amino acid derivative (for type and amount, see Table II) and, if applicable, hydroxybenzotriazole (see Table II) in DMF (2 ml) is mixed with diisopropylcarbodiimide (DIPCDI) (for amount, see Table II), and the mixture is added to the resin. The suspension is agitated until the Kaiser ninhydrin test[94] gives a negative result (2–12 hr). Then the resin is washed 10 times with DMF (2–4 ml), and deprotection is carried out again.

[94] V. Sarin, S. Kent, J. Tam, and R. Merrifield, *Anal. Biochem.* **117,** 147 (1981).

After the last coupling, a final deprotection is effected (50% morpholine in DMF), then the resin is washed 5 times with DMF and 10 times with dichloromethane. Cleavage is effected, with the resin still in the reaction vessel, by repetitive treatment with 1% trifluoroacetic acid in dichloromethane (2-ml portions, at least 5 times, or until the resin has taken on a constant magenta color). The collected filtrates are transferred to a round-bottomed flask and are then evaporated under reduced pressure on a rotatory evaporator. The residue is taken up in methanol (8 ml) containing methanolic sodium methoxide (0.5 *M*, 2 ml). After 1 hr, the mixture is neutralized with acetic acid (2 ml), transferred to a Parr hydrogenation apparatus, and then hydrogenated at 60 psi hydrogen pressure at room temperature over Pd/C (0.8 g). After 18 hr, the mixture is filtered, and the filtrate is evaporated. The residue is dissolved in aqueous ammonium bicarbonate (0.1 *M*, 3 ml) and applied to a Sephadex G-15 gel-filtration column (diameter 40 mm, length 500 mm). Elution with 0.1 *M* aqueous ammonium bicarbonate (UV monitoring) gives a major pool, which is further purified by reversed-phase FPLC (fast protein liquid chromatography), using a Waters (Milford, MA) RCM 25 $\times$ 30 radial compression unit and a μBondapak 10 μm/125A Guardpak precolumn followed by a μBondapak C_{18} 10 μm/125A column. The sample (10-mg portions) is dissolved in water containing 0.05% trifluoroacetic acid, and the solution is injected and eluted (6 ml/min) with a gradient of 0–40% acetonitrile in water, containing 0.05% trifluoroacetic acid. The UV absorbance of the eluate is monitored (280 nm). Appropriate fractions are pooled, concentrated, redissolved in water, and lyophilized to give the glycopeptide **15** (Fig. 6) (63.4 mg, 0.061 mmol, 19%). The FAB-MS spectrum (*m*/*z* 1038, M + 1) as well as the ^{1}H NMR and carbon NMR spectra[86] indicated a pure, homogeneous compound with the correct structure.

[7] Regeneration of Sugar Nucleotide for Enzymatic Oligosaccharide Synthesis

By YOSHITAKA ICHIKAWA, RUO WANG, and CHI-HUEY WONG

Introduction

Glycosyltransferase-mediated oligosaccharide synthesis is of current interest in synthetic carbohydrate chemistry.[1] Although the enzymatic method is regio- and stereoselective and does not require multiple protection and deprotection steps, it is still hampered by the unavailability of transferases, the problem of product inhibition in stoichiometric reactions, and the tedious preparation of the donor substrate sugar nucleotides when large-scale processes are needed. One way to overcome such drawbacks is to use multiple enzymatic systems with *in situ* cofactor regeneration, in which the sugar nucleotides are continuously regenerated from readily available monosaccharides.

Although many glycosyltransferases are involved in the biosynthesis of oligosaccharides, these enzymes utilize only eight sugar nucleotides (Scheme 1) as donor substrates in mammalian systems. Development of regeneration systems for each of these sugar nucleotides would therefore make possible the practical synthesis of most complex oligosaccharides, provided that glycosyltransferases are available. Efforts in this regard have led to the development of regeneration systems for uridine 5′-diphosphoglucose (UDP-Glc),[2,3] UDP-galactose (UDP-Gal),[2,3] UDP-glucuronic acid (UDP-GlcUA),[4] cytidine 5′-monophospho-*N*-acetyl-

[1] A. T. Beyer, J. E. Sadler, J. I. Rearick, J. C. Paulson, and H. L. Hill, *Adv. Enzymol.* **52,** 23 (1981); E. J. Toone, E. S. Simon, M. D. Bednarski, and G. M. Whitesides, *Tetrahedron* **45,** 5365 (1989); S. David, C. Auge, and C. Gautheron, *Adv. Carbohydr. Chem. Biochem.* **49,** 175 (1991); Y. Ichikawa, G. C. Look, and C.-H. Wong, *Anal. Biochem.* **202,** 215 (1992).

[2] C.-H. Wong, S. L. Haynie, and G. M. Whitesides, *J. Org. Chem.* **47,** 5416 (1982). For other GalT-catalyzed synthesis with *in situ* regeneration of UDP-Gal, see C. Auge, C. Mathieu, and C. Merienne, *Carbohydr. Res.* **151,** 147 (1986); J. Thiem and T. Wiemann, *Angew. Chem., Int. Ed. Engl.* **30,** 1163 (1991); J. Thiem and T. Wiemann, *Synthesis,* 141 (1992).

[3] C.-H. Wong, R. Wang, and Y. Ichikawa, *J. Org. Chem.* **57,** 4343 (1992).

[4] D. Gygax, P. Spies, T. Winkler, and V. Pfaar, *Tetrahedron* **47,** 5119 (1991).

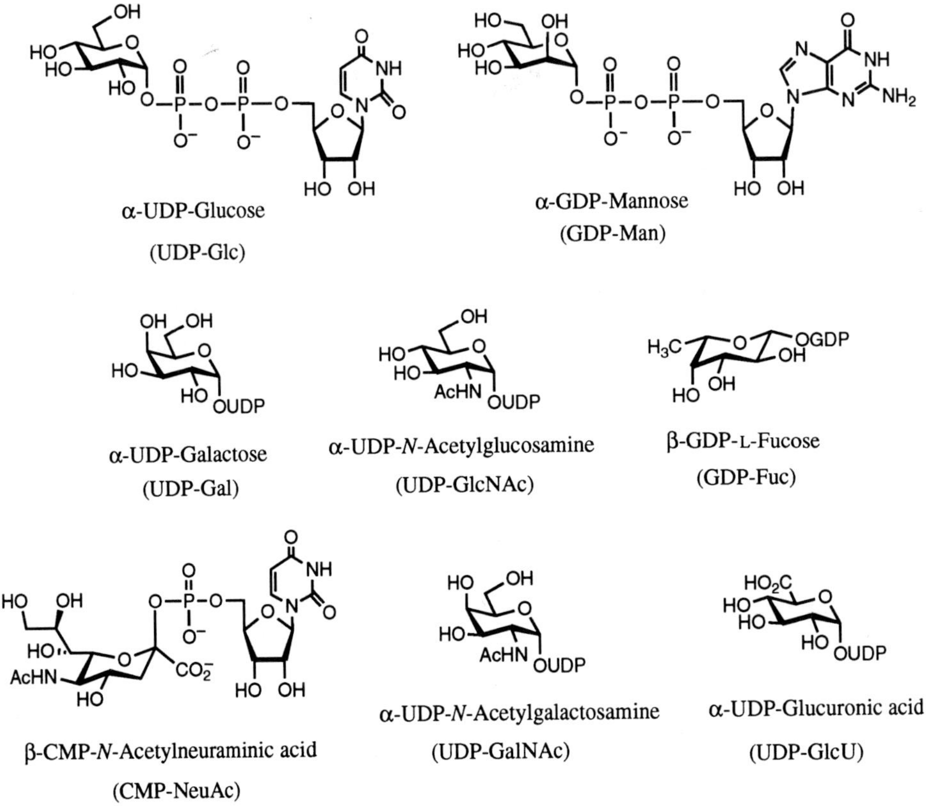

α-UDP-Glucose (UDP-Glc)

α-GDP-Mannose (GDP-Man)

α-UDP-Galactose (UDP-Gal)

α-UDP-*N*-Acetylglucosamine (UDP-GlcNAc)

β-GDP-L-Fucose (GDP-Fuc)

β-CMP-*N*-Acetylneuraminic acid (CMP-NeuAc)

α-UDP-*N*-Acetylgalactosamine (UDP-GalNAc)

α-UDP-Glucuronic acid (UDP-GlcU)

SCHEME 1. Sugar nucleotides as donor substrates for glycosyltransferases.

neuraminic acid (CMP-NeuAc),[5,6] and guanosine 5′-diphosphomannose (GDP-Man)[7] in large-scale oligosaccharide synthesis, and GDP-fucose (GDP-Fuc),[6] UDP-*N*-acetylglucosamine (UDP-GlcNAc),[8] and UDP-*N*-acetylgalactosamine (UDP-GalNAc)[8] in analytical scale synthesis. The regeneration systems for the last three sugar nucleotides have not been

[5] Y. Ichikawa, G.-J. Shen, and C.-H. Wong, *J. Am. Chem. Soc.* **113,** 4698 (1991).

[6] Y. Ichikawa, Y.-C. Lin, D. P. Dumas, G.-J. Shen, E. Garcia-Junceda, M. A. Williams, R. Bayer, C. Ketcham, L. E. Walker, J. C. Paulson, and C.-H. Wong, *J. Am. Chem. Soc.* **114,** 9283 (1992).

[7] P. Wang, G.-J. Shen, Y.-F. Wang, Y. Ichikawa, and C.-H. Wong, *J. Org. Chem.* **58,** 3985 (1993).

[8] G. C. Look, Y. Ichikawa, G.-J. Shen, P.-W. Chen, and C.-H. Wong, *J. Org. Chem.* **58,** 4326 (1993).

well developed mainly owing to the difficulty in the preparation of enzymes for cofactor (e.g., GDP-Fuc) regeneration or the lack of transferases (e.g., GlcNAc and GalNAc transferases) for use in large-scale processes. The problems, however, may be solved when overexpression systems become available for preparation of the enzymes.

Illustrated in Schemes 2–8 are representative examples for the regeneration of each of the eight sugar nucleotides used in enzymatic oligosaccharide synthesis. It has also been demonstrated that two different transferases coupled with the corresponding sugar nucleotide regeneration systems can be combined in one vessel for the formation of two glycosidic bonds (Scheme 9).[9] Furthermore, the monosaccharide sialic acid can be generated *in situ* from *N*-acetylmannosamine and pyruvate catalyzed by sialic acid aldolase. The aldolase substrate pyruvate is the by-product of phosphoenolpyruvate, which is the phosphorylation source utilized in sugar nucleotide regeneration.

Alternatively, a glycosidase (or transglycosidases) can be used in a kinetic mode to form a saccharide which is used *in situ* as a substrate for a glycosyltransferase (Scheme 10).[10] This system has the advantage that the product is no longer subject to the glycosidase-catalyzed hydrolysis, therefore improving the yield. The reaction is usually carried out under conditions in which the transferase reaction is faster than the glycosidase-catalyzed reaction, allowing the maximum conversion of the intermediate to the final product.

The following procedures describe experimental details regarding the use of sugar nucleotide regeneration in glycosyltransferase-catalyzed synthesis. It should be noted, however, that many functionally important complex oligosaccharides are derived from further modification of the initially synthesized oligosaccharides. Sulfation, methylation, and acylation are typical modification processes. All these enzymatic reactions require cofactors (e.g., 3′-phosphoadenosyl 5′-phosphosulfate, *S*-adenosylmethionine, and acetyl-CoA), and the enzymes as well as those required for regeneration of these cofactors have not been available for practical synthesis.[11]

[9] Y. Ichikawa, J. L.-C. Liu, G.-J. Shen, and C.-H. Wong, *J. Am. Chem. Soc.* **113,** 6300 (1991).

[10] G. F. Herrmann, Y. Ichikawa, C. Wandrey, F. C. A. Gaeta, J. C. Paulson, and C.-H. Wong, *Tetrahedron Lett.* **34,** 3091 (1993).

[11] H. K. Chenault, E. S. Simon, and G. M. Whitesides, *Biotechnol. Genet. Eng. Rev.* **6,** 221 (1988).

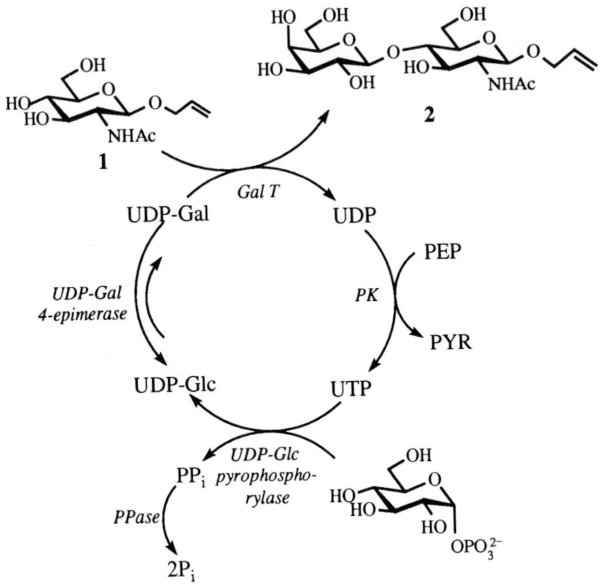

SCHEME 2. Synthesis of LacNAc with *in situ* cofactor regeneration employing UDP-Glc pyrophosphorylase and UDP-Gal-4-epimerase.

Methods

Galactosylation

Two multienzyme systems for the synthesis of *N*-acteyllactosamine (LacNAc) have been developed with *in situ* cofactor regeneration. One starts with glucose 1-phosphate (Glc-1-P) and uses UDP-Glc pyrophosphorylase (EC 2.7.7.9, UDPGP, UTP–glucose-1-phosphate uridylyltransferase) and UDP-Gal 4-epimerase (EC 5.1.3.2, UDPGE) (Scheme 2).[2] UDP-Galactose is generated from UDP-Glc with UDPGE; however, this equilibrium favors the formation of UDP-Glc, and Glc-1-P has to be prepared separately. The other procedure uses Gal instead of Glc-1-P as a donor precursor, and UDPGP, galactokinase (EC 2.7.1.6, GK), and Gal-1-P uridylyltransferase (EC 2.7.7.12, Gal-1-P UT) (Scheme 3).[3] The enzyme GK is specific for galactose, allowing the direct production of Gal-1-P, which is converted to UDP-Gal with Gal-1-P UT and UDP-Glc. The latter system has been used in the preparation of [1-^{13}C]Gal-LacNAc.

The multienzyme system in Scheme 3 started with [1-^{13}C]Gal, GlcNAcβ-*O*-allyl,[12] phosphoenolpyruvate (PEP), and catalytic amounts of

[12] R. T. Lee and Y. C. Lee, *Carbohydr. Res.* **37,** 193 (1974).

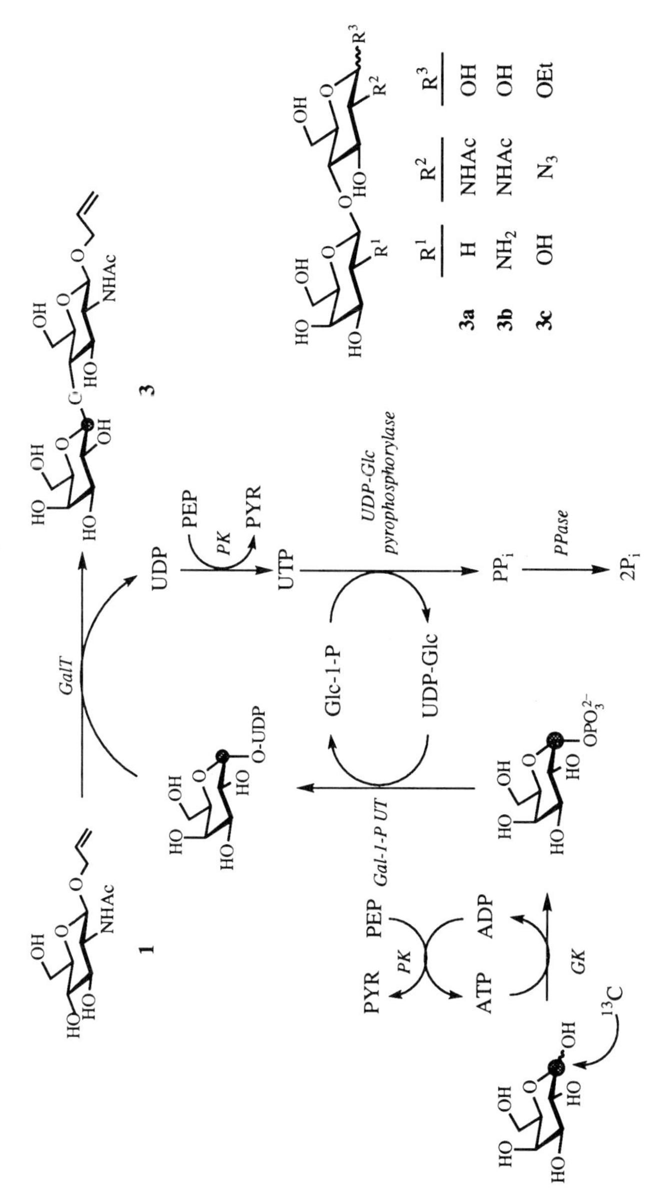

SCHEME 3. Synthesis of LacNAc and analogs with *in situ* cofactor regeneration employing galactokinase (GK) and Gal-1-Puridylyltransferase (Gal-1-P UT).

Glc-1-P, ATP, and UDP. UDP is converted to UTP with pyruvate kinase (EC 2.7.1.40, PK) and PEP, and UTP reacts with Glc-1-P catalyzed by UDPGP to produce UDP-Glc. The by-product inorganic pyrophosphate (PP_i) is decomposed by inorganic pyrophosphatase (EC 3.6.1.1, PPase). With Gal-1-P UT, UDP-Glc reacts with [^{13}C]Gal-1-P, generated from [^{13}C]Gal and ATP in the presence of GK, to give UDP[^{13}C]Gal and Glc-1-P. The [^{13}C]Gal of UDP[^{13}C]Gal is transferred to the acceptor (GlcNAcβ-*O*-allyl) by galactosyltransferase (EC 2.4.1.22, GalT, lactose synthase) to give [1-^{13}C]Gal containing Lac-NAc. The produced UDP is again converted to UTP with PK and PEP, which reacts with the released Glc-1-P to regenerate UDP-Glc. Using this multienzyme system, [1-^{13}C]Gal-LacNAc is obtained in 54% yield. The same procedure is also used in the preparation of unlabeled LacNAc and analogs **3a–3c** (Scheme 3). As indicated, this system also allows the regeneration of UDP-2-deoxy-D-galactose and UDPgalactosamine.

*LacNAcβ-O-allyl (**2**).* For the preparation of LacNAcβ-*O*-allyl (**2**, Scheme 2), a mixture of **1**[12] (2.0 g, 7.65 mmol), Glc-1-P (2.74 g, 7.65 mmol), PEP (K^+ salt, 1.6 g, 7.65 mmol), NAD^+ (193 mg, 0.25 mmol), $MnCl_2 \cdot 4H_2O$ (79.2 mg, 0.4 mmol), $MgCl_2 \cdot 6H_2O$ (162.6 mg, 0.8 mmol), dithiothreitol (DTT) (306 mg, 2 mmol), KCl (1.04 g, 15 mmol), NaN_3 (20 mg, 0.31 mmol), and UDP (90 mg, 0.19 mmol) in *N*-(2-hydroxyethyl)piperazine-*N*′-(2-ethanesulfonic acid) (HEPES) buffer (100 m*M*, pH 7.5; 200 ml) is adjusted with 10 *N* and 1 *N* NaOH to pH 7.5, and the enzymes UDPGE (10 U), UDPGP (20 U), PK (100 U), GalT (5 U), PPase (100 U) are added to the solution. The mixture is gently stirred under an argon atmosphere at room temperature (25°) for 5 days. The mixture is concentrated and chromatographed on silica gel with $CHCl_3$–ethyl acetate-methanol (5 : 2 : 2 to 5 : 2 : 3, v/v/v), to give a disaccharide, which is further purified with Sephadex G-25, with water, to give **2** (1.7 g, 50%); ^{1}H nuclear magnetic resonance (NMR) (D_2O): δ 2.00 (3 H, s, NHAc), 3.49 (1H, dd, J = 7.84, 9.97 Hz, H2 of Gal), 3.52–3.57 (1 H, m, H5 of GlcNAc), 3.63 (1H, dd, J = 3.31, 10.04 Hz, H3 of Gal), 3.65–3.75 (8 H, m), 3.79 (1 H, dd, J = 5.10, 12.27 Hz, H6a of GlcNAc), 3.88 (1H, br d, J = 3.32 Hz, H4 of Gal), 3.95 (1 H, dd, J = 2.14, 12.27 Hz, H6b of GlcNAc), 4.43 (1H, d, J = 7.81 Hz, H1 of Gal), 4.55 (1H, d, J = 8.28 Hz, H1 of GlcNAc), 5.21–5.29 (2 H, m, allylic), 5.82–5.90 (1H, m, allylic); ^{13}C NMR (D_2O): δ 22.6, 55.5, 60.5, 61.5, 69.0, 70.9, 71.4, 72.9, 75.2, 75.8, 78.8, 100.4, 103.3, 118.6, 133.7.

*Compound **3**.* To prepare **3** (Scheme 3), a solution of **1** (1.15 g, 4.4 mmol), [1-^{13}C]Gal (800 mg, 4.4 mmol), PEP K^+ salt (1.82 g, 8.8 mmol), UDP (90 mg, 0.19 mmol), ATP (100 mg, 0.18 mmol), cysteine (116 mg, 0.96 mmol), DTT (183 mg, 1.2 mmol), $MgCl_2 \cdot 6H_2O$ (244 mg, 12 mmol), $MnCl_2 \cdot 4H_2O$ (118 mg, 0.6 mmol), KCl (179 mg, 2.4 mmol), and

Glc-1-P (77 mg, 0.22 mmol) in HEPES buffer (100 m*M*, pH 7.5; 120 ml) is adjusted with 10 *N* and 1 *N* NaOH to pH 7.5, and the enzymes GK (10 U), PK (200 U), PPase (10 U), Gal-1-P UT (10 U), UDPGP (10 U), and GalT (10 U) are added to the solution. The mixture is gently stirred under an argon atmosphere at room temperature (~25°) for 3 days. The mixture is concentrated *in vacuo*, and the residue is chromatographed on silica gel, with ethyl acetate–methanol (2 : 1, v/v), to give a disaccharide, which is further purified on a column of Sephadex G-25, with water, to give **3** (1.06 g, 57%); ^{1}H NMR (D_2O): δ 2.00 (3H, s, NHAc), 3.48–3.52 (1H, m, H2 of Gal), 4.43 (1H, dd, $J_{H1,H2}$ = 8.32, $J_{H1,^{13}C1}$ = 162.33 Hz, H1 of Gal), 4.54 (1H, d, J = 8.32 Hz, H1 of GlcNAc); high-resolution mass spectrometry (HRMS): calculated for $^{12}C_{16}$ $^{13}CH_{29}NO_{11}Na$ (M + Na^+) 447.1672, found 447.1681.

*2-Deoxy-D-galactopyranosyl-β(1,4)-2-acetamido-2-deoxyglucopyranose (**3a**).* Compound **3a** (Scheme 3) is obtained in a yield of 36%. Both the ^{1}H NMR spectrum of its heptaacetate and the ^{13}C NMR spectrum of **3a** are in good agreement with those reported.[13]

*2-Amino-2-deoxy-D-galactopyranosyl-β(1,4)-2-acetamido-2-deoxyglucopyranose (**3b**).* Compound **3b** (Scheme 3) is obtained in 12% yield; ^{1}H NMR for HCl salt (D_2O): δ 2.022, 2.024 (s, NHAc of α and β anomer of GlcNAc), 3.17–3.23 (1H, m, H2 of GalN), 4.67 (d, J = 7.53 Hz, H1β of GlcNAc), 5.13 (d, J = 1.54 Hz, H1α of GlcNAc); HRMS: calculated for $C_{14}H_{26}N_2O_{10}Na$ (M + Na^+) 405.1485, found 405.1489. The ^{1}H NMR spectrum of the acetate form is in good agreement with that reported.[14]

*Ethyl D-Galactopyranosyl-β(1,4)-2-Azido-2-deoxy-D-glucopyranoside (**3c**).* In the case of synthesis of **3c** (Scheme 3), DTT was eliminated, since the 2-azido group is reduced to the corresponding amine with DTT (15%); ^{1}H NMR β anomer (D_2O): δ 1.22 (1H, t, J = 7.80 Hz, OCH_2CH_3), 3.27 (1H, J = 8.33, 9.64 Hz, H2 of $GlcN_3$), 4.40 (1H, d, J = 7.81 Hz, H1 of Gal), 4.55 (1H, J = 8.24 Hz, H1 of $GlcN_3$); HRMS: calculated for $C_{14}H_{25}N_3O_{10}Na$ (M + Na^+) 418.1438, found 418.1438.

The acceptor ethyl 2-azido-2-deoxy-D-galactopyranoside is prepared as follows. Triacetyl-D-glucal is azidonitrated[14] [NaN_3 and ammonium cerium (IV) nitrate $Ce(NH_4)_2(NO_3)_6$ in CH_3CN] and acetolyzed (sodium acetate in acetic acid) to give 2-azido-1,3,4,6-tetra-*O*-acetyl-2-deoxy-D-glucopyranose, which is treated with $TiBr_4$ in CH_2Cl_2 and ethyl acetate, giving a glycosyl bromide, and then glycosylated with ethanol in the presence of silver trifluoromethane sulfonate (AgOTf) and 4 Å molecular seives (MS4Å) in CH_2Cl_2 to give, after O-deacetylation with sodium methoxide

[13] J. Thiem and T. Wiemann, *Angew. Chem., Int. Ed. Engl.* **30,** 1163 (1991).

[14] R. U. Lemieux and R. M. Ratcliffe, *Can. J. Chem.* **57,** 1244 (1979).

in methanol, ethyl 2-azido-2-deoxy-D-glucopyranoside (22% overall yield) as a 1 : 1.5 mixture of α and β anomers; ^{1}H NMR (D_2O): δ 1.21 (t, J = 7.80 Hz, OCH_2CH_3 of β anomer), 1.22 (t, J = 7.80 Hz, OCH_2CH_3), 2.99 (dd, J = 7.43, 9.83 Hz, H2 of β anomer), 5.11 (d, J = 3.58 Hz, H1 of α anomer); HRMS: calculated for $C_8H_{15}N_3O_5Cs$ (M + Cs^+) 366.0066, found 366.0066.

Sialylation

The multienzyme system for sialylation is similar to that reported previously[5] except that α-2,3-sialyltransferase[6] is used in the sialylation. It starts with NeuAc, [1-^{13}C]Gal-LacNAc, PEP, and catalytic amounts of ATP and CMP (Scheme 4). CMP is converted to CDP by nucleoside monophosphate kinase (EC 2.7.4.4, NMK) in the presence of ATP, which is regenerated from its by-product ADP in a reaction catalyzed by PK in the presence of PEP, then to CTP with PEP by PK. The CTP formed then reacts with NeuAc catalyzed by CMP-NeuAc synthase (EC 2.7.7.43, *N*-acetylneuraminate cytidylyltransferase) to produce CMP-NeuAc. The

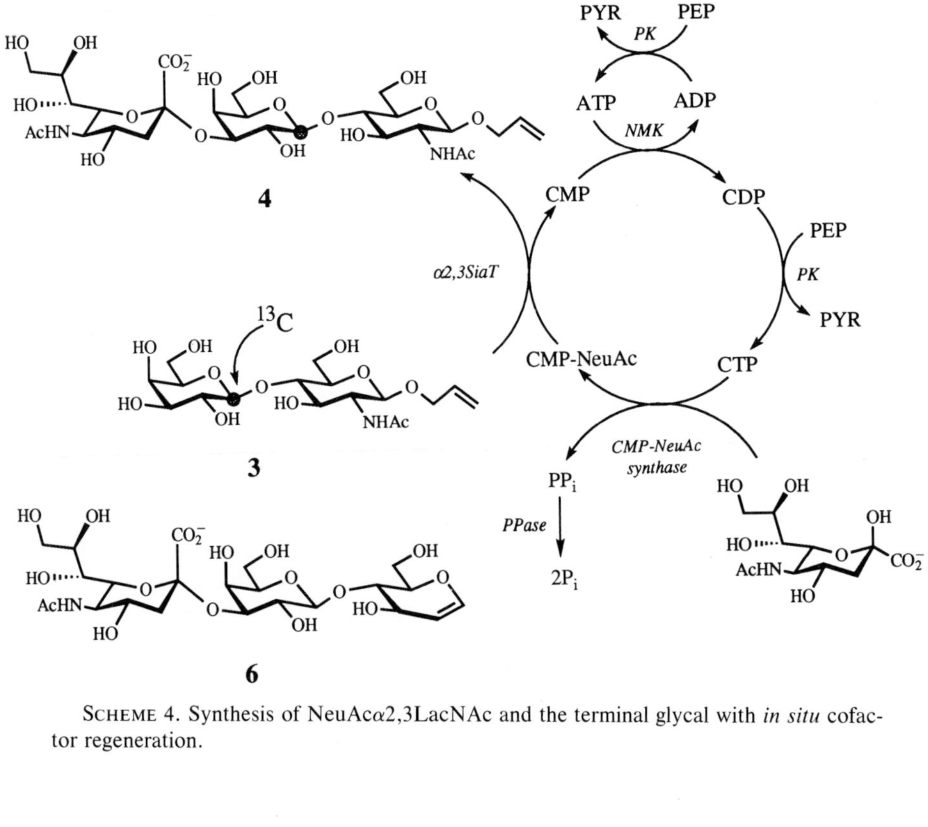

SCHEME 4. Synthesis of NeuAcα2,3LacNAc and the terminal glycal with *in situ* cofactor regeneration.

by-product PP_i is hydrolyzed to inorganic phosphate (P_i) by PPase. Sialylation of LacNAc is accomplished with CMP-NeuAc and α-2,3-sialyltransferase (EC 2.4.99.6, *N*-acetyllactosaminide α-2,3-sialyltransferase). The released CMP is again converted to CDP, to CTP, and finally to CMP-NeuAc. Using this system, [1-^{13}C]Gal-NeuAcα2,3Galβ-*O*-allyl (**4**) as well as the unlabeled trisaccharide is prepared. Interestingly, lactal (Galβ1,4Glucal) is also a good substrate for α-2,3-sialyltransferase, allowing NeuAcα2,3Galβ1,4Glucal (**6**) to be synthesized in 21% yield. Lactal is prepared either chemically[15] or enzymatically using GalT and glucal.[16] The glycal-containing oligosaccharides such as **6** may be converted to other sialyl Le^x derivatives.[6]

Compound 4. To prepare compound **4** (Scheme 4), a solution of **3** (210 mg, 0.50 mmol), NeuAc (160 mg, 0.52 mmol), PEP trisodium salt (120 mg, 0.51 mmol), $MgCl_2 \cdot 6H_2O$ (20 mg, 0.1 mmol), $MnCl_2 \cdot 4H_2O$ (4.9 mg, 25 μmol), KCl (7.5 mg, 0.10 mmol), CMP (16 mg, 50 μmol), ATP (2.7 mg, 5 μmol), and mercaptoethanol (0.34 μl) in HEPES buffer (200 m*M*, pH 7.5; 3.5 ml) is adjusted with 1 *N* NaOH to pH 7.5, and the enzymes NMK (5 U), PK (100 U), PPase (10 U), CMP-NeuAc synthase (0.4 U), and α-2,3-sialyltransferase (0.1 U) are added to the solution. The mixture is gently stirred under an argon atmosphere at room temperature (25°) for 3 days. The mixture is concentrated, and the residue is chromatographed on silica gel, with ethyl acetate–2-propanol–water (2 : 2 : 1, v/v/v), to give a trisaccharide, which is further purified with BioGel P-2 (Bio-Rad, Richmond, CA), with water, to give **4** (88 mg, 24%); ^{1}H NMR (D_2O): δ 1.81 (1H, br t, J = 12.02 Hz, $H3_{ax}$ of NeuAc), 2.04 (6 H, s, NHAc of GlcNAc and NeuAc), 2.76 (1H, dd, J = 4.57, 12.33 Hz, $H3_{eq}$ of NeuAc), 3.96 (1H, br d, J = 3.10 Hz, H4 of Gal), 4.13 (1H, dd, J = 3.09, 9.94 Hz, H3 of Gal), 4.56 (1H, dd, $J_{H1,H2}$ = 7.83, $J_{H1,^{13}C1}$ = 162.78 Hz, H1 of Gal), 4.58 (1H, d, J = 8.32 Hz, H1 of GlcNAc); HRMS: calculated for $C_{27}H_{44}N_2O_{19}Cs_2$ ($M - H^+ + 2Cs^+$) 980.0759, found 980.0720.

NeuAcα2,3-Lactal (6). Compound **6** (Scheme 4) is prepared in a yield of 82 mg; ^{1}H NMR (D_2O, 320 K): δ 1.84 (1H, br t, J = 12.18 Hz, $H3_{eq}$ of NeuAc), 2.08 (3H, s, NHAc of NeuAc), 2.82 (1H, dd, J = 4.46, 12.32 Hz, $H3_{eq}$ of NeuAc), 4.01 (1H, br d, J = 2.50 Hz, H4 of Gal), 4.16 (1H, dd, J = 2.50, 9.50 Hz, H3 of Gal), 4.43 (1H, dt, J = 1.18, 6.46 Hz, H3 of glucal), 4.65 (1H, d, J = 7.86 Hz, H1 of Gal), 4.88 (1H, dd, J = 2.63, 6.07 Hz, H2 of glucal), 6.51 (1H, dd, J = 1.45, 6.08 Hz, H1 of glucal);

[15] W. N. Haworth, E. L. Hirst, M. M. T. Plant, and R. J. W. Reinolds, *J. Chem. Soc.*, 2644 (1930).

[16] C.-H. Wong, Y. Ichikawa, T. Krach, C. Gautheron-Le Narvor, D. P. Dumas, and G. C. Look, *J. Am. Chem. Soc.* **113,** 8137 (1991).

HRMS: calculated for $C_{23}H_{35}NO_{17}NaCs_2$ $(M - H^+ + 2Cs^+)$ 864.0092, found 864.0066.

Fucosylation

The cloned human enzyme fucosyltransferase (FucT) is used for stoichiometric fucosylation with GDP-Fuc[17] (Scheme 5). Fucosylation of sialyl LacNAc (**4**) gives sialyl Le^x (**5**) after silica gel and BioGel P-2 column chromatographies. LacNAc (**3**) and the sialyl glycal **6** are also fucosylated to give Le^x trisaccharide (**7**) and sialyl Le^x glycal (**8**), respectively. These labeled and unlabeled fucosylated oligosaccharides are used for the conformational study employing NMR techniques.[6] One interesting observation is that α1,3-FucT and α1,3/4-FucT accept Galβ1,4-(5-thioGlc) to give a (5-thioGlc)-Le^x analog. Galβ1,4-deoxynojirimycin is, however, an inhibitor of both enzymes.[6]

As for the *in situ* regeneration of GDP-Fuc, we first examine the conversion of mannose 1-phosphate (Man-1-P) to GDP-Fuc via GDP-Man based on the biosynthetic pathway of GDP-Fuc in microorganisms. We use microbial enzymes because of the ease of access. Furthermore, this system allows regeneration of GDP-Man. GDPmannose pyrophosphorylase (GDP-Man PP) has been found in yeast,[18] and GDP-Fuc-generating enzymes are known to exist in the bacterium *Klebsiella pneumonia*.[19] In this regeneration, GTP is generated from GDP in the presence of PEP and PK. Man-1-P reacts with GTP to give GDP-Man by GDP-Man PP from dried yeast cells. GDP-Man is transformed to GDP-Fuc in the presence of NADPH and GDP-Fuc-generating enzymes partially purified from the bacterium. The oxidized $NADP^+$ is converted back to NADPH by *Thermoanaerobium brockii* alcohol dehydrogenase (TADH) (EC 1.1.1.2) and 2-propanol. Production of GDP-Man and GDP-Fuc is confirmed by high-performace liquid chromatography (HPLC), and fucosylation of LacNAc and **4** to give **7** and **5** in yields of 5–10 mg is accomplished.[6]

An alternative method is to start with fucose 1-phosphate (Fuc-1-P), which is converted to GDP-Fuc catalyzed by GDP-Fuc pyrophosphorylase (GDP-Fuc P). We have partially purified GDP-Fuc P from porcine thyroids and have demonstrated that the regeneration system depicted in Scheme 5 is functional on an analytical scale for the synthesis of sialyl Le^x.

*Compounds **5**, **7**, and **8**.* For fucosylation (Scheme 5), a solution of FucT (20 mU; 2 ml) is added to a solution of **4** (23 mg, 31 μmol) and GDP-

[17] Y. Ichikawa, M. M. Sim, and C.-H. Wong, *J. Org. Chem.* **57,** 2943 (1992).

[18] A. Munch-Peterson, this series, Vol. 5, p. 171; E. S. Simon, S. Grabowski, and G. M. Whitesides, *J. Org. Chem.* **55,** 1834 (1990).

[19] V. Ginsburg, this series, Vol. 8, p. 293.

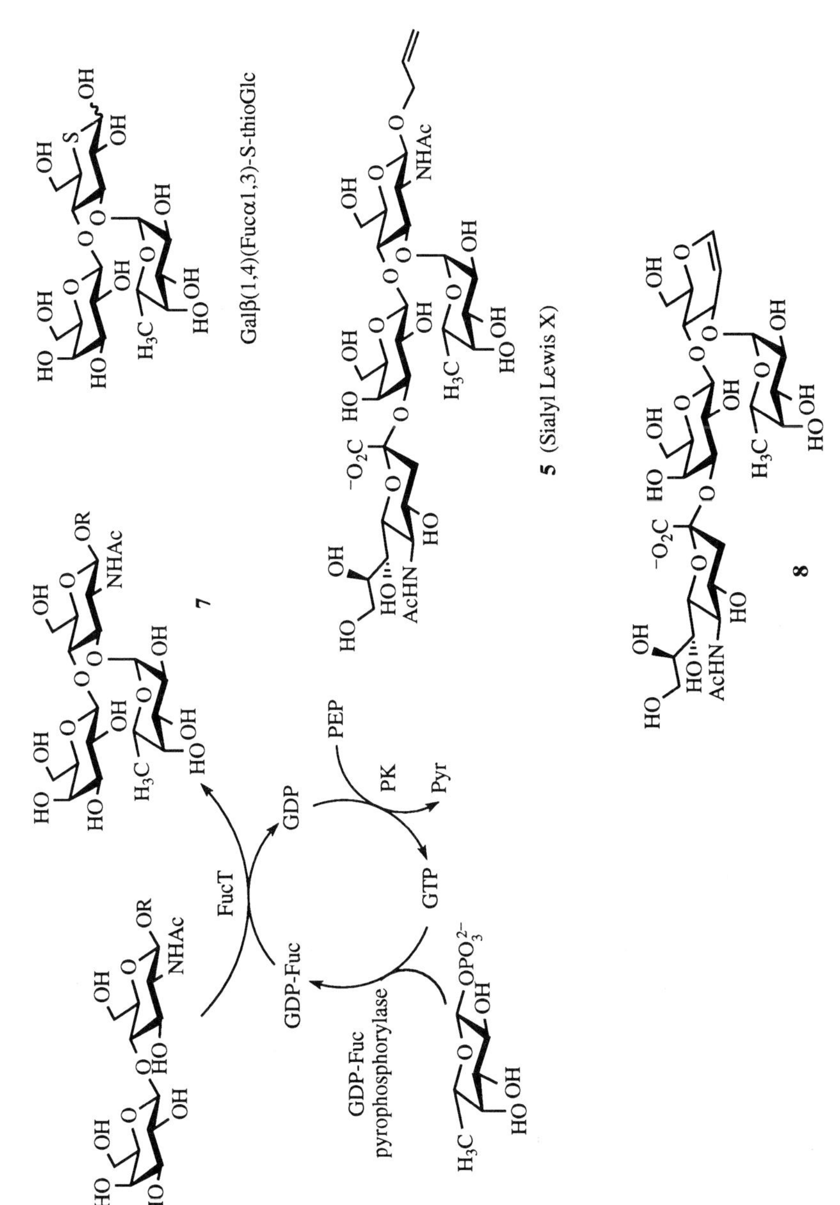

SCHEME 5. GDP-Fuc regeneration employing GDP-Fuc pyrophosphorylase.

Fuc[17] (24 mg, 36 μmol) in HEPES buffer (3 ml; 200 mM, pH 7.5) containing 5 mM ATP, 20 mM Mn^{2+}, and the mixture is gently stirred under an argon atmosphere for 5 days at room temperature (25°). The mixture is concentrated and chromatographed on silica gel, with ethyl acetate–2-propanol–water (2 : 2 : 1, v/v/v), to give a tetrasaccharide, which is further purified with BioGel P-2, with water. The eluant is passed through a column of Dowex 50W-X8 (H^+), eluted with water, neutralized with 1 N NaOH, and lyophilized to give **5** (18 mg). Similarly, compounds **7** (42 mg) and **8** (51 mg) are prepared. For complete NMR (^{1}H and ^{13}C) assignments for **5** and **7**, see the reported data.[6] For compound **8**, ^{1}H NMR (D_2O): δ 1.15 (3 H, d, J = 6.61 Hz, 6-CH_3 of Fuc), 1.76 (1H, br t, J = 12.00 Hz, $H3_{ax}$ of NeuAc), 1.98 (3 H, s, NHAc of NeuAc), 2.71 (1H, dd, J = 4.52, 12.38 Hz, $H3_{eq}$ of NeuAc), 4.44 (1H, br q, J = 7.50 Hz, H5 of Fuc), 4.58 (1H, d, J = 8.0 Hz, H1 of Gal), 4.95 (1H, dd, J = 2.5, 6.0 Hz, H2 of glucal), 5.00 (1H, d, J = 3.98 Hz, H1 of Fuc), 6.45 (1H, d, J = 6.0 Hz, H1 of glucal).

Galβ(1,4)(Fucα1,3)-5-thio-Glc. A solution of Galβ(1,4)(5-thioGlc) (30 mg, 84 μmol), GDP-Fuc (60 mg, 84 μmol), and α1,3/4-FucT (0.5 U)[20] in sodium cacodylate buffer (5.4 ml; 50 mM, pH 6.2) containing 5 mM ATP and 20 mM $MnCl_2$ is stirred for 2 days at room temperature. The R_f values of the starting material and the product are 0.39 and 0.31, respectively, in ethyl acetate–acetic acid–water (3 : 2 : 1, v/v/v) on silica thin-layer chromatography (TLC). The reaction mixture is applied directly to a column of Sephadex G-25 Superfine (1.5 × 30 cm) and eluted with water. The fractions containing the product are pooled and successively passed through columns of QAE-Sephadex and Dowex 50-X8 (H^+) with water. The effluent is pooled and lyophilized (21 mg); ^{1}H NMR (D_2O): δ 1.13 (3H, d, J = 6.7 Hz, 6-CH_3 of Fuc), 3.40 (1H, dd, J = 6.4 and 11.7 Hz), 3.60 (1H, dd, J = 3.6 and 11.7 Hz), 4.52 (1H, d, J = 7.9 Hz), 4.95 (1H, J = 2.6 Hz), 5.34 (1H, d, J = 3.8 Hz).

Purification of GDP-Fucose Pyrophosphorylase. Porcine thyroid glands (376 g) are homogenized in ice-cold 10 mM 3-(N-morpholino)propanesulfonic acid (MOPS) (752 ml), pH 7.5, with 1 μg/ml each antipain, aprotinin, pure chymotrypsin, leupeptin, and pepstatin, in a Waring blendor (five 15-sec bursts on high setting). Cell debris is removed by centrifugation at 8000 g for 20 min at 4°. To the supernatant fraction is added 188 ml of a 2% solution of protamine sulfate. The mixture is stirred for 15 min, and the precipitate is removed by centrifugation as above. Solid ammonium sulfate is slowly added to the supernatant fraction to

[20] J. F. Fukowska-Latallo, R. D. Larson, R. P. Nair, and J. B. Lowe, *Gene Dev.* **4,** 1288 (1990).

50% saturation (0.291 g/ml at 0°). After centrifugation as described above, the precipitate is collected and resuspended in 500 ml of 1.2 *M* ammonium sulfate. The sample is mixed with a slurry of phenyl-Sepharose (50 ml) that has been equilibrated in 1.2 *M* ammonium sulfate. The resin with the bound enzyme is washed with 1.2 *M* ammonium sulfate (200 ml) and the enzyme activity eluted with 0.4 *M* ammonium sulfate (200 ml). Throughout the purification, GDP-Fuc pyrophosphorylase is assayed according to the method of Ishihara and Heath.[21] One unit of activity is defined as the incorporation of 1 μmol of inorganic [^{32}P]pyrophosphate into GTP per minute.

GDP-Fucose Regeneration Employing GDP-Fucose Pyrophosphorylase: Synthesis of Sialyl Lewis x. A solution of MOPS, pH 7.5 (50 m*M*), Fuc 1-P (10 m*M*), GDP (1 m*M*), PEP (10 m*M*), KF (5 m*M*), Mg^{2+} (10 m*M*), Mn^{2+} (10 m*M*), PK (5 U), sialyl-[^{3}H]LacNAcβ-*O*-$(CH_2)_6CO_2Me$ (10 m*M*), α1, 3-FucT (0.1 U), inorganic pyrophosphatase (5 U), and GDP-Fuc pyrophosphorylase (0.1 U) is mixed in a volume of 100 μl. The reaction is incubated on a tube turner at 37° for 64 hr. The sample is extracted with 10 volumes of methanol, dried by evaporation under reduced pressure, resuspended in water, and analyzed by thin-layer chromatography on silica gel plates with 2-propanol–1 *M* ammonium acetate (6 : 1, v/v) as solvent. Sialyl Lewis x is formed with a yield of 47%, as determined by scintillation counting.

α1,2-Mannosyltransferase

Substrate Specificity. Previous studies on the native α1,2-mannosyltransferase (ManT) have showed that the enzyme transfers mannose to methyl α-D-mannopyranoside and mannobiose.[22–24] The active domain of the enzyme prepared in this study also has a similar substrate specificity as the native membrane-bound enzyme from yeast.[23,24] It accepts methyl α-D-mannopyranoside, mannose, and Manα(1,2)ManαOMe as substrates with K_m values of 57, 193, and 28 m*M*, respectively. For comparison, the K_m value of mannose for the native enzyme is 100 m*M*. The C-6 modified methyl α-D-mannopyranosides such as 6-deoxy-ManαOMe, 6-azido-6-deoxy-ManαOMe, and 6-amino-6-deoxy-ManαOMe are poor substrates. *p*-Nitrophenyl α-D-mannopyranoside and other monosaccharides with an S-configuration at the 2 position, such as D-altrose, D-idose, D-talose,

[21] R. Prohaska and H. Schenkel-Brunner, *Anal. Biochem.* **69,** 536 (1975).
[22] L. Lehle and W. Tanner, *Biochim. Biophys. Acta* **350,** 225 (1974).
[23] M. S. Lewis and C. E. Ballou, *J. Biol. Chem.* **266,** 8255 (1991).
[24] A. Hausler and P. W. Robbins, *Glycobiology* **2,** 77 (1992).

D-arabinose, and D-lyxose, are not substrates. The mannosidase inhibitor, 1-deoxymannojirimycin, is neither a substrate nor an inhibitor of ManT.[7]

O-Mannosylpeptides are found to be good substrates for the ManT enzyme. For example, Cbz-Thr(α-Man)-Val-*O*-Me (**9**) is comparable with ManαOMe as an acceptor, but has a lower K_m value. The N-terminal deprotected compound and other longer peptide analogs are also good substrates. Interestingly, the *O*-glycopeptide Boc-Tyr-Thr(α-Man)-Val-*O*-Me[25] has a K_m value of 0.7 m*M* for ManT. This value is 10 times smaller than that of compound **9** and 80 times smaller than that of ManαOMe. This result suggests that ManT prefers certain peptide sequences in *O*-mannosylpeptides.

Oligomannose is the backbone structure component of N-linked and O-linked glycoproteins in yeast and mammalian cells. Such oligosaccharides are constructed by the highly ordered addition of monosaccharide units to the growing oligosaccharide chain. Studies on the biosynthesis of O-linked carbohydrate chains in *Saccharomyces cerevisiae* have suggested that the native α-1,2-mannosyltransferase is responsible for the transfer of the third mannose to the growing O-linked carbohydrate chains.[26] However, comparing the kinetic data of compounds **7** and **10**, the recombinant active domain of ManT seems to be more active on monomannosyl *O*-glycopeptides than on dimannosyl glycopeptides. The reason for this observation is not clear yet.

Synthesis of Mannose-Containing Oligosaccharides and Glycopeptides with Regeneration of GDP-Man. As illustrated in Scheme 6, the multienzyme system starts with mannose 1-phosphate.[17] Mannose 1-phosphate reacts with GTP catalyzed by GDPmannose pyrophosphorylase (EC 2.7.7.22; mannose-1-phosphate guanylyltransferase) from yeast cells[18] to form GDP-Man. GDP-Man is consumed by ManT to give the mannosyl oligosaccharide or glycopeptide, and the released GDP is again converted to GTP by PK (EC 2.7.7.9) and PEP. The inorganic pyrophosphate resulting from the reaction is hydrolyzed to inorganic phosphate by inorganic pyrophosphatase to shift the equilibrium and to avoid enzyme inhibition.

In a representative synthesis of Cbz-Thr(α-Man-1,2α-Man)-Val-*O*-Me (**10**),[7] a reaction mixture (2 ml, 100 m*M* Tris, pH 7.5, 5% acetone, 10 m*M* $MgCl_2$, 10 m*M* $MnCl_2$, 5 m*M* NaN_3, 1 m*M* ATP, 10 μ*M* theophylline, 30 μ*M* 2,3-dimercaptopropanol, and 50 μ*M* phenylmethylsulfonyl fluoride) containing Cbz-Thr(α-Man)-Val-*O*-Me (**9**) (100 mg, 95 m*M*), PEP (47 mg,

[25] W. Tanner, *Eur. J. Biochem.* **196,** 185 (1991).

[26] A. Hausler, L. Ballou, C. E. Ballou, and P. W. Robbins, *Proc. Natl. Acad. Sci. U.S.A.* **89,** 6846 (1992).

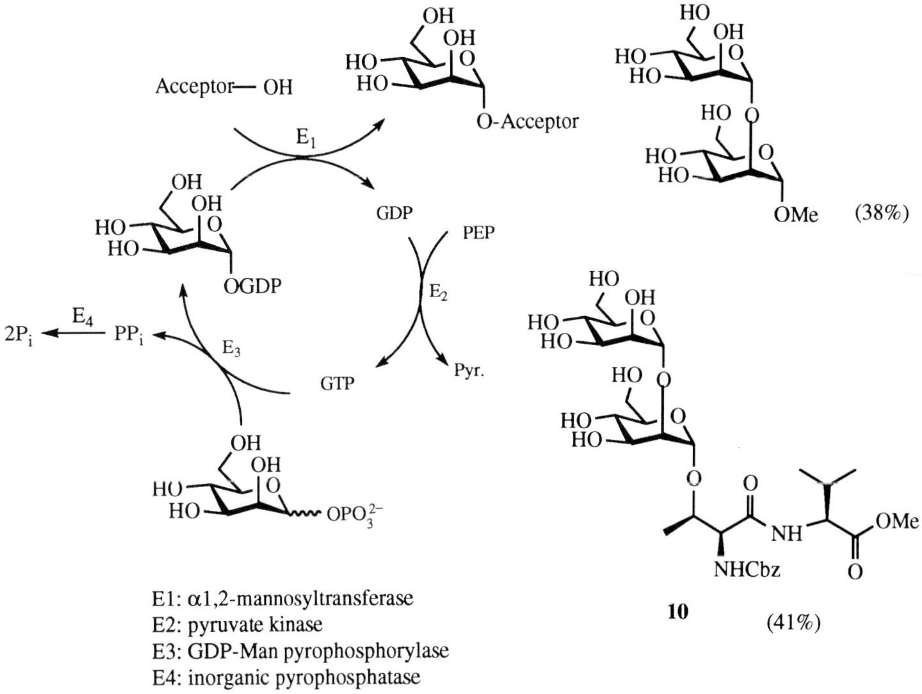

E1: α1,2-mannosyltransferase
E2: pyruvate kinase
E3: GDP-Man pyrophosphorylase
E4: inorganic pyrophosphatase

SCHEME 6. Synthesis of mannosyloligosaccharides with regeneration of GDP-Man.

100 m*M*), PK (50 U), dried yeast cells (50 mg), ManT (0.4 U), and inorganic pyrophosphatase (1 U) is slightly stirred at room temperature for 60 hr and then centrifuged. The supernatant is lyophilized and extracted with methanol. After removal of the methanol, the product is purified by silica gel column chromatography ($CHCl_3$–methanol–water, 6 : 3 : 0.5, v/v/v) to afford **10** (31 mg) in 41% overall yield based on consumed Cbz-Thr(α-Man)-Val-*O*-Me. The disaccharide Manα1,2ManαOMe is prepared in 38% yield in a similar manner.

Manα(1,2)ManαOMe. A reaction mixture (100 ml, 100 m*M* Tris, pH 7.5, 10 m*M* $MgCl_2$, 10 m*M* $MnCl_2$, 5 m*M* EDTA, 5 m*M* NaN_3, 1 m*M* ATP, 10 μ*M* theophylline, and 30 μ*M* 2,3-dimercaptopropanol) containing α-methyl D-mannopyranoside (0.4 g, 20 m*M*), Man-1-P (22 m*M*), PEP (10 g, 42.7 m*M*), GDP (30 mg, 1 m*M*), dried yeast cells (0.4 g), PK (300 U), ManT (1.2 U), and inorganic pyrophosphorylase (2 U) is slightly stirred at room temperature for 72 hr and then centrifuged. The supernatant is lyophilized and extracted with methanol. After removal of the methanol, the product is purified by silica gel column chromatography ($CHCl_3$–meth-

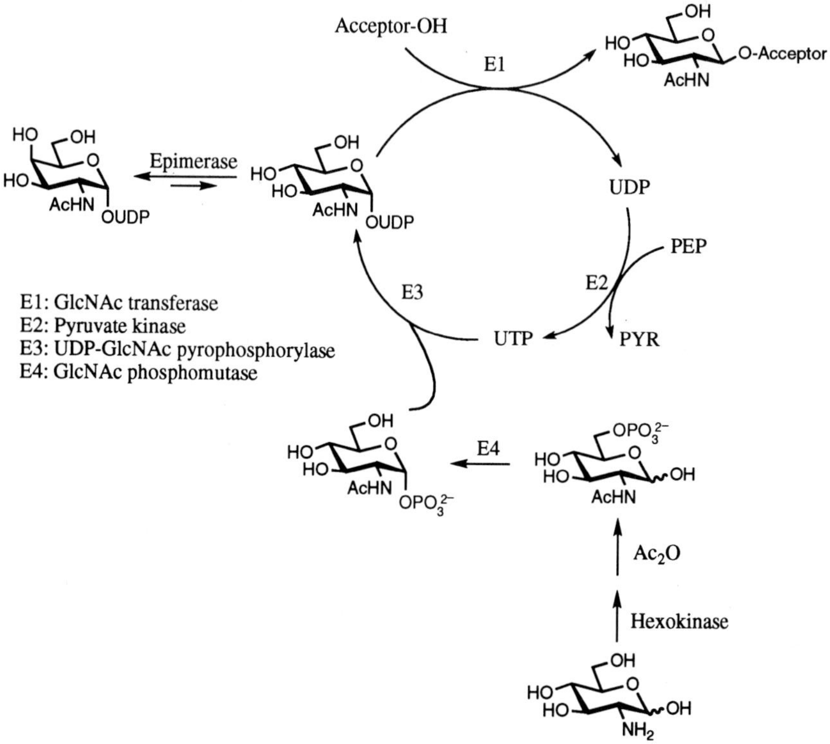

SCHEME 7. Regeneration of UDP-GlcNAc and UDP-GalNAc.

anol–water, 6 : 3 : 0.5, v/v/v) to afford Manα1,2ManαOMe (96 mg) in 38% overall yield based on consumed α-methyl D-mannopyranoside; ^{1}H NMR (D_2O): δ 3.41 (s, 3H), 3.60–3.80 (m, 6H), 3.86 (dd, 1H, J = 3.5, 10 Hz), 3.88 (dd, 1H, J = 3.5, 10 Hz), 3.90 (d, 1H, d, J = 1.5 Hz), 3.92 (d, 1H, J = 1.5 Hz), 3.97 (dd, 1H, J = 2, 3.5 Hz), 4.08 (dd, 1H, J = 2, 3.5 Hz), 5.02 (d, 1H, d, J = 2 Hz), 5.04 (d, 1H, J = 2.0 Hz); ^{13}C {^{1}H} NMR (D_2O): δ 54.90, 61.01, 61.24, 66.98, 67.03, 70.02, 70.29, 70.37, 72.63, 73.40, 78.62, 99.40, 102.43; HRMS: calculated for $C_{13}H_{24}O_{11}Na^+$ (M + Na$^+$) 379.1216, found 379.1220.

Cbz-Thr[αManα(1,2)Man]-Val-O-Me (**10**). The synthetic procedure is described above; ^{1}H NMR (500 MHz, D_2O): δ 0.83 (m, 6H), 1.21 (d, 3H, J = 4.5 Hz), 2.04 (m, 1H), 3.55–3.62 (m, 15H), 4.05 (m, 1H), 4.16 (m, 1H), 4.24 (m, 1H), 5.00 (s, 2H), 7.16–7.27 (m, 5H); ^{13}C {^{1}H} NMR (D_2O): δ 17.63, 17.68, 18.41, 30.10, 52.88, 58.68, 61.14, 61.21, 66.91, 67.19, 67.52, 70.03, 70.13, 70.47, 73.30, 76.47, 79.31, 99.84, 102.40, 127.87, 127.92,

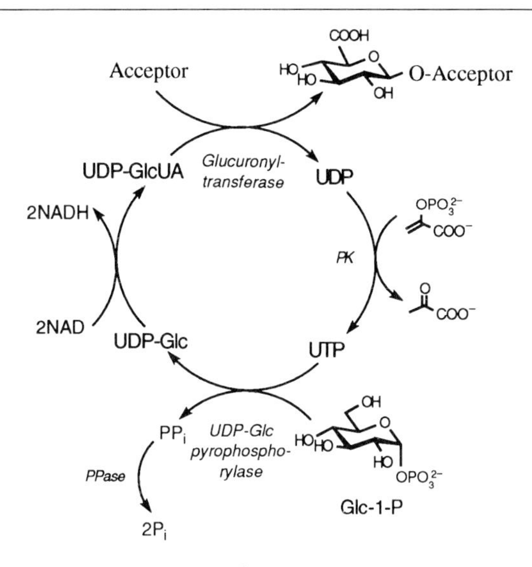

SCHEME 8. Regeneration of UDP-glucuronic acid.

128.60, 129.00, 172.82, 173.80; HRMS: calculated for $C_{30}H_{46}N_2O_{16}Cs^+$ ($M + Cs^+$) 823.1902, found 823.1928.

Coupling of Two Glycosyltransferase Reactions

The first procedure involves the *in situ* generation of NeuAc catalyzed by NeuAc aldolase coupled with regeneration of CMP-NeuAc (parts A and B in Scheme 9). The second procedure involves the *in situ* generation of NeuAc and LacNAc coupled with regeneration of UDP-Glc, UDP-Gal, and CMP-NeuAc (parts A, B, and C in Scheme 9). The two multienzyme systems operate very efficiently without problems of product inhibition.

The enzymatic aldol reaction (Scheme 9B) is first introduced to the previously investigated regeneration system (Scheme 9A): ManNAc is converted to NeuAc catalyzed by NeuAc aldolase[27] (EC 4.1.3.3, *N*-acetylneuraminate lyase; from Toyobo, New York, N.Y.) in the presence of pyruvic acid. Although NeuAc aldolase also catalyzes the reverse reaction (NeuAc to ManNAc and pyruvate), the produced NeuAc is irreversibly incorporated into cycle A (Scheme 9) via CMP-NeuAc catalyzed by CMP-sialate synthase[28] coupled with inorganic pyrophosphatase

[27] M.-J. Kim, W. J. Hennen, H. M. Sweers, and C.-H. Wong, *J. Am. Chem. Soc.* **110,** 6481 (1988).

[28] J. L.-C. Liu, G.-J. Shen, Y. Ichikawa, J. F. Rutan, G. Zapata, W. F. Vann, and C.-H. Wong, *J. Am. Chem. Soc.* **114,** 3901 (1992).

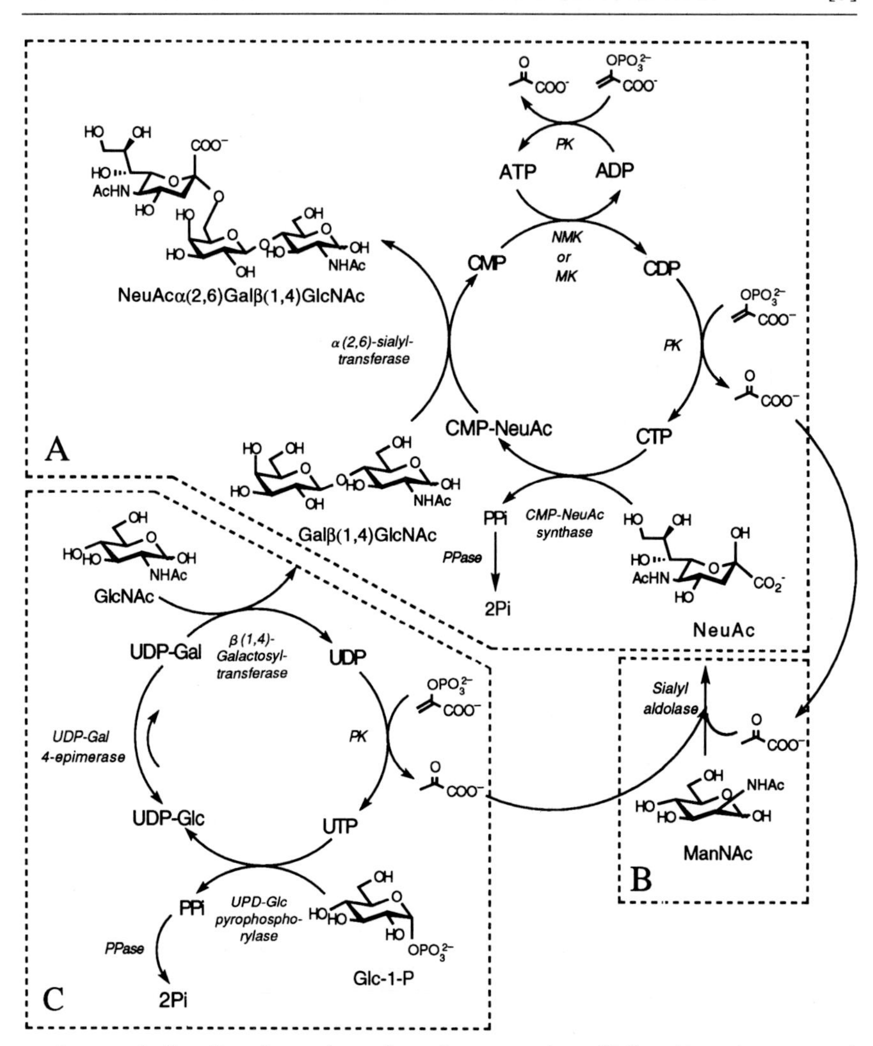

SCHEME 9. Coupling of two glycosyltransferase reactions. Sialic acid can be generated *in situ* by the sialic acid aldolase reaction.

(PPase)-catalyzed decomposition of the released inorganic pyrophosphate. The sialyl-LacNAc is obtained in 89% yield after BioGel P-2 column chromatography. The experimental procedure is as follows.

To 1.65 ml of HEPES buffer (200 m*M*, pH 7.5) are added ManNAc (43 mg, 180 μmol), LacNAc (22 mg, 60 μmol), CMP (2.0 mg, 6 μmol), ATP (0.32 mg, 0.6 μmol), PEP sodium salt (56 mg, 240 μmol), $MgCl_2 \cdot 6H_2O$ (12.2 mg, 60 μmol), $MnCl_2 \cdot 4H_2O$ (3.0 mg, 16 μmol), KCl (4.4 mg, 60 μmol), pyruvic acid sodium salt (33 mg, 300 μmol), NeuAc aldolase (EC 4.1.3.3; 45 U), myosin kinase (EC 2.7.4.3, MK; 100 U), PK (EC 2.7.1.40; 120 U), PPase (EC 3.6.1.1; 6 U), mercaptoethanol (0.22 μl), CMP-NeuAc synthase (EC 2.7.7.43; 0.3 U in 1 ml of 0.1 *M* Tris buffer, pH 9), and α-(2,6)-sialyltransferase (EC 2.4.99.1; 80 mU). The final volume of the reaction mixture is 3 ml. The reaction is conducted at room temperature for 2 days under argon. After disappearance of the starting material as judged by TLC on silica [R_f for LacNAc, 0.63; NeuAc, 0.31; sialyl-Lac-NAc, 0.30; CMP-NeuAc, 0.19 in 1 : 2.4 (v/v) 1 *M* ammonium acetate–2-propanol], the reaction mixture is directly applied on BioGel P-2 (200–400 mesh) column (2 × 36 cm) and eluted with water. The trisaccharide-containing fractions are pooled and lyophilized to give sialyl-LacNAc (37 mg, 89%) with ^{1}H NMR data identical with those reported.[6]

We then combine the LacNAc synthesizing cycles (Scheme 9C) and the above-mentioned cycle (Scheme 9A,B). This system requires only three monosaccharide components, GlcNAc, ManNAc, and Glc-1-P, for the sialyl-LacNAc synthesis. UDP-Glucose pyrophosphorylase converts Glc-1-P to UDP-Glc, which is epimerized to UDP-Gal by UDP-Gal 4-epimerase.[5] The galactose moiety of UDP-Gal is coupled with GlcNAc to generate LacNAc, which is incorporated into cycle A to produce sialyl-LacNAc. The experimental procedure is as follows.

To 2.6 ml of HEPES buffer (200 m*M*, pH 7.5) are added ManNAc (43 mg, 180 μmol), GlcNAc (13.3 mg, 60 μmol), Glc-1-P (21.5 mg, 60 μmol), CMP (2.0 mg, 6 μmol), UDP (2.8 mg, 6 μmol), ATP (0.32 mg, 0.6 μmol), PEP sodium salt (75 mg, 320 μmol), $MgCl_2 \cdot 6H_2O$ (16.3 mg, 80 μmol), $MnCl_2 \cdot 4H_2O$ (4.0 mg, 20 μmol), KCl (6.0 mg, 80 μmol), pyruvic acid sodium salt (33 mg, 300 μmol), NeuAc aldolase (45 U), MK (100 U), PK (120 U), PPase (12 U), mercaptoethanol (0.33 μl), GalT (1 U), UDP-Glc pyrophosphorylase (EC 2.7.7.9; 1 U), UDP-Gal 4-epimerase (1 U), CMP-NeuAc synthase (0.3 U in 1 ml of 0.1 *M* Tris buffer, pH 9), and α(2,6)-sialyltransferase (80 mU). The final volume of the reaction mixture is 4 ml. The reaction is stopped in 2 days, and pure sialyl-LacNAc (9 mg; 22%) is isolated based on the above-mentioned procedure.

Coupling of Glycosidase and Glycosyltransferase Reactions

In a representative synthesis of NeuAcα(2,6)-LacNAc (Scheme 10), to 1.07 ml HEPES buffer (0.2 *M*, 20 m*M* $MgCl_2$, 5.3 m*M* $MnCl_2$, 20 m*M* KCl, pH 7.5) containing 12.3 mg NeuAc (20 m*M*), 180 mg Lac (250 m*M*), 265 mg GlcNAc (600 m*M*), 30.2 mg PEP (trisodium salt, 40 m*M*), CMP (2 m*M*), and ATP (0.2 m*M*) are added MK (6 U), PK (EC 2.7.1.40, 80 U), PPase (4 U), CMP-NeuAc synthase (0.32 U), α(2,6)-sialyltransferase (52

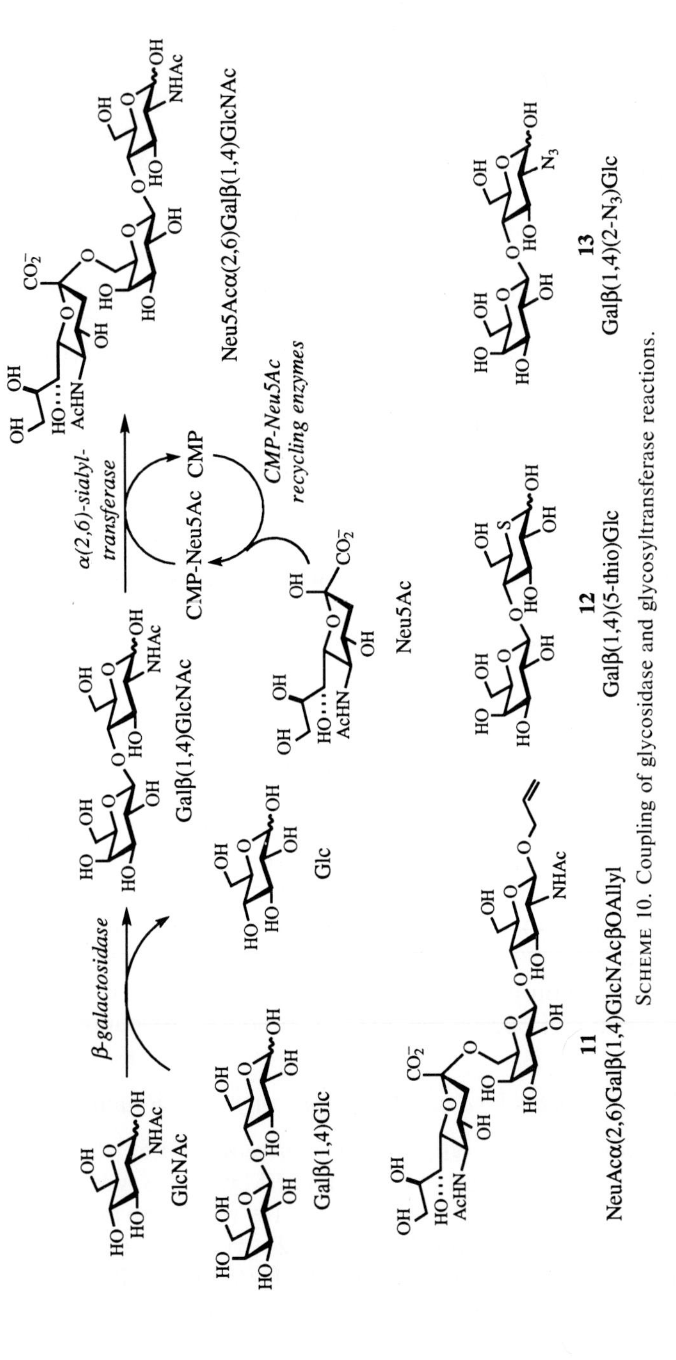

SCHEME 10. Coupling of glycosidase and glycosyltransferase reactions.

mU), and 1 mg of crude β-galactosidase from *Bacillus circulans* (EC 3.2.1.23, Daiwa Kasei KK, Osaka, Japan).[29] The total volume is adjusted to 2 ml. The reaction is conducted for 91 hr under argon at room temperature and is monitored by TLC on silica gel 60 [R_f for Lac, 0.16; LacNAc, 0.27; Gal, 0.29; Glc, 0.37; α(2,6)-sialyl-LacNAc, 0.45; NeuAc, 0.54; GlcNAc, 0.54 in 7:2:1 (v/v/v) 2-propanol–water–NH_4OH]. The reaction mixture is centrifuged, and the supernatant is directly applied to a BioGel P-2 column (200–400 mesh, 43 × 2 cm) with water as the eluent. The trisaccharide-containing fractions are pooled and lyophilized to give 6.8 mg of NeuAcα(2,6)Galβ(1,4)GlcNAc (26% yield); ^{1}H NMR: δ 1.677 (1H, t, J = 12.5 Hz, $H3_{ax}$ of NeuAc), 1.985 (3H, s, NHAc of GlcNAc), 2.022 (3H, s, NHAc of NeuAc), 2.628 (1H, dd, J = 5 Hz, 12.5 Hz, $H3_{eq}$ of NeuAc), 4.409 (1H, d, J = 8 Hz, H1 of Gal), 4.780 (0.5 H, d, J = 8 Hz, H1β of GlcNAc), 5.154 (0.5 H, d, J = 2.5 Hz, H1α of GlcNAc). No other sialylated product can be detected.

Neither Lac, one of the starting materials, nor Galβ(1,6)GlcNAc, a by-product of the β-galactosidase reaction, is a substrate of α(2,6)-sialyltransferase owing to the high K_m (390 mM) and low V_{max} (3%), respectively. Other kinetic parameters for the enzyme are as follows: Galβ(1,4)-GlcNAc,[30] V_{max} 1.00, K_m 12 mM; Galβ(1,6)GlcNAc,[30] V_{max} 0.03, K_m 140 mM; Galβ(1,4)Glc,[30] V_{max} 1.02, K_m 390 mM. Of several galactosides tested, lactose is found to be the best substrate for the galactosidase. The β-galactosidase accepts the following substrates with the indicated relative rate: Lac (100), *o*-nitrophenyl β-D-galactoside (10.6), methyl β-D-galactoside (1). In a similar manner, the sialylated saccharide **11** is also prepared. Compounds **12** and **13** are prepared separately in approximately 20% yield with the galactosidase; however, **12** is not a substrate for the sialyltransferase, and **13** is a very weak substrate.[10] Physical data are as follows: ^{1}H NMR NeuAcα(2,6)Galβ(1,4)GlcNAcβ-*O*-Allyl (**11**): δ 1.672 (1H, t, J = 12.5 Hz, H-3_{ax} of NeuAc), 1.984 (3H, s, NHAc of GlcNAc), 2.015 (3H, s, NHAc of NeuAc), 2.623 (1H, m, J = 4.5, 12.25 Hz, $H3_{eq}$ of NeuAc), 4.130 (1H, m, allyl), 4.294 (1H, m, allyl), 4.403 (1H, d, J = 8 Hz, H1 of Gal), 4.569 (0.5 H, d, J = 8.5 Hz, H1β of GlcNAc), 5.221 (1H, m, allyl), 5.268 (1H, m, allyl), 5.872 (1H, m, allyl); R_f 0.50 in 7:2:1 (v/v/v) 2-propanol–water–NH_4OH. The ^{1}H NMR spectrum of **12** is identical with that reported previously.[16] ^{1}H NMR Galβ(1,4)(2-N_3)Glc (**13**): 3.27 (dd, J = 8.32, 9.82 Hz, H2β), 4.42 (d, J = 7.75 Hz, H1′β), 4.69 (d, J = 8.72 Hz, H1β), 5.31 (d, J = 3.50 Hz, H1α).

[29] K. Sakai, R. Katsumi, H. Ohi, T. Usui, and Y. Ishido, *J. Carbohydr. Chem.* **11,** 553 (1992).
[30] J. C. Paulson, J. I. Rearick, and R. L. Hill, *J. Biol. Chem.* **252,** 2363 (1977).

[8] Chemical Synthesis of Core Structures of Oligosaccharide Chains of Cell Surface Glycans Containing Carba Sugars

By SEIICHIRO OGAWA

Introduction

The validamycin antibiotics (**1**), α-amylase inhibitor acarbose (**2**), and homologs were discovered more than 20 years ago. Their unique biological activities have been ascribed to the enzyme-inhibitory activities exerted by the essential core structures, validoxylamine A (**3**) and methyl acarviosin (**4**), respectively. Both compounds possess the 5a-carba sugar[1] validamine (**5**), a carbocyclic analog of hexose, and/or its unsaturated derivative, valienamine (**6**), which bind to the 5a-carba and/or true hexose residues by way of an imino linkage. In nature, the other type of linkage contained in the carba di-, tri-, and oligosaccharides has never been found so far.

5a-Carba-α- (**7**) and 5a-carba-β-D-glucopyranoses (**8**) are used as model compounds to establish indirectly the anomeric specificity of true α- and β-D-glucopyranoses. Thus, the α anomer but not the β anomer has been shown to inhibit both glucose-stimulated insulin release and islet glucokinase activity.[2] The α-stereospecificity reflects the preference for the α anomer of D-glucopyranose of both glucokinase and glucose-stimulated insulin release. Carba sugar analogs of naturally occurring oligosaccharides of biological interest have been used as model compounds for conformational analysis of true oligosaccharides or as substrate analogs and inhibitors for the study of enzymatic actions. The β anomer (**8**) has been shown to be the substrate of *Cellvibrio gilvuse* cellobioside phosphorylase to produce 5a-carba-cellobiose (**9**), suggesting indirectly that only the β anomer of D-glucose participates in the reverse reaction of this phosphorylase.[3]

Carba disaccharides are carbocyclic analogs of true disaccharides, in which one or both of the hexose (or pentose) residues is (are) replaced with a carba sugar. Therefore, for instance, three types of carba-maltoses are possible: type **A** is composed of two carba-D-glucopyanose units bonded by an ether linkage, and types **B** and **C** consist of one true and

[1] T. Suami and S. Ogawa, *Adv. Carbohydr. Chem. Biochem.* **48,** 21 (1990); S. Ogawa, *in* "Studies in Natural Products Chemistry" (Atta-ur-Rahman, ed.), p. 187. Elsevier, Amsterdam, London, New York, and Tokyo, 1993.

[2] I. Miwa, H. Hara, J. Okuda, T. Suami, and S. Ogawa, *Biochem. Int.* **11,** 809 (1985).

[3] M. Kitaoka, S. Ogawa, and H. Taniguchi, *Carbohydr. Res.* **247,** 355 (1993).

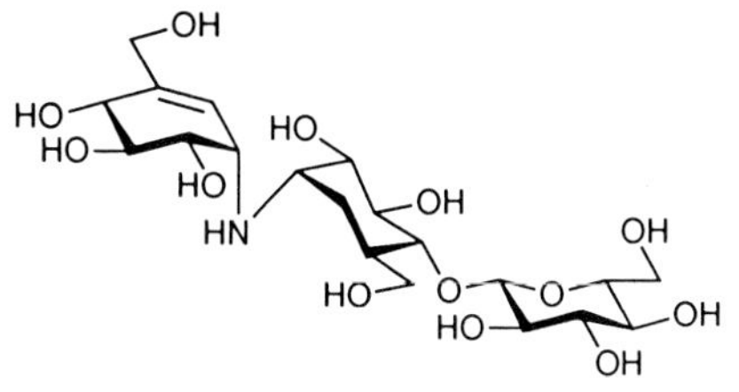

Validamycin A : **1**

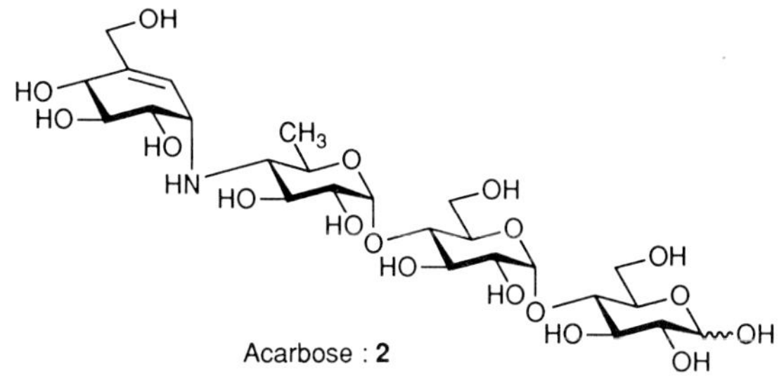

Acarbose : **2**

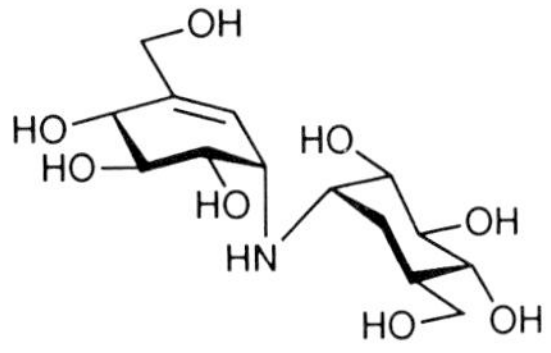

Validoxylamine A : **3**

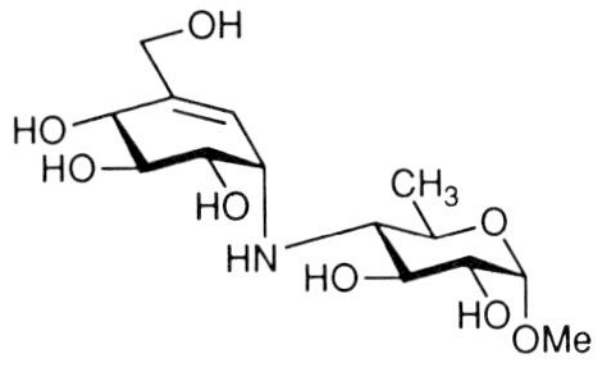

Methyl acarviosin : **4**

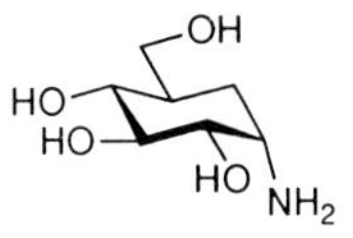

Validamine : **5**

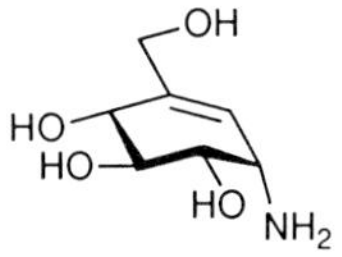

Valienamine : **6**

one carba-D-glucopyranose unit. The interunit linkage in the first two types (**A** and **B**) is ethereal, and that in the last (**C**) is glycosidic. Likewise, carbaoligosaccharides comprising such linkages as N-glycosidic or imino, S-glycosidic or thio ether, and C-glycosidic or methylene may be possible. Some 5a-carba di- and trisaccharides related to core structures common to cell surface glycans have been synthesized.[4]

[4] S. Ogawa and K. Nishi, *Carbohydr. Res.* **229,** 117 (1992).

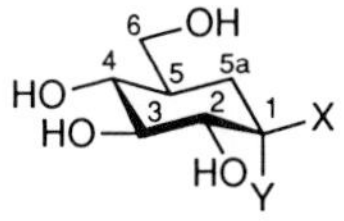

5a-Carba-α-D-glucopyranose, X = H, Y = OH : **7**
5a-Carba-β-D-glucopyranose, X = OH, Y = H : **8**

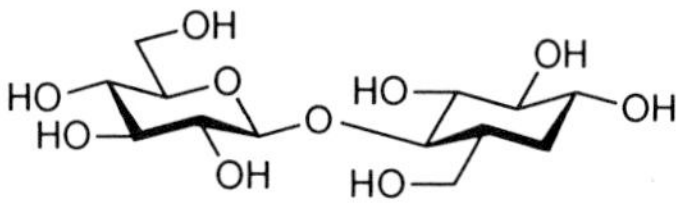

5a-Carba-β-cellobiose : **9**

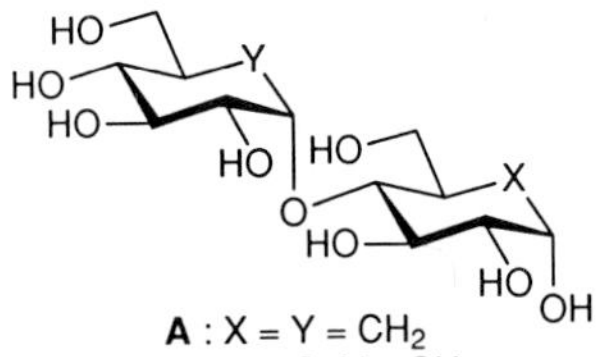

A : X = Y = CH_2
B : X = O, Y = CH_2
C : X = CH_2, Y = O

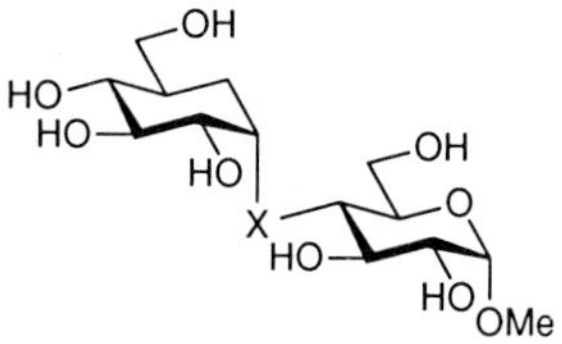

Carba-maltose: X = NH
X = O
X = S

General Considerations

In this chapter, the syntheses of some 5a-carba sugar analogs (types **B** and **C**) (**11–13**) of the branched mannotriose structure (**10**) containing the "trimannosyl core" common to glycoproteins are described. These analogs are expected to be utilized as model compounds for studies of conformational analysis of complex oligosaccharides and as substrates or inhibitors for elucidation of the enzymatic action of sugar hydrolases and some transferases, such as GlcNAc-transferases I–VI.

First, synthesis[4] of the trimannosyl core (type **C**, **11**) composed of 5a-carba-D-mannose residues by α-D-mannosylation of a 5a-carba glucal analog (**14**), (1*S*)-(1,3/2)-3-(hydroxymethyl)cyclohex-5-ene-1,2-diol, is described. Second, as a practical route to imino- and ether-linked 5a-carba oligosaccharides (type **B**) is described the synthesis of 5a-carba trisaccharide analogs (**12** and **13**) by use of 1,2-anhydro-3,4,6-tri-*O*-benzyl-5a-carba-β-D-mannopyranose (**15**) as the common 5a-carba-glycosyl donor.[5,6] Selective glycosylation of **14** with 3.5 molar equivalents of 2,3,4,6-tetra-

[5] S. Ogawa and T. Tonegawa, *Carbohydr. Res.* **204,** 51 (1990).
[6] S. Ogawa, S. Sasaki, and H. Tsunoda, *Chem. Lett.*, 1587 (1993).

O-benzoyl-α-D-mannopyranosyl bromide[7] (**16**) in benzene in the presence of mercury(II) cyanide affords the 3,6-di-*O*- (**17**, 36%) and the 3,4,6-tri-*O*-(α-D-mannopyranosyl) derivatives (**18**, 18%).

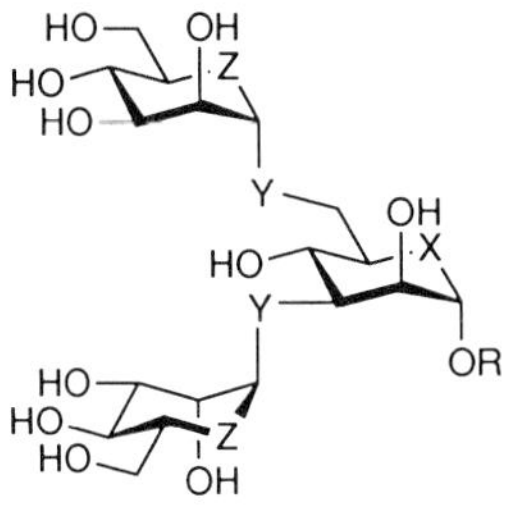

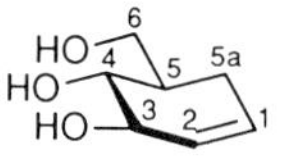

10 : R = Me, X = Y = Z = O
11 : R = H, X = CH_2, Y = Z = O
12 : R = Me, X = O, Y = NH, Z = CH_2
13 . R = Me, X = Y = O, Z = CH_2

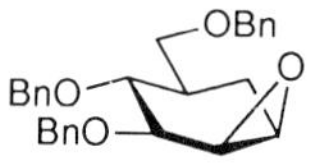

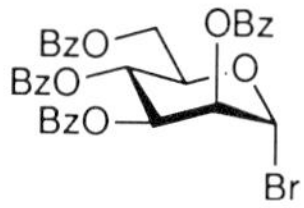

14

15

16

The protected precursor of the carba trisaccharide is a versatile synthetic intermediate for further transformation to various carba oligosaccharides of biological interest. Thus, epoxidation of **17** with *m*-chloroperbenzoic acid gives the crystalline α- (**19**, 33%) and β-epoxides (**20**, 49%). For instance, the β-epoxide is selectively opened at C-1 with an acetate ion to give the precursor of the 5a-carba-3,6-di-*O*-(α-D-mannopyranosyl)-α-D-mannopyranose (**11**).

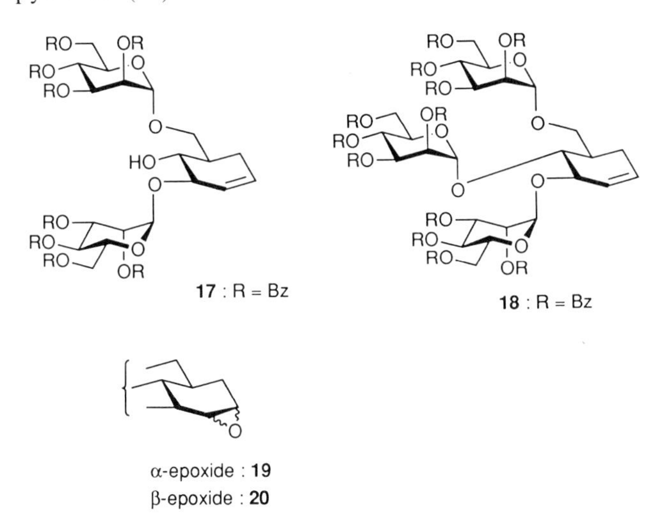

17 : R = Bz

18 : R = Bz

α-epoxide : **19**
β-epoxide : **20**

[7] P. K. Ness, H. G. Fletcher, Jr., and C. S. Hudson, *J. Am. Chem. Soc.* **72,** 2200 (1950).

Conventional epoxidation of the tri-*O*-benzyl derivative of **14** furnishes a versatile 5a-carba-α-D-mannosyl donor (**15**), which is selectively opened at C-1 by an amine or an oxonium ion generated from alcohol, affording the imino- or ether-linked carba disaccharide having the 5a-carba-glycosyl residues with the α-*manno* configuration. Thus, coupling of **15** and methyl 3-amino-3-deoxy-α-D-mannopyranoside[8] (**21**) in 2-propanol in a sealed tube at 120° for 9 days gives the condensate (**23**, 81%), which is deprotected to afford free carba disaccharide.[9] Furthermore, similar coupling of **15** with methyl 3,6-diamino-3,6-dideoxy-α-D-mannopyranoside[10] (**22**) gives mainly the protected carba trisaccharide **24** (59%), which can be readily hydrogenolyzed in the presence of 10% Pd–C, followed by purification over a column of Dowex 50W-X2 (H^+) resin, giving the free carba trisaccharide (**12**, 96%).[9] This compound shows a mild inhibitory activity [*I* of 43% at (100 μg/ml)] against α-D-mannosidase from Jack beans, which is comparable to that of methyl 4-(5a-carba-α-D-mannopyranosylamino)-4-deoxy-α-D-mannopyranoside.[11]

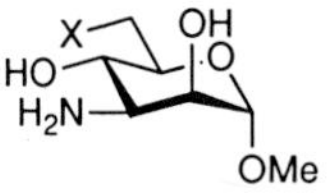

21 : X = OH
22 : X = NH_2

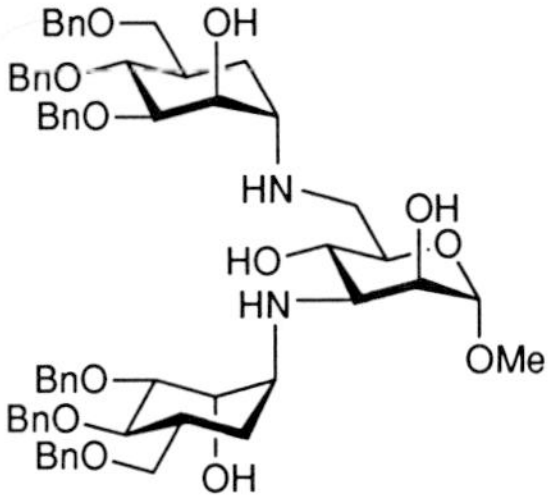

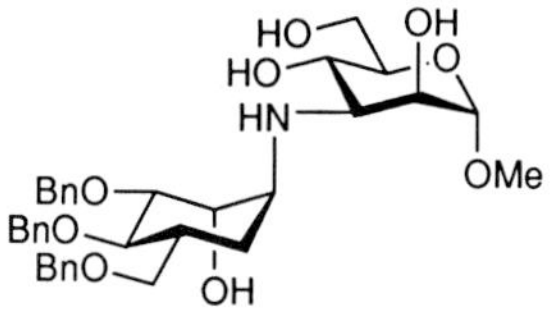

23

24

On the other hand, coupling of 1 molar equivalent of **15** and methyl 2-*O*-benzyl-4,6-*O*-benzylidene-α-D-mannopyranoside (**25**) in *N,N*-dimethylformamide (DMF) in the presence of sodium hydride and 15-crown-5 ether at 70° affords the carba disaccharide **26** (88%), which is benzylated and then reduced with diisobutylaluminum hydride (DIBAL) to give the 6-OH unsubstituted derivative (**27**), further treatment of which with **15** under similar conditions produces the ether-linked carba trisaccharide (**28**). Compound **28** is similarly deblocked to give the free carba trisaccha-

[8] H. H. Baer, *J. Am. Chem. Soc.* **84,** 83 (1962).
[9] S. Ogawa, S. Sasaki, and H. Tsunoda, *Carbohydr. Res.* submitted.
[10] S. Inouye, *Chem. Pharm. Bull.* **14,** 902 (1966).
[11] S. Ogawa and Y. Nakamura, *Carbohydr. Res.* **226,** 79 (1992).

ride (**13,** 44%). These ether-linked carba di- and trisaccharides do not show any inhibitory activity against α-mannosidase; however, they may be transformed into the versatile carba di- and trimannose donors, for instance, as the mannosyl halides or so, useful for further elaboration of biologically interesting oligosaccharides partially composed of 5a-carbamannosyl residues.

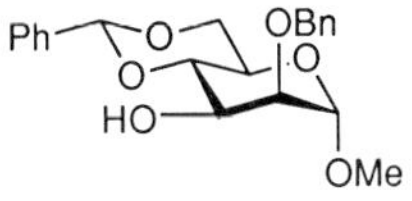

Ph O O OBn O HO OMe
25

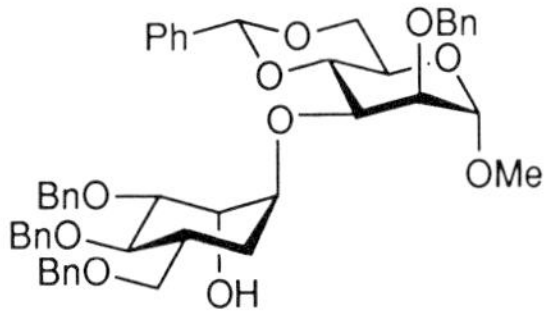

Ph O O OBn O O OMe BnO BnO BnO OH
26

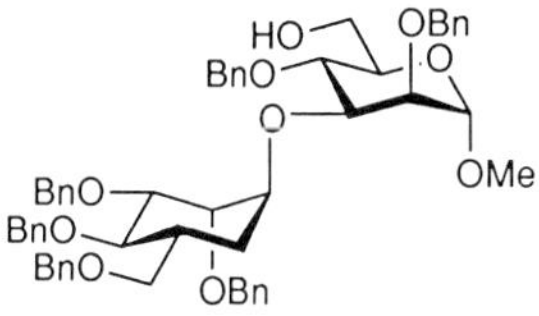

HO OBn BnO O O OMe BnO BnO BnO OBn
27

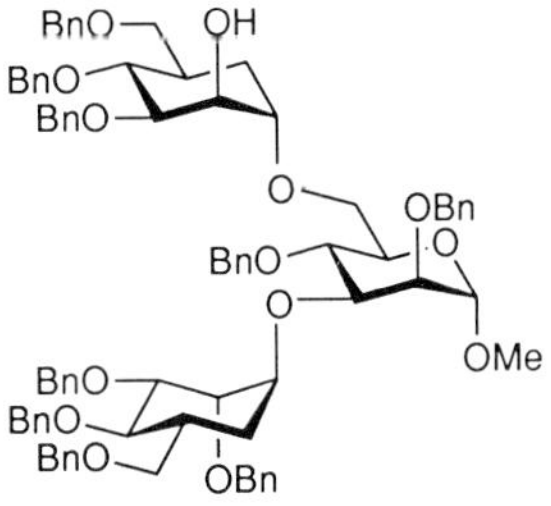

BnO OH BnO BnO O OBn BnO O O OMe BnO BnO BnO OBn
28

Procedures

General Methods

Thin-layer chromatography (TLC) is performed on silica gel 60 GF (E. Merck, Darmstadt, Germany), the components being detected by charring with H_2SO_4. Silica gel column chromatography is conducted on Wakogel C-300 (300 mesh) (Wako Pure Chemical Industries, Osaka, Japan). Organic solutions are dried over anhydrous sodium sulfate and evaporated below 50° under diminished pressure.

Syntheses

*Preparation of Pseudotrisaccharide (**11**); α-D-Mannosylation of (1R,2R,3R)-3-(Hydroxymethyl)cyclohex-5-ene-1,2-diol (**14**).* A mixture of **14** (25 mg, 0.17 mmol), $Hg(CN)_2$ (100 mg, 0.40 mmol), Drierite (100 mg), and benzene–nitromethane (1 : 1, v/v) (10 ml) is first heated at reflux to distill out benzene (~2 ml), and, after being cooled to 0°, 2,3,4,6-tetra-*O*-benzoyl-α-D-mannopyranosyl bromide (**16**, 400 mg, 0.60 mmol) is added and stirred for 25 hr at 0° in the dark. An insoluble material is removed by filtration through a bed of Celite, and the filtrate is evaporated. The residue is chromatographed on a column of silica gel (25 g) with butanone–toluene (1 : 15, v/v) containing triethylamine (~3%) as an eluent to give mainly the 3,6-di-*O*-(α-D-mannopyranosyl) derivative **17** (82 mg, 36%), mp 111°–113° (from ethanol), $[\alpha]_D^{23}$ −60° (*c* 0.83, chloroform), together with the 4,6-di- (3%) and the 3,4,6-tri-*O*-(α-D-mannopyranosyl) derivatives (**18**, 18%), mp 130°–132° (from ethanol), $[\alpha]_D^{23}$ −64° (*c* 1.4, chloroform).

*Epoxidation of Compound **17**.* To a solution of **17** (200 mg, 0.15 mmol) in 1,2-dichloroethane (3 ml) are added 1 *M* aqueous monosodium phosphate (1.5 ml), 1 *M*, aqueous disodium phosphate (1.5 ml), and *m*-chloroperbenzoic acid (53 mg, 0.31 mmol), and the mixture is vigorously stirred for 20 hr at 50°. After cooling, the mixture is diluted with chloroform (50 ml), washed with saturated aqueous $Na_2S_2O_3$ and water, dried, and evaporated. The residue is chromatographed on a column of silica gel (25 g) with butanone–toluene (1 : 10, v/v) as an eluent to give the α-epoxide (**19**, 33%), mp 128°–129° (from ethanol), $[\alpha]_D^{22}$ −50° (*c* 0.76, chloroform) and β-epoxide (**20**, 49%), mp 128°–130° (from ethanol), $[\alpha]_D^{22}$ −58° (*c* 1.2, chloroform).

*Preparation of 5a-Carba-3,6-di-O-(α-D-mannopyranosyl)-α-D-mannose (**11**).* A mixture of **20** (24 mg, 18 μmol), anhydrous sodium acetate (8 mg, 90 μmol), and 90% aqueous 2-methoxyethanol (2 ml) is stirred for 2 days at 120°, then evaporated. The residue is treated with acetic anhydride (3 ml) and pyridine (3 ml) overnight at room temperature. The mixture is evaporated, and the residual product is chromatographed on a column of silica gel (1 g) with acetone–toluene (1 : 4, v/v), as an eluent to give the undecaacetate (17 mg, 98%) of **11** as an amorphous powder $[\alpha]_D^{25}$ + 49° (*c* 1.0, chloroform). The compound is easily de-*O*-acetylated by treatment with methanolic sodium methoxide in methanol, giving the free carba trisaccharide (**11**) quantitatively.

*Preparation of Imino-Linked Carba Dimannoside (**23**).* A mixture of **15** (260 mg, 0.60 mmol), methyl 3-amino-3-deoxy-α-D-mannopyranoside (**21**, 94 mg, 0.49 mmol), and 2-propanol (1 ml) is heated in a sealed tube for 9 days at 120°, then evaporated. The residue is chromatographed on a column of silica gel (20 g) with acetone–toluene (1 : 1, v/v) as an eluent to give the pseudodisaccharide tribenzyl ether (**23**, 245 mg, 81%), $[\alpha]_D^{22}$ +20° (*c* 1.0, chloroform). Conventional acetylation of **23** gives the tetraacetate, $[\alpha]_D^{27}$ +16° (*c* 1.2, chloroform).

*Preparation of Imino-Linked Carba Trisaccharide (**12**).* A mixture of **15** (177 mg, 0.41 mmol), 3,6-diamino-3,6-dideoxy-α-D-mannopyranoside

(**22**, 36 mg, 0.19 mmol), and 2-propanol (1 ml) is heated in a sealed tube for 6 days at 120°, then evaporated. The residual product is processed, as described for the preparation of **23**, to give the pseudotrisaccharide hexabenzyl ether (**24**, 116 mg, 59%), $[\alpha]_D^{24}$ +33° (*c* 0.8, chloroform). Conventional acetylation of **24** gives the tetraacetate, $[\alpha]_D^{24}$ +29° (*c* 0.8, chloroform).

A solution of **24** (25 mg, 24 μmol) in ethanol (3 ml) is hydrogenated in the presence of a catalytic amount of 10% Pd–C under atmospheric pressure for 2 days at room temperature. The product is purified by elution from a column of Dowex 50W-X2 (H^+) resin with water and 1% ammoniacal methanol as an eluent to give the carba trisaccharide **12** (12 mg, 96%) as a syrup, $[\alpha]_D^{25}$ +29° (*c* 0.5, methanol).

*Preparation of Protected Ether-Linked Pseudodisaccharide (**26**).* To a solution of methyl 2-*O*-benzyl-4,6-*O*-benzylidene-α-D-mannopyranoside (**25**, 44 mg, 0.12 mmol) in DMF (1 ml) is added NaH (42 mg, 1.75 mmol, 15 equivalents) at 0° under argon, and the mixture is stirred for 30 min. The epoxide **15** (145 mg, 0.34 mmol, 2.8 equivalents) and 15-crown-5 ether (350 μl, 1.76 mmol) are added, and the mixture is stirred for 4 days at 70°. After cooling, the mixture is diluted with ethyl acetate (60 ml), washed with water, dried, and evaporated. The residue is chromatographed on a column of silica gel with acetone–hexane (1 : 5, v/v) as an eluent to give the pentabenzyl ether **26** (61 mg, 64%) as a syrup, $[\alpha]_D^{26}$ −0.9° (*c* 1.6, chloroform), together with **25** (27%) recovered.

*Preparation of Ether-Linked Carba Disaccharide Hexabenzyl Ether (**27**).* A solution of **26** (163 mg, 0.203 mmol) in DMF (3 ml) is treated with 60% sodium hydride (30 mg, 0.75 mmol) for 30 min, and then benzyl bromide (60 μl, 0.50 mmol) is added to it and the mixture stirred for 21 hr at room temperature. After addition of methanol at 0°, the mixture is diluted with ethyl acetate, washed with water thoroughly, dried, and evaporated. The residual product is chromatographed [silica gel (20 g), ethyl acetate–hexane (1 : 4, v/v)], to give the pentabenzyl ether (132 mg, 73%), $[\alpha]_D^{25}$ +6° (*c* 1.2, chloroform). Treatment of this compound (122 mg, 0.137 mmol) with DIBAL (1.5 *M* in toluene) (600 μl, 0.90 mmol) in CH_3Cl (2 ml) gives, after conventional processing and chromatography [silica gel (5 g), ethyl acetate–hexane (1 : 2, v/v)], compound **27** (108 mg, 88%) as a syrup, $[\alpha]_D^{24}$ +32° (*c* 0.90, chloroform).

*Preparation of Ether-Linked Pseudotrisaccharide (**13**).* Coupling of **15** (196 mg, 0.46 mmol) and **27** (138 mg, 0.154 mmol) is carried out, as in the preparation of **26**, to give **28** (90 mg, 44%), $[\alpha]_D^{24}$ +33° (*c* 1.0, chloroform), together with **27** (10%). A solution of **28** (10.8 mg, 815 μmol) in ethanol (2 ml) containing 1 *M* aqueous HCl (0.1 ml) is hydrogenated in the presence of a catalytic amount of palladium black under a hydrogen pressure of 3

kg/cm^2 for 1 day at room temperature. The filtered solution is evaporated and the residue is acetylated conventionally to give the decaacetate (6.2 mg, 82%) of **13**, $[\alpha]_D^{22}$ +26° (*c* 0.6, chloroform), which is de-*O*-acetylated by treatment with methanolic sodium methoxide to give, after chromatography [silica gel, acetonitrile–water (4 : 1, v/v)], compound **13**, $[\alpha]_D^{25}$ +7° (*c* 0.5, methanol), in quantitative yield.

[9] Chemical Synthesis of Glycosylamide and Cerebroside Analogs Composed of Carba Sugars

By SEIICHIRO OGAWA and HIDETOSHI TSUNODA

Introduction

Many kinds of glycosylamides, glycolipid analogs structurally related to glycosphingolipids and glycoglycerolipids, have been synthesized and subjected to several biological tests.[1] Some compounds have been found to be immunomodulators, and among them, *N*-(β-D-glucopyranosyl)-*N*-octadecyldodecanamide (**1**) produces a dose-dependent increase in the formation of antibodies against sheep erythrocytes *in vitro* in the concentration range 3–100 mg/ml.

These studies have stimulated us to synthesize some carbocyclic analogs of the glycosylamides,[2] namely, 5a-carba glycosylamides, which contain the 5a-carba-D-hexopyranose residues having *β-galacto, α-* and *β-gluco,* and *α-* and *β-manno* configurations. Interestingly, these compounds have been shown to possess biological activities similar to those shown by the corresponding true sugar analogs, thereby suggesting that 5a-carba sugar analogs may possibly be useful model compounds for studies of glycolipid chemistry and biochemistry. The synthesis of two 5a-carba glycosylamides having *α-gluco* (**2**) and *α-manno* configurations (**3**) is described here.

Although the biological functions of glycosphingolipids have been examined thoroughly, in order to gain further understanding of their biological functions it is necessary to provide isomerically pure glycosphingolipids as well as structural analogs in larger quantities by chemical synthesis.

[1] O. Lockhoff, *Angew. Chem., Int. Ed. Engl.* **30,** 1611 (1991).
[2] H. Tsunoda and S. Ogawa, *Liebigs Ann. Chem.* **103,** 1994.

We describe a method for a synthesis[3] of the imino-linked 5a-carba sugar analog **5** of glucoceramide (**4**). The method combines the 1,2-aziridinosphingosine derivative **12** with properly blocked 5a-carba-β-D-glucopyranosylamine (**22**).

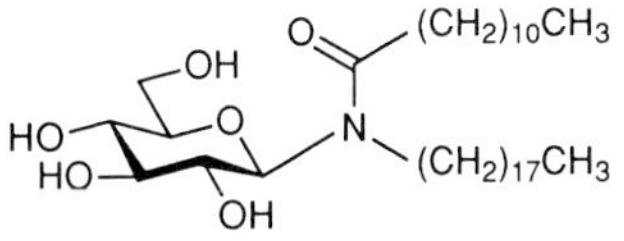

1

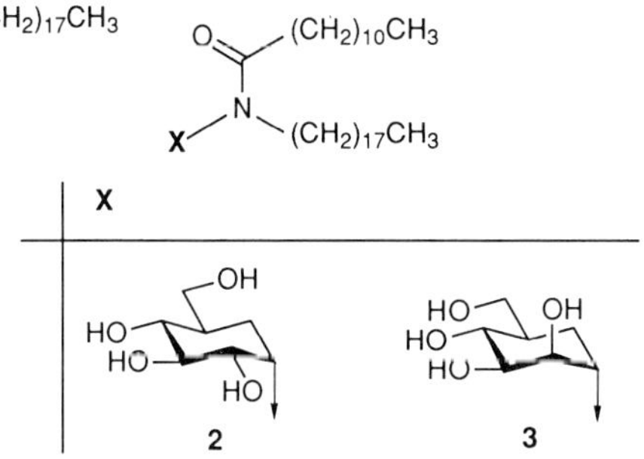

2

3

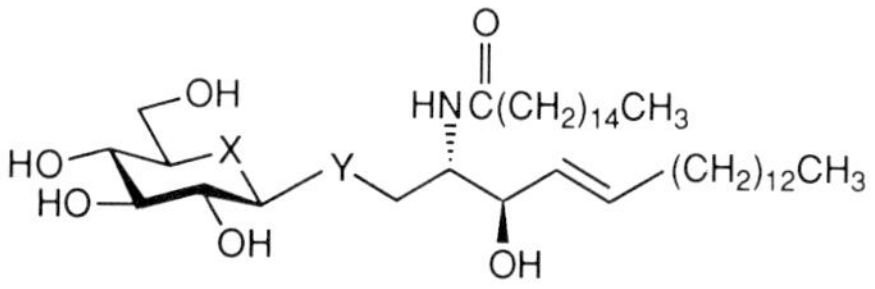

4 : X = Y = O
5 : X = CH_2, Y = NH

General Considerations

Preparation of 5a-Carba α-Glucosyl- and α-Mannosylamides

The synthesis employs a coupling of the anhydro derivatives of 5a-carba sugars with aliphatic amines and successive N-acylation with acyl chloride in the presence of 4,4′-dimethylaminopyridine (DMAP). This method may be applicable for the preparation of the homologous compounds. Thus, coupling of 1,2-anhydro-3-*O*-benzyl-4,6-*O*-benzylidene-5a-carba-β-D-mannopyranose[4] (**6**) with 1-octadecanamine for 4 days at 120° produces the sole secondary amine **7** (87%) having the *α-manno* configuration. Compound **7** is N,O-acylated with dodecanoyl chloride and DMAP

[3] H. Tsunoda and S. Ogawa, *Liebigs Ann. Chem.* submitted.

[4] S. Ogawa, T. Tonegawa, K. Nishi, and J. Yokoyama, *Carbohydr. Res.* **229,** 173 (1992).

to give the amide **8** (61%), which is de-O-acylated and successively deprotected to give the 5a-carba-α-D-mannosylamide (**3**) (69%). Alternatively, compound **8** is first de-O-acylated and the resulting alcohol **9** (79%) is treated with 1.2 molar equivalents of mesyl chloride in pyridine for 8 hr at room temperature to give the ester **10** (64%) with the *α-gluco* configuration. The initially formed mesylate is thought to be intramolecularly attacked by the neighboring amide carbonyl to give an intermediate oxazolinium ion, which is opened by a trace of water, giving **10**. Then compound **10** is N,O-acylated again with dodecanoyl chloride in CH_2Cl_2 to give the ester **11** (91%), which is successively de-O-isopropylidenated, de-O-acylated, and then hydrogenolyzed to give the 5a-carba-α-D-glucosylamide (**2**) (78%).

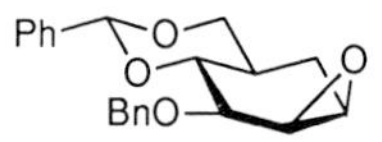

6

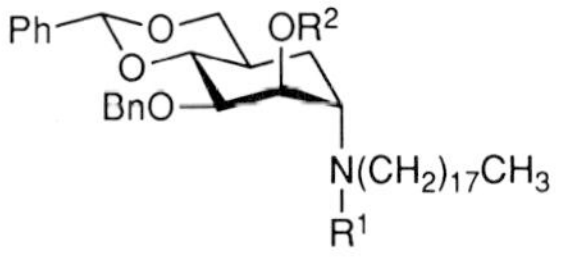

7 : $R^1 = R^2 = H$
8 : $R^1 = R^2 = C(O)(CH_2)_{10}CH_3$
9 : $R^1 = C(O)(CH_2)_{10}CH_3$, $R^2 = H$

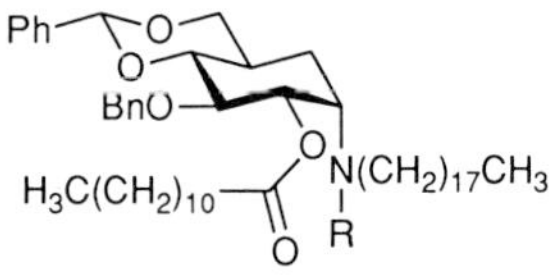

10 : R = H
11 : $R = C(O)(CH_2)_{10}CH_3$

Preparation of 5a-Carba-glucosyl-1-Azaceramide

Selective benzoylation of azidosphingosine[5] **12** gives the 1-*O*-benzoyl derivative **13** (65%), which is silylated followed by de-O-acylation to afford the 1-OH unprotected derivative **14** (73%). Reduction of **14** with triphenylphosphine and subsequent treatment with 2,4-dinitrofluorobenzene produces the protected aziridine **15** (35%) and the blocked amino alcohol **16** (45%). The latter compound can be transformed into **15** by the following sequence of reactions: iodination with triphenylphosphine, imidazole, and iodine, then successive treatment with silver fluoride in pyridine (83% overall yield).

First, 4,6-*O*-benzylidene-5a-carba-β-D-glucopyranosylamine (**23**) is prepared from a 5a-carba glucal derivative,[6] (1*R*,3*R*,6*R*,10*R*)-10-hydroxy-

[5] K. Koike, M. Numata, M. Sugimoto, Y. Nakahara, and T. Ogawa, *Carbohydr. Res.* **158,** 113 (1986).

[6] S. Ogawa, N. Nakamura, and T. Takagaki, *Bull. Chem. Soc. Jpn.* **59,** 2956 (1986).

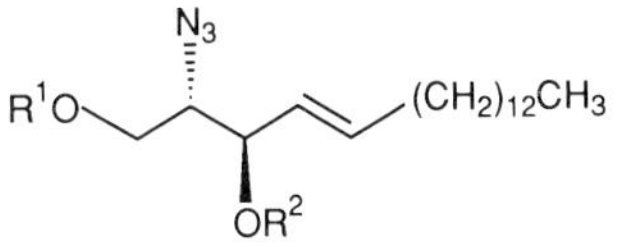

12 : $R^1 = R^2 = H$
13 : $R^1 = Bz, R^2 = H$
14 : $R^1 = H, R^2 = TBDMS$

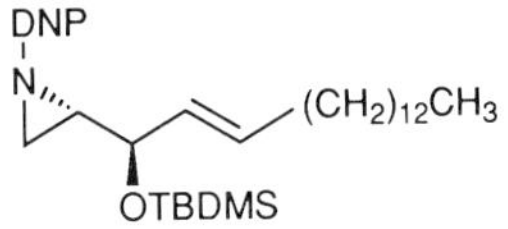

15

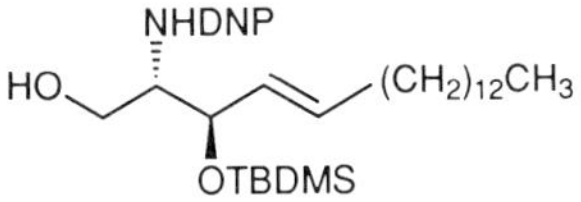

16

3-phenyl-2,4-dioxabicyclo[4.4.0]dec-8-ene. The 4,6-*O*-benzylidene derivative (**17**) is benzoylated to give **18** and then oxidized with osmium tetroxide in the presence of *N*-methylmorpholine *N*-oxide (NMO) to give selectively the 5a-carba glucopyranose derivative **19** (97%). After selective benzoylation of the equatorial 2-hydroxyl group (yielding **20**, 82%), the dibenzoate **20** is mesylated conventionally to the mesylate **21**, which is displaced by an azide ion via a direct S_N2 mechanism to give the sole azide **22** (87%) with a *β-gluco* configuration. Compound **22** is de-O-acylated with methanolic sodium methoxide and then hydrogenated in the presence of Raney nickel catalyst to give the amine **23** (90%).

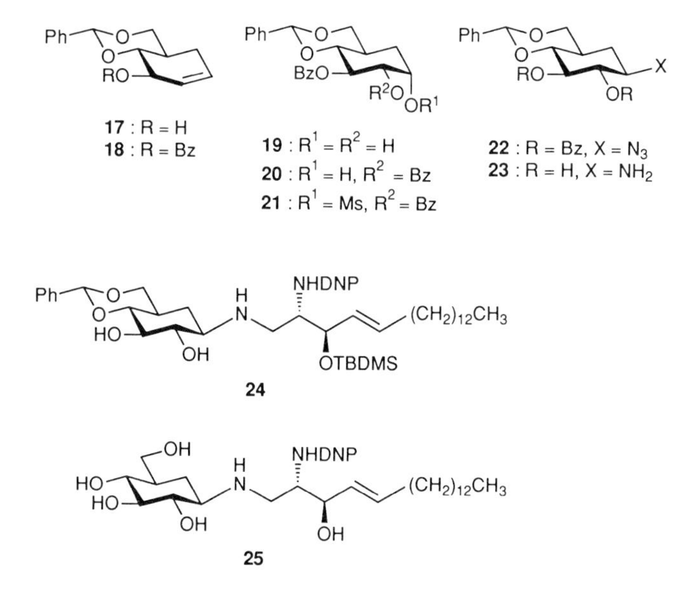

Coupling of 1 molar equivalent of **15** and **23** is carried out in 2-propanol in a sealed tube for 7 days at 120°, giving the condensate **24** in 52% yield. Treatment of **24** with tetrabutylammonium fluoride followed by de-O-benzylidenation with aqueous acetic acid gives the *N*-2,4-dinitrophenyl derivative **25** (93%). Removal of the N-protecting group with Amberlite IRA-400 (HO^-) resin and successive N-acylation with dodecanoyl chloride afford the 5a-carba-β-D-glucosyl-1-azaceramide (**5**) (31%).

Procedures

General Methods

Thin-layer chromatography (TLC) is performed on silica gel 60 GF (E. Merck, Darmstadt, Germany), the components being detected by charring with H_2SO_4. Silica gel column chromatography is conducted on Wakogel C-300 (300 mesh, Wako Pure Chemical Industries, Osaka, Japan). Unless otherwise noted, the syrupy compounds are purified by chromatography on silica gel. The organic solution is dried over anhydrous sodium sulfate and evaporated below 50° under diminished pressure.

Syntheses

Preparation of Secondary Amine **7**. A mixture of 1,2-anhydro-3-*O*-benzyl-4,6-*O*-benzylidene-5a-carba-β-D-mannopyranose (**6**) (276 mg, 0.81 mmol), 1-octadecanamine (273 mg, 0.98 mmol), and 2-propanol (1 ml) is heated in a sealed tube for 20 hr at 120°, then evaporated. The residual product is chromatographed on a column of silica gel (20 g) with ethyl acetate–toluene (1 : 8, v/v) as an eluent to give the amine **7** (432 mg, 87%), $[\alpha]_D^{25}$ +12° (*c* 1.0, chloroform).

Preparation of Amide **8**. A mixture of **7** (371 mg, 0.61 mmol), dodecanoyl chloride (845 μl, 3.7 mmol), and CH_2Cl_2 (10 ml) is stirred in the presence of DMAP (893 mg, 7.3 mmol) for 1 hr at room temperature and then evaporated. The residue is chromatographed on a column of silica gel (15 g) with ethyl acetate–hexane (1 : 10, v/v) as an eluent to give the amide **8** (363 mg, 61%), $[\alpha]_D^{25}$ +5° (*c* 1.1, chloroform).

Preparation of 5a-Carba-α-mannosylamide **3**. A mixture of **8** (267 mg, 0.27 mmol) and a solution of acetic acid–tetrahydrofuran (THF)–water (4 : 3 : 1, v/v/v) (5 ml) is stirred for 5 hr at 80°, then evaporated. The residue is dissolved in ethanol (15 ml), and the solution is hydrogenated in the presence of a catalytic amount of 10% Pd–C for 1 day at room temperature. The catalyst is removed by filtration, and the filtrate is evaporated. The residual product is treated with 0.1 *M* methanolic sodium methoxide (10 ml) for 1 hr at 0°. After neutralization with acetic acid, the mixture is evaporated and the residue chromatographed on a column of silica gel (6 g) with ethanol–toluene (1 : 10, v/v) as an eluent to give **3** (116 mg, 69%), $[\alpha]_D^{24}$ +6° (*c* 1.1, chloroform).

*Preparation of Amide **9**.* A mixture of **8** (490 mg, 0.50 mmol), 0.2 *M* methanolic sodium methoxide (5 ml), and CH_3Cl (5 ml) is stirred for 12 hr at room temperature. After neutralization with Amberite IR-120B (H^+) resin, the mixture is evaporated and the residual product chromatographed on a column of silica gel (15 g) with ethyl acetate–toluene (1 : 8, v/v) as an eluent to give **9** (317 mg, 79%), $[\alpha]_D^{23}$ +10° (*c* 0.89, chloroform).

*Preparation of Amine Ester **10**.* To a solution of **9** (317 mg, 0.40 mmol) in pyridine (10 ml) is added mesyl chloride (46 μl, 0.600 mmol, 1.5 equivalents) at 0°, and then the mixture is stirred for 12 hr at room temperature. The mixture is diluted with ethyl acetate (50 ml), and the solution is washed with water, dried, and evaporated. The residual product is chromatographed on a column of silica gel (20 g) with ethyl acetate–toluene (1 : 6, v/v) as an eluent to give **10** (203 mg, 64%), $[\alpha]_D^{26}$ +18° (*c* 1.0, chloroform).

*Preparation of Amide **11**.* Compound **10** (203 mg, 0.256 mmol) is treated with dodecanoyl chloride (121 μl, 0.51 mmol) in CH_2Cl_2 (6 ml) in the presence of DMAP (125 mg) for 5 hr at room temperature. The mixture is then processed as for the preparation of **8**. The product is chromatographed on a column of silica gel with ethyl acetate–hexane (1 : 8, v/v) as an eluent to give **11** (228 mg, 91%), $[\alpha]_D^{25}$ +12° (*c* 0.81, chloroform).

*Preparation of 5a-Carba-α-Glucosylamide (**2**).* Compound **11** (200 mg, 0.21 mmol) is deprotected, as for the preparation of **3**, to give the amide **2** (98 mg, 78%), $[\alpha]_D^{27}$ +21° (*c* 0.9, chloroform).

*Preparation of Aziridino Derivative of Sphingosine (**15**).* To a solution of the azide **12** (479 mg, 1.47 mmol) in pyridine (15 ml) is added benzoyl chloride (188 μl, 1.62 mmol) at −15°, and the mixture is stirred for 8 hr at room temperature. The product is chromatographed on a column of silica gel with ethyl acetate–hexane (1 : 10, v/v) as an eluent to give the benzoate **13** (409 mg, 65%), $[\alpha]_D^{25}$ −35° (*c* 0.92, chloroform).

*Protection of Hydroxyl Group of **13**.* To a solution of **13** (409 mg, 0.95 mmol) in $HCON(CH_2)_2$ (10 ml) is added *tert*-butyldimethylsilyl chloride (431 mg, 2.9 mmol) and imidazole (389 mg, 5.7 mmol), and the mixture is stirred for 20 hr at 60°. The mixture is diluted with ethyl acetate (50 ml), washed with water, dried, and evaporated. The residue is treated with methanolic sodium methoxide conventionally, and the product is chromatographed on a column of silica gel with ethyl acetate–hexane

(1 : 10, v/v) as an eluent to give the alcohol **14** (304 mg, 73%), $[\alpha]_D^{26}$ $-42°$ (*c* 0.92, chloroform) {literature value,[7] $[\alpha]_D^{22}$ $-40.5°$ (*c* 4.5, chloroform)}.

*Preparation of Aziridine **15***. A mixture of **14** (304 mg, 0.69 mmol), triphenylphosphine (200 mg, 0.76 mmol), and toluene (6 ml) is stirred for 20 min at 40°, then evaporated. The residue is dissolved in methanol (6 ml), and the solution is treated with 2,4-dinitrofluorobenzene (130 μl, 1.04 mmol) and triethylamine (193 μl, 1.38 mmol) for 12 hr at room temperature. The products are chromatographed on a column of silica gel with ethyl acetate–hexane (1 : 20, v/v) as an eluent to give the aziridine **15** (137 mg, 35%), $[\alpha]_D^{26}$ $-162°$ (*c* 1.2, chloroform), and the alcohol **16** (182 mg, 45%), $[\alpha]_D^{25}$ $-52°$ (*c* 1.1, chloroform).

Compound **16** (232 mg, 0.40 mmol) is treated with triphenylphosphine (210 mg, 0.80 mmol), imidazole (109 mg, 1.6 mmol) and iodine (203 mg, 0.80 mmol) in toluene (10 ml) for 30 min at room temperature. The iodide (269 mg, 97%) obtained is further treated with silver fluoride (64 mg, 0.51 mmol) for 2 hr at 0°, and the product is purified by chromatography to give **15** (188 mg, 86%).

*Preparation of Blocked Carba-β-Glucosylamine (**23**)*. A mixture of 4,6-*O*-benzylidene-5a-carba-glucal (**17**) (562 mg, 2.4 mmol) and benzoyl chloride (562 μl, 4.8 mmol) in pyridine (10 ml) is stirred for 1 hr at room temperature. The mixture is processed conventionally to give the benzoate **18** (743 mg, 91%), $[\alpha]_D^{24}$ $-38°$ (*c* 1.0, chloroform).

A solution of **18** (743 mg, 2.20 mmol) in acetone–water (4 : 1, v/v) (20 ml) is treated with a 50 m*M* solution (4.4 ml, 0.22 mmol) of OsO_4 in *tert*-butanol in the presence of NMO (772 mg, 6.6 mmol) for 2 hr at room temperature. A 920-mg portion (8.8 mmol) of $NaHSO_3$ is added to the mixture, which is stirred for 30 min at room temperature. The mixture is processed conventionally to give the diol **19** (795 mg, 97%), $[\alpha]_D^{23}$ $-25°$ (*c* 0.89, chloroform).

To a solution of **19** (321 mg, 0.864 mmol) in pyridine (6 ml) is added benzoyl chloride (120 μl, 1.04 mmol) at $-15°$, and the mixture is stirred for 1 hr at $-15°$ and then for 2 hr at room temperature. Conventional processing gives the dibenzoate **20** (337 mg, 82%), $[\alpha]_D^{23}$ $-20°$ (*c* 0.96, chloroform).

Compound **20** (869 mg, 1.8 mmol) is treated with mesyl chloride, as for the preparation of **9**, to give the mesylate **21** (910 mg, 90%), $[\alpha]_D^{24}$ $-16°$ (*c* 0.75, chloroform). A mixture of **21** (532 mg, 0.961 mmol), sodium azide (187 mg, 2.88 mmol), and aqueous 90% DMF (10 ml) is stirred for 12 hr at 120°, and then evaporated. The product is chromatographed on a column

[7] K. C. Nicolau, T. J. Caulfield, and H. Kataoka, *Carbohydr. Res.* **202,** 177 (1990).

of silica gel [ethyl acetate–hexane (1 : 5, v/v)] to give the azide **22** (404 mg, 87%), $[\alpha]_D^{26}$ +32° (*c* 1.2, chloroform). Compound **22** (361 mg, 0.747 mmol) is treated with 0.1 *M* methnaolic sodium methoxide in dichloromethane–methanol (1 : 1, v/v) (10 ml) in the usual manner. The diol (190 mg, 92%) obtained is hydrogenated in ethanol (5 ml) in the presence of Raney nickel catalyst under atmospheric pressure to give the amine **23** (155 mg, 90%), $[\alpha]_D^{25}$ +62° (*c* 1.0, methanol).

*Coupling of Aziridine **15** and Amine **23**; Preparation of Imino-Linked Carba Glucosylceramide (**5**).* A mixture of **15** (72 mg, 0.129 mmol) and **23** (48 mg, 0.19 mmol) in 2-propanol (1 ml) is heated in a sealed tube for 4 days at 120° and then evaporated. The residual product is chromatographed on a column of silica gel to give the coupled product **24** (62 mg, 52%) as a syrup, $[\alpha]_D^{26}$ +30° (*c* 0.83, chloroform).

Compound **24** (43 mg, 53 μmol) is treated with 1 *M* tetrabutylammonium fluoride (80 μl, 80 μmol) in THF (2 ml) for 1 hr at room temperature. The product is purified by use of preparative TLC [silica gel, ethanol–toluene (1 : 8, v/v)] to give the triol (36 mg, 97%), which is treated with a solution of acetic acid–water (4 : 1, v/v) (2 ml) for 1 hr at 90°. The product is chromatographed [silica gel, ethanol–toluene (1 : 30, v/v)] to give **25** (30 mg, 93%), $[\alpha]_D^{26}$ −5° (*c* 0.7, chloroform).

A mixture of **25** (19 mg, 31 μmol), Amberlite IRA-400 (OH^-) resin (0.2 ml), and acetone–methanol–water (3 : 5 : 20, v/v/v) (2 ml) is stirred for 1 day at room temperature. After filtration through a bed of Celite, the filtrate is evaporated, the residue is dissolved in THF (2 ml), and the solution is treated with hexadecanoyl chloride (4.0 μl, 31 μmol) and 30% aqueous sodium acetate (1 ml) for 20 min at 0°. The product is purified by chromatography on silica gel [chloroform–methanol (5 : 1, v/v)] to give the carba glucosylceramide **5** (8.0 mg, 31%), $[\alpha]_D^{25}$ −27° (*c* 0.40, chloroform).

[10] Galactosylation of Nucleosides at 5′-Position of Pentofuranoses

By JIRI J. KREPINSKY, DENNIS M. WHITFIELD, STEPHEN P. DOUGLAS, NICULINA LUPESCU, DAVID PULLEYBLANK, and FREDERICK L. MOOLTEN

Introduction

Several synthetic analogs of naturally occurring nucleosides, such as β-D-arabinofuranosylcytosine (araC) or 2′-deoxy-5-fluorouridine, are clinically useful anticancer or antiviral agents.[1] A practical problem with their use arises from the low selectivity of such agents: they kill normal cells, albeit more slowly, as well as neoplastic or virally infected cells. An increase in selectivity would substantially increase the curative potential of such drugs. For instance, if the cytotoxicity can be abolished by glycosylation of the drug, and if a suitable glycosidase were present only in a cancer cell, then this glycosylation would provide the needed increase in the selectivity.[2] Thus, if the glycosidase employed is β-galactosidase (the product of *lacZ* gene) introduced into cancer cells[2] by efficient gene insertion vectors, galactosylation of the nucleoside[3] would mask the toxicity of the drug until it is unmasked inside a cancer cell by the action of β-galactosidase.

The evaluation of biological activity requires absolutely pure compounds in relatively large quantities. Because glycosylation of carbohydrate primary hydroxyls usually proceeds smoothly,[4] in principle such galactosylated nucleosides should be obtainable by organic synthesis. However, the primary hydroxyl group 5′-OH of pentofuranoses in nucleosides is surprisingly unreactive.[5–7] Not only are orthoesters[8] formed when-

[1] N. D. Chkanikov and M. N. Preobrazhenskaya, *J. Carbohydr. Nucleosides Nucleotides* **8,** 391 (1981).

[2] F. L. Moolten, *Crit. Rev. Immunol.* **10,** 203 (1990).

[3] D. M. Whitfield, S. P. Douglas, T. H. Tang, I. G. Csizmadia, H. Y. S. Pang, F. L. Moolten, and J. J. Krepinsky, *Can. J. Chem.,* in press.

[4] For reviews, see H. Paulsen, *Angew. Chem., Int. Ed. Engl.* **21,** 155 (1982); H. Paulsen, *Angew. Chem., Int. Ed. Engl.* **29,** 823 (1990); K. Toshima and K. Tatsuta, *Chem. Rev.* **93,** 1503 (1993).

[5] L. M. Lerner, *in* "Chemistry of Nucleosides and Nucleotides" (L. B. Townsend, ed.), Vol. 2, p. 27. Plenum, New York and London, 1988.

[6] F. W. Lichtenthaler, Y. Sanemitsu, and T. Nohara, *Angew. Chem., Int. Ed. Engl.* **17,** 772 (1978).

using an ester protective group at C-2 (required to obtain β-anomeric specificity) under common Koenigs–Knorr reaction conditions, but even their formation is sluggish and low yielding.[9]

It has been shown[10] that a possible cause of this lack of reactivity is due to intramolecular hydrogen bonding between the 5′-hydroxyl group and the nucleoside base (in a nonpolar solvent). The Bredereck modification[11] of the Koenigs–Knorr reaction,[12] in which the hydrogen of the primary hydroxyl is replaced with a trityl group, is performed under acidic conditions, and it has been used to prepare 5′-glucosides and glucuronides in modest yields.[13] Another acidic condition under which such hydrogen bonds can be broken is the Schmidt glycosylation.[14] Using this reaction with several Lewis acid promoters, in particular with silver triflate,[15] the 5′-galactosylated nucleosides have been synthesized.[3]

The sequence of reactions leading to the desired products is portrayed in Scheme 1 using araC as an example. The acetylated nucleosides **1**,[16] **2**,[17] **3**, and **4**,[18] (see Scheme 1) having a free 5′-hydroxyl, are prepared as follows: N-acetylation of the free amino group (when present), temporary regioselective protection of the 5′-hydroxyl with the dimethoxytrityl (DMT) or trityl group, acetylation of the secondary hydroxyls, and removal of the DMT or trityl group. Attempts to react the 5′-hydroxyl selectively with DMT chloride without prior N-acetylation led to considerable substitution with DMT at the amino group. Some cleavage of nucleoside glycosidic bonds occurs during the removal of the DMT group with 1% dichloroacetic acid in CH_2Cl_2 necessitating a chromatographic purification.

[7] N. B. Hanna, R. K. Robins, and G. R. Revankar, *Carbohydr. Res.* **165,** 267 (1987).

[8] N. K. Kochetkov and A. F. Bochkov, *in* "Recent Developments in the Chemistry of Natural Carbon Compounds" (A. Bognár, V. Bruckner, and Cs. Szántay, eds.), **4,** 77, Akademiai Kiadó, Budapest, 1971.

[9] N. D. Chkanikov, V. N. Tolkachev, and M. N. Preobrazhenskaya, *Bioorg. Khim.* **4,** 1620 (1978).

[10] T. H. Tang, D. M. Whitfield, S. P. Douglas, J. J. Krepinsky, and I. G. Csizmadia, *Can. J. Chem.* **70,** 2434 (1992).

[11] H. Bredereck, A. Wagner, G. Faber, H. Ott, and J. Rauther, *Chem. Ber.* **92,** 1135 (1959).

[12] H. M. Flowers, this series, Vol. 50, 93.

[13] N. D. Chkanikov and M. N. Preobrazhenskaya, *Bioorg. Khim.* **6,** 67 (1980).

[14] R. R. Schmidt, *Angew. Chem., Int. Ed. Engl.* **25,** 212 (1986).

[15] S. P. Douglas, D. M. Whitfield, and J. J. Krepinsky, *J. Carbohydr. Chem.* **12,** 131 (1993).

[16] W. J. Wechter, *J. Med. Chem.* **10,** 762 (1967).

[17] A. M. Michelson and A. R. Todd, *J. Chem. Soc.*, 34 (1954).

[18] A. Seung-Ho, R. C. West, and C. I. Hong, *Steroids* **47,** 413 (1986).

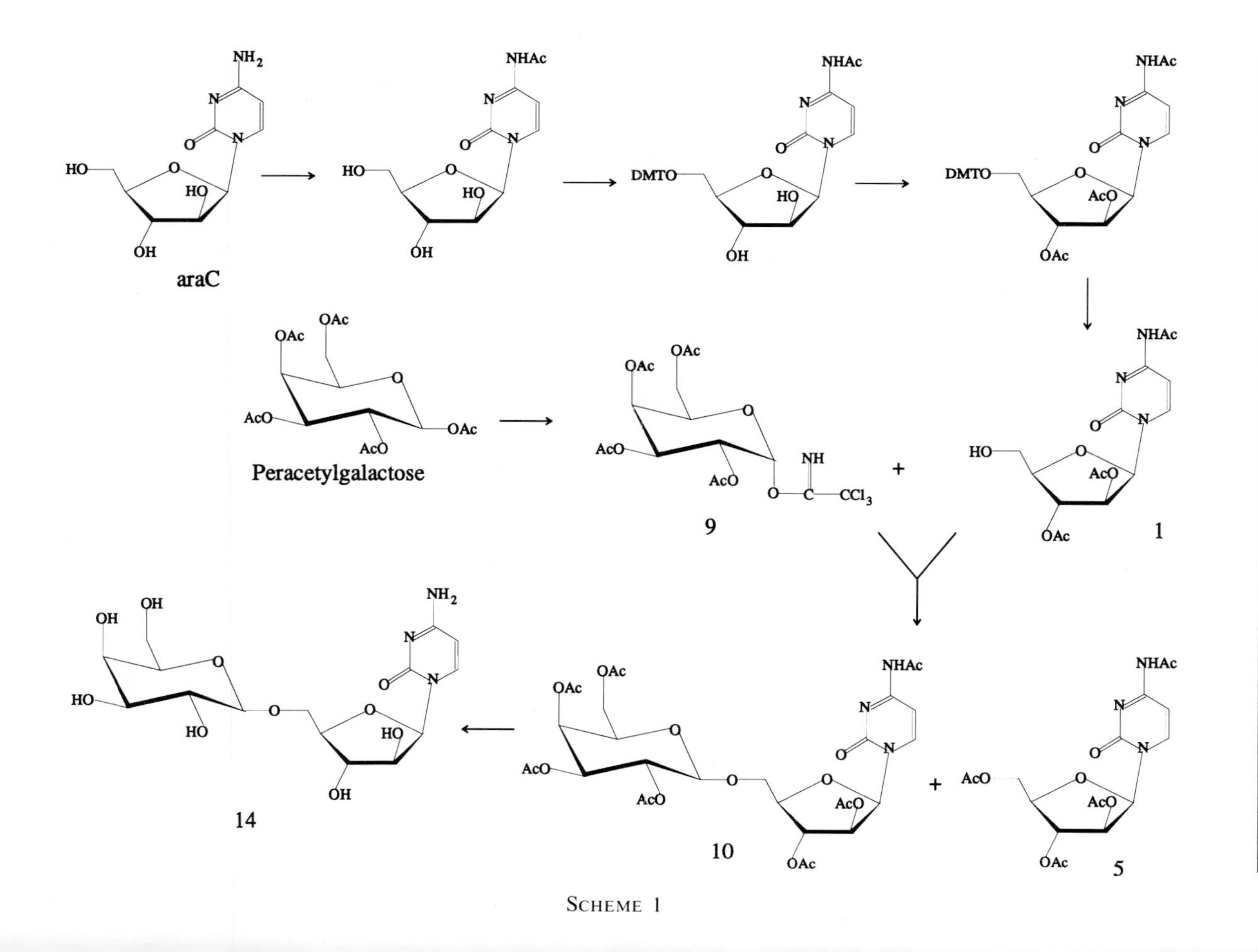
araC
Peracetylgalactose
9
1
10
5
14
DMTO
NHAc
AcO
OAc
HO
OH
NH_2
NH
CCl_3

SCHEME 1

It should be noted that acyl transfer from the 2-position of a glycosylating agent[19] to the OH of aglycon is a common side reaction. Thus galactosylation is always accompanied by the formation of nucleoside peracetates **5**,[20] **6**,[21] **7**,[22] and **8**[23] (see Scheme 1). Acyl transfer can be minimized using 2,3,4,6-tetra-*O*-acetyl-α-D-galactopyranosyl trichloroacetimidate **(9)**[24] with silver triflate in CH_2Cl_2 or in more acidic $CHCl_3$. The desired products are separated from those arising from acyl transfer by chromatography on silica gel followed by size-exclusion chromatography on Sephadex LH-20, and the acetate protective groups are removed via base-catalysed {1,8-diazabicyclo[5.4.0]undec-7-ene (DBU); K_2CO_3} transesterification.[3]

Materials and Methods

The starting nucleosides are available from a number of commercial sources (e.g., Sigma, St. Louis, MO; Aldrich, Milwaukee, WI). The 3′-acetylated nucleoside **3** is commercially available from Sigma. 1,2,3,4,6-Penta-*O*-acetyl-D-galactopyranose is obtained from Toronto Research Chemicals (Toronto, Canada). Reagents of highest purity available are used. All starting materials are dried overnight *in vacuo* (10^{-3} mm Hg) over KOH or P_2O_5 prior to use, and the solvents are distilled from appropriate drying agents. Solutions are concentrated at 10 mm Hg pressure in a rotary evaporator. Thin-layer chromatography (TLC) is performed on silica gel $60F_{254}$ (Merck, Darmstadt, Germany) plates and visualized by spraying with 50% aqueous sulfuric acid and heating at 200°. Silica gel (230–400 mesh, Toronto Research Chemicals) is used for flash chromatography. Sephadex LH-20 is obtained from Pharmacia (Uppsala, Sweden). Product identity and purity in each synthetic step should be determined by standard nuclear magnetic resonance (NMR) and mass spectroscopies. Details of ^{1}H and ^{13}C NMR spectra recorded at 500 (125 for ^{13}C) MHz

[19] P. J. Garegg, P. Konradsson, I. Kvarnström, T. Norberg, S. C. T. Svensson, and B. Wigilius, *Acta Chem. Scand. Ser. B* **39,** 569 (1985); T. Ziegler, P. Kováč, and C. P. J. Glaudemans, *Ann. Chem.* 613 (1990); N. I. Uvarova, G. I. Oshitok, and G. B. Elyakov, *Carbohydr. Res.* **27,** 79 (1973); A. Ya. Khorlin, V. A. Nesmeyanov, and S. E. Zurabyan, *Carbohydr. Res.* **43,** 69 (1975); R. U. Lemieux, *Chem. Can.* **16,** 14 (1964); J. Banoub and D. R. Bundle, *Can. J. Chem.* **57,** 2091 (1979); G. Wulff and G. Röhle, *Angew Chem., Int. Ed. Engl.* **13,** 157 (1974).

[20] A. P. Martinez, W. W. Lee, and L. Goodman, *J. Med. Chem.* **9,** 268 (1966).

[21] M. J. Robins, M. MacCoss, S. R. Naik, and G. Ramani, *J. Am. Chem. Soc.* **98,** 7381 (1976).

[22] A. M. Michelson and A. R. Todd, *J. Chem. Soc.*, 816 (1955).

[23] T. Kawaguchi, S. Fukushima, Y. Hayashi, and M. Nakano, *Pharm. Res.* **5,** 741 (1988).

[24] P. Zimmermann, R. Bommer, T. Bär, and R. R. Schmidt, *J. Carbohydr. Chem.* **7,** 435 (1988).

with a Bruker (Karlsruhe, Germany) AM 500 spectrometer or at 300 (75.5 for ^{13}C) MHz with a Bruker AM 300 spectrometer can be found in Ref. 3). Details of mass spectra recorded with a VG Analytical (Manchester, UK) ZAB-SE mass spectrometer in the fast atom bombardment mode (FAB-MS) using thioglycerol or glycerol–thioglycerol (3 : 1, v/v) matrices are described in Ref. 3.

Experimental Procedures

N^4-Acetyl-1-(2′,3′-di-O-acetyl-β-D-arabinofuranosyl)-cytosine ***(1)***. 1-(β-D-Arabinosyl)cytosine (3.0 g, 12.3 mmol; araC) is dissolved in dry methanol (300 ml) containing acetic anhydride (3.0 ml), and the mixture is refluxed under argon (see Scheme 1). Two additional portions of acetic anhydride (1.5 ml each) are added after 1 and 2 hr.[25] After heating for a total of 3 hr, the mixture is concentrated to about 10 ml and cooled on ice, and diethyl ether (100 ml) is added to precipitate a white solid. Mother liquors are filtered off by suction, the solid is dried and dissolved in dry pyridine (60 ml) containing 4,4′-dimethylaminopyridine (DMAP; 20 mg) under argon, dimethoxytrityl chloride (DMTCl; 6.27 g, 18.5 mmol) is added, and the mixture is stirred at room temperature for 24 hr. Then another portion (0.9 g) of DMTCl is added, and stirring is continued for another 8 hr. At this time TLC (1% CH_3OH–CH_2Cl_2) indicates that the reaction is complete. The flask is cooled in an ice bath, acetic anhydride (18 ml) is added dropwise by a syringe, and the mixture is stirred overnight (in the ice bath).

After warming to room temperature, the reaction mixture is evaporated under high vacuum, and the residue is subjected to chromatography on a short column of silica gel. Sequential elution with 500 ml each of CH_2Cl_2, 1% CH_3OH in CH_2Cl_2, 3% CH_3OH in CH_2Cl_2, and 5% CH_3OH in CH_2Cl_2 yields slightly impure solid peracetylated 5′-dimethoxytritylated compound (6.0 g). The latter is dried overnight, then dissolved in CH_2Cl_2 (150 ml), and to this stirred solution is added chloroacetic acid in CH_2Cl_2 (1% solution; 18.3 ml) dropwise under argon. After stirring has been continued for another 5 min, solid $NaHCO_3$ (1.6 g, 20 mmol) is added, and the reaction mixture is stirred for another 10 min. The solids are filtered off and rinsed with $CHCl_3$, and the combined filtrates are evaporated to dryness. The residue is at once subjected to chromatography on a short column of silica gel and eluted sequentially with 250 ml each of 1% CH_3OH in CH_2Cl_2, 3% CH_3OH in CH_2Cl_2, 5% CH_3OH in CH_2Cl_2, and 10% CH_3OH in CH_2Cl_2 to yield slightly impure **1** (3.5 g, 77%). After repeated chromatog-

[25] K. A. Watanabe and J. J. Fox, *Angew. Chem., Int. Ed. Engl.* **5**, 579 (1966).

raphy on silica gel with 5% $CH_3OH–CH_2Cl_2$, a pure specimen of **1** ($C_{15}H_{19}N_3O_8$) is obtained: $[\alpha]_D$ +96.3° (*c* 0.30, $CH_3OH–CH_2Cl_2$, 1 : 15, v/v).

N⁴-Acetyl-1-(3′-O-acetyl-2′-deoxy-β-D-ribofuranosyl)-cytosine ***(2)***. N^4-Acetyl-1-(3′-*O*-acetyl-2′-deoxy-β-D-ribofuranosyl)-cytosine **(2)** is prepared in an analogous manner in an overall yield of 40% (see Scheme 1); formula $C_{15}H_{19}N_3O_7$; $[\alpha]_D$ +54.9° (*c* 0.36, $CH_3OH–CH_2Cl_2$, 1 : 15, v/v).

1-(3′-O-Acetyl-2′-deoxy-β-D-ribofuranosyl)-5-fluorouracil ***(4)***. 1-(3′-*O*-Acetyl-2′-deoxy-β-D-ribofuranosyl)-5-fluorouracil ***(4)*** (Scheme 1) is prepared according to the procedure described[18] as follows (alternative methods have also been described[26]). To a mixture of 1-(2′-deoxy-β-D-ribofuranosyl)-5-fluorouracil (0.87 g, 3.53 mmol) and trityl chloride (1.47 g, 5.3 mmol) is added dry pyridine (9.0 ml), and the reaction mixture is stirred at room temperature for 1 hr, then at 50° for 18 hr. Then the reaction mixture is poured onto crushed ice (100 ml), and after separation of a precipitated gum the gum is rinsed with cold distilled water (2 times, 20 ml) and dissolved in acetone (18 ml). The insoluble material is filtered off, and the acetone is evaporated to dryness. The resulting foam dried *in vacuo* is dissolved in dry pyridine (18 ml), acetic anhydride (9.0 ml) is added, and the reaction mixture is stirred at room temperature for 18 hr. Then the mixture is poured into ice-cold water (20 ml); the emulsion formed is evaporated to dryness to yield an amorphous residue which after trituration with cold water (30 ml) turns into a white precipitate. The precipitate is filtered off and dissolved in aqueous acetic acid (80%, 27 ml), refluxed for 30 min, and allowed to cool to room temperature under stirring. Trityl alcohol precipitates out and is filtered off, and the mother liquors give on evaporation a residue which is subsequently dissolved in $CHCl_3$ (4.5 ml) and subjected to chromatography on silica gel (90 g). $CHCl_3$ (600 ml) elutes residual trityl alcohol, and methanol (200 ml) elutes impure 3′-*O*-acetyl-2′-deoxy-β-D-ribofuranosyl-5-fluorouracil **(4)**. The residue after evaporation of methanol is triturated with $CHCl_3$ to give a white precipitate (0.91 g, 89%) which after crystallization from methanol gives pure **4** (0.77 g, 76%).

2,3,4,6-Tetra-O-acetyl-α-D-galactopyranosyl Trichloroacetimidate ***(9)***. Anhydrous HBr gas is bubbled into dry CH_2Cl_2 (150 ml) at 0° until saturation (for 2 hr). To the saturated solution is added dry 1,2,3,4,6-pentaacetyl-D-galactopyranoside (10.3 g, 26 mmol), and the temperature

[26] H. J. Thomas and J. A. Montgomery, *J. Med. Pharm. Chem.* **5,** 24 (1962); W. Rode, T. Kulikowski, B. Kedzierska, and D. Sttugar, *Biochem. Pharmacol.* **36,** 203 (1987); K. Nozaki, A. Uemura, J. Yamashita, and M. Yasumoto, *Tetrahedron Lett.* **31,** 7327 (1990).

is maintained at 0° for 1 hr, after which the ice bath is removed. After 90 min the solution is cooled to 0° and resaturated with HBr. After another 30 min at 0°, TLC shows the reaction to be complete (R_f values: product, 0.80; starting material, 0.69; in ethyl acetate–hexane, 2 : 1, v/v). Dry argon is bubbled through the solution to remove excess HBr, and the reaction mixture is concentrated using a rotary evaporator. The amber residue is dissolved in acetone (250 ml) and water (65 ml) and then treated with Ag_2O (3.1 g, 13 mmol) under stirring overnight at room temperature. The solids are filtered off, and the filtrate is evaporated to dryness and further dried by coevaporation with toluene (2 times, 100 ml). The dry residue is dissolved in dry CH_2Cl_2 (150 ml), and to the stirred solution are added trichloroacetonitrile (5 ml; 50 mmol) and Cs_2CO_3 (0.425 g; 1.3 mmol) under argon at 0°. After stirring the reaction mixture at room temperature overnight, the solids are filtered off and rinsed with CH_2Cl_2, and the combined filtrate and washings are evaporated to dryness. The residue is subjected to chromatography on a silica gel column (hexane–ethyl acetate, 2 : 1, v/v). The α anomer **9** (Scheme 1) (9.0 g, 69%) elutes first, followed by the β anomer (3.0 g, 23%).

Syntheses of 5′-Galactosylated Nucleosides

N^4-Acetyl-1-[2′,3′-di-O-acetyl-5′-(2,3,4,6-tetra-O-acetyl-β-D-galactopyranosyl)-β-D-arabinofuranosyl]-cytosine ***(10)***. Alcohol **1** (500 mg, 1.37 mmol), trichloracetimidate **9** (Scheme 1) (945 mg, 1.92 mmol), and silver triflate (458 mg, 1.8 mmol) are dried together at high vacuum overnight. Chloroform (5 ml), freshly distilled from CaH_2, is added to the mixture of solids under an atmosphere of argon at room temperature. The reaction mixture is stirred for 24 hr; then diisopropylethylamine (5 drops) is added, and the solids are filtered off, rinsed with CH_2Cl_2 (2 times, 25 ml), the combined filtrate plus washings washed with cold 50 m*M* HCl (3 times, 50 ml) and cold saturated aqueous $NaHCO_3$ (1 time, 50 ml). The organic layer is dried with $MgSO_4$, filtered, and evaporated to dryness. The residue is dissolved in CH_2Cl_2 and subjected to flash chromatography on silica gel which on elution with 3% CH_3OH in CH_2Cl_2, followed by 5% CH_3OH in CH_2Cl_2 yields ultimately fractions containing a mixture of galactosylated nucleoside **10** and peracetylated nucleoside **5**. The mixture is further chromatographed on Sephadex LH-20 in CH_3OH–$CHCl_3$ (3 : 4, v/v) to give **10** (394 mg, 41%); formula $C_{29}H_{37}N_3O_{17}$; $[\alpha]_D$ +52.0° (*c* 0.59, CH_2Cl_2); UV, 303, 244 nm (CH_2Cl_2). FAB-MS shows m/z 700 (MH^+) and 331 (hexose). This fraction is followed by a fraction containing N^4-(2′,3′,5′-

tri-*O*-acetyl-β-D-arabinofuranosyl)-cytosine **(5)** (Scheme 1) [$C_{17}H_{21}N_3O_{17}$, $[\alpha]_D$ +118.2° (*c* 4.25, CH_2Cl_2)].

*N^4-Acetyl-1-[3′-O-acetyl-2′-deoxy-5′-(2,3,4,6-tetra-O-acetyl-β-D-galactopyranosyl)-β-D-ribofuranosyl]-cytosine **(11)***. Compound **11** (Scheme 1) {$C_{27}H_{35}N_3O_{15}$, $[\alpha]_D$ +6.93° (*c* 0.37, CH_2Cl_2); UV, 303, 243 nm (CH_2Cl_2)} is prepared as described for **10** (yield 35%). FAB-MS exhibits ions at *m/z* 642 (MH^+), 600 $(MH - CH_2CO)^+$, and 331 (peracetylated hexose). Exact mass measurements are as follows: observed 642.2128; calculated for $C_{27}H_{35}N_3O_{15}$, 642.2146. Compound **11** is accompanied by N^4-acetyl-1-[2′-deoxy-3′,5′-di-*O*-acetyl-β-D-ribofuranosyl]-cytosine **(6)** (Scheme 1) {$C_{15}H_{19}N_3O_7$, $[\alpha]_D$ +71.4 (*c* 12.4, CH_2Cl_2)}. Compounds **11** and **6** are separated by chromatography on silica gel and Sephadex LH-20.

*1-[3′-O-Acetyl-2′deoxy-5′-(2,3,4,6-tetra-O-acetyl-β-D-galactopyranosyl)-β-D-ribofuranosyl]-thymine **(12)***. Preparation of Compound **12** (Scheme 1) is an example of using another Lewis acid as a promoter. To a solution of 1-(3′-*O*-acetyl-2′-deoxy-β-D-ribofuranosyl)-thymine **(3)** (300 mg, 0.81 mmol) in CH_2Cl_2 (20 ml) containing activated molecular sieves 4Å (500 mg) is added a solution of trichloroacetimidate **9** (Scheme 1) (520 mg, 1.22 mmol) in CH_2Cl_2 (36.6 ml) at 5°, followed by boron trifluoride etherate (150 μl, 1.22 mmol). After stirring for 40 min, solid NH_4HCO_3 (100 mg) is added, and stirring is continued for another 10 min. The solids are filtered off and washed with CH_2Cl_2 (2 times, 25 ml), the filtrate and washings are combined, and the solvent is evaporated. The residue is subjected to chromatographic purifications as described in the procedure using silver triflate for preparation of **10** to give compound **12** {$C_{26}H_{34}N_2O_{15}$, $[\alpha]_D$ −13.2° (*c* 0.67, CH_2Cl_2); UV, 262 nm (CH_2Cl_2)} (yield 24%). FAB-MS exhibits ions at *m/z* 637 (MNa^+), 615 (MH^+), and 331 (peracetylated hexose). Compound **12** is accompanied by 1-(2′-deoxy-3′,5′-di-*O*-acetyl-β-D-ribofuranoxyl)-thymine **(7)** (Scheme 1).

*1-[2′-Deoxy-3′-O-acetyl-5′-(2,3,4,6-tetra-O-acetyl-β-D-galactopyranosyl)-β-D-ribofuranosyl]-5-fluorouracil **(13)***. 1-(3′-*O*-Acetyl-2′-deoxy-β-D-ribofuranosyl)-5-fluorouracil **(4)** (236 mg, 0.82 mmol) trichloracetimidate **9** (604 mg, 1.23 mmol), and silver triflate (315 mg, 1.22 mmol) are dried together at high vacuum overnight. Chloroform (10 ml), freshly distilled from CaH_2, is added to the mixture of solids under an atmosphere of argon at room temperature. The reaction mixture is stirred at 45°–50° for 1.5 hr (the reaction mixture has to be heated in order to increase solubility of reactants), then cooled to room temperature to form a cloudy solution which is filtered through a bed of celite. The celite bed is rinsed with CH_2Cl_2 (50 ml), and the combined filtrate and rinsings are washed with cold HCl (20 mM; 2 times, 50 ml) and saturated aqueous $NaHCO_3$ (50 ml),

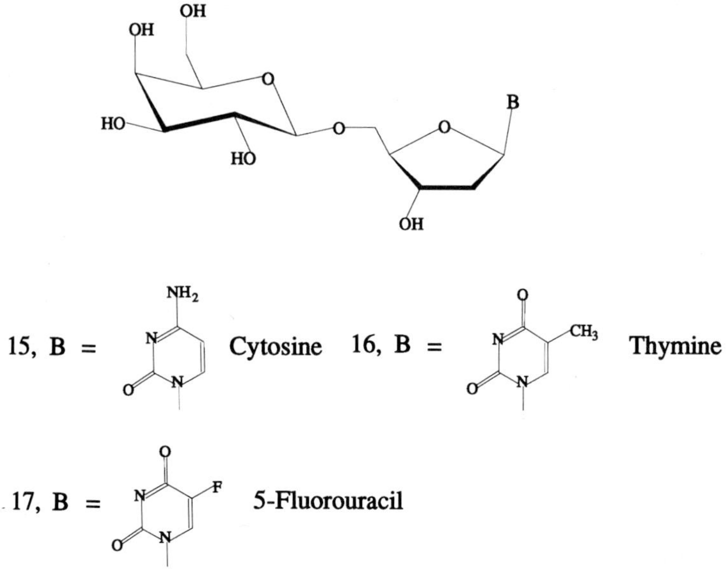

15, B = Cytosine 16, B = Thymine

17, B = 5-Fluorouracil

SCHEME 2

dried over Na_2SO_4, and evaporated to dryness. The residue is dissolved in CH_2Cl_2 and subjected to flash chromatography on a silica gel (90 g) column. The chromatography on elution with 3% CH_3OH in CH_2Cl_2 removes **9** together with other impurities and yields ultimately fractions containing a mixture of galactosylated nucleoside **13** and peracetylated nucleoside 2′-deoxy-3′,5′-di-*O*-acetyl-β-D-ribofuranosyl 5-fluorouracil **(8)** (total 311 mg). The two compounds are separated by subsequent chromatography on Sephadex LH-20 in CH_3OH–$CHCl_3$ (3 : 4, v/v) to give **13** (103 mg, 31%), followed by fractions containing **8** (76 mg). Minor remaining impurities in **13** can be removed by a repeated chromatography on Sephadex LH-20 followed by chromatography on silica gel in 2-propanol (2.5%) in CH_2Cl_2. Pure **13** {$C_{25}H_{31}N_2O_{15}F$; $[\alpha]_D$ −8.39° (*c* 0.43, $CHCl_3$)} shows in FAB-MS characteristic ions at *m/z* 619 (MH^+) and 331 (peracetylated hexose). The trans diaxial coupling constant ($J_{1,2}$ = 7.92 Hz) observed on the signal of Gal H-1 at δ 4.505 confirmed the β-anomericity of the newly formed O-glycosidic bond. The chemical shifts of both H5 (δ 3.79 and 3.30) in **4** changed to δ 4.24 and 3.77 in **13**, confirming that acetyl migration did not occur (H5-OAc would be found at much lower field).

1-(5′-β-D-Galactopyranosyl-β-D-arabinofuranosyl)-cytosine ***(14)***. Peracetylated galactosylated nucleoside **10** (120 mg) is dissolved in CH_3OH (20 ml) under argon, DBU (5 drops) is added, and the mixture is stirred overnight (Scheme 1). After an addition of toluene (15 ml), the reaction mixture is concentrated using a rotary evaporator until a precipitate appears. The solid is filtered off, rinsed with toluene and CH_2Cl_2, and then redissolved in CH_3OH and subjected to chromatography on Sephadex LH-20 using $CHCl_3$–CH_3OH (4 : 3, v/v). Pure **14** (62 mg, 89%) {$C_{15}H_{22}N_3O_{10}$, $[\alpha]_D$ +93.0° (*c* 0.13, CH_3OH)} is obtained. FAB-MS exhibits ions at *m/z* 406 (MH^+), 244 (MH^+ − hexose), and 112 (base + 2H).

1-(2′-Deoxy-5′-β-D-galactopyranosyl-β-D-ribofuranosyl)-cytosine ***(15).*** Compound **15** (Scheme 2) is obtained in 71% yield {$C_{15}H_{22}N_3O_9$, $[\alpha]_D$ +51.3° (*c* 0.37, CH_3OH); UV, 273, 230 nm (CH_3OH)} and is prepared as described for **14**. FAB-MS exhibits ions at *m/z* 390 (MH^+), 228 (MH^+ − hexose), and 112 (base + 2H).

1-(2′-Deoxy-5′-β-D-galactopyranosyl-β-D-ribofuranosyl)-thymine ***(16).*** Compound **16** (Scheme 2) {$C_{16}H_{24}N_2O_{10}$, $[\alpha]_D$ +11.2° (*c* 0.42, CH_3OH–water 3 : 2, v/v); UV, 264, 202 nm (CH_3OH)} is prepared as described for **14** (yield 96%). FAB-MS exhibits ions at *m/z* 427 (MNa^+), 405 (MH^+), 243 (MH^+ − hexose), and 127 (base + 2H).

1-(2′-Deoxy-5′-β-D-galactopyranosyl-β-D-ribofuranosyl)-5-fluorouracil ***(17).*** To a mixture of peracetylated galactosylated nucleoside **13** (12 mg, 20 μmol) and K_2CO_3 (13.4 mg; 97 μmol) is added CH_3OH (1.5 ml) under argon, and the mixture is stirred for 48 hr. The solid is filtered off and rinsed with CH_3OH, and the combined filtrates are evaporated to dryness. The residue is dissolved in water (1 ml), and passage through a BioGel P-2 (Bio-Rad, Richmond, CA) column (0.9 × 24 cm) gives pure **17** (Scheme 2) {$C_{15}H_{21}N_2O_{10}F$; $[\alpha]_D$ +12.32° (*c* 0.19, CH_3OH–water, 2 : 1, v/v); 7.1 mg, 90%}. Note that DBU could not be used because it is difficult to remove from the final product.

[11] Sialic Acid Analogs and Application for Preparation of Neoglycoconjugates

By REINHARD BROSSMER and HANS JÜRGEN GROSS

Introduction

Sialylated glycans attached to soluble or membrane-bound glycoconjugates are important ligand structures for a variety of receptor interactions, especially with regard to cellular adhesion phenomena, virus specificity for host cells, and half-lives of serum proteins.[1-5] However, evidence is increasing that only distinct sialoglycan sequences rather than global terminal sialic acids are recognized by many sialic acid-dependent receptor proteins. Modeling of glycoconjugates selectively in the sialic acid moiety provides an approach to gain further information on the structural features essential for such events. The chemical procedures for modification of glycosidically linked sialic acids possess some disadvantages owing to the chemical pretreatment, for example, oxidation, and to unspecific side reactions of the sialic acid molecule and/or the remainder of the glycoconjugate.[6-8] In addition, sialic acids are modified irrespective of the subterminal glycan, without the possibility of focusing on special sequences.

An approach to overcome such difficulties is by the synthesis of suitable sialic acids which may be transferred by a sialyltransferase from the respective CMP-glycoside to distinct glycan sequences determined by the well-defined enzyme specificity.[9-11] Thus, replacement of naturally occurring sialic acids by synthetic analogs endowed with special properties such as resistance against catabolic glycosidases or altered hydrophobic and steric characteristics can influence certain biological functions mediated by sialic acid. Previous kinetic studies performed with synthetic sialic acids differently substituted at the C-4, C-5, and C-9 positions and sialyltransferases specific for different acceptor sequences demonstrated a low enzyme specificity with respect to these molecular regions[9-15] and the feasibility of transferring such analogs to soluble as well as cell surface-bound glycoconjugates.[9-17]

[1] J. C. Paulson, G. N. Rogers, S. M. Caroll, H. H. Higa, T. Pritchet, G. Milks, and S. Sabesan, *Pure Appl. Chem.* **56,** 797 (1984).
[2] G. N. Rogers, G. Herrler, J. C. Paulson, and H.-D. Klenk, *J. Biol. Chem.* **261,** 5947 (1986).
[3] G. Ashwell and A. G. Morell, *Adv. Enzymol. Relat. Areas Mol. Biol.* **41,** 99 (1974).
[4] Y. Pilatte, J. Bignon, and C. R. Lambré, *Glycobiology* **3,** 201 (1993).
[5] T. A. Springer, *Cell (Cambridge, Mass.)* **76,** 301 (1994).
[6] S. Spiegel, this series, Vol. 138, p. 313.
[7] M. Wilchek and E. Bayer, this series, Vol. 138, p. 429.
[8] V. P. Bhavanandan, M. Murray, and E. A. Davidson, *Glycoconjugate J.* **5,** 467 (1988).
[9] H. S. Conradt, A. Bünsch, and R. Brossmer, *FEBS Lett.* **170,** 295 (1984).
[10] H. J. Gross, A. Bünsch, J. C. Paulson, and R. Brossmer, *Eur. J. Biochem.* **168,** 595 (1987).
[11] H. J. Gross and R. Brossmer, *Glycoconjugate J.* **4,** 145 (1987).
[12] H. J. Gross and R. Brossmer, *Eur. J. Biochem.* **177,** 583 (1988).
[13] H. J. Gross, U. Rose, J. M. Krause, J. C. Paulson, K. Schmid, R. E. Feeney, and R. Brossmer, *Biochemistry* **28,** 7386 (1989).
[14] H. J. Gross, U. Sticher, and R. Brossmer, *Anal. Biochem.* **186,** 127 (1990).
[15] U. Sticher, H. J. Gross and R. Brossmer, *Glycoconjugate J.* **8,** 45 (1991).
[16] H. J. Gross and R. Brossmer, *Glycoconjugate J.* **5,** 411 (1988).
[17] R. Kosa, R. Brossmer, and H. J. Gross, *Biochem. Biophys. Res. Commun.* **190,** 914 (1993).

This chapter describes the enzymatic synthesis of neoglycoconjugates derived from human plasma proteins and ganglioside G_{M1} [Galβ1,3GalNAcβ1,4(NeuAcα2,3)Galβ1,4Glc-ceramide] as well as the transfer of sialic acid analogs onto the surface of human cells. The procedure leads to a masking of the penultimate galactose in special glycan sequences by different synthetic sialic acids to an extent close to natural sialylation.

Synthesis of Sialic Acid Analogs

5-N-Acetyl-9-amino-9-deoxyneuraminic Acid

Principle. As starting material for the synthesis of 5-*N*-acetyl-9-amino-9-deoxyneuraminic acid, the methyl α-glycoside of *N*-acetylneuraminic acid methyl ester,[1,18–21] obtained from the peracetylated derivative[19–21] by Zemplén deacetylation (sodium methanolate), is employed. Treatment of **1** with *p*-toluenesulfonyl chloride in pyridine provides the 9-*O*-tosyl derivative **2,** which is converted to azide **3** by reaction with sodium azide in aqueous acetone. Saponification of the methyl ester of **3** under mild conditions yields the free acid **4** as a crystalline solid (yield 56%, based on peracetylated **1**). Subsequent careful hydrolysis of **4** with Dowex 50W-X8 (H^+) gives *N*-acetyl-9-azido-9-deoxyneuraminic acid **(5).**[18,22] Catalytic hydrogenation (PdO) of the azide function of **5** in weakly acidic solution affords amine **6** in an almost quantitative yield (Fig. 1). The nuclear magnetic resonance (NMR) spectra of all sialic acids described in this chapter are in agreement with the postulated structures.

Methyl 5-N-Acetyl-9-O-toluenesulfonyl-D-neuraminate Methyl α-Glycoside **(2).** To remove traces of water, compound **1** (2.0 g, 5.93 mmol) is coevaporated twice with dry pyridine, then dissolved in the same solvent (50 ml). The solution is cooled to 0°, and *p*-toluenesulfonyl chloride (1.13 g, 5.93 mmol) is added in small portions. The reaction mixture is allowed to warm to room temperature and kept overnight at 20°. After evaporation of the solvent *in vacuo,* ice water (4 ml) and ethyl acetate (40 ml) are added, and the organic phase is separated. The aqueous solution is extracted once again with ethyl acetate (40 ml), and the combined organic phases are concentrated. Addition of diethyl ether followed by *n*-hexane causes crys-

[18] R. Isecke and R. Brossmer, *Tetrahedron* **50,** 7445 (1994).
[19] R. Kuhn, P. Lutz, and D. L. MacDonald, *Chem. Ber.* **99,** 611 (1966).
[20] D. J. M. van der Vleugel, W. A. R. Heeswijk, and J. F. G. Vliegenthart, *Carbohydr. Res.* **102,** 121 (1982).
[21] P. Meindl and H. Tuppy, *Monatsh. Chem.* **97,** 990 (1966).
[22] R. Brossmer, U. Rose, D. Kasper, T. L. Smith, H. Grasmuk, and F. M. Unger, *Biochem. Biophys. Res. Commun.* **96,** 1282 (1980).

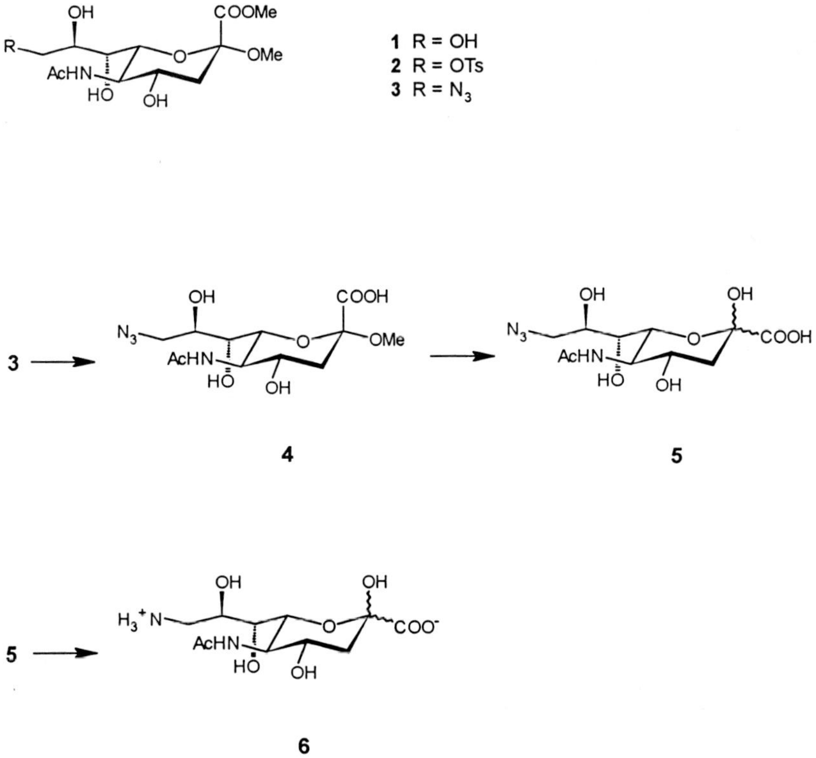

1 R = OH
2 R = OTs
3 R = N_3

FIG. 1. Structural diagrams of sialic acids **1–6**.

tallization of **2,** mp 150° (dec.). The yield is 2.28 g (78%); R_f 0.23 [cf. $R_{f,1}$ 0.05, 20 : 1 (v/v) ethyl acetate–methanol]; $[\alpha]_D^{20}$ +1.0° (*c* 0.5, methanol).

Methyl 5-N-Acetyl-9-azido-9-deoxy-D-neuraminate Methyl α-Glycoside (3). A mixture of **2** (1.84 g, 3.74 mmol), sodium azide (1.09 g, 16.8 mmol), water (8 ml), and acetone (24 ml) is refluxed for 40 hr. Thin-layer chromatography (TLC) shows complete conversion of **2** to **3** [$R_{f,3}$ 0.48, $R_{f,2}$ 0.58, 5 : 1 (v/v) chloroform–methanol]. The solvents are removed completely under reduced pressure, and the residue is taken up in 5 : 1 (v/v) chloroform–methanol (~12 ml), filtered, and concentrated. After column chromatography on silica gel [10 : 1 (v/v) chloroform–methanol], **3** is crystallized from methanol–diethyl ether, mp 168°–170°, in a yield of 1.11 g (82%); R_f 0.22 [12 : 1 (v/v) ethyl acetate–methanol; $[\alpha]_D^{20}$ −4.2° (*c* 0.5, methanol)].

5-N-Acetyl-9-azido-9-deoxy-D-neuraminic Acid Methyl α-Glycoside (4). Compound **3** (1.11 g, 3.06 mmol) is dissolved in methanol (90 ml), and 0.1 *N* NaOH (90 ml) is added. After being kept at room temperature for 1 hr, the solution is passed through a column of Dowex 50W-X8 (H^+) resin at 4°, then lyophilized. The residue crystallizes from methanol–diethyl ether–*n*-hexane, mp 159°–161° (dec.); yield 0.94 g (88%); R_f 0.44, [5 : 2 : 3 (v/v/v) *n*-butanol–acetic acid–water]; $[\alpha]_D^{20}$ +14.5° (*c* 0.5, water).

5-N-Acetyl-9-azido-9-deoxy-D-neuraminic acid **(5).** To a solution of **4** (0.40 g, 1.15 mmol) in water (120 ml) is added Dowex 50W-X8 (H^+) resin, and the mixture is heated at 80° with stirring for 1 hr. Analysis by TLC [2 : 3 : 1 (v/v/v) methanol–ethyl acetate—20% acetic acid] shows $R_{f,5}$ 0.56 (cf. $R_{f,4}$ 0.73). After removal of the resin by filtration, the solution is lyophilized and subsequently purified on a column of DEAE-Sephadex A-25 (HCO_3^-) (eluent 80 m*M* NH_4HCO_3). The fractions containing the product are combined, treated with Amberlite IR 120 (H^+) resin, and lyophilized after separation of the resin. The yield is 0.37 g (91%, for **5** · H_2O); R_f 0.38 [6 : 1 : 2 (v/v/v) *n*-propanol–25% ammonia–water]; $[\alpha]_D^{20}$ −16.3° (*c* 0.5, water).

5-N-Acetyl-9-amino-9-deoxy-D-neuraminic Acid **(6).** A solution of **5** · H_2O (0.37 g, 1.05 mmol) in water (30 ml) is brought to approximately pH 2 by the addition of acetic acid and shaken with PdO (50 mg) under a hydrogen atmosphere for 1 hr at ambient temperature. After removal of the catalyst by filtration, the solution is lyophilized. Crystallization from water–acetone affords 330 mg (91%, for **6** · $2H_2O$), mp approximately 190° (dec.); R_f 0.24 [5 : 3 (v/v) *n*-propanol–water]; $[\alpha]_D^{20}$ −24.1° (*c* 0.5, water).

N-Acylation of 9-Amino-9-deoxy-D-neuraminic Acid

Principle. As an example of 9-N-acylated derivatives of **6,** the preparation of 9-acetamido-5-*N*-acetyl-9-deoxy-D-neuraminic acid (9-acetamido-NeuAc) is described. The synthesis involves treatment of **6** with the respective acid anhydride. Although **6** contains four unprotected hydroxyl groups, no side reactions are observed. The method can be extended to the preparation of other N-acylated derivatives needed for special applications. In addition, the synthesis of the influenza C virus receptor analog 5-*N*-acetyl-9-deoxy-9-thioacetamido-D-neuraminic acid (9-thioacetamido-NeuAc) is outlined.

9-Acetamido-5-N-acetyl-9-deoxy-D-neuraminic Acid. To the solution of 5-*N*-acetyl-9-amino-9-deoxyneuraminic acid **(6)** (77 mg, 0.25 mmol) in 2.5 ml methanol–water (4 : 1, v/v) are added 1 *M* sodium hydroxide (0.25 ml) and acetic anhydride (50 μl). After stirring for 1 hr at room temperature the 9-acetamido derivative is purified by chromatography on DEAE-Sephadex A-25 (HCO_3^-). Elution with ammonium hydrogen carbonate (80 m*M*) and treatment with IR-120 (H^+) resin afford, after lyophilization, 80 mg (92%) of the desired compound. Analysis by TLC [6 : 1 : 2 (v/v/v) *n*-propanol–concentrated ammonia–water] gives R_f 0.28; $[\alpha]_D^{20}$ −13.7° (*c* 0.5, water).

5-N-Acetyl-9-benzamido-9-deoxy-D-neuraminic Acid and 5-N-Acetyl-9-deoxy-9-hexanoyl-D-neuraminic Acid. The compounds 5-*N*-acetyl-9-benzamido-9-deoxy-D-neuraminic acid (9-benzamido-NeuAc) and 5-*N*-acetyl-9-deoxy-9-hexanoyl-D-neuraminic acid (9-hexanoylamido-NeuAc) are prepared in the same way from **6** as 9-acetamido-NeuAc using the respective acid anhydride.

5-N-Acetyl-9-deoxy-9-thioacetamido-D-neuraminic Acid Ammonium Salt. Compound **6** (80 mg, 0.26 mmol) is suspended in aqueous methanol [2.0 ml, 14 : 1 (v/v) methanol–water].[18,23] At 0°, triethylamine (0.25 ml, 1.8 mmol) is added followed by methyl dithioacetate[24] (0.25 ml, 2.3 mmol). The mixture is allowed to warm to room temperature, and after 14 hr TLC shows complete conversion to the thio compound. For purification, the reaction mixture is concentrated under reduced pressure and then dissolved in a small volume of methanol (~1 ml). Addition of ethyl acetate (~3 ml) causes precipitation of the crude compound. Further purification is accomplished by chromatography on DEAE-Sephadex A-25 (HCO_3^-). After rinsing with water, elution is done with 6 m*M* ammonium hydrogen carbonate. The fractions containing the product are pooled and lyophilized three times. The preparation yields 76 mg of a white powder; R_f 0.29 [8 : 1 (v/v) *n*-propanol–water]; $[\alpha]_D^{20}$ −23.3° (*c* 0.5, water).

5-N-Acetyl-9-deoxy-9-(N′-Fluoresceinyl)thioureido-D-neuraminic Acid. To a solution of **6** (47 mg, 0.15 mmol) in 0.5 *M* sodium carbonate (2.2 ml) and methanol (3 ml) is added fluorescein isothiocyanate isomer I (115 mg, 0.30 mmol). After 2 hr the mixture is diluted with water (20 ml) and lyophilized. The residue is purified on a column (2.3 × 20 cm) of silica gel RP-2 (Merck, Darmstadt, Germany), which is eluted first with 5 : 1 (v/v) chloroform–methanol and then with 1 : 1 (v/v) of the same solvent. The main fraction is concentrated under reduced pressure and further purified on a Sep-Pak C_{18} cartridge. After washing with acetonitrile–water (1 : 5, v/v), solvents are removed and the residue is dissolved in water and precipitated with 2.5% formic acid. The yield is 66.8 mg (61%); R_f 0.31 [1 : 4 : 1 (v/v/v) methanol–ethyl acetate–acetic acid].

5-N-Aminoacetyl-D-neuraminic Acid

Principle. Because 5-*N*-fluoresceinated neuraminic acid substituted directly at the amino group is difficult to prepare, probably for steric

[23] R. Brossmer, R. Isecke, and G. Herrler, *FEBS Lett.* **323,** 96 (1993).

[24] J. Meijer, P. Vermeer, and L. Brandsma, *Recl. Trav. Chim. Pays-Bas* **92,** 601 (1973).

reasons, glycine is introduced as a C_2 spacer. N-Unsubstituted neuraminic acid benzyl α-glycoside **(10)**, which is obtained by treatment of the peracetylated benzyl α-glycoside of methyl *N*-acetylneuraminate **(9)** with barium hydroxide at reflux, serves as starting material. Reaction of **10** with *Z*-glycine *p*-nitrophenyl ester followed by removal of the protecting groups applying catalytic hydrogenation produces 5-*N*-aminoacetyl-D-neuraminic acid **(12)**. In contrast to *N*-deacetyl neuraminic acid, **12** is a stable analog which can be activated to the corresponding CMP-glycoside (Fig. 2).

Methyl N-Acetyl-D-neuraminate **(7)**. To the solution of *N*-acetylneuraminic acid (5.0 g, 16.18 mmol) in 500 ml dry methanol is added Dowex 50W-X8 (H^+) resin (25 ml) which has been carefully washed with dry methanol. After stirring at room temperature for 15–20 hr, TLC [5 : 2 : 3 (v/v/v) *n*-butanol–acetic acid–water] shows complete conversion to the

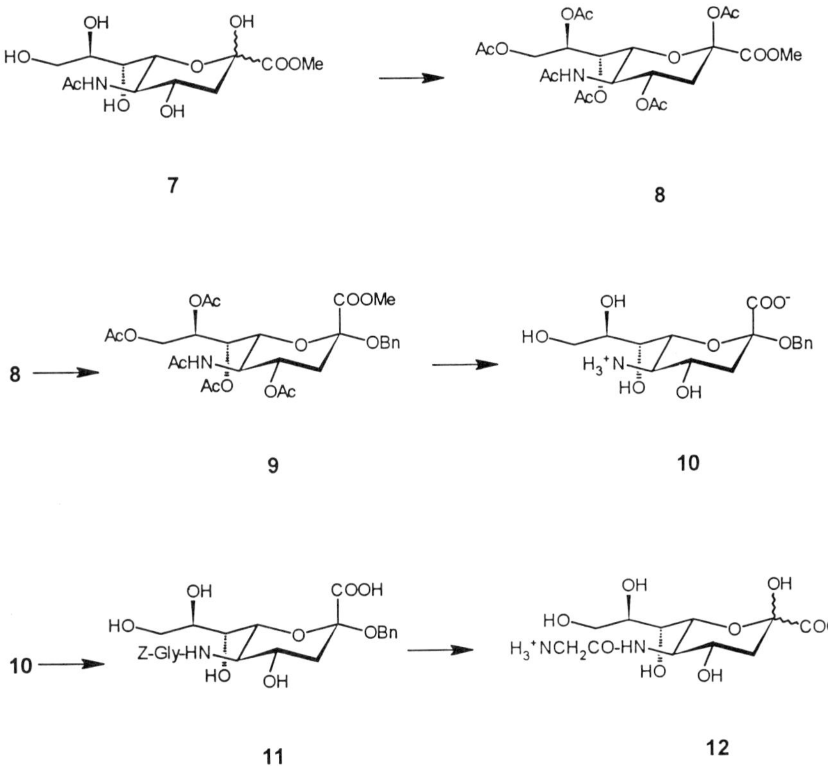

(Bn = $C_6H_5CH_2$)

(Z-Gly = Z-aminoacetyl = $C_6H_5CH_2OCO\text{-}NHCH_2CO$)

FIG. 2. Structural diagrams of sialic acids **7–12**.

ester. The resin is separated and stirred 3 times with dry methanol for 5 min each. After removal of the combined solvents under reduced pressure an amorphous solid remains. The yield is nearly quantitative; R_f 0.53 [solvent: 5:2:3 (v/v/v) *n*-butanol-acetic acid-water).

*Peracetylated Methyl N-Acetyl-*D*-neuraminate* ***(8)***. Compound **7** (5.0 g, 15.48 mmol) is twice evaporated with dry pyridine (10 ml). To this is added 50 ml of the same solvent and dropwise acetic anhydride (20 ml) under cooling with ice. After 24 hr at 4°, TLC [5:2:3 (v/v/v) *n*-butanol–acetic acid–water] shows the reaction to be complete; R_f 0.76. After stirring with ice water (200 ml) for 1 hr at room temperature the mixture is twice concentrated under reduced pressure, adding water each time. Treatment with Dowex 50W-X8 (H^+) resin removes remaining pyridine. Lyophilization affords 7.5 g (91%) amorphous white solid.

*Peracetylated Methyl N-Acetyl-*D*-neuraminate Benzyl α-Glycoside* ***9***. To the solution of the peracetylated methyl ester **8** (5.0 g, 9.38 mmol) in acetic acid (100 ml) is added acetic anhydride (50 ml). The mixture is cooled in ice, saturated with dry HCl, and kept overnight at 4°. After removal of solvents and evaporation twice with approximately 10 ml toluene at 1 torr the residue is treated with 50 ml benzyl alcohol and 3 g powdered molecular sieves (3 Å) overnight at room temperature. Analysis by TLC (ethyl acetate) shows R_f 0.37. The solution is filtered, extracted with aqueous saturated $NaHCO_3$, and then extracted with saturated NaCl. The dried (Na_2SO_4) organic layer is concentrated under reduced pressure, and the residue is purified by flash chromatography on silica gel 60, particle size 0.040–0.063 mm (9 × 12 cm) (Merck). Benzyl alcohol is removed by washing with hexane (bp 40°–60°)–ethylacetate, 2:1 (v/v) (~2.5 liters). After elution with ethyl acetate, **9** is obtained as a colorless foam (4.6 g, 84%).

D*-Neuraminic Acid Benzyl α-Glycoside* ***(10)***. The peracetylated benzyl glycoside **9** (1.0 g, 1.72 mmol) is refluxed in a saturated water–methanol (1:1, v/v) solution of barium hydroxide (40 ml) for ~18 hr. Analysis by TLC shows complete conversion to **10**, $R_{f,10}$ 0.37 [5:1 (v/v) *n*-propanol–water], ninhydrin positive. The mixture is cooled to room temperature, then filtered, and the filtrate is brought to approximately pH 7 by the addition of small pieces of dry ice. After separation of the copious precipitate, the solution is cooled to 0° and adsorbed on Amberlite IR 120 (H^+) resin (60 ml). After washing with water, **10** is eluted with 1 *M* aqueous ammonia (60 ml) to afford, after freeze-drying, 522 mg (85%), which may be crystallized from methanol or ethanol–water, yielding 399 mg (65%); mp >200°; $[\alpha]_D^{20}$ −41.9° (*c* 0.5, water).

*5-N-(Z-Aminoacetyl)-*D*-neuraminic Acid Benzyl α-Glycoside* ***(11)***. To the solution of the sodium salt of **10** (265 mg, 0.7 mmol) in 3 ml dry dimethylformamide (DMF) is added *N*-*Z*-glycine nitrophenyl ester (417 mg, 1.26 mmol) in 4 ml dry DMF. The solvent is removed under reduced pressure (~1 torr), and the residue is extracted with diethyl ether–water. The material from the freeze-dried water phase is purified by chromatography on DEAE-Sephadex A-25 (HCO_3^-). Elution with 0.1 *M* ammonium hydrogen carbonate and treatment with IR 120 (H^+) resin give, after freeze-drying, 358 mg (93%) **11;** R_f 0.52 [1 : 4 : 1 (v/v/v) methanol–ethyl acetate–acetic acid]; $[\alpha]_D^{20}$ −3° (*c* 0.5, methanol).

*5-N-Aminoacetyl-D-neuraminic Acid **(12)**.* Compound **11** (358 mg, 0.653 mmol) is dissolved in 20 ml methanol–water (1 : 1, v/v) and hydrogenated using 150 mg palladium oxide. After 4 hr the filtered suspension is concentrated under reduced pressure and lyophilized (202 mg, 95%); R_f 0.36 [4 : 1 : 2 (v/v/v) *n*-propanol–concentrated ammonia–water); $[\alpha]_D^{20}$ −23.5° (*c* 0.5, water).

The NMR spectra of compounds **7–12** are in accordance with the respective structures.

5-N-Thioacetyl-D-neuraminic Acid as Ammonium Salt

Principle. The N-unsubstituted neuraminic acid (Neu) methyl α-glycoside[18] is N-thioacetylated. Enzymatic cleavage of the glycoside then affords 5-*N*-thioacetyl-D-neuraminic acid (5-*N*-thioacetyl-Neu).[18]

Triethylammonium 5-N-Thioacetyl-D-neuraminate Methyl α-Glycoside. The peracetylated methyl α-glycoside of methyl *N*-acetylneuraminate (1.0 g, 1.98 mmol) is refluxed in a saturated aqueous solution of barium hydroxide (40 ml) for 16 hr, cooled to room temperature, and filtered, and the filtrate is brought to approximately pH 7 by the addition of small pieces of dry ice. After removal of the precipitate by filtration, the solution is cooled to 0°, and Amberlite IR 120 (H^+) cation-exchange resin is added until Ba^{2+} can no longer be detected. The resin is removed by filtration and washed exhaustively with water, and the collected solutions are lyophilized to give a first fraction of de-N-acetylated neuraminic acid methyl α-glycoside (320 mg, 58%). Because a considerable amount of the compound adheres to the resin, it is finally rinsed with 0.5 *M* aqueous ammonia (40 ml) to afford, after freeze-drying, a second fraction (170 mg, 31%). An analytical sample is obtained by crystallization from ethanol–water; mp 195°; R_f 0.21 [5 : 1 (v/v) *n*-propanol–water]; $[\alpha]_D^{20}$ −12.4° (*c* 0.5, water).

A suspension of the compound (100 mg, 0.36 mmol) in water (0.2 ml) is diluted with methanol (2.5 ml). After cooling to 0°, triethylamine (0.40 ml, 2.9 mmol) and methyl dithioacetate[24] (0.31 ml, 2.9 mmol) are added.

The reaction mixture is allowed to warm to room temperature and is stirred overnight. Analysis by TLC [1 : 1 (v/v) ethyl acetate–methanol] then shows complete conversion to the strongly UV positive product. Volatile material is removed under reduced pressure, and the residue is taken up in 1 : 1 (v/v) ethyl acetate–methanol (~3 ml). After filtration, addition of ethyl acetate followed by diethyl ether causes crystallization. Recrystallization from methanol–ethyl acetate–diethyl ether affords colorless crystals, mp 169°. The yield is 136 mg (86%); R_f 0.41 [1 : 1 (v/v) ethyl acetate–methanol]; $[\alpha]_D^{20}$ −51.5° (*c* 0.5, water).

*Ammonium N-Thioacetyl-*D*-neuraminate.* Hydrolysis of the *N*-thioacetylneuraminate methyl glycoside with dilute hydrochloric acid results in extensive decomposition. However, the glycoside can be cleaved enzymatically. Methyl glycoside (50 mg, 0.11 mmol) is dissolved in water (0.5 ml) which has been brought to pH 5.0 with the addition of dry ice. (*Note:* Results obtained in this way do not differ from those obtained when sodium acetate buffer is used.) After addition of *Arthrobacter ureafaciens* sialidase (EC 3.2.1.18; 0.2 U) (Boehringer, Mannheim, Germany) and incubation of the solution for 10 hr at 37°, more sialidase (0.1 U) is added and the incubation continued for 14 hr. The mixture is transferred onto a column of DEAE-Sephadex A-25 (HCO_3^-), washed with water, and eluted with 6 m*M* NH_4HCO_3. The fractions containing the desired product are collected and freed from NH_4HCO_3 by lyophilization repeated three times. The procedure yields 34 mg of a white powder; R_f 0.41 [5 : 1 (v/v) *n*-propanol–water]; $[\alpha]_D^{20}$ +4.6° (*c* 0.5 water).

*5-N-Formyl-*D*-neuraminic Acid.* 5-*N*-Formyl-D-neuraminic acid (5-*N*-formyl-Neu) is prepared according to Brossmer and Nebelin.[25]

*5-N-Acetyl-4-deoxy-*D*-neuraminic Acid.* 5-*N*-Acetyl-4-deoxy-D-neuraminic acid is prepared according to Hagedorn and Brossmer.[26]

Note: The N-unsubstituted methyl α-glycoside of 5-*N*-acetyl-D-neuraminic acid (NeuAc) has been obtained by heating the corresponding methyl ester with tetramethyl ammonium hydroxide.[27] The respective N-unsubstituted benzyl α-glycoside has been prepared by heating the peracetylated methyl ester with 2 *M* sodium hydroxide in a sealed tube at 110°.[28]

[25] R. Brossmer and E. Nebelin, *FEBS Lett.* **4,** 335 (1969).

[26] H. W. Hagedorn and R. Brossmer, *Helv. Chim. Acta* **69,** 2127 (1986).

[27] W. Schmid, L. Z. Avila, K. W. Williams, and G. M. Whitesides, *Bioorg. Med. Chem. Lett.* **3,** 747 (1993).

[28] N. E. Byramova, A. B. Tuzikov, and N. V. Bovin, *Carbohydr. Res.* **237,** 161 (1992).

Synthesis, Purification, and Characterization of CMP-Activated Sialic Acid Analogs

Principle

Synthetic sialic acids differently substituted at the C-9 or C-5 position can be enzymatically converted by 40–90% to the corresponding CMP-glycosides employing CMP-sialic acid synthase (EC 2.7.7.43, acylneuraminate cytidylyltransferase) from bovine brain. As in previous studies, enzyme specificity with respect to these positions turned out to be low.[10–12] Additionally, a procedure is demonstrated for direct acylation of the amino group at C-9 of cytidine-5′-monophospho-9-amino-9-deoxy-NeuAc (CMP-9-amino-NeuAc) without decomposition of the labile glycosidic linkage. The CMP-activated products are purified by high-performance liquid chromatography (HPLC) and gel filtration and characterized with regard to their composition (CMP moiety/sialic acid moiety) and contaminations (CMP, respective sialic acid analog, CTP) employing analytical HPLC systems.[10–13]

Preparative Synthesis of CMP-Activated Sialic Acid Analogs

A typical preparative reaction mixture[10,11] (14 ml) contains 160 m*M* Tris, pH 9, 40 m*M* $MgCl_2$, 2 m*M* dithioerythritol, 15 m*M* CTP, and 7–10 m*M* sialic acid analog (9-amino-, 9-azido-, 9-acetamido-, 9-benzamido-, 9-hexanoylamido-, or 9-thioacetamido-NeuAc, 5-*N*-aminoacetyl-, 5-*N*-formyl-, or 5-*N*-thioacetyl-Neu, or 4-deoxy-NeuAc). The reaction is started by the addition of 3–4 U CMP-sialic acid synthase (3–4 U/ml) partially purified as described[10]; after 2 hr additional enzyme (1–1.5 U) and CTP (5 m*M* final concentration) is added (1.5 ml total). The CMP activation during synthesis can be monitored by analytical HPLC at 275 nm (see below).

After 4–5 hr, synthesis is stopped by the addition of 29 ml cold acetone, the tube is cooled on ice for at least 30 min, and the mixture is centrifuged at 6000 *g* for 15 min. The pellet is reextracted at least twice by the addition of 4 ml twice-distilled water and 5 ml cold acetone, and the mixture is processed as before. Supernatants are combined and acetone evaporated, yielding a final preparation volume of about 3–4 ml. The CMP-glycoside preparation is purified by HPLC as outlined below.

Structures of enzymatically synthesized CMP-activated sialic acid analogs substituted at positions C-5 and C-9 of the NeuAc moiety are outlined in Figs. 3 and 4 (for the structure of CMP-4-deoxy-NeuAc, see Ref. 11.) Final conversions of 60–80% are achieved except for 9-acetamido-NeuAc (40%); the conversion of parent NeuAc under identical conditions reaches

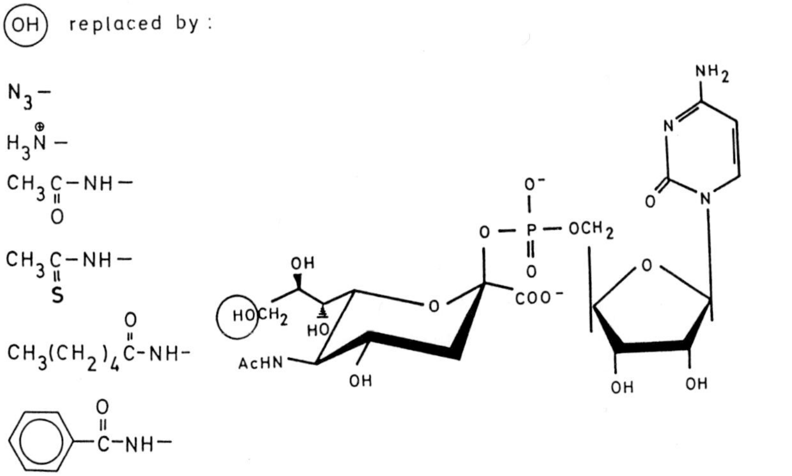

FIG. 3. Structure of sialic acid analogs modified at the C-9 position. The hydroxyl group at C-9 was replaced by an azido, amino, acetamido, thioacetamido, hexanoylamido, or benzamido group.

70–75%. Cytidine-5′-monophospho-5-*N*-acetyl-9-deoxy-9-(*N*′-fluoresceinyl)thioureido-D-neuraminic acid (CMP-9-fluoresceinyl-NeuAc) (for structure, see Ref. 12 and this volume, [12]) may be synthesized enzymatically starting from 9-deoxy-9-N-fluoresceinyl-NeuAc, but in low conversion yield (25%).[12]

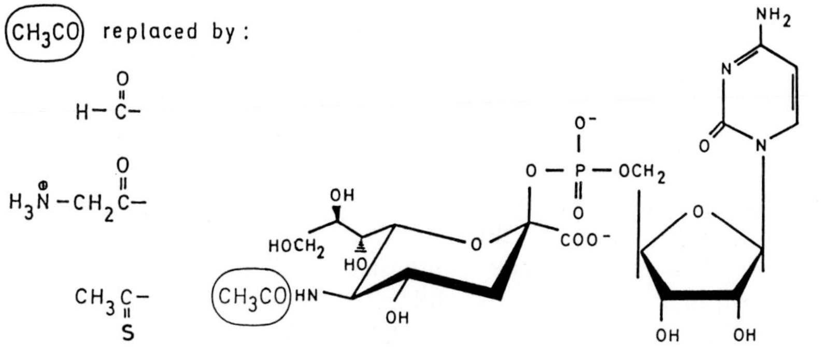

FIG. 4. Structure of sialic acid analogs modified at the C-5 position. The acetyl group at C-5 was replaced by a formyl, aminoacetyl, or thioacetyl group.

Preparative Chemical Synthesis of CMP-Activated 9-Substituted Sialic Acid Analogs

A new method (this volume, [12][29]) which proceeds by direct coupling of reactive fluorescent dyes to the amino group of CMP-9-amino-NeuAc affords CMP-9-fluoresceinyl-NeuAc in superior yield (90–95%) compared to the enzymatic procedure. The CMP-9-acetamido-NeuAc can be prepared by direct acylation of CMP-9-amino-NeuAc, taking advantage of the novel approach developed for the synthesis of fluorescent and photoactivatible CMP-glycosides. Synthesis starts from CMP-9-amino-NeuAc synthesized enzymatically and purified as outlined below; it is dissolved in 0.1 *M* $NaHCO_3$, pH 9.5, at a concentration of 10–15 m*M*. Acetic anhydride is added in 40-fold molar excess, and the mixture maintained above pH 7.5. The acetylation can be monitored by analytical HPLC at 275 nm as described below, which differentiates the starting compound CMP-9-amino-NeuAc from the CMP-9-acetamido-NeuAc formed. The reaction is complete after 15 min at 37°. The conversion yield is markedly improved compared to the enzymatic CMP-activation (95 versus 40%), and decomposition of the CMP-glycoside during the reaction remains below 5%.

Purification of CMP-Activated Sialic Acid Analogs

Purification of each CMP-activated product is performed as described,[10–12] employing semipreparative HPLC on an aminopropyl-phase column (0.7 × 25 cm, 5 μm particle size) isocratically operated with 15 m*M* KH_2PO_4–acetonitrile (45/55 v/v). The contaminating inorganic phosphate remaining after HPLC separation is precipitated for at least 1 hr at 4° by addition of ethanol in 5-fold excess. After the solution is centrifuged, ethanol is carefully evaporated from the supernatant. Final purification and desalting are achieved by gel filtration on a column (2.5 × 100 cm) of BioGel P-2 (200–400 mesh, Bio-Rad, Richmond, CA) run with twice-distilled water.

Overall yields of pure CMP-NeuAc analogs amount to 35–55%, except for CMP-9-acetamido- and CMP-9-fluoresceinyl-NeuAc[10,12] which are obtained in 20 and 17% yield, respectively, owing to the low *enzymatic* conversion. Both CMP-glycosides can be obtained in about 55% overall yield by *chemical* derivatization of CMP-9-amino-NeuAc and subsequent purification, carried out as described above in the case of CMP-9-acetamido-NeuAc and slightly modified in case of the fluorescent CMP-glyco-

[29] R. Brossmer and H. J. Gross, European Patent 91906420.4-2110; patent pending in Japan and the United States.

side (see [12], this volume). Purified CMP-glycosides may be stored lyophilized or in a neutral solution above pH 7.5 at $-20°$ and $-80°$ for several months without significant decomposition.

Characterization of CMP-Activated Sialic Acid Analogs

The synthetic CMP-glycosides are characterized by analytical HPLC systems after acid hydrolysis (1 *N* HCl, 60 min, room temperature).[10-12] The HPLC system consists of a Consta Metric III pump, a Spectro Monitor III and a Shimadzu C-R6A chromatointegrator.[10] The CMP-glycosides and CMP liberated are measured at 275 nm employing an aminopropylphase column (0.4 × 5.0 cm; 5 μm particle size) operated isocratically with 15 m*M* KH_2PO_4–acetonitrile (50/50 v/v) at 3–4 ml/min; sialic acid analogs released from CMP-glycosides or from neoglycoproteins prepared (see below) can be identified at 200 nm according to the retention time and quantified with respect to corresponding external standards in the effluent of a Spherisorb-NH_2 column (0.4 × 25 cm, 5 μm particle size) run isocratically with 15 m*M* KH_2PO_4–acetonitrile (25–30/75–70 v/v) at 2–3 ml/min.[10,13] After acid hydrolysis of each CMP-NeuAc analog, the molar ratio of CMP to the respective NeuAc analog proves to be equivalent. The concentration of CMP-glycoside is calculated on the basis of the CMP liberated.

Preparations of CMP-glycosides are contaminated by 2–6% CMP and 4–7% sialic acid analog as determined by the HPLC systems. As CTP is known to be a potent inhibitor of sialyltransferases, this nucleotide is determined by analytical HPLC employing a DEAE column (0.4 × 12.5 cm, 5 μm particle size), operated with 35 m*M* KH_2PO_4; the CTP content of CMP-glycosides is always less than 0.2%. The inorganic phosphate in each preparation measured according to Ames[30] is below 10%. The CMP-NeuAc analogs synthesized may be futher analyzed for impurities by TLC.[31]

Independently, the chemical structure of CMP-9-amino- and CMP-9-acetamido-NeuAc is confirmed by 500 MHz ^{1}H NMR spectroscopy.[10]

Preparation of Soluble Neoglycoconjugates by Enzymatic Transfer of Sialic Acid Analogs

Principle

After enzymatic desialylation of several human plasma glycoproteins corresponding neoglycoproteins are obtained by subsequent sialylation of

[30] B. N. Ames, this series, Vol. 8, p. 115.

[31] H. H. Higa and J. C. Paulson, *J. Biol. Chem.* **260,** 8838 (1985).

complex type glycans with synthetic sialic acid in α2,6-linkage employing purified rat liver Galβ1,4GlcNAc α-2,6-sialyltransferase (EC 2.4.99.1 CMP-*N*-acetylneuraminate–β-galactoside α-2,6-sialyltransferase).[10–13,16] *In vitro* sialylation of the ganglioside acceptor G_{M1} by Galβ1,3GalNAc α-2,3-sialyltransferase (EC 2.4.99.4, CMP-*N*-acetylneuraminate–β-galactoside α-2,3-sialyltransferase) purified from porcine liver and synthetic CMP-glycosides affords neoglycolipids. To achieve a high resialylation the transfer assay is kinetically optimized predominantly by increasing the amount of enzyme, incubation time, and molar excess of donor over acceptor substrate. As a result, the galactose sites of glycoproteins are masked to an extent of 60–90% by synthetic sialic acid analogs, differently substituted at the C-9, C-5, or C-4 positions, compared to a sialylation yield of 70–80% with the parent NeuAc. Enzymatic conversion of G_{M1} to the corresponding G_{D1a} carrying a terminal 5-aminoacetyl-Neu succeeds in about 40% yield.

The substituents introduced specifically modify certain biological and chemical properties of the sialic acid, such as the susceptibility to cleavage by sialidases and *O*-acetylesterase[16,32] as well as the hydrophobicity, space-filling, and hydrogen-bonding capacity. Transfer of 9-amino-NeuAc is of special biological interest as it renders glycan chains of neoglycoproteins resistant to sialidase catabolism.

Enzymatic Desialylation of Acceptor Glycoprotein

Enzymatic desialylation[16] is preferred over acidic hydrolysis to preserve physiological protein conformations. The reaction mixture containing 50 m*M* sodium acetate, pH 5.5, 7 m*M* $CaCl_2$, 0.01% sodium azide, 50 mg of α_1-acid glycoprotein purified from human serum, and 1 U *Clostridium perfringens* sialidase immobilized on agarose (1 U per 1 ml agarose) is dialyzed at 37° for 24 hr against a buffer consisting of 50 m*M* sodium acetate, pH 5.5, 7 m*M* $CaCl_2$, 0.01% sodium azide (two buffer changes) to remove the NeuAc released. Subsequently, the mixture is dialyzed for 24 hr at 4° against twice-distilled water and lyophilized. The asialoglycoprotein is then dissolved in twice-distilled water. The other asialoglycoproteins shown in Table II were prepared accordingly, but on a smaller scale.

The asialoglycoprotein preparations contain less than 5% bound NeuAc as determined by the thiobarbituric acid method.[33] Acceptor sites are given in terms of the galactose content of the respective asialoglycoprotein

[32] G. Herrler, H. J. Gross, A. Imhof, R. Brossmer, G. Milks, and J. C. Paulson, *J. Biol. Chem.* **267,** 12501 (1992).

[33] L. Warren, *J. Biol. Chem.* **234,** 1971 (1959).

or ganglioside preparation determined after acid hydrolysis (1 *N* HCl for 3 and 4 hr at 100°) using the galactose dehydrogenase assay.[34]

Procedure for Analytical Resialylation

To investigate the resialylation efficiency of rat liver α-2,6-sialyltransferase for different sialic acid analogs, the analytical transfer assay is modified compared to kinetic standard assays[35] in order to favor complete saturation of the glycan acceptor. Compared to kinetic procedures the pH value of the reaction mixture is raised from pH 6.5 to 6.7 to enhance the stability of CMP-glycosides, the amount of acceptor glycoprotein is reduced thus gaining a 24-fold molar excess of the donor CMP-glycoside over the galactose acceptor sites, a high enzyme concentration is applied (22 mU/ml), and the incubation time is extended to at least 17 hr.

The reaction mixture (160 μl) contains 55 m*M* sodium cacodylate, pH 6.7, 0.16 mg Triton X-100, 1 mg bovine serum albumin (BSA), 22 μg asialo α_1-acid glycoprotein (9.5 nmol galactose acceptor sites) or 80–90 μg of another asialoglycoprotein (see Table II) (9–10 nmol galactose acceptor sites), and 1.5 m*M* of the respective CMP-glycoside. The reaction starts after addition of 3.5 mU α-2,6-sialyltransferase from rat liver, which may be purified[36,37] or purchased commercially. The assay mixture for the enzymatic synthesis of neoglycolipid is composed as above but contains 0.1 mg G_{M1} (47 nmol terminal galactose) as acceptor, 0.16 mg BSA, 0.2 mU Galβ1,3GalNAc α-2,3-sialyltransferase (1.25 mU/ml) purified from porcine liver as described.[38] Corresponding reference solutions are composed identically, but lack the respective sialyltransferase.

The tubes are incubated for 17 hr at 37° and the glycoprotein or the ganglioside is precipitated subsequently by the addition of 1.3 ml cold 1% phosphotungstic acid in 0.5 *N* HCl. The sediment is washed twice with phosphotungstic acid and finally dissolved in 100 μl of 0.2 *M* NaCl as described earlier.[10,13]

To determine the resialylation efficiency, sialic acid analogs transferred to asialoglycoprotein or ganglioside are released by acid hydrolysis (0.1 *N* HCl, 60 min, 80°) and subsequently quantified by the thiobarbituric acid assay[33] as well as by analytical HPLC at 200 nm with respect to appropriate external standards.[10–13] For the exact quantification of sialyl transfer, standards of each analog are also exposed to acid hydrolysis to consider the

[34] K. Wallenfels and G. Kurz, this series, Vol. 9, p. 112.
[35] J. Weinstein, U. de Souza-e-Silva, and J. C. Paulson, *J. Biol. Chem.* **257,** 13845 (1982).
[36] J. Weinstein, U. de Souza-e-Silve, and J. C. Paulson, *J. Biol. Chem.* **257,** 13835 (1982).
[37] U. Sticher, H. J. Gross, and R. Brossmer, *Biochem. J.* **253,** 577 (1988).
[38] U. Sticher, Ph.D. Thesis, Universität Heidelberg (1992).

derivative-specific decomposition during this procedure. The acid-labile 4-deoxy-NeuAc can be released only enzymatically and is quantified by the analytical HPLC system at 200 nm.[11]

Comments

Neoglycoprotein Glycans Terminated by Synthetic Sialic Acid Analogs. Asialo α_1-acid glycoprotein serves to establish the preparation of a neoglycoprotein bearing complex-type lactosamine glycans terminated to the highest possible extent by synthetic sialic acids. The resialylation values obtained applying the donor substrates shown in Figs. 3 and 4 and the Galβ1,4GlcNAc α-2,6-sialyltransferase from rat liver are summarized in Table I. The galactose sites of asialo α_1-acid glycoprotein are saturated to the same extent of about 75–85% by incorporation of NeuAc, 9-acetamido-NeuAc, 9-thioacetamido-NeuAc, and 9-hexanoylamido-NeuAc, 5-formyl-Neu, 5-aminoacetyl-Neu, 5-thioacetyl-Neu, and 4-deoxy-NeuAc. Sialyl transfer of 9-azido-NeuAc and 9-benzamido-NeuAc reaches even about 90%, whereas 9-amino-NeuAc is transferred by 15% lower than the value for parent NeuAc. A higher excess of donor CMP-glycoside (48 nmol donor/nmol galactose sites) does not further improve

TABLE I
ANALYTICAL SCALE RESIALYLATION OF ASIALO α_1-ACID GLYCOPROTEIN BY SIALYLTRANSFERASE[a]

Donor substrate	Resialylation (%)
CMP-NeuAc[b]	80–83
CMP-9-azido-NeuAc[b]	91
CMP-9-amino-NeuAc[b]	66
CMP-9-acetamido-NeuAc[b]	82
CMP-9-thioacetamido-NeuAc	76
CMP-9-hexanoylamido-NeuAc[b]	84
CMP-9-benzamido-NeuAc[b]	95
CMP-5-*N*-formyl-Neu	78
CMP-5-*N*-aminoacetyl-Neu	86
CMP-5-*N*-thioacetyl-Neu	75
CMP-4-deoxy-NeuAc[c]	82

[a] The resialylation value is given as the percentage of total galactose acceptor sites and represents the average of two to five independent experiments using rat liver Galβ1,4GlcNAc α-2,6-sialyltransferase.
[b] Values taken from Ref. 13.
[c] Value taken from Ref. 11.

the resialylation efficiency.[13] For comparison, the sialylation state of native α_1-acid glycoprotein is in the range of 85–90%.

The high resialylation of the asialoglycoprotein by 9-benzamido-NeuAc, even though the analog carries a large substituent, is favored by the low donor K_m value of 30 μM.[13] Resialylation by the sialidase-resistant 9-amino-NeuAc proceeds to a saturation value of 66% despite the very low enzyme affinity (K_m 720 μM; for NeuAc is 45 μM[13]). Notably, 5-aminoacetyl-Neu, which represents another zwitterionic derivative resistant to *Vibrio cholerae* neuramidase, is transferred to the same extent as parent NeuAc.

Resialylation of asialo α_1-acid glycoprotein has also been studied by isoelectric focusing, which allows one to monitor qualitatively the sialylation pattern.[13] The band pattern observed with NeuAc or 9-acetamido-NeuAc is very similar to that of the native glycoprotein with regard to distribution and relative intensity, proving again the effective reconstitution of the native sialylation state.

Neoglycoproteins terminating in special synthetic sialic acids can also be prepared from other biologically interesting asialoglycoproteins using the rat liver α-2,6-sialyltransferase. To this end, resialylation of sialidase-resistant 9-amino-NeuAc is demonstrated, as termination of glycans by this derivative renders the glycoprotein resistant toward sialidases and thus blocks turnover of the glycan moiety. Table II compares the resialylation of several human plasma proteins with NeuAc and 9-amino-NeuAc; whereas remodeling of the asialo-glycoproteins by NeuAc yields a masking

TABLE II
COMPLEX-TYPE GLYCANS OF DESIALYLATED HUMAN PLASMA GLYCOPROTEINS RESIALYLATED BY SIALYLTRANSFERASE[a]

	Resialylation (%) by	
Asialoglycoprotein	NeuAc	9-Amino-NeuAc
α_1-Acid glycoprotein[b]	80	64
Antithrombin III (AT III)[c]	78	70
Zinc α_2-glycoprotein[b]	72	58
Tissue plasminogen activator[d]	85	60

[a] Resialylation is given as the percentage of total galactose acceptor sites and represents the average of three independent experiments using Galβ1,4GlcNAc α-2,6-sialyltransferase from rat liver.

[b] Kindly provided by Dr. K. Schmid (Boston, Massachusetts).

[c] Kindly provided by Behring Werke (Marburg, Germany).

[d] Kindly provided by Knoll AG (Ludwigshafen, Germany).

of 72–85% of total galactose sites, 9-amino-NeuAc is always 10–30% less effective.[38] Although these proteins bear N-linked glycans of different antennary structure, the resialylation yield is not markedly influenced by these steric features.

Neoglycolipid Glycans Terminated by Sialic Acids Analogs. A model neoglycolipid is obtained by enzymatic sialylation of the terminal galactose of the ganglioside G_{M1} with 9-amino-NeuAc and 5-aminoacetyl-Neu. About 40% of the acceptor G_{M1} is converted to the corresponding G_{D1a} neoglycolipid carrying 5-aminoacetyl-Neu; sialylation with the parent NeuAc to G_{D1a} proceeds to an extent of 50%. In contrast, less than 5% G_{D1a} sialylated by 9-amino-NeuAc can be obtained, a result which is in accord with the very low acceptance of the porcine submaxillary gland Galβ1,3GalNAc α-2,3-sialyltransferase for CMP-9-amino-NeuAc observed in kinetic studies.[13] Corresponding G_{D1a} neoglycolipids carrying other 9- and 5-substituted sialic acid analogs can be prepared accordingly (not shown). Separation of starting G_{M1} from the products may be easily achieved by TLC (silica gel, $CHCl_3/CH_3OH/0.25\%$ $CaCl_2$, 54/43/10, v/v/v).

Procedure for Semipreparative Resialylation

Asialo α_1-acid glycoprotein as acceptor also serves to exemplify the preparation of neoglycoproteins sialylated at the terminal galactose by synthetic analogs on a milligram scale.[16] To achieve economically a high resialylation by the Galβ1,4GlcNAc α-2,6-sialyltransferase from rat liver, the different K_m values determined previously for CMP-NeuAc and CMP-9-amino-NeuAc (45 and 750 μM, respectively[13]) must be considered. Hence, although V_{max} values of both CMP-glycosides are identical, the 20-fold lower affinity for CMP-9-amino-NeuAc is responsible for the increasing inhibition of sialyl transfer by the reaction product CMP; inhibition is especially pronounced at donor concentrations below 1 mM.[13] By applying saturating concentrations of both donor substrates (about 10 times the K_m value), a neoglycoprotein can be prepared from asialo α_1-acid glycoprotein by incorporation of NeuAc and 9-amino-NeuAc exclusively α2,6-linked to terminal galactose according to the transferase specificity. The concentration of CMP-NeuAc is not optimized here to achieve a high resialylation but to kinetically compare transfer of both sialic acids.

The reaction mixture (1 ml) contains 50 mM sodium cacodylate, pH 6.7, 1 mg Triton X-100, 1 mg BSA, 1.5 mg asialo α_1-acid glycoprotein (650 nmol galactose sites), and 0.8 mM CMP-NeuAc or 6.0 mM CMP-9-amino-NeuAc. The reaction is started by the addition of 19.5 mU Galβ1,4GlcNAc α2,6-sialyltransferase (19.5 mU/ml). The time course of the resialylation can be monitored as follows: after appropriate incubation times at 37°, aliquots (100 μl) are withdrawn from the reaction mixture, and the glycoprotein is precipitated as described for the analytical assay. The BSA is included in the reaction mixture only to improve acid precipitation.

After 17 hr at 37° the resialylated glycoprotein may be separated from CMP-glycoside by gel filtration on Sephadex G-50 (column 0.8 × 15 cm, 10 mM NH_4HCO_3 as buffer). The glycoprotein is lyophilized and redissolved in twice-distilled water.[16] The gel-filtration step completely separates donor CMP-glycoside and neoglycoprotein, with preservation of the physiological protein conformation. Sialyl transfer can be calculated as outlined for the analytical assay after acid hydrolysis either by the thiobarbituric acid method or by analytical HPLC at 200 nm, which allows the identification of NeuAc and 9-amino-NeuAc by the retention times.[10]

Comments

The time courses for resialylation of asialo α_1-acid glycoprotein with NeuAc and 9-amino-NeuAc are compared in Fig. 5. Both compounds show an identical efficiency of incorporation: after 7 hr about 65% of available galactose sites are sialylated. Prolonged incubation of up to 24 hr results in a slow increase of NeuAc incorporation to 75%, but no further transfer of 9-amino-NeuAc can be observed. The difference may

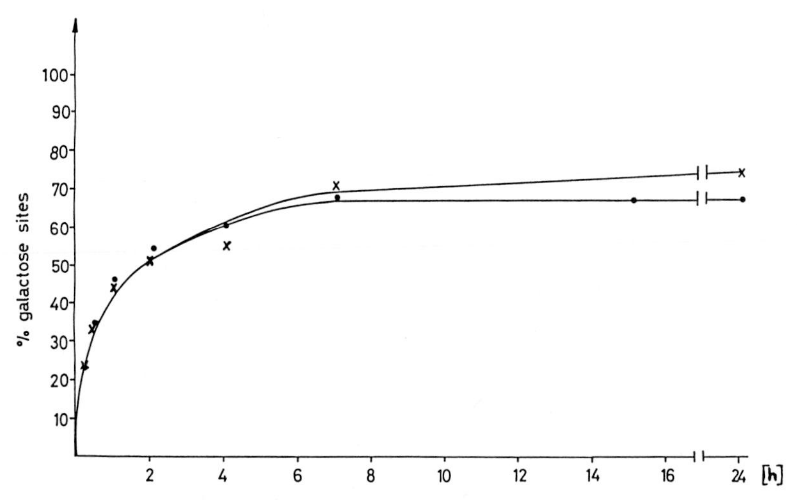

FIG. 5. Time course of semipreparative sialylation of asialo α_1-acid glycoprotein by 9-amino-NeuAc (●) and NeuAc (x), employing rat liver Galβ1,4GlcNAc α-2,6-sialyltransferase.

be explained by the low affinity for the synthetic CMP-glycoside, which strongly favors product inhibition by the increase in CMP resulting from enzyme action. Higher saturation values might be achieved employing a larger, albeit uneconomical, molar excess of donor CMP-glycoside versus acceptor substrate. For the kinetic reasons explained, 9-amino-NeuAc represents the most difficult derivative for effective enzymatic sialyl transfer. Resialylation of asialoglycoproteins by other synthetic sialic acids which do not suffer from low donor affinity of the respective sialyltransferase proceeds at least to the value achieved with the parent NeuAc applying 1–1.5 m*M* donor substrate in the assay described. For example, final resialylation of asialo α_1-acid glycoprotein with 9-fluoresceinyl-NeuAc by the rat liver enzyme amounts to 65%; the corresponding value obtained with NeuAc is 61%.[12]

Sialylation of Cell Surface Glycoconjugates by Enzymatic Transfer of Sialic Acid Analogs

Principle

The approach to remodel soluble glycoconjugates by enzymatic sialylation has been applied to the plasma membrane of native cells. Synthetic sialic acids are transferred from respective CMP-glycosides to surface-exposed lactosamine glycans by rat liver Galβ1,4GlcNAc α-2,6-sialyltransferase.[17,39,40] The principle of enzymatic surface sialylation is exemplified for a human B-cell line and CMP-9-fluoresceinyl-NeuAc; determination of membrane-bound fluorescence at the level of a single cell can be accomplished by flow cytometry.

Procedure

The human B-cell line IM-9 [American Type Culture Collection (ATCC), Rockville, MD, CCL 159] serves as a cellular model to establish surface sialylation of human hematopoietic cells by synthetic sialic acids.[17,39,40] Prior to enzymatic sialylation IM-9 cells are washed 2 times and resuspended in CIB, which is composed of PBS (phosphate-buffered saline), pH 6.8, supplemented with 5 mg/ml BSA and 1 mg/ml glucose. Asialo IM-9 cells are obtained by enzymatic desialylation of 10^7 cells suspended in 1 ml CIB for 30 min at 37° using 200 mU sialidase from *Vibrio cholerae;* subsequently cells are washed 3 times in PBS before sialylation to remove sialidase.

[39] C. Bollheimer, R. Kosa, A. Sauer, R. Schwartz-Albiez, and H. J. Gross, *Biol. Chem. Hoppe-Seyler* **374,** GC 1a (1993).

[40] R. Kosa, M.D. Thesis, Universität Heidelberg (1994).

The sialylation assay (100 μl) is performed in CIB containing 2.5 × 10^5 untreated or desialylated IM-9 cells, 50 *μM* CMP-9-fluoresceinyl-NeuAc, 200 *μM* 2,3-anhydro-NeuAc to specifically block remaining sialidase, and 3.5 mU of rat liver α-2,6-sialyltransferase (35 mU/ml).[35,37] Prior to use the enzyme preparations containing 0.1–0.5% Triton X-100 or Triton CF-54 are depleted of detergent.[17] To this end, a suspension of Bio-Beads SM-2 (Bio-Rad; washed with CIB) preequilibrated at a ratio of 1 mg (wet weight) per 15 μl in CIB is added to the sialyltransferase preparation (2 U/ml) at a volume ratio of 15 : 1, and the mixture is set on ice for 45–60 min. After sedimentation of the beads, about 80% of the enzyme activity is recovered in the supernatant. Detergent is thereby sufficiently depleted; cells remain viable to about 90% during the 90-min incubation at 37° as checked by Trypan Blue exclusion.

To monitor the time course of sialyl transfer, aliquots are withdrawn after appropriate incubation times, cells are washed twice in PBS, and surface-bound fluorescence is measured by flow cytometry using a standard flow cytometer equipped with an argon laser (488 nm) and an appropriate software program.

Comments

The time course for transfer of 9-fluoresceinyl-NeuAc to the surface of untreated or desialylated IM-9 cells is depicted in Fig. 6. The amount of transferred fluorescent NeuAc is expressed by the mean fluorescence

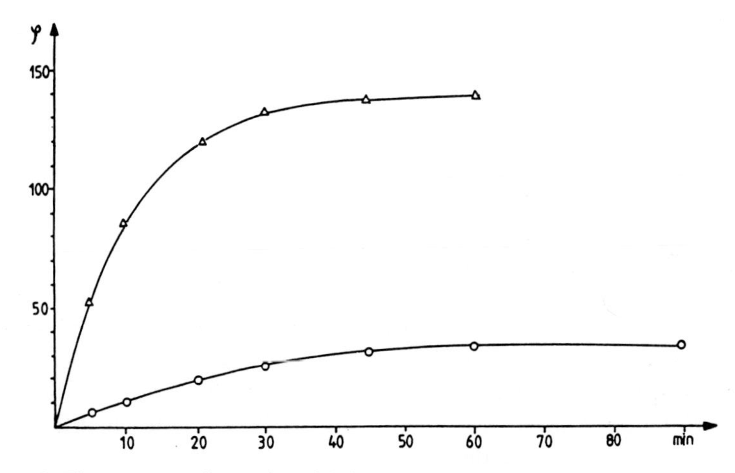

FIG. 6. Time course of transfer of 9-fluoresceinyl-NeuAc onto the surface of intact untreated (○) or desialylated (△) IM-9 cells by rat liver Galβ1,4GlcNAc α-2,6-sialyltransferase.

value per single cell (φ) calculated by the cytometer software. Sialylation of asialo cells leads to 4 times higher final surface-bound fluorescence compared to untreated cells and is restricted to the cell surface, as 85–90% of the fluorescence incorporated can be released after subsequent treatment of labeled cells with sialidase from *Vibrio cholerae* (200 mU/ml in CIB, 90 min, 37°).[40]

The enzymatic labeling of cell surface glycans by fluorescent sialic acid analogs according to the acceptor specificity of a certain sialyltransferase represents a novel approach to study plasma membrane glycosylation by flow cytometry at the level of a single cell. The method further allows simultaneous viability staining and detection of distinct marker antigens by suitably labeled monoclonal antibodies.[39]

In addition, lactosamine glycans at the surface of human native erythrocytes or IM-9 cells, untreated or desialylated, can be remodeled by enzymatic introduction of 9-azido-, 9-amino-, and 9-acetamido-NeuAc[17,40]; furthermore, a receptor analog for influenza C virus can be restored on desialylated Madin–Darby canine kidney (MDCK) cells via transfer of 9-acetamido-NeuAc by Galβ1,4GlcNAc α-2,6-sialyltransferase from rat liver[32]; such remodeled MDCK cells allow both virus infection and propagation.[32,41] In contrast, although MDCK cells resialylated with 9-thioacetamido-NeuAc accept influenza C virus, infection does not take place.[23]

Concluding Remarks

Neoglycoproteins carrying terminal 9-amino-NeuAc in α2,6-linkage have been derived from several human plasma glycoproteins by enzymatic remodeling. Termination of glycoconjugate glycans by the sialidase-resistant sialic acid analog deserves special interest as it blocks catabolism of glycoconjugate glycans by bacterial, viral, or mammalian sialidases. Among several purified sialyltransferases studied, only the lactosamine-specific α-2,6-sialyltransferases effectively transfer the zwitterionic analog.

Neoglycoproteins have been prepared from asialo α_1-acid glycoprotein by resialylation with four 9-N-acylated sialic acid analogs (Fig. 3). The chemical structure of 9-acetamido-NeuAc and 9-thioacetamido-NeuAc closely resembles that of 9-*O*-acetyl-NeuAc, the natural cell surface receptor determinant for influenza C virus. However, in contrast to the natural receptor, the analogs resist cleavage by the viral receptor-destroying *O*-acetylesterase.[23,32] 9-Benzamido-NeuAc and 9-hexanoylamido-NeuAc carry a bulky hydrophobic substituent. Notably, despite distinctly differ-

[41] G. Herrler, H. J. Gross, and R. Brossmer, *Biol. Chem. Hoppe-Seyler* **371,** 791 (1990).

ent shapes, both compounds can be enzymatically incorporated to the same extent as the parent NeuAc. Based on these results it is feasible to prepare soluble or membrane-attached glycoproteins containing an outer shell of hydrophobic groups in the oligosaccharide chains. Enzymatic activation and transfer of 4-deoxy-NeuAc prove the minor importance of a hydroxy group at C-4 for substrate recognition.[11] The substitutions at position C-5 (Fig. 4) leading to 5-*N*-aminoacetyl-, 5-*N*-formyl-, and 5-*N*-thioacetyl-neuraminic acid modify the acetyl group of sialic acid which, as is now well established, represents a major structural determinant for a number of receptor recognitions. One interesting example is the study with the hemagglutinin of the influenza A virus X-31 strain which shows 5-thioacetyl-Neu methyl α-glycoside to bind more strongly compared to the natural receptor.[42]

The results reveal that, within the specificity limitations of sialyltransferases, enzymatic synthesis is a suitable procedure for introducing sialic acid analogs into defined sequences of soluble or plasma membrane-bound glycoproteins or glycolipids according to the respective acceptor specificity. In particular, sialic acid analogs specifically modified at the C-4, C-5, or C-9 positions may thus be tailored to influence a distinct recognition event.[23,32,42–50]

[42] D. Machytka, I. Kharitonenkov, R. Isecke, P. Hetterich, R. Brossmer, R. A. Klein, H.-D. Klenk, and H. Egge, *FEBS Lett.* **334,** 117 (1993).

[43] H. J. Gross and R. Brossmer, *Clin. Chim. Acta* **197,** 237 (1991).

[44] T. J. Pritchett, R. Brossmer, U. Rose, and J. C. Paulson, *Virology* **160,** 502 (1987).

[45] R. Brossmer, M. Wagner, and E. Fischer, *J. Biol. Chem.* **267,** 8752 (1992).

[46] E. Fischer, N. Q. Khang, G. Letendre, and R. Brossmer, *Glycoconjugate J.* **11,** 51 (1994).

[47] G. T. J. van der Horst, U. Rose, R. Brossmer, and F. W. Verheijen, *FEBS Lett.* **277,** 42 (1990).

[48] G. T. J. van der Horst, G. M. S. Mancini, R. Brossmer, U. Rose, and F. W. Verheijen, *J. Biol. Chem.* **265,** 10801 (1990).

[49] S. Kelm, J. Paulson, U. Rose, R. Brossmer, W. Schmid, B. Bandgar, E. Schreiner, M. Hartmann, and E. Zbiral, *Eur. J. Biochem.* **205,** 147 (1992).

[50] R. Brossmer, H. J. Gross, U. Rose, U. Sticher, P. Mirelis, F. Gradel, A. Imhof, and A. Kovac, *in* "GBF Monographs Vol. 15, Protein Glycosylation: Cellular, Biotechnological and Analytical Aspects" (H. S. Conradt, ed.), p. 209 (VCH, 1991).

[12] Fluorescent and Photoactivatable Sialic Acids

By REINHARD BROSSMER and HANS JÜRGEN GROSS

Introduction

Structures of sialoglycans and expression of sialyltransferases are of growing interest in cell biology since important receptor interactions and cell adhesion events are mediated by distinct sialylated glycan sequences.[1-5] Detection of sialoglycans by labeling the sialic acid moiety via incorporation of fluorescent dyes requires chemical pretreatment.[6,7] The method most frequently used for this purpose is periodate oxidation of glycolipids or glycoproteins, which affords sialic acids with a shortened side chain (at C-8 or C-7) with concomitant generation of a reactive aldehyde group. A serious drawback of this method is that the aldehyde also leads to unspecific side reactions, and some important biological functions of sialoglycoconjugates may be lost by truncating the exocyclic glyceryl side chain. In addition, labile glycoproteins may suffer and intact cells may be damaged during periodate oxidation. For many analytical applications, however, a major disadvantage is the lack of selectivity of the chemical procedure by which sialic acids are labeled irrespective of the subterminal glycan sequence.

Photoactivatable sialic acids may be covalently tethered to biologically active glycoconjugates or membrane-bound receptors by enzymatic transfer followed by photoinduced cross-linking with the respective ligand which may lead to more insight into sialic acid-dependent recognition events. Use of chemical procedures for the synthesis of such neoglycoconjugates may lead to the same problem as outlined above for the synthesis of the fluorescent pendants.[6,7]

Sialyltransferases are Golgi-located enzymes which terminate the synthesis of glycoconjugate glycans by transferring sialic acids from the corresponding donor CMP-glycoside in α-glycosidic linkage (α2,3, α2,6, or α2,8) to a nonreducing Gal-, GalNAc-, GlcNAc-, or sialic acid residue in

[1] J. C. Paulson, G. N. Rogers, S. M. Caroll, H. H. Higa, T. Pritchet, G. Milks, and S. Sabesan, *Pure Appl. Chem.* **56,** 797 (1984).
[2] G. N. Rogers, G. Herrler, J. C. Paulson, and H.-D. Klenk, *J. Biol. Chem.* **261,** 5947 (1986).
[3] G. Ashwell and A. G. Morell, *Adv. Enzymol. Relat. Areas Mol. Biol.* **41,** 99 (1974).
[4] Y. Pilatte, J. Bignon, and C. R. Lambré, *Glycobiology* **3,** 201 (1993).
[5] T. A. Springer, *Cell (Cambridge, Mass.)* **76,** 301 (1994).
[6] S. Spiegel, this series, Vol. 138, p. 313.
[7] M. Wilchek and E. Bayer, this series, Vol. 138, p. 429.

the glycan acceptor. In the past, detection of sialyltransferase activity required radiometric assays employing ^{3}H- or ^{14}C-labeled cytidine-5′-monophospho-5-*N*-acetyl-D-neuraminic acid (CMP-NeuAc). In general, the activity of the enzymes is very low in crude tissue preparations, cellular extracts, or body fluids. Therefore, comparative studies at saturating concentrations of the donor CMP-NeuAc require radiolabeling with high specific activity. Studies performed at subsaturating donor levels, frequently used to reduce consumption of radiolabeled CMP-glycoside, may lead to kinetically invalid data.[8,9]

As an approach to overcome the methodological difficulties inherent in common sialyltransferase assays and in chemical labeling of sialoglycans, fluorescent and photoactivatable sialic acids have been synthesized which could be enzymatically transferred from the respective CMP-glycoside to soluble or membrane-bound acceptor glycoconjugates.[8,10] Kinetic studies have proved the general suitability of CMP-sialic acid analogs to act as donor substrates for purified sialyltransferases.[8,10–13] Because of the defined acceptor specificity, these enzymes in combination with fluorescent or photoactivatable donor CMP-glycosides are excellent tools for selective introduction of a fluorescence or photoactivatable substituent to a distinct oligosaccharide sequence.[8,10,13] Further, the fluorescent CMP-sialic acids allow to establish a very sensitive fluorometric sialyltransferase assay.[8]

This chapter describes the synthesis, kinetic properties, and biochemical applications of fluorescent and photoactivatable sialic acids developed for a specific glycan labeling mediated by sialyltransferases.

Synthesis, Purification, and Characterization of Fluorescent CMP-Activated Sialic Acids

Principle

Several CMP-activated fluorescent sialic acids substituted at the C-9 (Fig. 1) or C-5 positions (Fig. 2) of the sialic acid moiety by different fluorescent dyes are prepared by combining enzymatic and chemical syn-

[8] H. J. Gross, U. Sticher, and R. Brossmer, *Anal. Biochem.* **186,** 127 (1990).

[9] M. M. Weiser, W. D. Klohs, D. K. Podolsky, and J. R. Wilson, *in* "The Glycoconjugates IV" (M. J. Horowitz, ed.), p. 301. 1982. Academic Press, New York.

[10] H. J. Gross and R. Brossmer, *Eur. J. Biochem.* **177,** 583 (1988).

[11] H. J. Gross, A. Bünsch, J. C. Paulson, and R. Brossmer, *Eur. J. Biochem.* **168,** 595 (1987).

[12] H. J. Gross, U. Rose, J. M. Krause, J. C. Paulson, K. Schmid, R. E. Feeney, and R. Brossmer, *Biochemistry* **28,** 7386 (1989).

[13] R. Brossmer and H. J. Gross, European Patent 91906420.4-2110; patent pending in Japan and the United States.

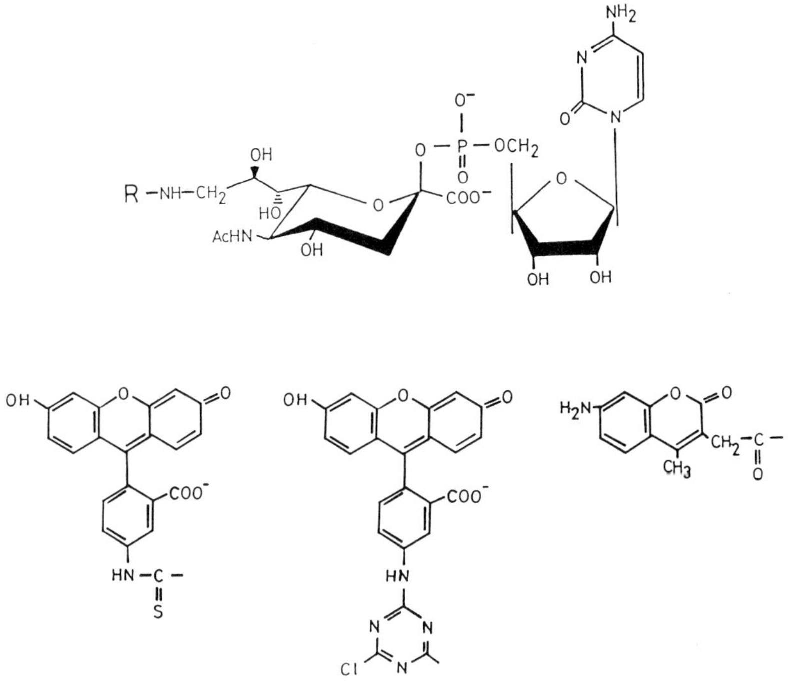

FIG. 1. Structure of CMP-sialic acids substituted by different fluorescent dyes (R) at C-9 of the sialic acid moiety. Substitution with R represents, from left to right, CMP-9-fluoresceinyl-NeuAc, CMP-9-(fluoresceinylaminomonochlorotriazinyl)amino-NeuAc (CMP-9-MTAF-NeuAc), and CMP-9-(7-amino-4-methylcoumarinyl)acetamido-NeuAc (CMP-9-AMCA-NeuAc).

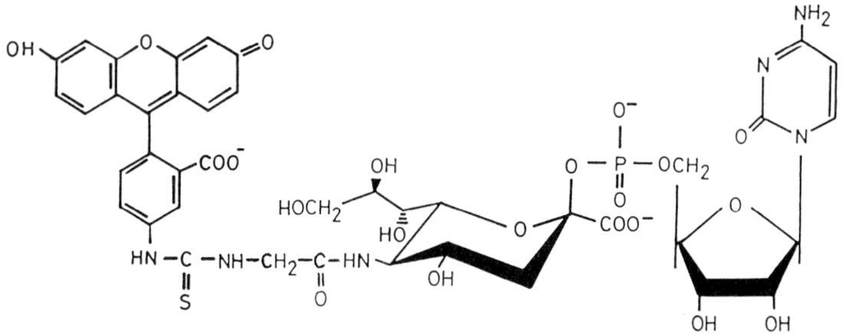

FIG. 2. Structure of CMP-5-(*N*-fluoresceinyl)thioureido-acetyl-D-neuraminic acid (CMP-5-fluoresceinyl-Neu).

thesis. Starting from enzymatically synthesized cytidine-5′-monophospho-5-*N*-acetyl-9-amino-9-deoxy-D-neuraminic acid (CMP-9-amino-NeuAc) or cytidine-5′-monophospho-5-*N*-aminoacetylneuraminic acid (CMP-5-*N*-aminoacetyl-Neu) (for structure, synthesis, and purification, see this volume, [11]), the chemical step then involves direct coupling of the deprotonated amino group of the sialic acid moiety with amino-reactive fluorescent dyes at alkaline pH. The conversion yield ranges from 75 to 95%. The procedure does not affect the amino group of cytosine and avoids cleavage of the labile glycosidic linkage.

Fluorescent CMP-sialic acids are purified by semipreparative high-performance liquid chromatography (HPLC) and are characterized, with respect to concentration (via the CMP moiety) and contaminants (CMP, respective fluorescent sialic acid), by analytical HPLC as outlined for other synthetic CMP-sialic acid analogs (see [11] in this volume).[10–12]

Synthesis of Fluorescent CMP-Sialic Acids

The starting compounds CMP-9-amino-NeuAc and CMP-5-*N*-aminoacetyl-Neu are prepared by enzymatic conversion of the respective sialic acid analog employing CMP-sialic acid synthase (EC 2.7.7.43, acylneuraminate cytidylyltransferase) from bovine brain, and the compounds are purified as outlined elsewhere in this volume (see [11]).[11] Despite chemical lability, fluorescent CMP-glycosides can be obtained by direct coupling of suitable amino-reactive dyes to CMP-9-amino-NeuAc or CMP-5-*N*-aminoacetyl-Neu. Fluorescent dyes "activated" to contain isothiocyanate, *N*-hydroxysuccinimide ester, or dichlorotriazinyl groups may be purchased or prepared chemically from the respective parent dyes (see, e.g., section on preparation of photoactivatable CMP-sialic acid).

Chemical conversion starts from purified CMP-9-amino-NeuAc or CMP-5-*N*-aminoacetyl-Neu dissolved just before coupling in 0.15 *M* sodium hydrogen carbonate, pH 9.0–9.5, at a concentration of 10–15 m*M*. A solution of the amino-reactive fluorescein isothiocyanate (FITC, isomer I), 7-amino-4-methylcoumarin-3-acetic acid *N*-hydroxysuccinimide ester (AMCA-NHS), or dichlorotriazinylaminofluorescein (DTAF) in dry *N,N*-dimethylformamide (DMF; analytical grade) is added to achieve a 5- to 10-fold molar excess over the starting CMP-glycoside. The final DMF concentration in the reaction mixture should not exceed 35% to avoid precipitation of the CMP-glycoside. For an efficient reaction, the solution must be maintained at pH 9.0–9.5, since only a deprotonated amino group at position C-5 or C-9 is reactive. The conversion is monitored by analytical HPLC at 275 nm as described below, which differentiates the initial CMP-9-amino-NeuAc or CMP-5-*N*-aminoacetyl-Neu from the fluorescent CMP-glycosides formed. The reaction is complete after 1–2 hr at 37°, except for coupling of DTAF which requires at least 15 hr at 37°. Before HPLC separation excess ethanolamine is added to the reaction mixture in order to block remaining "activated" dye.

The products are obtained in 90–95% yield when applying isothiocyanate- or *N*-hydroxysuccinimide ester-activated dyes, whereas conversion in the case of DTAF reaches only 75%.[13] Decomposition of CMP-glycosides during the reaction remains below 5%, and the amino group of the cytosine moiety is not affected. Enzymatic synthesis of the corresponding CMP-glycoside by CMP-sialic acid synthase starting from 5-*N*-acetyl-9-deoxy-9-(*N'*-fluoresceinyl)thioureido-D-neuraminic acid (9-fluoresceinyl-NeuAc) achieves only 17% overall yield.[10]

The procedure affords CMP-9-fluoresceinyl-NeuAc, CMP-9-(fluoresceinylaminomonochlorotriazinyl)amino-NeuAc (CMP-9-MTAF-NeuAc), CMP-9-(7-amino-4-methylcoumarinyl)acetamido-NeuAc (CMP-9-AMCA-NeuAc), and CMP-5-fluoresceinyl-Neu. The corresponding structures are shown in Figs. 1 and 2; the fluorescent substituents are not drawn to scale with respect to the CMP-sialic acid moiety.

Purification of Fluorescent CMP-Sialic Acids

Purification of each CMP-glycoside synthesized is performed essentially as described for other CMP-sialic acid analogs (see [11], this volume)[10–12] employing primarily semipreparative HPLC on an aminopropyl-phase column (0.7 × 12.5 cm, 5 μm particle size) isocratically operated with 40–45 : 60–55 (v/v) 15 m*M* KH_2PO_4–acetonitrile. The peak of the fluorescent CMP-glycoside is collected from the effluent monitored at 275 nm, and acetonitrile is evaporated under reduced pressure at 30° while maintaining the solution at pH 7.0–7.5. After addition of ethanol (analytical grade) in 6-fold (v/v) excess, inorganic phosphate is precipitated for 24 hr at 4° and then the mixture is centrifuged. To dissolve coprecipitated fluorescent CMP-glycoside, the sediment is extracted by a mixture of 1 : 6 (v/v) water–ethanol. Phosphate is again precipitated for 24 hr at 4° followed by centrifugation. Ethanol is carefully removed from the combined supernatants under reduced pressure. To further reduce the phosphate content, precipitation and extraction of the sediment are repeated once more in the same way; pH 7.0–7.5 must be maintained during the whole procedure. Final purification and desalting by gel filtration on BioGel P-2 (see [11], this volume) are omitted as fluorescent compounds markedly adsorb to the gel.

If necessary, to further reduce contaminating phosphate a final purification step may be included as follows. The preparation of fluorescent

CMP-glycoside obtained after ethanol evaporation (0.2–2 ml) is diluted by an equal volume of acetone (analytical grade), stored at −20° for 24 hr, and then centrifuged. Finally, acetone is carefully removed under reduced pressure.

The overall yield of purified fluorescent CMP-glycosides amounts to 50–70% relative to the starting CMP-9-amino-NeuAc or CMP-5-*N*-aminoacetyl-Neu. The purified fluorescent CMP-glycosides may be stored lyophilized or in less than 15 m*M* solution at pH 7.5 and at −20° to −80° for several months without significant decomposition.

Note: Careful pH control (pH 7.0–7.5) is necessary during the entire purification procedure to avoid decomposition of the CMP-glycosides.

Characterization of Fluorescent CMP-Sialic Acids

Synthetic CMP-glycosides are characterized by analytical HPLC after acid hydrolysis (1 *N* HCl, 60 min, room temperature) essentially as outlined for other synthetic CMP-glycosides (see [11], this volume).[10–12] The CMP-glycosides and CMP liberated are measured at 275 nm employing an aminopropyl-phase column (0.4 × 5.0 cm; 5 μm particle size) operated isocratically with 50 : 50 (v/v) 15 m*M* KH_2PO_4–acetonitrile at 3–4 ml/min. Fluorescent sialic acids released can be differentiated according to the retention times (monitored at 200–250 nm) in the effluent of a Spherisorb-NH_2 column (0.4 × 25 cm, 5 μm particle size) run isocratically with 25–30 : 75–70 (v/v) 15 m*M* KH_2PO_4–acetonitrile at 2–3 ml/min.[8,10] As the fluorescent substituent causes an increase in extinction (fluoresceinyl 3.2-fold, MTAF 4.5-fold, and AMCA 1.2-fold) compared to CMP-NeuAc, the concentration of CMP-glycoside is calculated on the basis of the CMP liberated.[10] The retention times of CMP-9-fluoresceinyl-NeuAc, CMP-9-MTAF-NeuAc, and especially CMP-5-fluoresceinyl-Neu are markedly enhanced compared to CMP-NeuAc or other synthetic CMP-glycosides.[10,11] In general, preparations of CMP-glycosides are contaminated with 2–6% CMP and 4–7% sialic acid analog as determined by analytical HPLC. The level of inorganic phosphate in each preparation is measured according to Ames.[14] The fluorescent CMP-sialic acids synthesized may also be analyzed for impurities by thin-layer chromatography (TLC).[15]

Enzymatic Transfer of Fluorescent Sialic Acids

Principle

The purified fluorescent CMP-sialic acids summarized in Figs. 1 and 2 have been examined as donor substrates for several purified sialyltrans-

[14] B. N. Ames, this series, Vol. 8, p. 115.

[15] H. H. Higa and J. C. Paulson, *J. Biol. Chem.* **260,** 8838 (1985).

ferases. For this purpose a fluorometric enzyme assay has been developed based on the introduction of fluorescent sialic acids to a glycoconjugate acceptor and subsequent separation of the labeled acceptor from fluorescent CMP-glycoside by gel filtration and quantification in the column effluent.[8,13] The fluorescent CMP-glycosides differently substituted at position C-9 (Fig. 1) or C-5 (Fig. 2) are suitable substrates for sialyltransferases independent of their respective linkage and acceptor specificity [K_m values 2- to 40-fold lower and V_{max} of comparable order (10–100%) to that of CMP-NeuAc].

Fluorometric Sialyltransferase Assay

The standard reaction mixture (20–30 μl) contains 62.5 m*M* sodium cacodylate, pH 6.5, 0.1–0.5% Triton X-100, 25–100 μM fluorescent CMP-sialic acid, and appropriate amounts of sialyltransferase.[8] To improve detection of a certain sialyltransferase, the composition of the mixture, especially the detergent, may be modified.[8,12,16]

Acceptor asialoglycoproteins are obtained either by enzymatic desialylation (see this volume [11]) or by mild acid hydrolysis.[10] Asialofetuin (4 mg/ml) carrying both N- and O-linked glycans represents an acceptor suitable for measuring global sialyltransferase activity. Acceptor specificity can be differentiated employing asialo α_1-acid glycoprotein (1.6 mg/ml) containing only N-linked complex-type glycans, bovine asialomucin (3.7 mg/ml), or antifreeze glycoprotein (1.7 mg/ml), both containing only O-linked mucin-type disaccharides. Acceptor saturation of a certain sialyltransferase may require higher concentrations. The specificity toward gangliosides is determined employing G_{M1} [Galβ1,3GalNAcβ1,4(NeuAcα2,3)Galβ1,4Glc-ceramide], G_{D1a}, or G_{D1b} (0.2–0.5 mg/ml). Application of other gangliosides may be hindered by their low solubility in aqueous buffer, especially considering the gel-filtration system for detection described below.

The reaction starts after addition of the respective sialyltransferase preparation (routinely 5–20 μU) and is terminated after an appropriate incubation time (routinely 20–40 min) at 37° by addition of 0.5 *M* CTP (4 μl). Corresponding references are performed without sialyltransferase, with enzyme in the presence of 0.5 *M* CTP (4 μl) to block enzyme activity, or with enzyme in the absence of exogenous acceptor to measure only activity toward the endogenous acceptor present in the sialyltransferase preparation. The latter control is especially important when investigating sialyltransferase activity of tissue preparations or body fluids.[8,16] For each novel sialyltransferase it is recommended that one examine the linearity

[16] H. J. Gross and R. Brossmer, *Clin. Chim. Acta* **197**, 237 (1991).

of the reaction rate with incubation time, as inhibition by the sialylated glycoconjugate may occur, especially above 20% donor substrate consumption.[8]

Fluorometric Detection System

Aliquots (routinely 5–20 μl) of the reaction mixture are applied to a chromatographic system equipped with an HPLC or a peristaltic pump, a fluorescence spectrophotometer (excitation 490 nm/emission 520 nm for CMP-9-fluoresceinyl-NeuAc, CMP-9-MTAF-NeuAc, and CMP-5-fluoresceinyl-Neu; excitation 345 nm/emission 445 nm for CMP-9-AMCA-NeuAc), and a chromatointegrator.[8] Labeled glycoproteins or gangliosides are separated from the donor substrate on a column (0.7 × 16 cm) of Sephadex G-50 operated isocratically with 0.1 *M* Tris-HCl, pH 8.5, 80 m*M* NaCl, 0.2% Nonidet P-40 or Triton X-100 at a flow rate of 1 ml/min. Detergent may be omitted if only glycoprotein acceptors are employed. Glycoconjugates sialylated with fluorescent sialic acids separate well from corresponding donor CMP-glycosides up to at least 20 μl sample. The amount of acceptor-bound fluorescent sialic acid is quantified from the effluent and calculated with respect to an external standard (20–100 pmol) of the corresponding fluorescent donor CMP-glycoside.[8] Incorporation values measured fluorometrically are confirmed by analytical HPLC after hydrolytic release of bound fluorescent sialic acid.[8] When samples are protected from light, the fluorescence emission is consistent for at least 2 days at 4°.

Comments

Kinetic Properties of Sialyltransferases for Fluorescent CMP-Sialic Acids. To prove the general applicability of the fluorescent donor CMP-glycosides with regard to their different structures (Figs. 1 and 2), kinetic properties of sialyltransferases specific for a distinct acceptor sequence [Galβ1,4GlcNAc, Galβ1,4(3)GlcNAc, Galβ1,3GalNAc, or GalNAc] as well as for the type of glycosidic linkage formed (α2,3 or α2,6) are assessed. To this end, sialyltransferases with defined specificity purified from various origins[17–22] are employed using the fluorometric assay system developed.

[17] U. Sticher, H. J. Gross, and R. Brossmer, *Glycoconjugate J.* **8,** 45 (1991).
[18] J. Weinstein, U. de Souza-e-Silva, and J. C. Paulson, *J. Biol. Chem.* **257,** 13835 (1982).
[19] J. E. Sadler, J. I. Rearick, J. C. Paulson, and R. L. Hill, *J. Biol. Chem.* **254,** 4434 (1979).
[20] J. E. Sadler, J. I. Rearick, and R. L. Hill, *J. Biol. Chem.* **254,** 5934 (1979).
[21] U. Sticher, Ph.D. Thesis, Universität Heidelberg (1992).
[22] U. Sticher, H. J. Gross, and R. Brossmer, *Biochem. J.* **253,** 577 (1988).

The kinetic studies[8,10,13] are exemplified by results depicted in Table I which were obtained with two commercially available purified sialyltransferases. Each enzyme acts on a specified glycan sequence. The α-2,6-sialyltransferase from rat liver (EC 2.4.99.1, CMP-*N*-acetylneuraminate–β-galactoside α-2,6-sialyltransferase) sialylates the lactosamine sequence,[23] and the α-2,3-sialyltransferase from porcine liver (EC 2.4.99.4, CMP-*N*-acetylneuraminate–β-galactoside α-2,3-sialyltransferase) the O-linked mucin-type disaccharide[21]; kinetic data of the corresponding α-2,3-sialyltransferase from porcine submaxillary glands,[19] also commercially available, are very similar.[8,13]

The fluorescent CMP-glycosides summarized in Figs. 1 and 2 are well-suited donor substrates for each sialyltransferase examined so far. The kinetic data differ significantly depending on the structure of the fluorescent residue and its position in the sialic acid moiety, as well as on the acceptor and linkage specificity of the sialyltransferase. Relative to the kinetic data of CMP-NeuAc, the affinity of each enzyme for the fluorescent donor substrates is 6-10- or 2-40-fold higher, and V_{max} varies within a range of 10–100% (see Table I).[8,13,21]

For CMP-5-fluoresceinyl-Neu and, even more pronounced, for CMP-9-fluoresceinyl-NeuAc, transferases catalyzing α-2,3-sialylation show a markedly lower reaction rate than α-2,6-sialyltransferases (see Table I).[8,13] Notably, kinetic properties (K_m, V_{max}/K_m) of sialyltransferases for CMP-5-fluoresceinyl-Neu and for CMP-9-fluoresceinyl-NeuAc are more favorable than for CMP-NeuAc (Table I). Furthermore, the 5-substituted sialic acid analog incorporated in glycoconjugate glycans resists cleavage by sialidases[24]; CMP-5-fluoresceinyl-Neu is thus especially useful in suppressing interference by sialidase cleavage when measuring the sialyltransferase activity of crude tissue preparations. In contrast, the reaction rate determined for CMP-9-MTAF-NeuAc, which carries the largest fluorescent residue (Fig. 1), is barely dependent on the linkage specificity of the transferase (Table I).[13]

The CMP-9-AMCA-NeuAc analog shows a higher K_m value compared to compounds substituted by the fluoresceinyl residues (Table I), but an advantage is that the fluorescence quenching of this dye in organic milieu is low and the Stokes shift (100 nm) is markedly higher than that of fluorescein (~25 nm). Acceptor K_m values of rat liver α-2,6-sialyltransferase for several human plasma glycoproteins carrying different *N*-glycan structures (bi-, tri-, and tetraantennary) determined with CMP-9-fluoresceinyl-NeuAc are not significantly influenced by the fluorescent residue; obviously, the large-sized substituent does not exert steric hindrance.[8]

[23] J. Weinstein, U. de Souza-e-Silve, and J. C. Paulson, *J. Biol. Chem.* **257,** 13845 (1982).
[24] H. J. Gross, *Eur. J. Biochem.* **203,** 269 (1992).

TABLE I
APPARENT KINETIC CONSTANTS[a] OF PURIFIED SIALYLTRANSFERASES FOR FLUORESCENT AND PHOTOACTIVATABLE CMP-GLYCOSIDES

Sialyltransferase	Donor	Acceptor[b]	K_m[c] (μM)	V_{max} (rel.)[d]	V_{max}/K_m (1/mM)
Galβ1,4GlcNAc	CMP-NeuAc[e]	AORM	45	1.0	22
α-2,6-sialyltransferase	CMP-9-F-NeuAc[f]	AORM	7	0.95	143
from rat liver	CMP-5-F-Neu[g]	AORM	4.5	1.15	255
	CMP-9-MTAF-NeuAc	AORM	5	0.5	100
	CMP-9-AMCA-NeuAc	AORM	15	0.8	54
	CMP-9-BB-NeuAc[h]	AORM	68	0.8	12
	CMP-9-ASA-NeuAc[i]	AORM	80	1.1	14
	CMP-9-AB-NeuAc[j]	AORM	66	0.9	14
	CMP-5-ABA-NeuAc[k]	AORM	161	0.5	3
Galβ1,3GalNAc	CMP-NeuAc[e]	AFG	20	1.0	50
α-2,3-sialyltransferase	CMP-9-F-NeuAc[f]	AFG	1.5	0.15	100
from porcine liver		AF	2.5	0.25	100
		AM	4.0	0.15*	37*
		G_{M1}	3.0	0.1	33
	CMP-5-F-Neu[g]	AF	0.5	0.35	700
		AM	0.5	0.2*	400*
		G_{M1}	0.5	0.15	300
	CMP-9-MTAF-NeuAc	AF	1.5	0.6	400
	CMP-9-AMCA-NeuAc	AF	7.5	0.6*	80*
		AM	9.5	0.55*	58*

[a] Duplicate kinetic assays were performed at five different donor concentrations near the respective K_m, and the kinetic parameters were calculated from Hanes plots[8]; values represent means from at least three independent experiments except those marked with an asterisk, which are only approximate.

[b] Asialo α_1-acid glycoprotein (AORM) serves as acceptor for rat liver Galβ1,4GlcNAc α-2,6-sialyltransferase, and antifreeze glycoprotein (AFG), asialofetuin (AF), asialomucin (AM), or ganglioside G_{M1} as acceptor for porcine liver Galβ1,3GalNAc α-2,3-sialyltransferase.

[c] Values represent an apparent Michaelis constant for the donor substrate at a fixed acceptor concentration, and thus vary depending on the acceptor employed.

[d] Values are related to the V_{max} determined with parent CMP-NeuAc (1.0).

[e] Kinetic data from Refs. 8 and 21 were determined by a radiometric assay.

[f] CMP-9-fluoresceinyl-NeuAc.

[g] CMP-5-fluoresceinyl-Neu.

[h] CMP-9-benzoylbenzamido-NeuAc(*5*).

[i] CMP-9-azidosalicoylamido-NeuAc(*4*).

[j] CMP-9-azidobenzoylamido-NeuAc(*3*).

[k] CMP-5-N-(azidobenzoyl)-aminoacetyl-Neu(*6*).

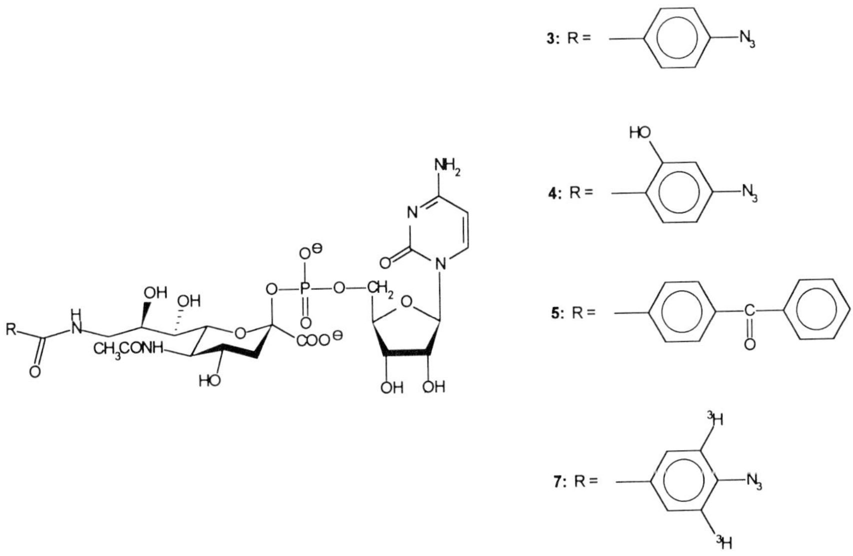

FIG. 3. Structure of CMP-sialic acids substituted by a photoactivatable group at C-9 of the sialic acid moiety. Bound to the carbonyl, the substituent R for compound **3** is the 4-azidobenzoyl group; **4,** 4-azidosalicoyl; **5,** 4-benzoylbenzoyl; **7,** 4-azido[^{3}H]benzoyl.

Synthesis of Photoactivatable Sialic Acids

Principle

Synthesis of the photoreactive CMP-glycosides with an aryl azide or benzophenone moiety in the sialic acid molecule (Figs. 3 and 4) is per-

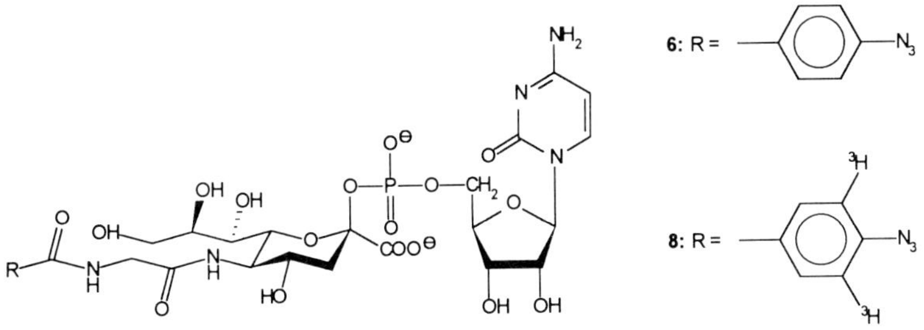

FIG. 4. Structure of CMP-sialic acids substituted by a photoactivatable group at C-5 of the sialic acid moiety. Bound to the carbonyl, the substituent R for compound **6** is the 4-azidobenzoyl group; **8,** 4-azido[^{3}H]benzoyl.

formed by chemical coupling of CMP-9-amino-NeuAc **(1)** or CMP-5-*N*-aminoacetyl-Neu **(2)** (for structure, synthesis, and purification of these compounds, see [11], this volume) with the respective *N*-hydroxysuccinimide ester (NHS) in DMF. Purification of the CMP-glycoside (Figs. 3 and 4) thus obtained is achieved by HPLC, ethanol precipitation, and BioGel P-2 (Bio-Rad, Richmond, CA) gel filtration.[25]

Caution: Care should be taken to avoid direct exposure to light during the syntheses.

General Procedure

To a solution of **1** (12.3 mg, 20 μmol) in water (100 μl) are added DMF (1.5 ml), 0.5 *M* sodium hydrogen carbonate (250 μl), and the NHS ester (40 μmol) of either 4-azidobenzoic acid (Pierce Chemical Co., Rockford, IL), 4-azido-2-hydroxybenzoic acid (Pierce), or benzoylbenzoic acid.[25] The mixture is kept at room temperature for 1.5 hr, and the synthesis is monitored by analytical HPLC on an aminopropyl-phase column [0.4 $\times$ 12.5 cm; 5 μm particle size; 50 : 50 (v/v) 15 m*M* KH_2PO_4–acetonitrile] at 275 nm. After completion of the reaction and removal of the solvents under reduced pressure at ambient temperature, the CMP-glycoside (final volume 0.2–2.0 ml) is separated by semipreparative HPLC [aminopropyl-phase column (0.7 $\times$ 12.5 cm; 5 μm particle size; 35–50 : 65–50 (v/v) 15 m*M* KH_2PO_4–acetonitrile]. Inorganic phosphate is precipitated by the

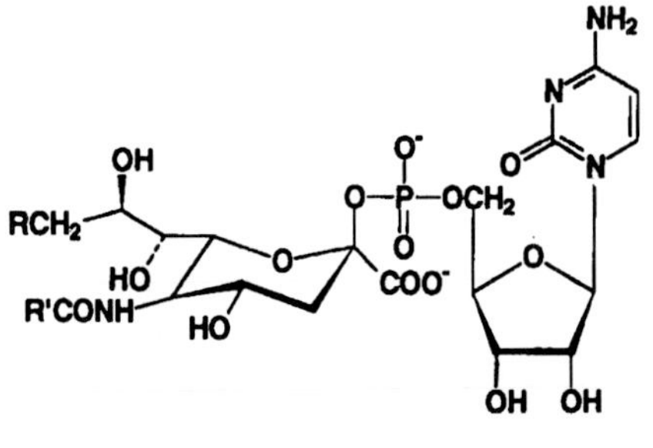

	R	R'
1	NH_2	CH_3
2	OH	NH_2CH_2

SCHEME 1. Structural diagrams of compounds **1** and **2**.

addition of 5 volumes of cold ethanol, and the mixture is centrifuged (600 *g*, 15 min) after cooling on ice for 30 min. The supernatant is decanted, and the pellet is extracted twice with a cold mixture (6 ml) of 5 : 1 (v/v) ethanol–water and treated as outlined above. After concentration of the combined supernatants, further purification is achieved by gel filtration on BioGel P-2 (200–400 mesh; 2.5 × 80 cm) eluting with twice-distilled water (pH 7.0–7.5). Fractions containing the CMP-glycoside are collected and concentrated under reduced pressure to 0.2 ml at ambient temperature. Yields are as follows: **3,** 16.4 μmol (82%); **4,** 16.0 μmol (80%); **5,** 17.4 μmol (87%).[25]

Note: The NHS ester of 4-benzoylbenzoic acid is prepared by the reaction of 4-benzoylbenzoic acid with *N*-hydroxysuccinimide and dicyclohexylcarbodiimide in dry acetone.[26] The compound is also commercially available.

CMP-5-N-(4-Azidobenzoyl)-aminoacetyl-D-neuraminic Acid ***(6)***

Reaction of CMP-5-aminoacetyl-Neu **(2)** (12.6 mg, 20 μmol) (for structure and synthesis see [11], this volume) with the NHS ester of 4-azidobenzoic acid proceeds under the same conditions as described for the synthesis of **3** and affords after identical purification compound **6.** The yield is 15 μmol (75%).[25]

CMP-5-N-Acetyl-9-(4-azido-[^{3}H]benzamido)-9-deoxy-D-neuraminic Acid ***(7)*** *and CMP-5-N-(4-Azido-[^{3}H]benzoyl)-aminoacetyl-D-neuraminic Acid* ***(8)***

Photoactivatable CMP-glycosides containing an arylazido group and a radiolabel are prepared from **1** or **2** (see above) employing tritiated 4-azidobenzoic acid NHS ester (Figs. 3 and 4).[25]

CMP-9-(4-Azido-[^{3}H]benzamido)-9-deoxy-NeuAc **(7).** To the solution of **1** (0.98 mg, 1.6 μmol) in water (50 μl) are added dimethyl sulfoxide (analytical grade; 150 μl), 0.5 *M* sodium hydrogen carbonate (25 μl), and the NHS ester of 4-azido-[^{3}H]benzoic acid (50 μCi, 48.8 Ci/mmol, NEN, Boston, MA) dissolved in 2-propanol (50 μl). The mixture is allowed to stand at room temperature for 1.5 hr, and the turnover of the reaction is determined by analytical HPLC (see above) coinjecting unlabeled **7** as a standard. Purification is performed by analytical HPLC and BioGel P-2 gel filtration (see above) except that ethanol precipitation of inorganic phosphate is omitted. The procedure yields 24.5 μCi of **7** (49% relative to the starting succinimide ester; specific radioactivity 48.6 Ci/mmol).

[25] V. Mirelis, Ph.D. Thesis, Universität Heidelberg (1994).
[26] T. H. Ji and I. Ji, *Anal. Biochem.* **121,** 286 (1982).

CMP-5-N-(4-Azido-[^{3}H]benzoyl)-aminoacetyl-D-neuraminic Acid **(8)**. To the solution of CMP-5-*N*-aminoacetyl-Neu **(2)** (0.71 mg, 1.13 μmol) (for structure and synthesis, see [11], this volume) in water (180 μl) are added dimethyl sulfoxide (350 μl), 0.5 *M* sodium hydrogen carbonate (50 μl), and the NHS ester of 4-azido-[^{3}H]benzoic acid (70 μCi, 48.6 Ci/mmol, NEN) dissolved in 2-propanol (70 μl). After 2 hr at room temperature, the reaction mixture contains 60% of the radiolabeled, photoactivatable CMP-glycoside. Quantification and purification are performed as described for the preparation of **7.** The yield is 27.1 μCi (39% relative to the starting succinimide ester; specific radioactivity 48.6 Ci/mmol).

5-N-Acetyl-9-(4-azido-2-hydroxybenzamido)-9-deoxy-D-neuraminic acid [9-(4-azidosalicoyl)amido-9-deoxy-NeuAc]. To the solution of 5-*N*-acetyl-9-amino-9-deoxy-D-neuraminic acid[27] (100 mg, 0.325 mmol) (for synthesis see [11], this volume) in 1 ml water are added 0.5 *M* sodium hydrogen carbonate (1 ml) and the NHS ester of 4-azidosalicylic acid (150 mg, 0.54 mmol) dissolved in 1 : 1 (v/v) methanol–ethyl acetate (3 ml). After 1 hr at room temperature, 0.5 *M* sodium hydrogen carbonate (0.5 ml) and a solution of the NHS ester (80 mg) in 1 : 1 (v/v) methanol–ethyl acetate (2 ml) are again added. After 3 hr, solvents are removed under reduced pressure, the mixture is dissolved in water, and insoluble material is removed by centrifugation. Purification is performed by chromatography on DEAE Sephadex A-25 (4 × 10 cm). After washing with water (200–300 ml), the column is eluted with 0.5 *M* ammonium hydrogen carbonate. The fractions positive to the Ehrlich reaction are concentrated, and the aqueous solution of the residue is extracted with diethyl ether. Treatment of the water phase with IR 120 (H^+) cation-exchange resin and lyophilization yield 85.5 mg (~60%); R_f 0.52 [5 : 2 (v/v) *n*-propanol–water]; $[\alpha]_D^{20}$ $-13°$ (*c* 0.3, water).

Comments

The structure of each CMP-glycoside is proved after acid hydrolysis, by HPLC (see [11], this volume). Contaminating CMP (<2%) and free photoreactive sialic acid (<3%) are quantified by analytical HPLC, and inorganic phosphate (<3%) is determined colorimetrically.[14] The photolabile compounds should be handled with care and stored lyophilized or in solution at −20° to −80° protected from light.

To improve yields, for the synthesis of nonlabeled photoreactive CMP-glycosides the respective NHS ester is applied in excess.[25] In contrast, preparation of labeled compounds requires an excess of the starting CMP-

[27] R. Isecke and R. Brossmer, *Tetrahedron* **50,** 7445 (1994).

sialic acid analog. It is to be noted that the respective photoreactive group carried by the CMP-sialic acid remains stable during synthesis and purification.

Analogous to the synthesis of 9-(4-azidosalicoylamido)-NeuAc, additional sialic acids carrying a photoreactive group are obtained, for example, 9-(4-benzoylbenzamido)-9-deoxy-NeuAc and 9-(4-azidobenzoyl)-9-deoxy-NeuAc. The kinetic data for CMP-sialic acid synthase determined with the latter compounds are of the same order as for the parent NeuAc and, therefore, allow the conversion of the photoactivatable sialic acids to the corresponding CMP-glycoside. In most cases, however, it is preferable to prepare the CMP-glycosides by chemical synthesis employing CMP-9-amino-NeuAc as the starting compound as described above.

Owing to the favorable kinetic parameters, the photoreactive CMP-sialic acids are transferred by rat liver α-2,6-sialyltransferase. Results are shown using asialo α_1-acid glycoprotein as acceptor (Table I).

Concluding Remarks: Biochemical Applications of Fluorescent and Photoactivatable CMP-Sialic Acids

Due to favorable kinetic parameters (compare V_{max}/K_m values in Table I), the fluorescent CMP-glycosides (Figs. 1 and 2) are best suited to establish a very sensitive fluorometric sialyltransferase assay based on the detection system described above.[8] By employing a set of acceptor glycoproteins with distinct N- or O-linked-glycan sequences and gangliosides, the assay permits characterization of the acceptor specificity.[8,16] The fluorometric assay is more favorable than the common radiometric method owing to the high sensitivity, the ease of assay procedure, and the lack of precautions which are otherwise necessary in handling radiolabeled compounds. Furthermore, enzyme activity may be determined at saturating donor concentrations, facilitating reliable comparative studies on sialyltransferase activity of different sources.

The decrease in the reaction rate of α-2,3-sialyltransferase measured for CMP-5-fluoresceinyl-Neu and especially for CMP-9-fluoresceinyl-NeuAc (V_{max} 10–35% relative to V_{max} of CMP-NeuAc; Table I)[8,13] is sufficiently compensated for by the high sensitivity. The fluorometric assay reaches a detection limit as low as 0.03 pmol at a ratio of 1.5 to 1.0 for enzymatic transfer to nonspecific adsorption. Thus, using CMP-5-fluoresceinyl-Neu or CMP-9-fluoresceinyl-NeuAc, at least 1 nU of rat liver α-2,6-sialyltransferase and 3–6 nU of porcine liver or porcine submaxillary gland α-2,3-sialyltransferase can be detected.[8]

Human plasma sialyltransferase activity in healthy persons has been investigated employing CMP-9-fluoresceinyl-NeuAc and the fluorometric detection system described.[16] Despite the very low level of enzyme activity (below 50 μU/ml), the assay allows characterization of acceptor specificity and kinetic properties. The study proved the presence of distinct plasma enzymes acting either on N-linked complex-type glycans or on O-linked GalNAc residues. In the past, studies on human plasma sialyltransferase as a clinical marker gave conflicting results which in most cases arose from radiometric assays performed under subsaturating conditions and from restriction to global sialyltransferase determination.[9,16] Prognostic and diagnostic significance may now be reevaluated, focusing on the different acceptor specificities of human plasma sialyltransferases.[28]

Fluorescent CMP-sialic acids have served to develop a sialyltransferase-dependent procedure for selective labeling of glycoconjugate glycans. In contrast to the chemical methods[6,7] to attain the same goals, specific oligosaccharide sequences contained in soluble or membrane-bound glycoconjugates can be specifically labeled with fluorescent CMP-sialic acids and sialyltransferases with defined acceptor specificity. For example, galactose sites of asialo α_1-acid glycoprotein were enzymatically sialylated to terminate in α2,6-linked- and α2,3-linked 9-fluoresceinyl-NeuAc by 65 and 10%, respectively; antifreeze glycoprotein was modified by incorporation of α2,6-linked 9-fluoresceinyl-NeuAc to 11% of the total GalNAc sites.[10] Thus, transfer of fluorescent sialic acid analogs to each acceptor glycoprotein exceeded the level necessary for efficient labeling. Labeled glycoproteins can be obtained in native conformation after separation of the fluorescent CMP-sialic acid by gel filtration as described for the preparation of neoglycoproteins (see [11], this volume).[10]

Transferring of fluorescent sialic acids with the commercially available specific sialyltransferases (Table I) allows the evaluation of the expression of either lactosamine- or mucin-type glycans on the surface of intact cells. Quantification of surface-bound fluorescence at the level of a single cell is easily accomplished by flow cytometry (see [11], this volume).[29,30]

The specificity of the Golgi carrier for CMP-sialic acids has been studied in permeabilized cells employing CMP-9-fluoresceinyl-NeuAc and CMP-5-fluoresceinyl-Neu. Despite the bulky fluoresceinyl residue, both CMP-glycosides were efficiently translocated from the cytosol into the Golgi lumen and thus are well-suited to investigate the lumenal sialylation within the intact organelle.[24]

[28] E.-M. Rothe, H. J. Gross, R. Raedsch, J. M. Krause, R. Brossmer, and B. Kommerell, *Eur. J. Clin. Invest.* **22,** A17 (1992).

[29] C. Bollheimer, R. Kosa, A. Sauer, R. Schwartz-Albiez, and H. J. Gross, *Biol. Chem. Hoppe-Seyler* **374,** GC 1a (1993).

[30] R. Kosa, Ph.D. Thesis, Universität Heidelberg (1994).

CMP-9-(4-azidosalicoyl)-NeuAc can be readily iodinated without decomposition by employing the Iodogen method (Pierce). Thus, after enzymatic transfer, the photoactivatable sialic acid carrying a photoreactive group and a radiolabel is incorporated in certain glycan chains of glycoconjugates according to the acceptor specificity of the sialyltransferase applied. Obviously, this approach may be extended to other photoreactive substituents provided the respective CMP-glycosides possess reasonably good substrate properties toward sialyltransferases.

For certain studies it may be sufficient to transfer enzymatically a sialic acid analog containing only a radiolabel. For example, ^{125}I-labeled immunoglobulin G (IgG) antibodies were obtained using iodinated CMP-9-deoxy-9-salicylamido-NeuAc under mild conditions.[31] Another application is the study of the active center of enzymes and lectins recognizing sialic acid. In this context, a bacterial and a mammalian sialidase were labeled in the active center by applying the strong inhibitor *N*-acetyl-2,3-anhydroneuraminic acid carrying a ^{125}I-labeled 4-azidosalicoylamido group at C-9.[32,33]

[31] U. Schwartz, G. Wunderlich, and R. Brossmer, *FEBS Lett.* **337,** 213 (1994); ibid. **343,** 181 (1994).

[32] G. T. J. van der Horst, U. Rose, R. Brossmer, and F. W. Verheijen, *FEBS Lett.* **277,** 42 (1990).

[33] G. T. J. van der Horst, G. M. S. Mancini, R. Brossmer, U. Rose, and F. W. Verheijen, *J. Biol. Chem.* **265,** 10801 (1990).

[13] Glycosyl Phosphites as Glycosylation Reagents

By SHIN AOKI, HIROSATO KONDO, and CHI-HUEY WONG

Introduction

The development of new methods for the efficient construction of glycosidic bonds is an objective of current research in carbohydrate synthesis.[1] The practical and stereocontrolled synthesis of oligosaccharides

[1] For representative chemical glycosylations, see R. U. Lemieux, *Chem. Soc. Rev.* **7,** 423 (1978); H. Paulson, *Angew. Chem., Int. Ed. Engl.* **21,** 155 (1982); R. R. Schmidt, *Angew. Chem., Int. Ed. Engl.* **25,** 212 (1986); H. Kunz, *Angew. Chem., Int. Ed. Engl.* **26,** 294 (1987); K. Okamoto and T. Goto, *Tetrahedron* **45,** 5835 (1989); K. C. Nicolaou, T. Caufield, H. Kataoka, and P. Kumazawa, *J. Am. Chem. Soc.* **110,** 7910 (1988); D. R. Mootoo, P. Konradsson, U. Udodong, and B. Fraser-Reid, *J. Am. Chem. Soc.* **110,** 5583 (1988); R. L. Halcomb and S. J. Danishefsky, *J. Am. Chem. Soc.* **111,** 6661 (1989); R. W. Friesen and S. J. Danishefsky, *J. Am. Chem. Soc.* **111,** 6656 (1989); A. G. M. Barrett, B. C. B.

METHODS IN ENZYMOLOGY, VOL. 247

containing sialic acid, for example, remains of particular interest because of their important roles in biological recognition and cellular communication.[2] However, the steric hindrance of the anomeric center and the ease of elimination during activation and glycosylation make the sialylation reaction particularly difficult to execute in a high-yield and stereoselective manner.

A novel and high-yield sialylation using sialyl phosphite as donor[3] and trimethylsilyltrifluoromethane sulfonate (TMSOTf) as catalyst has been reported. We also have reported the preparation of several other dibenzylglycosyl phosphites, which can be easily converted to glycosyl phosphates and sugar nucleotides[4] (Scheme 1). We here describe the application of glycosyl phosphites to the synthesis of different types of glycosides.

Experimental Methods

Dibenzyl N,N-Diethylphosphoramidite

The procedure for the synthesis of dibenzyl *N,N*-diethylphosphoramidite (DDP) is essentially the same as that described previously.[5] To a dry

Bezuidenhoudt, A. F. Gasiecki, A. R. Howell, and M. A. Russell, *J. Am. Chem. Soc.* **111,** 1392 (1989); K. Briner and A. Vesella, *Helv. Chim. Acta* **72,** 1371 (1989); D. Kahne, S. Walker, Y. Cheng, and D. Van Engen, *J. Am. Chem. Soc.* **111,** 6881 (1989); K. Suzuki, H. Maeta, and T. Matsumoto, *Tetrahedron Lett.* **36,** 4853 (1989); Y. Ito and T. Ogawa, *Tetrahedron* **46,** 89 (1990); A. Hasegawa, T. Nagahama, H. Ohki, H. Hotta, and M. Kiso, *J. Carbohydr. Chem.* **10,** 493 (1991); A. Marra and P. Sinay, *Carbohydr. Res.* **195,** 303 (1990); H. Lönn and K. Stenvall, *Tetrahedron Lett.* **33,** 115 (1992); E. Kirchner, F. Thiem, R. Dernick, J. Heukeshoven, and J. Thiem, *J. Carbohydr. Chem.* **7,** 453 (1988); F. Barres and O. Hindsgaul, *J. Am. Chem. Soc.* **113,** 9376 (1991); G. Stork and G. Kim, *J. Am. Chem. Soc.* **114,** 1087 (1992); P. J. Garegg, *Acc. Chem. Res.* **25,** 575 (1992); K. Toshima and K. Tatsuta, *Chem. Rev.* **93,** 1503 (1993). For reviews of enzymatic glycosylation, see Y. Ichikawa, G. C. Look, and C.-H. Wong, *Anal. Biochem.* **202,** 215 (1992); D. G. Drueckhammer, W. J. Hennen, R. L. Pederson, C. F. Barbas, C. M. Gautheron, T. Krach, and C.-H. Wong, *Synthesis* **7,** 499 (1991); E. J. Toone, E. S. Simon, M. D. Bednarski, and G. M. Whitesides, *Tetrahedron* **45,** 5365 (1989); S. David, C. Auge and C. Gautheron, *Adv. Carbohydr. Chem.* **49,** 175 (1991).

[2] R. W. Ledeen, R. K. Yu, M. M. Rapport, and K. Suzuki, "Ganglioside Structure, Function and Biomedical Potential." Plenum, New York, 1984; M. L. Phillips, E. Nudelman, F. C. A. Gaeta, M. Perez, A. K. Singhal, S. Hakomori, and J. C. Paulson, *Science* **250,** 1130 (1990); A. Varki, *Glycobiology* **3,** 97 (1993).

[3] For sialylations with TMSOTf as catalyst, see H. Kondo, Y. Ichikawa, and C.-H. Wong, *J. Am. Chem. Soc.* **114,** 8748 (1992); T. J. Martin and R. R. Schmidt, *Tetrahedron Lett.* **33,** 6123 (1992). For the reaction using $ZnCl_2$ as Lewis acid, see Y. Watanabe, C. Nakamoto, and S. Ozaki, *Synlett,* 115 (1993). For synthesis of CMP-NeuAc, see T. J. Martin and R. R. Schmidt, *Tetrahedron Lett.* **34,** 1765 (1993).

[4] M. M. Sim, H. Kondo, and C.-H. Wong, *J. Am. Chem. Soc.* **115,** 2260 (1993).

[5] R. L. Pederson, J. Esker, and C.-H. Wong, *Tetrahedron* **47,** 2643 (1991).

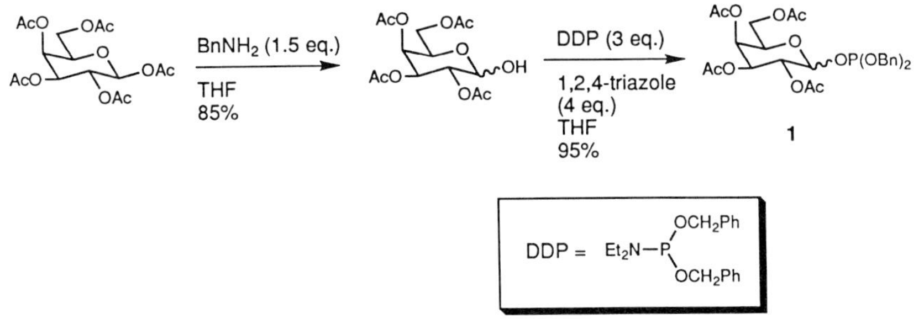

SCHEME 1. Preparation of galactosyl phosphite donor.

3-necked 5-liter round-bottomed flask equipped with an overhead stirrer is added 4.2 liters of anhydrous ether and 137.3 g (1.0 mol) of phosphorus trichloride. The solution is cooled to 0°, under dry N_2, and 73.1 g (1.0 mol) diethylamine and 101.2 g (1.0 mol) triethylamine are added dropwise. The mixture is stirred at room temperature for 24 hr (efficient stirring is required for good yields). The solution is again cooled to 0°, and 216.3 g (2.0 mol) of benzyl alcohol and 220.0 g (2.17 mol) of triethylamine are added dropwise. The solution is stirred at room temperature for 36 hr. The triethylamine hydrochloride salt is removed by filtration, and the ether is removed under reduced pressure. The remaining residue is vacuum distilled, $bp_{0.3}$ 154°–156°, to yield 196.5 g (0.619 mol, 62%) DDP. The compound is stable and can be kept for several months without any decomposition or disproportionation. The ^1H nuclear magnetic resonance (NMR) (200 mHz, $CDCl_3$) data are as follows: δ 1.1 (2t, 6H, 2 CH_3, J = 7.1 Hz), 3.1 (2q, 4H, 2NCH_2, J = 7.1 Hz), 4.75, 4.8 (2d, 4H, 2 CH_2Ph), 7.3 (m, 10H, 2 Ph); ^{13}C NMR (50 MHz, $CDCl_3$): δ 17.5 (CH_3), 38.7, 39.1 (NCH_2), 62.5, 62.9 (CH_2Ph), 123.0, 123.2, 124.1 138.8 (aromatic).

Preparation of Glycosyl Phosphites

Compounds **1** and **13–16** are prepared according to the procedure described previously.[4] Compound **10** is prepared according to the procedure of Garegg *et al.*[6]

Dibenzyl 3,4,6-Tri-O-acetyl-2-deoxy-2-phthalimido-β-D-glucopyranosyl Phosphite (**2**). As described previously,[7] dibenzyl *N,N*-diethylphosphoramidite (146 gm, 0.46 mmol) is added to a solution of 3,4,6-tri-*O*-acetyl-2-deoxy-2-phthalimido-β-D-glucopyranose[8] (0.1 g, 0.23 mmol) and 1*H*-tetrazole (65 mg, 0.93 mmol) in dry tetrahydrofuran (THF) (3 ml) under

[6] P. J. Garegg, H. Hultberg, and S. Walling, *Carbohydr. Res.* **108,** 97 (1982).
[7] T. Sugai, H. Ritzen, and C.-H. Wong, *Tetrahedron: Asymmetry* **4,** 1051 (1993).
[8] H. Paulsen and B. Helpap, *Carbohydr. Res.* **216,** 289 (1991).

Donors

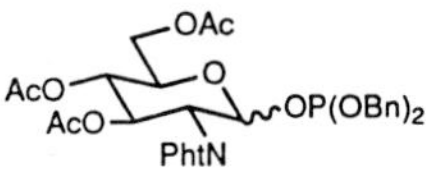

AcO, AcO, OAc, O, PhtN, OP(OBn)2

2

AcO, AcO, OAc, O, TrocHN, OP(OBn)2

3

AcO, AcO, COOMe, O, OAc, OP(OBn)2

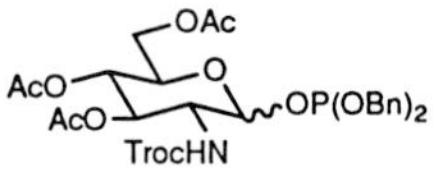

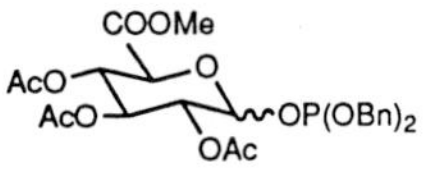

4

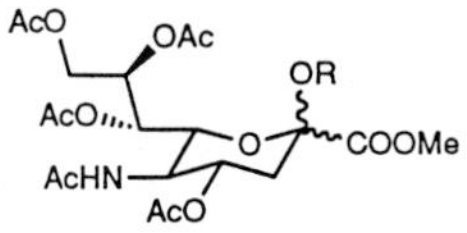

AcO, OAc, OR, AcO, AcHN, AcO, O, COOMe

5 : R = H
6 : R = P(OBn)2

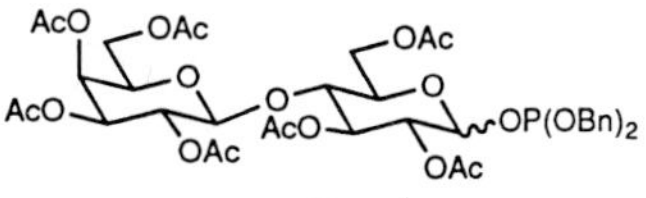

AcO, OAc, O, AcO, OAc, O, AcO, OAc, O, OAc, OP(OBn)2

7

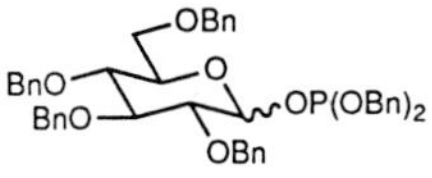

BnO, BnO, OBn, O, OBn, OP(OBn)2

8

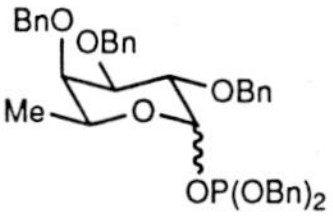

BnO, OBn, Me, O, OBn, OP(OBn)2

9

Acceptors

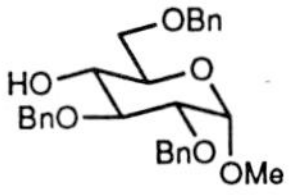

HO, BnO, OBn, O, BnO, OMe

10

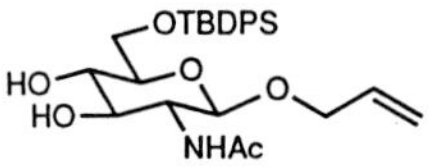

HO, HO, OTBDPS, O, NHAc, O

11

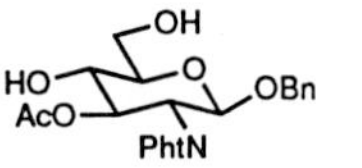

HO, AcO, OH, O, PhtN, OBn

12

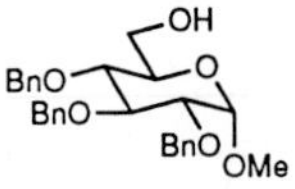

BnO, BnO, OH, O, BnO, OMe

13

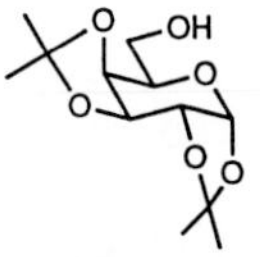

O, O, OH, O, O, O

14

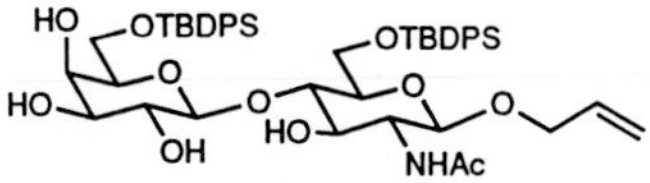

HO, OTBDPS, HO, O, OH, O, OTBDPS, HO, O, NHAc, O

15

HO, OTBDPS, TBDPSO, O, OH, O, OTBDPS, HO, O, NHAc, O

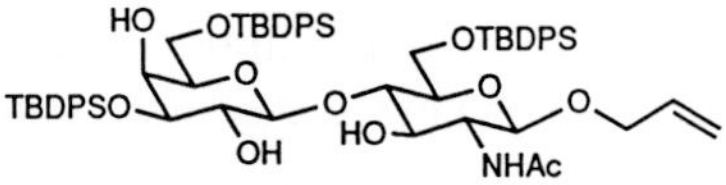

16

an argon atmosphere at room temperature. The mixture is allowed to stir at room temperature for 2 hr. The reaction mixture is diluted with CH_2Cl_2, then washed with ice-cold saturated $NaHCO_3$, saturated NaCl, and water. The organic layer is separated, dried over anhydrous Na_2SO_4, and concentrated *in vacuo*. The residual syrup is chromatographed on silica gel with ethyl acetate–hexane (1 : 3, v/v) to give **2** (140 mg, 88%). ^{1}H NMR (400 MHz, $CDCl_3$): δ 5.97 (1H, t, J = 8.4 Hz, H-1), 5.89 (1H, dd, J = 9.2, 10.7 Hz, H-3), 5.22 (1H, dd, J = 9.3, 10.0 Hz, H-4), 4.46 (1H, dd, J = 8.4, 10.7 Hz, H-2).

*Dibenzyl 3,4,6-Tri-O-acetyl-2-deoxy-2-(2,2,2-trichloroethoxycarbonylamino)-D-glucopyranosyl Phospite (**3**).* As described previously,[7] dibenzyl *N,N*-diethylphosphoramidite (527 mg, 0.46 mmol) is added to a solution of 3,4,6-tri-*O*-acetyl-2-deoxy-2-(2,2,2-trichloroethoxycarbonylamino)-β-D-glucopyranose[9] (0.2 g, 0.42 mmol) and 1*H*-tetrazole (116 mg, 1.66 mmol) in dry THF (3 ml) under an argon atmosphere at room temperature. The mixture is allowed to stir at room temperature for 2 hr. The reaction mixture is distilled with CH_2Cl_2, then washed with ice-cold saturated $NaHCO_3$, saturated NaCl, and water. The organic layer is separated, dried over anhydrous Na_2SO_4, and concentrated *in vacuo*. The residual syrup is chromatographed on silica gel with ethyl acetate–hexane (1 : 2, v/v) to give **3** as an anomeric mixture (225 mg, 67%). ^{1}H NMR ($CDCl_3$) of α-**3**: δ 2.01 (3H, s, CH_3CO), 2.02 (6H, 2, 2 CH_3CO), 3.89 (1H, dd, J = 2.2, 12.4 Hz, H-6), 4.00 (1H, m, H-5), 4.08 (1H, dt, J = 3.4, 10.4 Hz, H-2), 4.13 (1H, dd, J = 4.1, 12.5 Hz, H-6′), 4.53 (1H, d, J = 12.0 Hz), 4.76 (1H, d, J = 11.7 Hz), 4.87–4.96 (4H, m), 5.09 (1H, t, J = 9.8 Hz, H-4), 5.16–5.25 (2H, m, H-3, NH), 5.59 (1H, dd, J = 3.4, 7.7 Hz, H-1).

*Methyl 2,3,4-tri-O-acetyl-1-O-dibenzylphosphityl Glucopyranuronate (**4**).* Dibenzyl *N,N*-diethylphosphoramidite (1.8 g, 5.67 mmol) is added to a solution of methyl 2,3,4-tri-*O*-acetylglucopyranuronate (0.76 g, 2.27 mmol) and 1,2,4-triazole (0.66 g, 9.56 mmol) in dry THF (5 ml) under an argon atmosphere at room temperature. The mixture is allowed to stir at room temperature for 2 hr. The reaction mixture is diluted with CH_2Cl_2, then washed with ice-cold saturated $NaHCO_3$ and saturated NaCl. The organic layer is dried and concentrated *in vacuo*. The residual syrup is chromatographed on silica gel with ethylacetate–hexane (1 : 1, v/v) to give **4** (0.83 g, 65%) as a 1 : 1 mixture of α and β anomers. ^{1}H NMR (400 MHz, $CDCl_3$): δ 1.85, 1.89, 2.01, 2.02, 2.02, 2.03 (3H, each s, OCH_2CH_3), 3.68, 3.70 (3H, each s, $COOCH_3$), 4.07 (1H, d, J = 9.6 Hz), 4.10 (1H, q, J = 7.2 Hz), 4.46 (1H, d, J = 10.2 Hz, H-5), 4.82–5.00 (10H, m), 5.12 (1H, dd, J = 2.0, 5.2 Hz), 5.20 (1H, dd, J = 9.7, 9.6 Hz), 5.26–5.28 (1H, m),

[9] A. Banaszek, X. B. Cornet, and A. Zamojski, *Carbohydr. Res.* **144,** 342 (1985).

5.60 (1H, t, J = 9.7 Hz), 5.83 (1H, dd, J = 3.8, 8.3 Hz, H-1), 7.27–7.35 (10H, m, phenyl protons).

*Methyl 5-Acetamido-4,7,8,9-tetra-O-acetyl-2-(dibenzylphosphityl)-3,5-dideoxy-β-D-glycero-D-galacto-2-nonulopyranosonate (**6**).* DDP (0.25 g, 0.78 mmol) is added dropwise to a solution of **5**[10] (0.166 g, 0.34 mmol) and 1*H*-tetrazole (0.10 g, 1.43 mmol) in THF (5 ml) under a nitrogen atmosphere, and the mixture is stirred for 4 hr at room temperature. The mixture is extracted with CH_2Cl_2 (10 ml), and the organic phase is washed with ice-cold dilute HCl, aqueous $NaHCO_3$, and ice–water, dried over anhydrous Na_2SO_4, and evaporated *in vacuo* to give a crude material, which is chromatographed on a silica gel column with ethyl acetate–hexane (5 : 1, v/v) to give **6** (0.17 g, 68%) as a colorless syrup. ^{1}H NMR (400 MHz, D_2O): δ 2.39 (dd, $J_{3eq,4}$ = 4.9 Hz, $J_{3eq,3ax}$ = 13.00 Hz, H-3eq), 4.83–4.91 (m, H-4), 3.99 (ddd, $J_{5,NH}$ = $J_{4,5}$ = 10.4 Hz, H-5), 3.74 (dd, $J_{5,6}$ = 10.60 Hz, $J_{6,7}$ = 2.00, H-6), 5.14 (dd, $J_{7,8}$ = 2.40 Hz, H-7), 4.83–4.91 (m, H-8), 4.10 (dd, $J_{8,9a}$ = 7.40 Hz, $J_{9a,9b}$ = 12.4 Hz, H-9a), 4.57 (dd, $J_{8,9b}$ = 2.00 Hz, H-9b), 3.73 (s, CO_2CH_3), 1.96, 2.01, 2.06, 2.07 (s, CH_3CO^-), 1.79 (s, CH_3CON^-). ^{13}C NMR (100 MHz, $CDCl_3$): δ 20.7, 20.8, 20.9, 21.0, 23.1, 36.0, 49.5, 53.5, 62.6, 67.3, 67.4, 67.8, 69.3, 70.8, 94.8, 128.0, 128.7, 135.5, 141.8, 153.0, 169.1, 170.2, 170.4, 170.8. High-resolution mass spectrometry (HRMS): Calculated for $C_{34}H_{42}NO_{15}PCs$ (M + Cs^+), 868.1346. Found, 868.1346.

*Dibenzyl 2,3,4,6-Tetra-O-acetyl-β-D-galactopyranosyl-(1,4)-2,3,6-tri-O-acetyl-D-glucopyranosyl Phosphite (**7**).* A solution of peracetylated lactose (0.5 g, 0.737 mmol) and benzylamine (125 μl) in THF (5 ml) is stirred at room temperature for 38 hr. The mixture is added to chloroform (100 ml) and washed 2 times with 50 ml of 0.1 *N* HCl followed by 50 ml saturated $NaHCO_3$. The organic layer is dried with $MgSO_4$, filtered, and concentrated *in vacuo*. The residue is chromatographed (ethyl acetate–*n*-hexane, 1 : 1 to 7 : 3, v/v) to give 2,3,4,6-tetra-*O*-acetyl-β-D-galactopyranosyl-β-(1,4)-2,3,6-tri-*O*-acetylglucopyranose (415 mg, 89%) as a mixture of anomers. ^{1}H NMR (400 MHz, $CDCl_3$): δ 1.97, 2.03, 2.05, 2.06, 2.07, 2.08, 2.12, 2.13, 2.16 (each s, OCH_2CH_3), 3.62–3.69 (m), 3.78 (t, J = 9.4 Hz), 3.89 (t, J = 7.0 Hz), 4.04–4.22 (m), 4.42 (d, J = 5.7 Hz), 4.45–4.53 (m), 4.74 (t, J = 7.5 Hz), 4.82 (dd, J = 3.5, 10.0 Hz), 4.97 (dd, J = 3.3, 10.4 Hz), 5.08–5.17 (m), 5.23 (t, J = 9.2 Hz), 5.32–5.39 (m), 5.53 (t, J = 9.6 Hz).

A solution of 2,3,4,6-tetra-*O*-acetyl D-galactopyranosyl-β(1,4)-2,3,6-tri-*O*-acetylglucopyranose (144 mg, 0.212 mmol), 1,2,4-triazole (103 mg), and dibenzyl *N,N*-diethylphosphoramidite (200 μl) in dichloromethane (3 ml) is stirred for 2 hr at room temperature. The mixture is added to

[10] A. Marra and P. Sinay, *Carbohydr. Res.* **190,** 317 (1989).

dichloromethane (20 ml) and washed with 10 ml saturated $NaHCO_3$ and water (10 ml). The organic layer is dried with Na_2SO_4, filtered, and concentrated under vacuum. The residue is chromatographed (hexane–ethylacetate, 7 : 3 to 1 : 1, v/v) to give **7** (172 mg, 92%) as a mixture of anomers. ^1H NMR (400 MHz, $CDCl_3$): δ 1.88, 1.92, 1.98, 2.04, 2.05, 2.07, 2.18 (each s, OCH_2CH_3), 3.68 (ddd, J = 1.8, 4.9, 9.9 Hz), 3.74–3.91 (m), 4.02–4.20 (m), 4.43–4.52 (m), 4.80–5.16 (m), 5.22 (t, J = 9.1 Hz), 5.35 (dd, J = 0.8, 3.4 Hz), 5.54 (t, J = 10.1 Hz, H-1β), 5.69 (dd, J = 3.5, 8.5 Hz, H-1α), 7.23–7.45 (m). Fast atom bombardment (FAB) HRMS: 1013 (M^+).

*Dibenzyl 2,3,4,6-Tetra-O-benzyl-D-glucopyranosyl Phosphite (**8**).* A solution of dibenzyl *N,N*-diethylphosphiramidite (1.10 g, 3.5 mmol) in CH_2Cl_2 (10 ml) is added dropwise to a solution of 2,3,4,6-tetra-*O*-benzyl D-glucopyranose (0.80 g, 1.4 mmol) and 1*H*-tetrazole (0.16 g, 2.3 mmol) in anhydrous CH_2Cl_2 (15 ml) under nitrogen at room temperature. After stirring for 2 hr, the mixture is diluted with CH_2Cl_2, washed with saturated aqueous $NaHCO_3$ and saturated NaCl, dried over anhydrous Na_2SO_4, filtered, and concentrated under reduced pressure. The crude product is purified further by silica gel column chromatography (ethyl acetate–hexane–triethylamine, 10 : 300 : 5, v/v/v) to provide a 83 : 17 mixture of the α and β anomers of **8** (1.08 g, 95% yield). This phosphite has been found to be stable at −5° over 2 months after its preparation.

The ^1H NMR ($CDCl_3$) data are as follows (only the chemical shifts of anomeric protons are shown): δ 5.03 (1H, dd, $J_{H\text{-}H}$ = 8.0, $J_{H\text{-}P}$ = 8.0 Hz, H-1 of β anomer), 5.65 (1H, dd, $J_{H\text{-}H}$ = 3.2, $J_{H\text{-}P}$ = 8.3 Hz, H-1 of α anomer). ^{13}C NMR ($CDCl_3$): δ 64.1 (d, $J_{C\text{-}P}$ = 8.7 Hz), 64.3 (d, $J_{C\text{-}P}$ = 7.2 Hz), 64.6 (d, $J_{C\text{-}P}$ = 8.9 Hz), 68.1, 68.6, 71.5, 72.9, 73.4, 73.5, 75.4, 75.7 (d, $J_{C\text{-}P}$ = 8.2 Hz), 77.1, 77.5, 79.9, 79.9, 81.5, 82.7, 82.7, 84.6, 92.1 (d, $J_{C\text{-}P}$ = 17.9 Hz), 97.2 (d, $J_{C\text{-}P}$ = 15.2 Hz), 127.4, 127.5, 127.6, 127.7, 127.7, 127.8, 127.9, 128.0, 128.3, 137.8, 137.9, 138.0, 138.1, 138.2, 138.5, 138.7. ^{13}P NMR ($CDCl_3$): δ 135.1 (α), 135.7 (β). Infrared spectroscoy (IR) (neat) 3063, 3030, 2868, 1951, 1811, 1734, 1606, 1496, 1453, 1363, 1209, 1071, 997, 914, 790, 732, 695 cm^{-1}.

*Dibenzyl 6-Deoxy-2,3,4-tri-O-benzyl-L-galactopyranosyl Phosphite (**9**).* A solution of DDP (1.00 g, 3.15 mmol) in THF (8 ml) is added dropwise to a solution of 6-deoxy-2,3,4-tri-*O*-benzyl-L-galactopyranose (0.76 g, 1.7 mmol) and 1*H*-tetrazole (0.2 g, 2.9 mmol) in THF (8 ml) at room temperature. After stirring for 2 hr at room temperature, the reaction mixture is quenched with saturated aqueous $NaHCO_3$ (20 ml) and extracted with CH_2Cl_2 (30 ml, 3 times). The combined organic layers are washed with saturated aqueous NaCl (20 ml, 2 times), dried over Na_2SO_4, filtered, and concentrated under reduced pressure. The remaining residue is purified by silica gel chromatography (ethyl acetate–hexane–triethylamine,

1 : 10 : 0.5, v/v/v) to give a 38 : 62 mixture of the α and β anomers of **9** (0.88 g, 74% yield).

Physical data for α-**9** are as follows: R_f 0.32 (ethyl acetate–*n*-hexane, 1 : 5, v/v). ^{1}H NMR (400 MHz, $CDCl_3$): δ 1.07 (3H, d, J = 6.5 Hz, CH_3), 3.64 (1H, dd, J = 1.7, 1.7 Hz, H-4), 3.92 (1H, dd, J = 2.8, 10 Hz, H-3), 4.00 (1H, br dq, J = 6.5 Hz, 1.7 Hz, H-5), 4.10 (1H, dd, J = 3.4, 10 Hz, H-2), 4.65 (1H, d, J = 11.6 Hz, one of benzylic protons), 4.69–4.93 (8H, 8 d, benzylic protons), 4.99 (1H, d, J = 11.6 Hz, one of benzylic protons), 5.66 (1H, dd, $J_{H\text{-}H}$ = 3.4 Hz, $J_{H\text{-}P}$ = 8.1 Hz, H-1), 7.25–7.38 (25 H, m, PhH). ^{13}C NMR (100 MHz, $CDCl_3$): δ 16.61, 64.01 (d, $J_{C\text{-}P}$ = 8.9 Hz, PO*C*H$_2$Ph), 64.21 (d, $J_{C\text{-}P}$ = 7.6 Hz, PO*C*H$_2$Ph), 67.68, 72.98, 73.06, 74.85, 76.29 (d, $J_{C\text{-}P}$ = 4.0 Hz, C-2), 77.51, 78.73, 92.89 (d, $J_{C\text{-}P}$ 17.4 Hz, anomeric C-1), 127.38, 127.44, 127.52, 127.61, 127.86, 128.13, 128.20, 128.26, 128.34, 128.39, 138.40, 138.49, 138.76. ^{31}P {^{1}H} NMR (162 MHz, $CDCl_3$): δ 135.43. HRMS: Calculated for $C_{41}H_{43}O_7PCs$ (M + Cs^+), 567.1148. Found, 567.1146.

Physical data for β-**9** are as follows: R_f 0.27 (ethyl acetate–hexane, 1 : 5, v/v). ^{1}H NMR (400 MHz, $CDCl_3$): δ 1.18 (3H, d, J = 6.4 Hz, CH_3), 3.51–3.60 (3H, m, H-3,4,5), 3.94 (1H, dd, J = 7.7, 9.4 Hz, H-2), 4.53–5.02 (11H, m, H-1 and benzylic protons), 7.24–7.39 (25H, m, PhH). ^{13}C NMR (100 MHz, $CDCl_3$): δ 16.72, 64.20 (d, $J_{C\text{-}P}$ = 7.5 Hz, PO*C*H$_2$Ph), 64.45 (d, $J_{C\text{-}P}$ = 8.4 Hz, PO*C*H$_2$Ph), 71.09, 73.11, 74.61, 75.20, 76.11, 79.74 (d, $J_{C\text{-}P}$ = 4.8 Hz, C-2), 82.62, 97.41 (d, $J_{C\text{-}P}$ = 15.7 Hz, anomeric C-1), 127.34, 127.46, 127.52, 127.57, 127.87, 127.93, 128.18, 128.22, 128.24, 128.30, 128.39, 128.46, 138.09, 138.19, 138.24, 138.41, 138.50, 138.54. ^{31}P {^{1}H} NMR (162 MHz, $CDCl_3$): δ 135.05. HRMS: Calculated for $C_{41}H_{43}O_7PCs$ (M + Cs^+), 567.1148. Found, 567.1130.

To prepare 6-deoxy-2,3,4-tri-*O*-benzyl-L-galactopyranose, acetyl chloride (0.3 ml, 3.2 mmol) is added to a solution of L-fucose (5.0 g, 30.5 mmol) in methanol (100 ml) and heated at reflux temperature overnight. The reaction mixture is concentrated under reduced pressure. The crude methyl 6-deoxy-L-galactopyranoside is used for the next step without purification.

Sodium hydride (5.4 g, purity 80%, 0.18 mol) is washed with THF (20 ml), dried, and suspended in dimethylformamide (DMF) (30 ml). A solution of crude methyl 6-deoxy-L-galactopyranoside described above in DMF (40 ml) is added to the suspension of NaH in DMF at 60°. The reaction mixture is stirred for 30 min at the same temperature, and benzyl bromide (20.6 g, 0.12 mol) is added. After stirring at 40°–50° overnight, the reaction mixture is quenched with methanol (3 ml) and concentrated under reduced pressure. The remaining residue is diluted with ethyl acetate (200 ml), washed with saturated aqueous $NaHCO_3$ (50 ml, 2 times) and saturated aqueous NaCl (50 ml, 2 times), dried over Na_2SO_4, filtered, and concentrated under reduced pressure. The crude product is suspended in acetic acid (120 ml) and 3 *M* sulfuric acid (15 ml), then stirred at 90° for 20 min. To the reaction mixture are added benzene (200 ml) and ice-cold water (100 ml), and the aqueous layer is separated. The organic layer is washed with saturated aqueous $NaHCO_3$ (50 ml, 2 times) and saturated aqueous NaCl (50 ml, 2 times) dried over Na_2SO_4, filtered, and concentrated under reduced pressure. The remaining residue is purified with silica gel column chromatography (ethyl acetate–hexane, 1 : 30, v/v) to give the desired product (5.0 g, 38% for three steps) as a 65 : 35 ($\alpha : \beta$) anomeric mixture.

Physical data for the product are as follows: ^1H NMR (500 MHz, $CDCl_3$): δ 1.14 (3H, d, J = 6.5 Hz, CH_3 of α anomer), 1.20 (3H, d, J = 6.5 Hz, CH_3 of β), 3.53 (1H, dd, J = 6.5 Hz, H-5 of β), 3.55 (1H, dd, J = 3.0, 9.5 Hz, H-3 of β), 3.59 (1H, dd, J = 0.5, 3.0 Hz, H-4 of β), 3.67 (1H, dd, J = 0.5, 2.5 Hz, H-4 of α), 3.73 (1H, dd, J = 7.5, 9.5 Hz, H-2 of β), 3.89 (1H, dd, J = 3.0, 10.0 Hz, H-3 of α), 4.04 (1H, dd, J = 3.5, 9.5 Hz, H-2 of α), 4.10 (1H, q, J = 6.5 Hz, H-5 of α), 4.63 (1H, dd, J = 4.0, 7.5 Hz, H-1 of β), 4.65–4.99 (12 d, benzylic protons), 5.26 (1H, dd, J = 2.0, 4.0 Hz, H-1 of α), 4.65–4.99 (benzylic protons), 7.26–7.36 (m, PhH). HRMS: Calculated for $C_{27}H_{30}O_5Cs$ (M + Cs^+), 567.1148. Found, 567.1130.

*Allyl 2-Acetamido-2-deoxy-6-O-tert-butyldiphenylsilyl-β-D-glucopyranoside (**11**).* A solution of allyl 2-acetamido-2-deoxy-β-D-glucopyranoside (7.0 g, 26.8 mmol), imidazole (2.0 g, 29.5 mmol), and *tert*-Bu$(Ph)_2$SiCl (8.10 g, 29.5 mmol) in DMF (150 ml) is stirred for 10 hr at room temperature. Water (2 ml) is added to the cooled mixture, and the mixture is concentrated. The residue is chromatographed on silica gel with chloroform–ethyl acetate–methanol (6 : 3 : 1, v/v/v) to give **11** (88%). ^1H NMR ($CDCl_3$): δ 1.03 (9H, s, ^{t}Bu), 2.02 (3H, s, $NHCH_2Cl_3$), 3.41 (1H, ddd, J = 3.41, 5.58, 9.31 Hz, H-5), 3.50–3.54 (1H, m, H-2), 3.54 (1H, t, J = 9.20 Hz, H-4), 3.69 (1H, dd, J = 8.65, 10.05 Hz, H-3), 3.87 (1H, dd, J = 5.58, 10.98 Hz, H-6), 3.96 (1H, dd, J = 3.28, 10.98 Hz, H-6′), 4.50 (1H, d, J = 8.31 Hz, H-1), 6.23 (1H, d, J = 6.07 Hz, NH). ^{13}C NMR ($CDCl_3$): δ 19.2, 23.6, 26.7, 57.6, 64.0, 69.4, 72.0, 75.1, 75.8, 99.3, 117.9, 127.6, 127.7, 129.7, 133.2, 133.3, 133.6, 135.6, 135.7, 172.4.

*Benzyl 3-O-Acetyl-2-deoxy-2-phthalimido-β-D-glucopyranoside (**12**).* A solution of benzyl 3,4,6-tri-*O*-acetyl-2-deoxy-2-phthalimido-β-D-glucopyranoside (15.68 g, 2.98 mmol) in methanol (50 ml) is treated with sodium methoxide (1.6 ml of a 25% solution in methanol). After 1 hr the solution is neutralized with Dowex 50 H^+ resin, filtered, and concentrated. The residue is dissolved in acetone (180 ml) and 2,2-dimethoxypropane (180 ml), and a catalytic amount of *p*-toluenesulfonic acid is added. Stirring of the solution continues overnight. The precipitate is filtered off, and to the filtrate is added aqueous $NaHCO_3$. The mixture is stirred, filtered, and concentrated under reduced pressure. The residue is diluted with chloroform, washed with brine, dried with Na_2SO_4, filtered, concentrated, and crystallized with chloroform–toluene to give benzyl 2-deoxy-4,6-*O*-isopropylidene-2-phthalimido-β-D-glucopyranoside (10.41 g, 79%), mp 205°.

Physical data for the product are as follows: ^{1}H NMR ($CDCl_3$): δ 1.42, 1.52 (3H, each s, CH_3C), 3.45 (1H, dt, J = 5.4, 9.8 Hz, H-5), 3.65 (1H, t, J = 9.2 Hz, H-4), 3.86 (1H, t, J = 10.4 Hz, H-6), 4.00 (1H, dd, J = 5.5, 10.8 Hz, H-6′), 4.23 (1H, dd, J = 8.5, 10.4 Hz, H-2), 4.45 (1H, dd, J = 8.9, 10.4 Hz, H-3), 4.49 (1H, d, J = 12.3 Hz, $PhCH_2$), 4.81 (1H, d, J = 12.3 Hz, $PhCH_2$), 5.22 (1H, d, J = 8.4 Hz, H-1), 7.00–7.08 (5H), 7.68–7.76 (4H). HRMS: Calculated for $C_{24}H_{25}NO_7Cs$ ($M + Cs^+$), 572.0685. Found, 572.0691.

Benzyl 2-deoxy-4,6-*O*-isopropylidene-2-phthalimido-β-D-glucopyranoside (1.0 g, 2.28 mmol) is stirred in pyridine–acetic anhydride (30 ml, 2 : 1, v/v) for 2.5 hr. The solution is evaporated *in vacuo*, coevaporated twice with toluene, dissolved in 60% aqueous acetic acid (40 ml), and stirred at 60° for 20 min. The solution is cooled, evaporated, and subjected to column chromatography (ethyl acetate) to give **12** (739 mg, 72%). ^{1}H NMR (400 MHz, $CDCl_3$): δ 1.94 (3H, s, OCH_2CH_3), 2.13 (1H, m, 6-OH), 2.96 (1H, d, J = 5.2 Hz, 2-OH), 3.63 (1H, m, H-5), 3.80 (1H, dd, J = 4.9, 9.3 Hz, H-4), 3.88 (1H, m, H-6), 3.99 (1H, m, H-6), 4.28 (1H, dd, J = 8.5, 10.7 Hz, H-2), 4.57 (1H, d, J = 12.2 Hz, $PhCH_2$), 4.82 (1H, d, J = 12.3 Hz, $PhCH_2$), 5.41 (1H, d, J = 8.5 Hz, H-1), 5.65 (1H, dd, J = 8.9, 10.7 Hz, H-3), 7.00–7.14 (5H), 7.71–8.80 (4H).

*Allyl 6-O-(tert-Butyldiphenylsilyl)-β-D-galactopyranosyl-(1,4)-2-acetamido-2-deoxy-6-O-(tert-butyldiphenylsilyl)-β-D-glucopyranoside (**15**).* To a cooled mixture of β-*O*-allyl-*N*-acetyllactosamine (890 mg, 2.1 mmol) and imidazole (315 mg, 4.62 mmol) in DMF (30 ml) is added dropwise *tert*-Bu(Ph)$_2$SiCl (1.21 g, 4.41 mmol; 1.15 ml) at 0°–5°, and the mixture is stirred for 10 hr at room temperature. Water (2 ml) is added to the cooled mixture, and the mixture is stirred for 30 min at room temperature and concentrated. The residue is chromatographed on silica gel with chloroform–ethyl acetate–methanol (12 : 7 : 1, v/v/v) to give **15** (871 mg, 46%) and **16** (152 mg, 7%). Physical data for **15** are as follows: ^{1}H NMR ($CDCl_3$): δ 1.017 (9H, s, ^{t}Bu), 1.023 (9H, s, ^{t}Bu), 1.95 (3H, s, $NHCH_2CH_3$), 3.37–3.40 (1H, m, H-5), 3.47 (1H, dd, J = 2.93, 9.17 Hz, H-3′), 3.65 (1H, br t, J = 8.4 Hz, H-2′), 3.96 (1H, br d, J = 2.6 Hz, H-4′), 4.46 (1H, d, J = 7.8 Hz, H-1′), 4.62 (1H, d, J = 8.3 Hz, H-1). HRMS: Calculated for $C_{49}H_{65}NO_{11}Si_2Na$ $(M + Na)^+$, 1032.4096. Found, 1032.0439.

Glycosidations

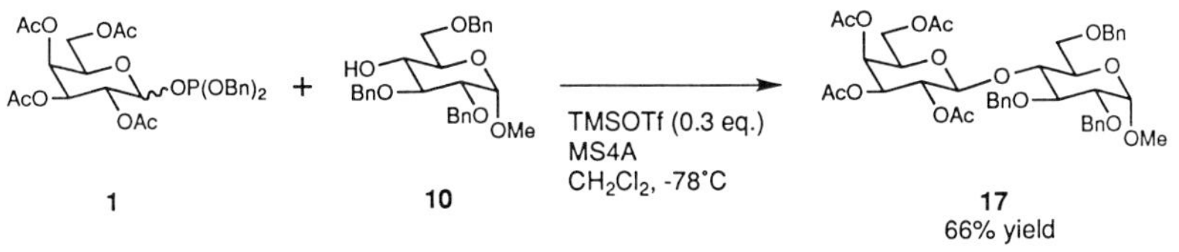

AcO OAc O OP(OBn)2 + HO OBn O BnO BnO OMe → TMSOTf (0.3 eq.) MS4A CH_2Cl_2, -78°C

1 10 17 66% yield

SCHEME 2. Glycosidations.

Acetylation of **15** with acetic anhydride and pyridine gives the corresponding acetate quantitatively: ^{1}H NMR ($CDCl_3$): δ 1.03 (9H, s, ^{t}Bu), 1.07 (9H, s, ^{t}Bu), 1.78, 1.79, 1.95, 2.00, 2.02 (5 3H, s, 4 OCH_2CH_3, NCH_2CH_3), 3.30–3.34 (1H, m, H-5), 3.53 (1H, *t*, *J* = 8.8 Hz, H-6′a), 3.62–3.68 (1H, m, H-5′), 3.71–3.78 (1H, m, H-6′b), 3.87 (1H, dd, *J* = 2.9, 11.27 Hz, H-6a), 3.92 (1H, dd, *J* = 2.5, 11.27 Hz, H-6b), 4.02–4.13 (3H, m, allylic, H-2,4), 4.27–4.34 (1H, m, allylic), 4.40 (1H, d, *J* = 7.6 Hz, H-1), 4.68–4.71 (1H, m, H-1′), 4.94 (1H, t, *J* = 9.2 Hz, H-3), 5.01–5.03 (2H, m, H-2′,3′), 5.57 (1H, d, *J* = 1.3 Hz, H-4′), 5.68 (1H, d, *J* = 9.5 Hz, NH).

Glycosidation Reactions

Scheme 2 depicts an example of glycosidation using glycosyl phosphites. Compounds **18–26** can be prepared by a similar strategy.

*Methyl 2,3,4,6-Tetra-O-acetyl-β-D-galactopyranosyl-(1,4)-2,3,6-tri-O-benzyl-α-D-glycopyranoside (**17**).* A 10% solution of TMSOTf in CH_2Cl_2 (74 ml, 38 μmol) is added dropwise to a mixture of dibenzyl 2,3,4,6-tetra-*O*-acetyl-D-galactopyranosyl phosphite (**1**, 2.75 mg, 0.13 mmol), methyl 2,3,6-tri-*O*-benzyl-α-D-glucopyranoside (**10**, 15.69 mg, 0.13 mmol), and molecular sieves 4 Å in anhydrous CH_2Cl_2 (1.6 ml) at −78° under argon (Scheme 2). After stirring for 30 min at the same temperature, the reaction is quenched with saturated aqueous $NaHCO_3$, warmed to room temperature, and extracted with CH_2Cl_2. The combined organic layers are washed with saturated aqueous NaCl, dried over anhydrous Na_2SO_4, filtered, and concentrated under reduced pressure. The remaining residue is purified by silica gel column chromatography (ethyl acetate–*n*-hexane) to provide **17** as colorless syrup (66 mg, 66% yield).

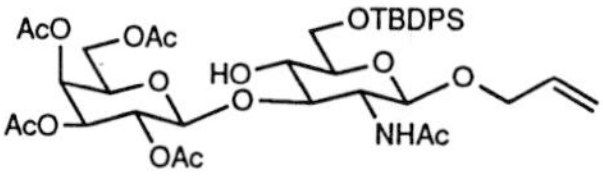

18

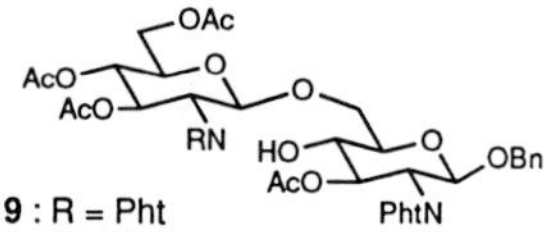

19 : R = Pht
20 : R = TrocH

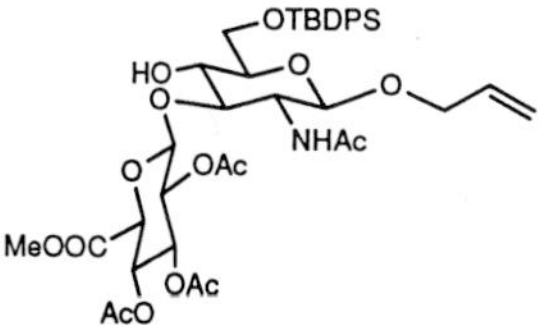

21

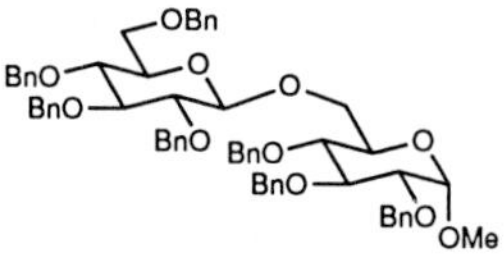

22

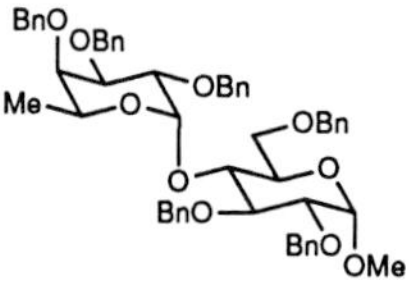

α-**23**

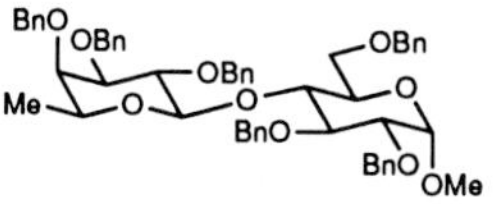

β-**23**

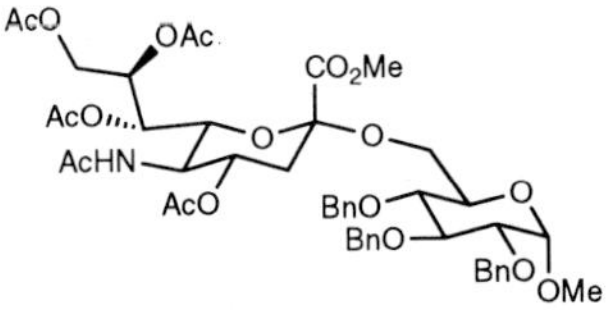

23

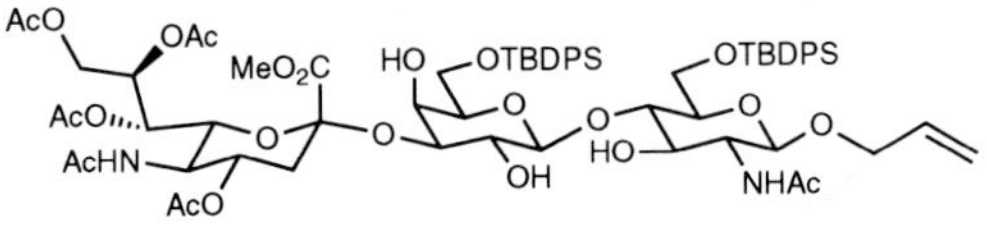

24

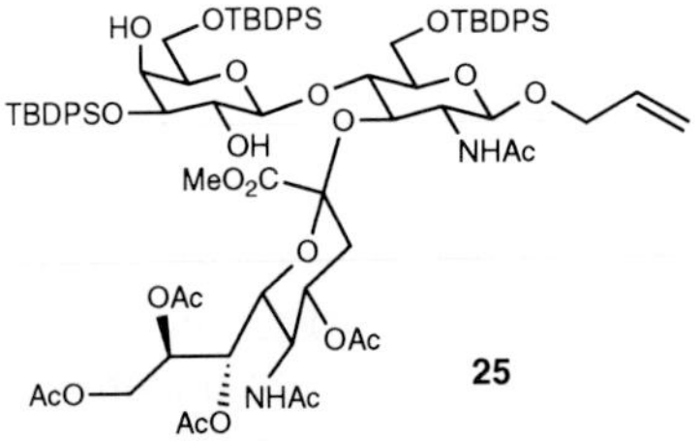

25

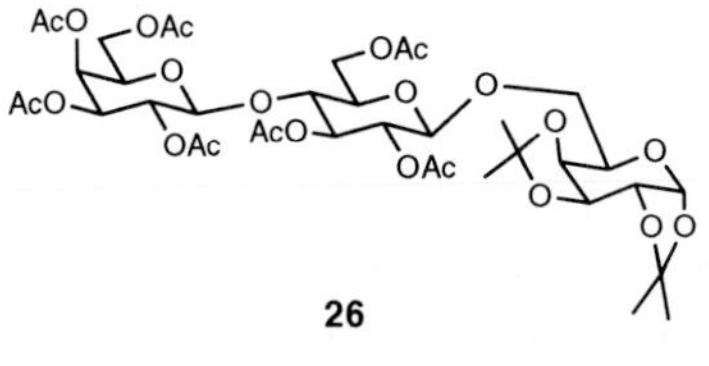

26

Physical data for **17** are as follows: ^{1}H NMR (400 MHz, $CDCl_3$): δ 1.95, 1.96, 2.00, 2.09 (3H, each s, O_2CCH_3), 3.37 (3H, s, OCH_3), 3.51 (1H, dd, J = 3.6, 9.6 Hz, Glu H-2), 3.50 (1H, dd, J = 9.0, 9.5 Hz, Glu H-4), 3.59 (1H, dd, J = 1.9, 10.5 Hz, Glu H-6), 3.63 (1H, brddd, Glu H-5), 3.74 (1H, dd, J = 2.8, 10.5 Hz, Glu H-6), 3.86 (1H, dd, J = 9.5, 9.6 Hz, Glu H-3), 3.89 (2H, m, Gal H-5), Gal H-6), 3.96 (1H, dd, J = 8.2,

11.2 Hz, Gal H-6), 4.41 (1H, d, J = 12.1 Hz, one of OCH_2Ph), 4.46 (1H, d, J = 8.0 Hz, Gal H-1), 4.59 (1H, d, J = 3.6 Hz, Glu H-1), 4.63 (1H, d, J = 12.4 Hz, one of OCH_2Ph), 4.75 (1H, dd, J = 3.4, 10.4 Hz, Gal H-3), 4.75 (1H, d, J = 12.9 Hz, two of OCH_2Ph), 4.82 (1H, d, J = 10.8 Hz, one of OCH_2Ph), 4.95 (1H, d, J = 10.9 Hz, one of OCH_2Ph), 5.08 (1H, dd, J = 8.0, 10.4 Hz, Gal H-2), 5.24 (1H, dd, J = 3.4, 0.8 Hz, Gal H-4), 7.25–7.42 (15H, m, aromatic protons). ^{13}C NMR (100 MHz, $CDCl_3$): δ 20.6, 20.6, 20.8, 29.7, 55.4, 60.6, 66.7, 67.5, 69.6, 70.0, 70.3, 71.1, 73.5, 73.7, 75.2, 78.9, 79.7, 98.4, 100.1, 127.3, 127.5, 127.8, 128.1, 128.2, 128.4, 128.6, 137.6, 138.3, 139.22, 169.1, 170.1, 170.2, 170.2. IR (neat): 3088, 3030, 29.32, 2869, 1748, 1497, 1454, 1369, 1223, 1048, 912, 735, 699 cm^{-1}. HRMS: Calculated for $C_{42}H_{50}O_{15}Cs$ (M + Cs^+), 927.2204. Found, 927.2211. $[\alpha]_D^{25}$ −3 (c 0.5, $CHCl_3$).

*Allyl 2,3,4,6-Tetra-O-acetyl-β-D-galactopyranosyl-(1,3)-2-acetamido-2-deoxy-6-tert-butyldiphenylsilyl-β-D-glucopyranoside (**18**).* A solution of **1** (64 mg, 0.128 mmol), **9** (76 mg, 0.128 mmol), and molecular sieves 3 Å in CH_2Cl_2 (2 ml) is cooled to −78°, and TMSOTf (14 mg, 63 μmol) is added. After stirring for 1 hr at −78°, the reaction is quenched with triethylamine and washed with saturated $NaHCO_3$. The organic layer is dried over Na_2SO_4, filtered, and concentrated under reduced pressure. The residue is chromatographed on silica gel ($CHCl_3$–methanol, 15 : 1, v/v) to give **18** (66 mg, 62%) as colorless plates: mp 179°–181°.

Physical data for **18** are as follows: ^1H NMR ($CDCl_3$): δ 1.08 (9H, s, ^{t}Bu), 1.72, 1.99, 2.03, 2.06, 2.16 (3H, each s, OCH_2CH_3 and $NHCH_2CH_3$), 3.42 (1H, bd, J = 8.7 Hz, GlcN H-5), 3.54 (1H, ddd, J = 7.9, 8.0, 9.5 Hz, GlcN H-2), 3.80–3.85 (3H, m, GlcN H-4, GLcN H-6, Gal H-6), 3.89–3.95 (2H, m, GlcN H-6′ and Gal H-6′), 4.00 (1H, dd, J = 9.12, 9.44 Hz, GlcN H-3), 4.06 (1H, dd, J = 6.0, 12.8 Hz, allylic proton), 4.14 (1H, bd, J = 6.56 Hz, GalN H-5), 4.32 (1H, dd, J = 5.0, 12.8 Hz, allylic proton), 4.70 (1H, d, J = 8.0 Hz, Gal H-1), 4.77 (1H, d, J = 8.0 Hz, GlcN H-1), 4.97 (1H, dd, J = 3.4, 10.4 Hz, Gal H-3), 5.189 (1H, dd, J = 1.16, 10.4 Hz, allylic proton), 5.19 (1H, dd, J = 8.0, 10.4 Hz, Gal H-2), 5.26 (1H, dd, J = 1.6, 17.0 Hz, allylic proton), 5.36 (1H, bd, J = 3.4 Hz, Gal H-4), 5.64 (1H, d, J = 8.0 Hz, NH), 5.84–5.94 (1H, m, allylic proton), 7.37–7.43 (6H, m, phenyl protons), 7.71–7.75 (4H, m, phenyl protons). ^{13}C NMR ($CDCl_3$): δ 19.31, 20.31, 20.50, 20.55, 20.60, 23.64, 26.79, 55.66, 61.18, 61.87, 66.85, 68.79, 69.26,.70.75, 71.22, 71.26, 74.52, 79.81, 98.87, 100.90, 117.42, 127.63, 127.65, 127.86, 129.84, 129.89, 133.88, 135.48, 135.93, 169.17, 169.90, 170.10, 170.38, 170.46. HRMS: Calculated for $C_{41}H_{55}NO_{15}SiCs$ (M + Cs^+), 962.2395. Found, 962.2380.

*Benzyl 3,4,6-Tri-O-acetyl-2-deoxy-2-phthalimido-β-D-glucopyranosyl-(1,6)-2-deoxy-2-phthalimido-3-O-acetyl-β-D-glucopyranoside (**19**).* A solution of **2** (140 mg, 0.22 mmol) in CH_2Cl_2 (1 ml) is added to a solution of **12** (99 mg, 0.23 mmol), TMSOTf (22 μl, 0.12 mmol), and molecular sieves 3 Å in CH_2Cl_2 (2 ml) at $-78°$. After 3 hr when the solution reaches room temperature, the reaction is quenched with saturated aqueous $NaHCO_3$, diluted with CH_2Cl_2, and washed with saturated $NaHCO_3$. The organic solvent is dried over Na_2SO_4, filtered, and concentrated under reduced pressure. The residue is chromatographed on silica gel (hexane–ethyl acetate, 2 : 3, v/v) to give **19** (97 mg, 49%) as a colorless syrup.

Physical data for **19** are as follows: $[\alpha]_D^{25}$ $-21.2°$ (*c* 0.8, $CHCl_3$). ^1H NMR (400 MHz, $CDCl_3$): δ 1.84, 1.88, 2.04, 2.13 (3H, each s, OCH_2CH_3), 3.01 (1H, OH), 3.47 (1H, t, *J* = 9.3 Hz, H-4), 3.65–3.71 (1H, m, H-5), 3.79–3.91 (2H, H-5′ and H-6), 4.10–4.35 (4H, H-6, 2 H-6′, benzyl protons), 4.12 (1H, dd, *J* = 8.5, 10.7 Hz, H-2), 4.41 (1H, dd, *J* = 8.5, 10.7 Hz, H-2′), 4.56 (1H, d, *J* = 12.2 Hz, benzyl protons), 5.21 (1H, t, *J* = 9.6 Hz, H-4′), 5.23 (1H, d, *J* = 8.6 Hz, H-1), 5.55 (1H, dd, *J* = 8.8, 10.6 Hz, H-3), 5.56 (1H, d, *J* = 8.5 Hz, H-1′), 5.82 (1H, dd, *J* = 9.2, 10.6 Hz, H-3′), 6.97–7.11 (5H, phenyl protons), 7.62–7.79 (8H, phthalimido protons). ^{13}C NMR (100 MHz, $CDCl_3$): δ 20.4, 20.6, 20.7, 54.4, 61.7, 68.7, 70.2, 70.5, 70.6, 71.9, 73.2, 74.9, 96.6, 98.3, 123.4, 136.8, 167.5, 169.4, 170.1, 170.8, 171.2. HRMS: Calculated for $C_{43}H_{42}N_2O_{17}Cs$ ($M + Cs^+$), 991.1538. Found, 991.1546.

*Benzyl 3,4,6-Tri-O-acetyl-2-deoxy-2-(2,2,2-trichloroethoxycarbonylamido)-β-D-glucopyranosyl-(1,6)-2-deoxy-2-phthalimidio-3-O-acetyl-β-D-glucopyranoside (**20**).* A solution of **3** (124 mg, 0.17 mmol) in CH_2Cl_2 (1 ml) is added to a solution of **12** (74 mg, 0.17 mmol), TMSOTf (17 μl, 90 μmol), and molecular sieves 3 Å in CH_2Cl_2 (2 ml) at $-78°$. After 3 hr when the solution reaches room temperature the reaction is quenched with saturated aqueous $NaHCO_3$ and extracted with CH_2Cl_2 (2 times, 20 ml). The organic layer is washed with saturated $NaHCO_3$ and then saturated NaCl, dried over Na_2SO_4, filtered, and concentrated under reduced pressure. The residue is chromatographed on silica gel (hexane–ethyl acetate, 2 : 3, v/v) to give **20** (97 mg, 63%) as a colorless syrup.

Physical data for **20** are as follows: $[\alpha]_D^{25}$ $-24.7°$ (*c* 0.8, $CHCl_3$). ^1H NMR (400 MHz, $CDCl_3$): δ 1.85, 2.02, 2.05 (3H, each s, OCH_2CH_3), 3.00–3.11 (2H), 3.63–3.80 (4H), 3.92 (1H, dd, *J* = 4.2, 11.4 Hz, H-3), 4.16–4.32 (4H), 4.55 (1H, d, *J* = 12.4 Hz, benzyl proton), 4.66 (1H, d, *J* = 12.4 Hz, CH_2CCl_3), 4.79 (1H, d, *J* = 12.4 Hz, CH_2CCl_3), 4.84 (1H, d, *J* = 12.4 Hz, benzyl proton), 5.08 (1H, t, *J* = 9.6 Hz, H-4′), 5.26 (1H, t, *J* = 9.9 Hz, H-3′), 5.35 (2H, d, *J* = 8.4 Hz, H-1, 1′), 5.63 (1H, dd, *J* = 8.3, 10.7 Hz, H-3), 7.05–7.12 (5H, phenyl protons), 7.70–7.82 (4H, phenyl protons). ^{13}C NMR (100 MHz, $CDCl_3$): δ 20.6, 20.7, 54.6, 55.9, 61.9, 68.5, 69.8, 71.1, 71.7, 72.0, 73.2, 74.4, 74.8, 95.3, 97.3, 101.1,

123–137.0, 154.3, 167.6, 169.4, 170.6, 170.8, 171.1. HRMS: Calculated for $C_{38}H_{41}C_{13}N_2O_{17}Cs$ (M + Cs^+), 1035.0525. Found, 1035.0558.

*Allyl [Methyl (2,3,4-Tri-O-acetyl-β-D-glucopyranosyl)uronate]-(1,3)-2-acetamido-2-deoxy-6-O-tert-butyldiphenylsilyl-β-glucopyranoside (**21**).* A solution of **4** (0.29 g, 0.5 mmol), **11** (0.25 g, 0.5 mmol), and molecular sieves 3 Å in CH_2Cl_2 (5 ml) is cooled to −78°, and TMSOTf (56 mg) is added. After stirring for 1 hr at −78°, the reaction is quenched with triethylamine and washed with saturated $NaHCO_3$. The organic solvents are dried over Na_2SO_4, filtered, and concentrated under reduced pressure. The residue is chromatographed on silica gel ($CHCl_3$–methanol, 15 : 1, v/v) to give **21** (0.07 g, 18%) as colorless oil.

Physical data for **21** are as follows: ^{1}H NMR (400 MHz, $CDCl_3$): δ 1.07 (9H, s, ^{t}Bu), 2.02, 2.03, 2.038, 2.40 (3H, each s, OCH_2CH_3 and $NHCH_2CH_3$), 3.40 (1H, bd, J = 9.0 Hz, GlcN H-5), 3.51 (1H, ddd, J = 8.0, 8.2, 9.0 Hz, GlcN H-2), 3.75 (3H, s, $COOCH_3$), 3.80–3.92 (3H, m, GlcN H-4, GlcN H-6, and GlcUA H-5), 4.04–4.15 (3H, m, GlcN H-3, GlcN H-6′, and allyl proton), 4.33 (1H, dd, J = 5.10, 12.8 Hz, allyl proton), 4.79 (1H, d, J = 8.0 Hz, GlcN-H-1), 4.81 (1H, d, J = 7.0 Hz, GlcUA H-1), 5.02 (1H, dd, J = 8.2, 8.8 Hz, GlcUA H-3), 5.16–5.30 (4H, m, GlcUA H-2, GlcUA H-4, and allyl protons), 5.82 (1H, d, J = 7.8 Hz, NH), 5.85–5.95 (1H, m, allyl proton), 7.36–7.46 (6H, m, phenyl protons), 7.73–7.77 (4H, m, phenyl protons). ^{13}C NMR (100 MHz, $CDCl_3$): δ 19.3, 20.2, 20.4, 20.5, 23.7, 26.7, 53.2, 56.9, 61.6, 68.8, 69.3, 70.5, 71.2, 71.7, 71.9, 74.3, 80.2, 98.9, 100.2, 117.4, 127.6, 127.9, 129.8, 129.9, 133.9, 135.4, 135.9, 166.6, 168.9, 169.4, 169.8, 170.6. HRMS: Calculated for $C_{40}H_{53}NO_{15}Cs$ (M + Cs^+), 948.2239. Found, 948.2206.

*Methyl 2,3,4,6-Tetra-O-benzyl-β-D-glucopyranosyl-(1,6)-2,3,4-tri-O-benzyl-α-D-glucopyranoside (**22**).* A 10% solution of TMSOTf in CH_2Cl_2 (50 μl, 26 μmol) is added dropwise to a mixture of **8** (68 mg, 87 μmol), methyl 2,3,4-tri-*O*-benzyl-α-D-glucopyranose (**13**) (41 mg, 88 μmol), and molecular sieves 4 Å in anhydrous CH_2Cl_2 (1.2 ml) under argon at −78°. After stirring for 1 hr at the same temperature, the reaction is quenched with saturated aqueous $NaHCO_3$, warmed to room temperature, and extracted with CH_2Cl_2. The combined organic layers are washed with saturated aqueous NaCl, dried over anhydrous Na_2SO_4, filtered, and concentrated under reduced pressure. The remaining residue is purified by silica gel column chromatography (ethyl acetate–*n*-hexane) to provide **22** as a colorless powder (57 mg, 66% yield), mp 191°–121°, $[\alpha]_D^{25}$ +23 (c 0.9, $CHCl_3$). All ^{1}H NMR ($CDCl_3$), ^{13}C NMR ($CDCl_3$), IR, and FAB-MS data are in agreement with those reported previously.[11]

[11] S.-I. Hashimoto, T. Honda, and S. Ikegami, *J. Chem. Soc., Chem. Commun.*, 685 (1989).

*Methyl 6-Deoxy-2,3,4-tri-O-benzyl-α-L-galactopyranosyl-(1,4)-2,3,6-tri-O-benzyl-α-D-glucopyranoside and Methyl 6-Deoxy-2,3,4-tri-O-benzyl-β-L-galactopyranosyl-(1,4)-2,3,6-tri-O-benzyl-α-D-glucopyranoside (α-**23** and β-**23**).* A 10% (v/v) solution of TMSOTf in ether (53 μl, 27 μmol) is added dropwise to a mixture of β-**9** (43.6 mg, 94 μmol), **10** (62.4 mg, 92 μmol), and molecular sieves 4 Å in ether (3 ml) at −78° under argon. The reaction mixture is stirred for 30 min at −78° and quenched with triethylamine (0.1 ml) and saturated aqueous $NaHCO_3$ (5 ml). The mixture is extracted with CH_2Cl_2 (30 ml, 3 times), and the combined organic layer is washed with saturated aqueous NaCl (20 ml, 2 times) dried over Na_2SO_4, filtered, and concentrated under reduced pressure. The remaining residue is purified by silica gel chromatography in ethyl acetate–*n*-hexane (1 : 10, v/v) to give 9.0 mg of β-**23** (11%), 27.4 mg of α-**23** (34%), and 19 mg of recovered **10** (43%).

Physical data for α-**23** are as follows: R_f 0.55 (ethyl acetate–*n*-hexane, 1:2, v/v). ^{1}H NMR (400 MHz, $CDCl_3$): δ 0.66 (3H, d, J = 6.4 Hz, CH_3 of fucose), 3.34 (3H, s, OCH_3), 3.34 (1H, brs, Glc H-6), 3.57 (1H, dd, J = 3.6 Hz, 9.4 Hz, Glc H-2), 3.64 (1H, brs, Fuc H-4), 3.78 (1H, brs, Glc H-6), 3.79 (1H, dd, J = 9.8, 9.8 Hz, Glc H-4), 3.84 (1H, dd, J = 2.7, 10.3 Hz, Fuc H-3), 3.90 (1H, dd, J = 9.2, 9.2 Hz, Glc H-3), 3.97 (1H, dd, J = 3.6, 10.3 Hz, Fuc H-2), 3.98–4.00 (1H, m, Fuc H-5), 4.32 (1H, d, J = 12.2 Hz, benzylic proton), 4.37 (1H, d, J = 12.2 Hz, benzylic proton), 4.53–4.62 (4H, 4 d, benzylic protons), 4.57 (1H, d, J = 4.0 Hz, Glc H-1), 4.68–4.78 (5H, 5 d, benzylic protons), 4.94 (1H, d, J = 11.6 Hz, benzylic proton), 4.99 (1H, d, J = 3.6 Hz, Fuc H-1), 5.05 (1H, d, J = 10.9 Hz, benzylic proton), 7.21–7.41 (30 H, m, PhH). ^{13}C NMR (100 MHz, $CDCl_3$): δ 16.21, 55.00, 66.62, 68.53, 70.28, 72.70, 73.18, 73.29, 73.68, 74.33, 74.70, 75.56, 76.28, 77.52, 79.45, 80.19, 80.51, 97.58, 97.71, 127.32, 127.50, 127.54, 127.58, 127.65, 127.78, 127.92, 128.01, 128.10, 128.17, 128.21, 128.25, 128.31, 128.38, 128.42. HRMS: Calculated for $C_{55}H_{66}O_{10}Cs$ (M + Cs^+), 1013.3241. Found, 1013.3258.

Physical data for β-**23** are as follows: R_f 0.59 (ethyl acetate–*n*-hexane, 1:2, v/v). ^{1}H NMR (400 MHz, $CDCl_3$): δ 1.05 (3H, d, J = 6.4 Hz, CH_3 of fucose), 3.29 (1H, dq, J = 6.2, 2.9 Hz, Fuc H-5), 3.41 (1H, dd, J = 3.0 Hz, 9.7 Hz, Fuc H-3), 3.42 (3H, s, OCH_3), 3.49 (1H, br d, J = 2.9 Hz, Fuc H-4), 3.51 (1H, dd, J = 3.5, 9.2 Hz, Glc H-2), 3.63 (1H, dd, J = 5.2, 10.8 Hz, Glc H-6), 3.72 (1H, dd, J = 7.8, 9.6 Hz, Fuc H-2), 3.72–3.75 (1H, m, Glc H-5), 3.84 (1H, dd, J = 8.7, 10.9 Hz, Glc H-4), 3.87 (1H, dd, J = 1.9, 10.7 Hz, Glc H-6), 3.96 (1H, dd, J = 8.9, 9.2 Hz, Glc H-3), 4.53 (1H, d, J = 12.1 Hz, benzylic proton), 4.60–4.66 (3H, 3 d, benzylic protons), 4.62 (1H, d, J = 3.6 Hz, Glc H-1), 4.72–4.85 (6H, 6 d, benzylic protons), 4.57 (1H, d, J = 3.6 Hz, Glc H-1), 4.80 (1H, d,

J = 7.4 Hz, Fuc H-1), 4.99 (1H, d, J = 11.8 Hz, benzylic proton), 5.00 (1H, d, J = 11.3 Hz, benzylic proton), 7.20–7.39 (30 H, m, PhH). ^{13}C NMR (100 MHz, $CDCl_3$): δ 16.67, 55.28, 69.09, 69.76, 70.27, 73.09, 73.35, 73.38, 73.82, 74.70, 75.31, 75.39, 77.34, 79.71, 79.94, 82.21, 82.30, 97.86, 102.93, 127.21, 127.32, 127.45, 127.49, 127.57, 127.92, 127.96, 128.02, 128.13, 128.17, 128.38, 128.46, 128.54, 138.10, 138.66, 138.81, 138.85. HRMS: Calculated for $C_{55}H_{66}O_{10}Cs$ (M + Cs^+), 1013.3241. Found, 1013.3260.

*Methyl [Methyl (5-Acetamido-4,7,8,9-tetra-O-acetyl-3,5-dideoxy-α-D-glycero-D-galacto-2-nonulopyranosyl)onate]-(2,6)-2,3,4-tri-O-benzyl-α-D-glucopyranoside (**24**).* A solution of phosphite **6** (51 mg, 70 μmol), methyl glucopyranoside **13** (48 mg, 0.10 mmol), and molecular sieves 3 Å in CH_3CN (1.5 ml) is cooled to −42°, and TMSOTf (3.0 mg, 10 μmol) is added. After the mixture is stirred for 30 min at −42°, the reaction is quenched with saturated aqueous $NaHCO_3$ and the mixture warmed to room temperature. The mixture is added to ethyl acetate and washed with saturated $NaHCO_3$. The organic solvents are dried over Na_2SO_4, filtered, and concentrated under reduced pressure. The residue is chromatographed on silica gel (CH_2Cl_2–methanol, 25 : 1, v/v) to provide **24α** (42.1 mg, 68%) and **24β** (7.9 mg, 12%) as colorless syrups.

The α Anomer of **24** shows the following: ^{1}H NMR ($CDCl_3$): δ 1.80, 1.85, 2.00, 2.01, 2.12 (3H each, s, OCH_2CH_3 and NCH_2CH_3), 1.96 (1H, t, J = 12.9, Hz, H-3ax), 2.65 (1H, dd, J = 4.76, 12.96 Hz, H-3eq), 3.35 (3H, s, OMe), 3.41 (1H, dd, J = 1.70, 10.6 Hz, Glc H-6), 3.51 (1H, dd, J = 3.52, 9.68 Hz, Glc H-2), 3.59 (1H, dd, J = 9.1, 10.0 Hz, Glc H-4), 3.74 (3H, s, $COOCH_3$), 3.73–3.79 (2H, m, Glc H-5 and NeuAc H-9), 3.94 (1H, dd, J = 9.2, 9.6 Hz, Glc H-3), 3.99 (1H, ddd, J = 10.4, 10.8, 9.8 Hz, NeuAc H-5), 4.03 (1H, dd, J = 2.6, 12.6 Hz, NeuAc H-9′), 4.09 (1H, dd, J = 2.1, 10.8 Hz, NeuAc H-6), 4.22 (1H, dd, J = 4.0, 10.6 Hz, Glc H-6′), 4.60 (1H, d, J = 3.6 Hz, Glc H-1), 4.74 (1H, d, J = 10.8 Hz, $PhCH_2$), 4.77 (1H, d, J = 10.7 Hz, $PhCH_2$), 4.79 (1H, d, J = 12.2 Hz, $PhCH_2$), 5.09 (1H, d, J = 9.8 Hz, NeuAc NH), 5.25 (1H, dd, J = 2.1, 9.4 Hz, NeuAc H-7), 5.32 (1H, ddd, J = 2.6, 4.7, 99.4 Hz, NeuAc H-8), 7.25–7.36 (15H, m, Ph). HRMS: Calculated for $C_{48}H_{59}NO_{18}Cs$ (M + Cs^+), 1070.2786. Found, 1070.2788.

*Allyl (Methyl (5-Acetamido-4,7,8,9-tetra-O-acetyl-3,5-dideoxy-α-D-glycero-D-galacto-2-nonulopyranosyl)onate]-(2,3)-[6-O-(tert-butyldiphenylsilyl)-β-D-galactosyl]-(1,4)-2-acetamido-2-deoxy-6-O-(tert-butyldiphenylsilyl)-β-D-glucopyranoside (**25**).* To a stirred solution of **15** (102 mg, 0.11 mmol) and molecular sieves 3 Å in dry CH_3CN (0.5 ml) is added TMSOTf (4 mg) at 0° under an argon atmosphere. The reaction mixture is cooled to −40°, and phosphite **6** (56 mg, 76 μmol) is added dropwise over 20 min. After the addition is over, the mixture is allowed to warm to −32° to −30° and is stirred for 1 hr at the same temperature. The reaction mixture is diluted with cold ethyl acetate and quenched with cold saturated aqueous $NaHCO_3$. The organic layer is separated and dried over anhydrous sodium sulfate. The solution is evaporated *in vacuo* to give a crude material, which is chromatographed on a silica gel column ($CHCl_3$–CH_3OH, gradient elution from 25 : 1 to 15 : 1, v/v) to give **25** (28 mg, 27%) as a colorless syrup and unreacted **15** (75 mg). For **25**, ^{1}H-NMR ($CDCl_3$): δ 1.04, 1.09 (9H each, s, ^{t}Bu), 1.52, 1.82, 1.83, 1.99, 2.06, 2.13 (3H each, s, OCH_2CH_3 and $NHCH_2CH_3$), 2.70 (1H, dd, J = 4.6, 13.1 Hz, H-3eq of NeuAc), 3.28–3.31 (2H, m), 3.40–3.48 (1H, m), 3.54–3.59 (1H, m), 3.69 (3H, s, $COOCH_3$), 3.75–3.98 (7H, m), 4.02–4.15 (3H, m), 4.18–4.30 (4H, m), 4.82–4.90 (2H, m), 4.92 (1H, ddd, J = 5.2, 10.7, 11.3 Hz, H-4 of NeuAc), 4.98 (1H, dd, J = 1.7, 12.3 Hz, H-9′ of NeuAc), 4.99 (1H, m), 5.13–5.32 (1H, m), 5.21–5.32 (2H, m), 5.45–5.50 (1H, m), 5.65–5.78 (2H, m), 6.55 (1H, bd), 7.03 (1H, bd), 7.27–7.48 (12H, m, phenyl protons), 7.57–7.62 (2H, m, phenyl protons), 7.70–7.78 (6H, m, phenyl protons). HRMS: Calculated for $C_{69}H_{92}N_2O_{23}Si_2Cs$ (M + Cs^+), 1505.4684. Found, 1505.4696.

*Allyl [3,6-Di(tert-butylphenylsilyl)-β-D-galactopyranosyl]-(1,4)-[methyl (5-acetamido-4,7,8,9-tetra-O-acetyl-3,5-dideoxy-D-glycero-α-D-galacto-2-nonulopyranosyl)onate]-(2,3)-2-acetamido-2-deoxy-6-O-(tert-butyldiphenylsilyl)-β-glucopyranoside (**26**).* To a stirred solution of **16** (36 mg, 32 μmol) and molecular sieves 3 Å in dry CH_3CN (0.5 ml) is added TMSOTf (2 mg) at 0° under an argon atmosphere. The reaction mixture is cooled at −40°, and phosphite **6** (28 mg, 38 μmol) is added dropwise over 20 min. After the addition is over, the mixture is allowed to warm to −32° to −30° and is stirred for 2 hr at the same temperature. The reaction mixture is diluted with cold ethanol acetate and quenched with cold saturated aqueous $NaHCO_3$. The organic layer is separated, dried over anhydrous Na_2SO_4, and evaporated *in vacuo* to give a crude material, which is chromatographed on a silica gel column ($CHCl_3$–CH_3OH, gradient elution from 25 : 1 to 15 : 1, v/v) to give **26** (27 mg, 44%) as a colorless syrup, and some unreacted **16** (9 mg) is recovered. For **26**, ^{1}H NMR ($CDCl_3$): δ 0.95, 0.97, 1.15 (9H each, s, ^{t}Bu), 1.81, 1.86, 2.01, 2.06, 2.16 (3H each, s, OCH_2CH_3 and $NHCH_2CH_3$), 2.67 (1H, dd, J = 4.9, 13.1 H-3eq of NeuAc), 2.78 (1H, bs), 3.26 (1H, m), 3.50–3.58 (2H, m), 3.62–3.67 (1H, m), 3.68 (3H, s, $COOCH_3$), 3.75–4.12 (11H, m), 4.26–4.30 (2H, m), 4.78–5.10 (5H, m), 5.45–5.55 (2H, m), 5.60 (1H, m, H-8 of NeuAc), 5.78–5.82 (1H, m), 6.99 (1H, d, J = 10.1 Hz), 7.04 (1H, d, J = 6.0 Hz), 7.16–7.42 (20H, m), 7.50 (2H, bd), 7.60 (2H, bd), 7.68 (4H, bd), 7.77 (2H, bd). HRMS: Calculated for $C_{85}H_{110}N_2O_{23}Si_3Cs$ (M + Cs^+), 1743.5862. Found 1743.5747. ^{1}H NMR ($CDCl_3$) for compound **16**: δ 0.98, 0.99, 1.10

(9H, s, $_t$Bu), 1.91 (3H, s, $NHCH_2CH_3$), 3.98–4.05 (2H, m), 4.25–4.33 (1H, m), 4.40–4.45 (1H, m), 4.72 (1H, d, J = 8.4 Hz, H-1), 5.14–5.18 (1H, m), 5.21–5.27 (1H, m), 5.41 (1H, d, J = 7.72 Hz, H-1′), 5.32–5.42 (1H, m).

*2,3,4,6-Tetra-O-acetyl-β-D-galactopyranosyl-(1,4)-2,3,6-tri-O-acetyl-β-D-glucopyranosyl-(1,6)-1,2:3,4-di-O-isopropylidene-α-D-galactopyranose (**27**).* A solution of **7** (172 mg, 0.195 mmol) and diisopropylidene-D-galactose **14** (74 mg, 0.285 mmol) and in CH_2Cl_2 (2 ml) is stirred with crushed molecular sieves 3 Å under argon for 45 min. The mixture is cooled to −42°, and TMSOTf (10 μl) is added. After stirring for 30 min at −42°, the mixture is warmed to 0° over 20 min and quenched with 1 ml saturated $NaHCO_3$. The organic solvents are dried over Na_2SO_4, filtered, and concentrated under reduced pressure. The residue is chromatographed on silica gel (hexane/ethyl acetate 6:4 to 3:7) to give **27** (69.2 mg, 40%). ^{1}H NMR (400 MHz, $CDCl_3$): δ 1.32 (6H, s, 2 CH_3), 1.45 (3H, s, CH_3), 1.50 (3H, s, CH_3), 1.98, 2.05, 2.08, 2.14, 2.17 (3H, each s, OCH_2CH_3), 3.58–3.70 (2H, m), 3.75–3.95 (3H, m), 3.99 (1H, dd, J = 3.6, 11.1 Hz), 4.05–4.20 (5H, m), 4.29 (1H, dd, J = 2.4, 4.9 Hz), 4.45–4.52 (2H, m), 4.56–4.62 (2H, m), 4.91 (1H, dd, J = 8.0, 9.6 Hz, H-2′), 4.95 (1H, dd, J = 3.5, 10.5 Hz, H-3′), 5.11 (1H, dd, J = 7.8, 10.4 Hz, H-2′), 5.21 (1H, t, J = 9.3 Hz, H-3′), 5.34 (1H, dd, J = 0.7, 3.3 Hz, H-4′), 5.50 (1H, t, J = 9.3 Hz, H-3′). HRMS: Calculated for $C_{38}H_{54}O_{23}Cs$ ($M + Cs^+$), 1011.2110. Found, 1011.2111.

Acknowledgments

This work was supported by the National Institutes of Health (GM44154). S. Aoki is a Glaxo Fellow.

Section II

Enzymatic and Affinity Methods

[14] Synthetic Neoglycoconjugates in Glycosyltransferase Assay and Purification

By MONICA M. PALCIC, MICHAEL PIERCE, and OLE HINDSGAUL

Introduction

Synthetic neoglycoconjugates are useful tools for both the isolation and assay of glycosyltransferase enzymes. The main advantage in using a synthetic approach is that a variety of natural as well as unnatural oligosaccharide structures can be prepared. These will be free of cross-reacting acceptors within the detection limits of chemical characterization, typically nuclear magnetic resonance (NMR) spectroscopy, high-performance liquid chromatography (HPLC), or mass spectrometry. In addition, incorporation levels can be controlled and a variety of proteins can be used for conjugation. In our laboratories, the most commonly used protein for conjugate preparation is the readily available bovine serum albumin (BSA).

In this chapter we review the use of synthetic neoglycoconjugates for the assay and isolation of *N*-acetylglucosaminyltransferase V [GnT-V; EC 2.4.1.155, α(1,3(6)-mannosylglycoprotein β-1,6-*N*-acetylglucosaminyl transferase], an enzyme involved in the branching of asparagine-linked oligosaccharides.[1,2] Biosynthetically, this enzyme catalyzes the transfer of *N*-acetylglucosamine (GlcNAc) from UDPGlcNAc to asparagine-linked acceptors with the minimum oligosaccharide structure **1**, converting them to **2** (Scheme 1). The smallest efficient synthetic acceptor for GnT-V is trisaccharide **3**, which the enzyme converts to tetrasaccharide **4**.[3-5] The 8-methoxycarbonyloctyl aglycon in acceptor **3** and product **4** can be covalently attached to BSA using the method of Pinto and Bundle,[6] giving neoglycoconjugates **5** and **6**, respectively. The acceptor conjugate **5** can be employed in radiochemical solution assays for GnT-V activity or as an immobilized acceptor in enzyme-linked immunosorbent (ELISA) assays.[7]

[1] R. D. Cummings, I. S. Trowbridge, and S. Kornfeld, *J. Biol. Chem.* **257,** 421 (1982).
[2] H. Schachter, *Biochem. Cell Biol.* **64,** 163 (1986).
[3] M. Pierce, J. Arango, S. H. Tahir, and O. Hindsgaul, *Biochem. Biophys. Res. Commun.* **146,** 679 (1987).
[4] O. Hindsgaul, S. H. Tahir, O. Srivastava, and M. Pierce, *Carbohydr. Res.* **173,** 263 (1988).
[5] M. M. Palcic, L. D. Heerze, M. Pierce, and O. Hindsgaul, *Glycoconjugate J.* **5,** 49 (1988).
[6] B. M. Pinto and D. R. Bundle, *Carbohydr. Res.* **124,** 313 (1983).
[7] S. C. Crawley, O. Hindsgaul, G. Alton, M. Pierce, and M. M. Palcic, *Anal. Biochem.* **185,** 112 (1990).

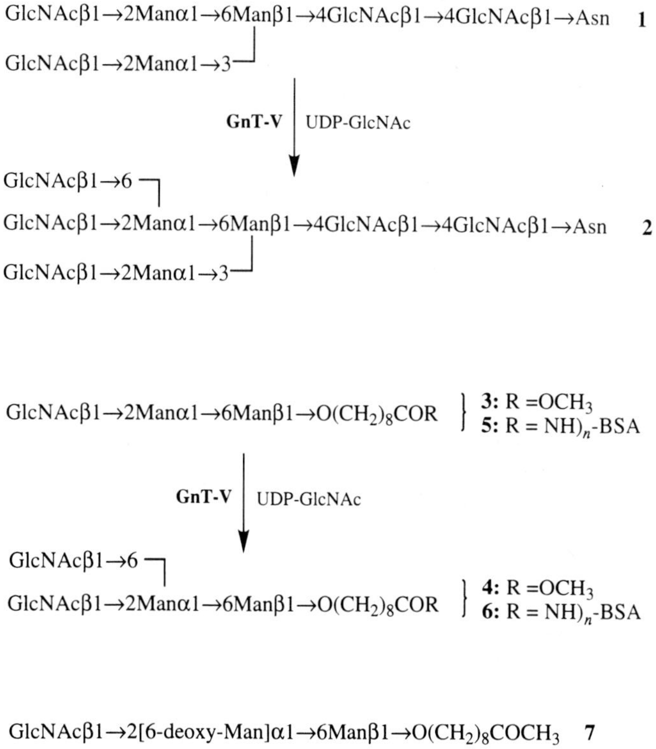

GlcNAcβ1→2Manα1→6Manβ1→4GlcNAcβ1→4GlcNAcβ1→Asn **1**
GlcNAcβ1→2Manα1→3 (to Manβ1)

↓ **GnT-V** | UDP-GlcNAc

GlcNAcβ1→6 (to Manα1)
GlcNAcβ1→2Manα1→6Manβ1→4GlcNAcβ1→4GlcNAcβ1→Asn **2**
GlcNAcβ1→2Manα1→3 (to Manβ1)

GlcNAcβ1→2Manα1→6Manβ1→O$(CH_2)_8$COR } **3:** R =OCH_3; **5:** R = NH$)_n$-BSA

↓ **GnT-V** | UDP-GlcNAc

GlcNAcβ1→6 (to Manα1)
GlcNAcβ1→2Manα1→6Manβ1→O$(CH_2)_8$COR } **4:** R =OCH_3; **6:** R = NH$)_n$-BSA

GlcNAcβ1→2[6-deoxy-Man]α1→6Manβ1→O$(CH_2)_8COCH_3$ **7**

SCHEME 1

When covalently coupled to Sepharose, the neoglycoconjugate **5** serves as an affinity ligand for column chromatography in the isolation of GnT-V. The 6′-deoxygenated analog **7** of the trisaccharide acceptor **3** is a competitive inhibitor of the *N*-acetylglucosaminyltransferase V with a K_i of 60 to 140 μM depending on the enzyme source.[8–10] The BSA conjugate of inhibitor **7** has also been covalently coupled to Sepharose and shown to be useful as an affinity matrix for the isolation of the transferase.[10] We here review the use of neoglycoconjugates in the assay and isolation of GnT-V from hamster kidney, rat kidney, and human serum.

[8] M. M. Palcic, J. Ripka, K. J. Kaur, M. Shoreibah, O. Hindgaul, and M. Pierce, *J. Biol. Chem.* **265,** 6759 (1990).
[9] O. Hindsgaul, K. J. Kaur, G. Srivastava, M. Blaszczyk-Thurin, S. C. Crawley, L. D. Heerze, and M. M. Palcic, *J. Biol. Chem.* **266,** 17858 (1991).
[10] M. G. Shoreibah, O. Hindsgaul, and M. Pierce, *J. Biol. Chem.* **267,** 2920 (1992).

Preparation of Albumin–Oligosaccharide Conjugates from Synthetic 8-Methoxycarbonyloctyl Glycosides

Principle. Synthetic oligosaccharides **3**, **4**, and **7** are prepared by multistep chemical synthesis as the 8-methoxycarbonyloctyl glycosides.[4,8,11] The methyl ester at the end of the hydrocarbon spacer is then quantitatively converted to the acyl hydrazide by reaction with hydrazine. The product hydrazide, in turn, is converted *in situ* to the acyl azide which specifically acylates the lysine amino groups of a protein such as BSA. The example given here is for the acceptor **3**, but identical procedures have been used for the remaining two oligosaccharides **4** and **7** with similar results. The procedure used is adapted from that of Pinto and Bundle.[6]

Reagents

- GlcNAcβ1→2Manα1→6Manβ1→$O(CH_2)_8COOCH_3$ (**3**),[4,11] 14.5 mg
- Hydrazine hydrate, 85% (v/v in water) (BDH, Poole, UK)
- Analytical silica thin-layer chromatography (TLC) plates (Merck, Darmstadt, Germany)
- GV filter, 0.2 μm (Millipore, Bedford, MA)
- C_{18} sample preparation cartridge (Waters, Milford, MA; Sep-Pak)
- HPLC-grade methanol
- Reagent grade dry dimethylformamide (DMF); immediately before use, the DMF is placed in a round-bottomed flask equipped with a stir bar and placed under water aspirator vacuum (~15 torr) while stirring to degas it through a drying tube containing Drierite
- Approximately 1.0 *M* solution of N_2O_4 in dichloromethane,[6] prepared by condensing gaseous N_2O_4 at low temperature, weighing it, and diluting with dichloromethane; the solution is stored sealed at −20°
- Bovine serum albumin (Pentex, crystalline, Miles Laboratories, Naperville, IL)
- 0.35 *M* $KHCO_3$, 80 m*M* $Na_2B_4O_7$ buffer, pH 8.9
- Ultrafiltration cell, 50 ml, with a PM10 membrane (Amicon, Danvers, MA)

Procedure. The methyl ester **3** is placed in screw-capped culture tube (13 × 100 mm, with a Teflon-lined screw cap) containing a stir bar. In a well-vented fume hood 95% ethanol (2.0 ml) is added, the solution is magnetically stirred, and then hydrazine hydrate (85%, 600 μl) is added. The tube is stoppered and the solution stirred overnight. The reaction can be monitored by TLC on silica gel using ethyl acetate–methanol–water (6 : 3 : 1, v/v/v) as the developing solvent and observing the starting ester **3** (R_f ≥0.5) converting to the product hydrazide with R_f near 0.1. The

[11] S. H. Tahir and O. Hindsgaul, *Can. J. Chem.* **64,** 1771 (1986).

absolute R_f values can vary significantly from experiment to experiment, but large differences are always observed between the starting material and product. The compounds are visualized by spraying the TLC plate with a solution of 5% sulfuric acid in ethanol and charring on a hot plate.

Within 16 hr, the reaction is complete, and the solvents are removed in the fume hood under a stream of nitrogen while keeping the tube in a bath at 40° (this can take up to 1 hr). Water (2 ml) is then added and evaporated 3 times using a nitrogen stream, and the final tube is placed under high vacuum overnight to ensure that all of the hydrazine (which interferes in the next step) is removed. Then DMF (200–300 μl) is added to the tube, which is cooled with stirring to −50° in an acetone bath cooled to that temperature with pieces of dry ice. Then the stock N_2O_4/CH_2Cl_2 solution (50 μl) is added using a Hamilton syringe which has been precooled in a freezer to at least −20°. The stirred reaction mixture is then allowed to warm slowly over 20 min to between −10 and −20°. Both N_2O_4 and dichloromethane are volatile, and the concentration of the stock solution can change significantly with time when stored. It is therefore important to monitor the formation of the azide by TLC using the same solvent system as above. Formation of the azide, which has a greater R_f (near 0.35 compared to about 0.1 for the hydrazide) is usually over 90% complete within 30 min. If not, more N_2O_4/CH_2Cl_2 (20 μl) is added until the apparent yield of the product is 90% by TLC.

When azide formation is complete, a solution of BSA (28.8 mg) in the bicarbonate–borate buffer (3 ml) cooled to 4° is added directly to the tube, which is capped and mixed end-over-end several times. The tube is placed in the refrigerator and left overnight. Water is then added, and the sample is transferred to the ultrafiltration cell where the contents are diluted to 50 ml with water. Ultrafiltration with stirring under pressure is continued until the sample volume reaches about 10 ml. Water is again added to 50 ml, and the procedure is repeated 4 more times (in total, 200 ml ultrafiltrate is collected). The final contents of the cell (~10 ml) are then removed, filtered through a Millipore filter attached to a plastic syringe, and lyophilized to dryness in a round-bottomed flask or similar container.

The degree of coupling can be estimated in various ways. On a relatively large scale such as this, the simplest method is to weigh a quantity of the conjugate (near 0.5 mg), then dissolve it in water and estimate the incorporation of sugar in the conjugate by the phenol–sulfuric acid method of Dubois *et al.*[12] Mannose is used as the standard, since the trisaccharide contains 2 mol of this sugar. The amount of protein in the conjugate can

[12] M. Dubois, K. A. Gilles, J. K. Hamilton, P. A. Rebers, and F. Smith, *Anal. Chem.* **31,** 296 (1956).

be independently estimated using a protein assay (such as the assay from Bio-Rad, Richmond, CA), if desired. The exact incorporation number is not critical to the outcome of the subsequent experiments, although incorporations above 8–10 are desirable. In the coupling just described, an incorporation of 13 mol of acceptor **3** per molecule of BSA is achieved.

Preparation of Affinity Chromatography Supports from Albumin–Oligosaccharide Conjugates

Principle. Glycosyltransferases bind to oligosaccharide acceptors with K_D values usually in the 10–1000 μM range. In the lower end of this range, immobilized acceptors have the potential to act as ligands for the affinity purification of the enzymes. Here, we immobilize the BSA–conjugate of a the synthetic trisaccharide acceptor for GnT-V and show its use in the purification of the enzyme. The corresponding inhibitor has already been shown to be useful for this purpose.[10] The very simple and reproducible immobilization procedure of Stults *et al.*,[13] involving the oxidation of ethylene glycol groups on Sepharose CL, followed by reductive amination, is used.

Reagents

Neoglycoconjugate of **5** (prepared as above)
Sepharose CL-6B (Pharmacia, Piscataway, NJ)
Sodium periodate
Ethylene glycol
Pyridine–borane (Aldrich, Milwaukee, WI)
0.2 *M* Sodium phosphate buffer, pH 7.0
1 *M* Ethanolamine hydrochloride buffer, pH 8.0
Sodium cacodylate buffer (50 m*M* sodium cacodylate buffer, 10 m*M* $MnCl_2$, 0.25% Triton X-100, pH 6.5)

Procedure. The Sepharose (3 ml) is washed with 30 ml of water in a sintered glass funnel and transferred to a 50-ml Erlenmeyer flask as a suspension with 15 ml water. Solid sodium periodate (0.12 g) is then added, and the flask is briefly swirled every 15 min for 45 min. Ethylene glycol (40 μl) is added, and the suspension is swirled every 5 min for 15 min. The gel is transferred to the sintered glass funnel, where it is washed with water (100 ml) followed by phosphate buffer (100 ml). The gel is then resuspended in phosphate buffer (4 ml) and transferred to a screw-capped tube containing the neoglycoconjugate of **5** (2.8 mg). Pyridine–borane (20 μl) is then added; the tube is sealed and rotated end-over-end at 4° over-

[13] N. L. Stults, L. M. Asta, and Y. C. Lee, *Anal. Biochem.* **180,** 114 (1989).

night (18 hr). On a sintered glass funnel, the gel is then washed with an initial 30 ml volume of water (which is reserved), then an additional 120 ml. The gel is transferred back to the screw-capped tube in 6 ml ethanolamine buffer, pyridine–borane (20 μl) is added, and end-over-end mixing is continued overnight. The gel is then washed with water (100 ml) and stored in the cold (4°) in cacodylate buffer, where it is stable indefinitely.

An estimate of the incorporation of the BSA conjugate onto the gel can be obtained by determining the protein content of the initial 30 ml aqueous wash from the coupling reaction. Before determination, the remaining coupling reagents are removed from the sample by ultrafiltration as is done for the initial attachment of the oligosaccharide described above. In the experiment described here, 33% of the conjugate is bound to the resin. This produces a loading of 0.3 mg of the BSA conjugate per milliliter of gel, or approximately 90 nmol of oligosaccharide per milliliter.

Radiochemical Assay for *N*-Acetylglucosaminyltransferase

Principle. The assay is based on the transfer of radiolabeled GlcNAc from UDPGlcNAc to the acceptor neoglycoprotein [Eq. (1)]. Unreacted donor is removed from radiolabeled product by gel-filtration chromatography and the amount of transfer quantitated by liquid scintillation counting. This also establishes whether the conjugate is suitable for immobilization in microtiter wells for use in the ELISA described below. The use of natural glycoconjugates as substrates for a variety of glycosyltransferase assays is reviewed by Sadler *et al.*[14]

$$\text{UDP}[^3\text{H}]\text{GlcNAc} + \text{GlcNAc}\beta1\rightarrow2\text{Man}\alpha1\rightarrow6\text{Man}\beta1\rightarrow\text{O(CH}_2)_8\text{CONH}]_n\text{—BSA} \xrightarrow{\text{GnT-V}} \text{GlcNAc}\beta1\rightarrow2[[^3\text{H}]\text{GlcNAc}\beta1\rightarrow6]\text{Man}\alpha1\rightarrow6\text{Man}\beta1\rightarrow\text{O(CH}_2)_8\text{COHN}]_n\text{—BSA} \quad (1)$$

Reagents

Stock neoglycoconjugate **5** (1 mg/ml in water), stored at −20°
50 m*M* Sodium cacodylate buffer, pH 6.5, containing 0.1% (w/v) Triton X-100, 20% glycerol, 10 m*M* EDTA, and 1 mg/ml BSA (Sigma, St. Louis, MO)
UDP[^{3}H]GlcNAc, 300,000 disintegrations/min (dpm) (American Radiolabeled Chemicals, St. Louis, MO, 25 Ci/mmol)

[14] J. E. Sadler, T. A. Beyer, C. L. Oppenheimer, J. C. Paulson, J. P. Prieels, J. I. Rearick, and R. L. Hill, this series, Vol. 83, p. 458.

UDPGlcNAc, 1 mM aqueous solution (Sigma)
Hamster kidney GnT-V, 100 microunits (μU), purified by extraction and UDP-hexanolamine chromatography[10]
Sephadex G-50 equilibrated with 0.2 M NaCl
Disposable plastic serological pipettes, 10 ml
0.2 M NaCl
EcoLite(+) liquid scintillation cocktail (ICN, San Diego, CA)

Procedure. Prior to assay, 10 μl of 1 mM unlabeled UDPGlcNAc and 300,000 dpm of UDP[^{3}H]GlcNAc are lyophilized together to dryness in 0.5-ml microcentrifuge tubes. Enzyme (30 μU) is added with 5 or 10 μg of conjugate **5** and assay buffer to a total volume of 100 μl. A control reaction with all components except for acceptor **5** is carried out in parallel. After 5 hr at 37° the solution is transferred to a 15-cm column of Sephadex G-50 packed in a 10-ml plastic serological pipette. The column is prepared by cutting the pipette to the appropriate mark with a hot razor and packing a small plug of glass wool at the bottom of the pipette to hold the resin. The resin is equilibrated with 0.2 M NaCl, and the column can be allowed to run dry before and after the application of the enzyme reaction mixture. Two rinses of 10 μl are used to transfer all of the solution from the microcentrifuge tube onto the column. The column is developed with 0.2 M NaCl using scintillation vials for the collection of 10-drop (~0.5 ml) fractions. After collecting 30 fractions, 10 ml of scintillation cocktail is added to each vial for counting.

A typical elution profile is shown in Fig. 1, where the radiolabeled product elutes in fractions 7 to 12, well separated from the unreacted labeled UDPGlcNAc. The control reaction lacking acceptor does not show an increase in radioactivity above background levels (25 dpm) in fractions 7 to 12. It can be seen in Fig. 1 that increasing product is obtained for increasing quantities of neoglycoconjugate acceptor. For hamster kidney enzyme, the apparent K_m for this acceptor conjugate is 10 μg acceptor in a reaction volume of 100 μl. The assay can also be scaled down to smaller volumes of 20 μl; in these cases the neoglycoconjugate is lyophilized with the donor prior to assay.

Enzyme-Linked Immunosorbent Assay for *N*-Acetylglucosaminyltransferase V

Principle. The assay makes use of the trisaccharide acceptor neoglycoconjugate **5**, immobilized in microtiter plates, which is converted to product **6** by the action of *N*-acetylglucosaminyltransferase V.[7] Product-specific antibodies are employed to detect and quantify the amount of transfer.

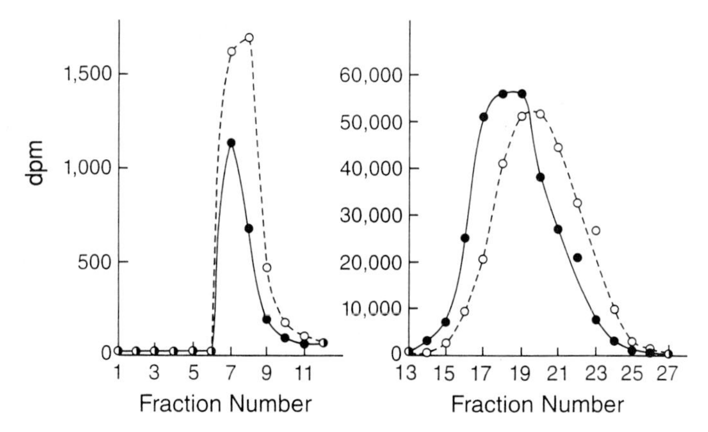

FIG. 1. Separation of radiolabeled GnT-V reaction product glycoconjugate **6** from unreacted UDP [^{3}H]GlcNAc by gel-filtration chromatography. The profiles shown are for incubation mixtures with 5 (●) and 10 (○) mg of acceptor conjugate **5** incubated with 30 μU of hamster kidney enzyme for 5 hr at 37°. Product elutes in fractions 7 to 12.

Coating of Plates

Reagents

Phosphate-buffered saline (PBS) solution, containing 7.8 m*M* Na_2HPO_4, 2.2 m*M* KH_2PO_4, 0.9% NaCl, and 15 m*M* NaN_3
PBST, PBS with 0.05% Tween 20 detergent
Stock neoglycoconjugate **5** (1 mg/ml in water), stored at −20°
50 m*M* Sodium phosphate buffer, pH 7.5, containing 5 m*M* $MgCl_2$ and 15 m*M* NaN_3
5% (w/v) Bovine serum albumin (Sigma) in PBS

Procedure. Microtiter plates are coated by the addition of 100 μl of neoglycoconjugate (20 μg/ml) in 50 m*M* sodium phosphate buffer, pH 7.5, containing 5 m*M* $MgCl_2$ and 15 m*M* NaN_3, and incubation at ambient temperature for 16 hr. The solution is removed by aspiration and replaced with 200 μl of 5% BSA. After 4 hr at ambient temperature, this solution is removed and the wells washed 3 times with PBS solution (200 μl/wash) and once with 200 μl of water. The wells are allowed to dry in air for 1 hr and then wrapped with transparent plastic wrap and stored at 4°. Plates are washed again with 200 μl of water immediately before use in enzyme assays.

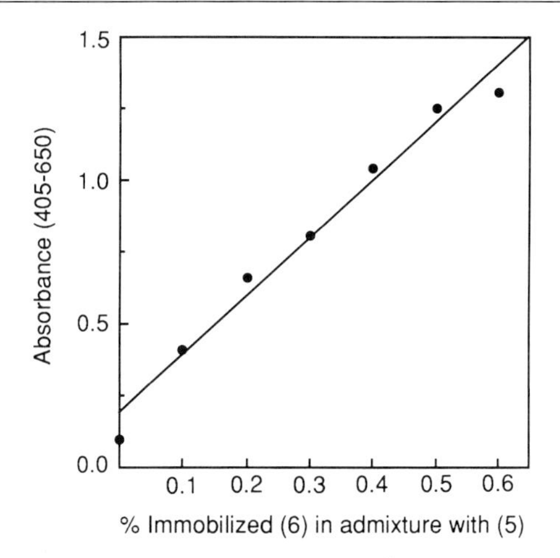

FIG. 2. Standard curve for ELISA response of wells coated with increasing quantities of product neoglycoconjugate **6** in admixture with acceptor conjugate **5**.

Assay for N-Acetylglucosaminyltransferase V

Reagents

UDPGlcNAc (Sigma)
Human serum sample as prepared in Ref. 7
30 m*M* 2-(*N*-Morpholino)ethanesulfonic acid (MES) buffer, pH 6.5
Polyclonal antisera specific for product conjugates[7]
Alkaline phosphatase conjugate of goat anti-rabbit immunoglobulin G (IgG) (Sigma)
p-Nitrophenyl phosphate substrate solution [1 mg tablet (Sigma)/ml of 1 *M* diethanolamine hydrochloride buffer, pH 9.8, containing 1% BSA and 500 μ*M* $MgCl_2$], freshly prepared

Procedure. Assays are carried out in 100 μl of 30 m*M* MES buffer, pH 6.5, with 1% Triton X-100 and 5.6 m*M* UDPGlcNAc and 0.22–0.88 μU of GnT-V enzyme (0.4–2.4 μl of serum) in the precoated microtiter plates. The plates are covered with plastic wrap and incubated at 37° for 60 or 90 min. The enzyme solution is removed by aspiration, and the wells are washed (2 times with 200 μl PBST followed by once with 200 μl water) and then incubated for 2 hr at ambient temperature with refined rabbit antiserum (100 μl of 1/8000 dilution in 1% BSA in PBST). The wells are aspirated, washed twice with 200 μl of PBST, and then incubated with alkaline phosphatase-conjugated goat anti-rabbit IgG (100 μl of 1/1000 dilution in 1% BSA in PBST) for 2 hr at ambient temperature. The second-

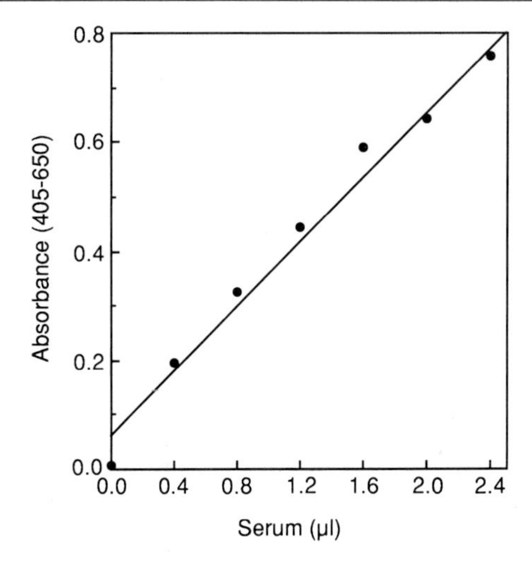

FIG. 3. Dependence of GnT-V ELISA response on concentration of serum.

ary antibody solution is removed by aspiration, the wells are washed 3 times with 200 μl PBST, once with 200 μl water, once with 300 μl water, and then 100 μl *p*-nitrophenyl phosphate solution is added. The increase in absorbance at 405 nm with background correction at 650 nm is monitored over time with a microtiter plate reader. Either kinetics or absolute absorbances after 30 or 60 min can be used. Mixtures of substrate and product conjugate are used to standardize the assay when day-to-day variability is a problem. Figure 2 shows that product detection in a standard curve is linear up to 0.6%. Figure 3 shows that the GnT-V assay in human serum is linear up to 2.4 μl of serum sample.

Applications of Enzyme Assay

Among other applications, the ELISA for GnT-V has been used to monitor the transfection of African green monkey kidney COS cells in the cloning of the enzyme[15] and the changes in GnT-V activity which accompanies chemical-induced differentiation of F9 teratocarcinoma cells.[16] Other ELISA methods for glycosyltransferases that quantify transfer to immobilized neoglycoconjugate acceptors include an $\alpha(1\rightarrow4)$-fuco-

[15] M. Shoreibah, G.-S. Perng, B. Adler, J. Weinstein, R. Basu, R. Cupples, D. Wen, J. K. Browne, P. Buckhaults, N. Fregien, and M. Pierce, *J. Biol. Chem.* **268,** 15381 (1993).

[16] M. Heffernan, R. Lotan, B. Amos, M. M. Palcic, R. Takano, and J. W. Dennis, *J. Biol. Chem.* **268,** 1242 (1993).

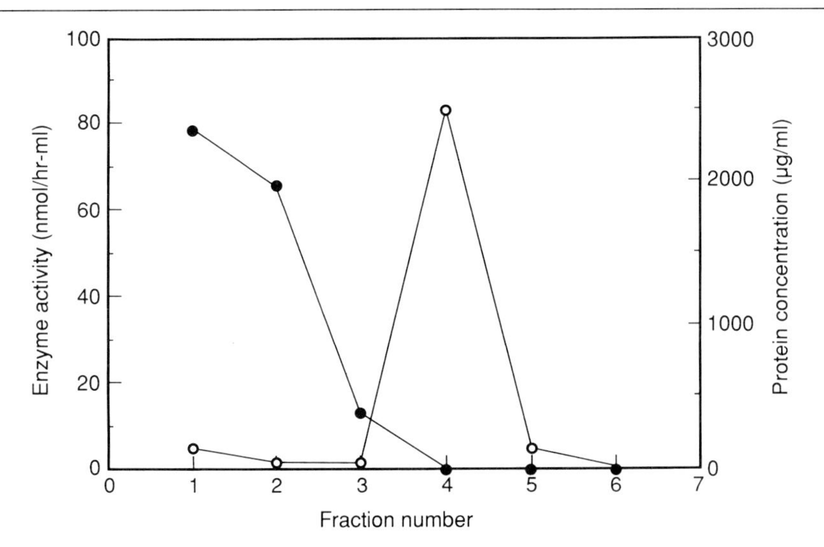

FIG. 4. Affinity chromatography using acceptor–BSA conjugate coupled to agarose. Partially purified GnT-V from rat kidney acetone powder was applied to a 1.5-ml column of agarose–conjugate at 4° and eluted with MES buffer containing Triton X-100, similar to the elution conditions described.[10] Activity was then eluted by warming the column to room temperature, including 0.5 *M* NaCl in the buffer, and raising the pH to 8.0. Activity was measured using the trisaccharide acceptor **3**.[4] Protein in the fractions before activity elution was assayed using the Bio-Rad protein assay. Protein in the salt-eluted fractions was determined by comparing intensities of silver staining after subjecting aliquots of the fractions and various concentrations of crystalline BSA to SDS–PAGE, then visualizing protein by silver staining. Plotted are enzyme activity (○) and protein concentration (●).

syltransferase,[17] blood group A and B glycosyltransferases,[18] a galactosyltransferase,[19] and an $\alpha(1\rightarrow3)$-fucosyltransferase.[20]

Purification of *N*-Acetylglucosaminyltransferase with Acceptor–Albumin Conjugate Columns

Principle. A variety of glycosyltransferases have been partially purified by taking advantage of their affinities for acceptor glycoproteins or glyco-

[17] M. M. Palcic, R. M. Ratcliffe, L. R. Lamontagne, A. H. Good, G. Alton, and O. Hindsgaul, *Carbohydr. Res.* **196,** 133 (1990).

[18] L. M. Keshvara, E. M. Newton, A. H. Good, O. Hindsgaul, and M. M. Palcic, *Glycoconjugate J.* **9,** 16 (1992).

[19] P. F. Zatta, K. Nyame, M.. J. Cormier, S. A. Madox, P. A. Prieto, D. F. Smith, and R. D. Cummings, *Anal. Biochem.* **194,** 185 (1991).

[20] T. Tachikawa, S. Yazawa, T. Asao, S. Shin, and N. Yanaihara, *Clin. Chem.* **37,** 2081 (1991).

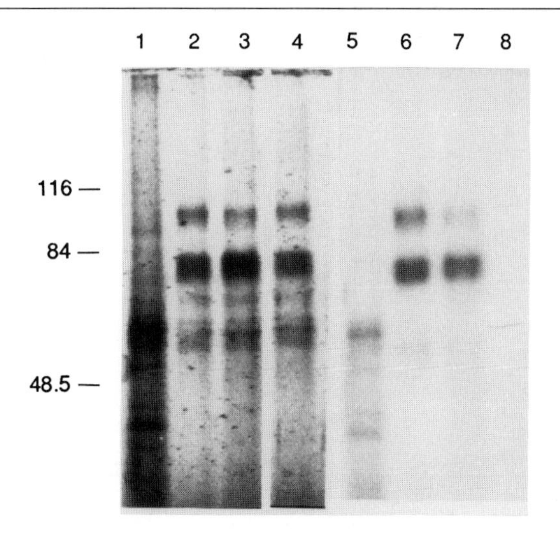

FIG. 5. Analysis by SDS–PAGE of samples from the rapid (48 hr) purification of GnT-V in the presence of protease inhibitor by acceptor by neoglycoconjugate affinity chromatography of rat kidney acetone powder detergent extract. Samples were reacted with an active site-specific radioactive photoaffinity ligand, 5′-[^{32}P]thiol-UDP[27] under various conditions prior to electrophoresis. After SDS–PAGE, the gel was silver stained, photographed, dried, and subjected to autoradiography. Lanes 1–4 show the silver-stained gel, whereas lanes 5–8 represent autoradiography of the gel. Lanes 1 and 5, affinity chromatography run-through fraction; lanes 2 and 6, affinity-purified sample; lanes 3 and 7, photoaffinity labeling performed in the presence of 10 m*M* UDPGlcNAc; lanes 4 and 8, photoaffinity labeling in 1 m*M* UDPGlcNAc. The gel was run as described in Ref. 10. Molecular weight markers are: 116, *E. coli* β-galactosidase; 84, fructose-6-phosphate kinase; 48.5, fumarase.

peptides derived therefrom. These include a sialyltransferase,[21] a galactosyltransferase,[22] *N*-acetylglucosaminyltransferases,[23,24] and a fucosyltransferases.[25,26] The use of an inhibitor neoglycoconjugate column for the isolation of GnT-V has been reported.[10] Here the acceptor conjugate is employed.

[21] J. Weinstein, U. de Souza-e-Silva, and J. C. Paulson, *J. Biol. Chem.* **257,** 13835 (1982).

[22] J. Mendicino, S. Sivakami, M. Davila, and E. V. Chandrasekaran, *J. Biol. Chem.* **257,** 3987 (1982).

[23] Y. Nishikawa, W. Pegg, H. Paulsen, and H. Schachter, *J. Biol. Chem.* **263,** 8270 (1988).

[24] A. Nishikawa, Y. Ihara, M. Hatakeyama, K. Kangawa, and N. Taniguchi, *J. Biol. Chem.* **267,** 18199 (1992).

[25] A. Mitsakos and F.-G. Hanisch, *Biol. Chem. Hoppe-Seyler* **370,** 239 (1989).

[26] J. A. Voynow, R. S. Keiser, T. F. Scanlin, and M. C. Glick, *J. Biol. Chem.* **266,** 21572 (1991).

Reagents

Buffer A: 50 m*M* MES, pH 6.5, 0.2% Triton X-100, 5 m*M* EDTA, and 0.05% NaN_3

Buffer B: 50 m*M* MES, pH 6.5, 0.1% Triton X-100, 20% glycerol, and 0.05% NaN_3 UDP (Sigma)

Rat kidney GnT-V, partially purified by extraction and UDP-hexanolamine-Sepharose chromatography,[10] equilibrated with buffer A

Econo-Pac column (Bio-Rad) packed with the BSA–acceptor-Sepharose resin (1.1 × 3 cm, 3 ml bed volume), equilibrated with buffer B

Procedure. Rat kidney enzyme is prepared by Triton X-100 extraction from an acetone powder followed by chromatography on UDP-hexanolamine-Sepharose.[10] The specific activity is 0.00195 μmol/mg protein-hr. The enzyme is dialyzed against buffer A, then brought to 1 m*M* UDP and 20% glycerol before application to the neoglycoconjugate column at 4°. After loading the enzyme, the column is washed with 20 ml of buffer B. The column is stopped, brought to room temperature (24°), and then eluted with the inclusion of 500 m*M* NaCl in buffer A after adjustment of the pH to 8.0. Figure 4 shows a profile of the chromatography. The specific activity of the eluted enzyme is 41 μmol/mg-hr, which translates to over 250,000-fold purification.

Sodium dodecyl sulfate–polyacrylamide gel electrophoresis (SDS–PAGE) with silver staining of the eluted enzyme shows two main bands at 75 and 69 kDa and a faint band at 95 kDa. Peptide sequences from proteolysis of the main bands have been used to design degenerate oligonucleoide probes for the screening of cDNA libraries by the polymerase chain reaction (PCR).[15] A cDNA encoding for full-length fibroblast GnT-V gives a molecular mass of 81.4 kDa for the protein predicted by the nucleotide coding sequence. There are six potential N-linked glycosylation sites, suggesting that the 95-kDa band corresponds to full-length GnT-V which is proteolyzed during purification. Rapid (48 hr) and small-scale isolation of the enzyme (data shown in Fig. 5)[27] with the inclusion of a 10-fold greater concentration of protease inhibitor cocktail to the acetone solubilization buffer yields enzyme with comparable specific activity and now an intense band at 95 kDa, along with the doublet at 75 and 69 kDa (Fig. 5).

Acknowledgments

We thank Ms. A. H. Good for the enzyme assay results using the conjugate acceptor. This work was supported through research grants from the Medical Research Council of Canada (M.M.P.), the Natural Sciences and Engineering Research Council of Canada (O.H.), and the U.S. National Institutes of Health (M.P.).

[27] R. S. Haltiwanger, G. D. Holt, and G. W. Hart, *J. Biol. Chem.* **265,** 2563 (1990).

[15] Affinity Chromatography of Oligosaccharides on *Psathyrella velutina* Lectin Column

By AKIRA KOBATA, NAOHISA KOCHIBE, and TAMAO ENDO

Introduction

Immobilized lectins have been used successfully for many years to isolate glycoconjugates with specific sugar structures. Researchers in a number of laboratories have begun to realize the potential of lectin columns for the analysis of sugar chains in glycoconjugates; serial lectin column chromatography in particular affords a simple and very sensitive method for fractionating various oligosaccharides.[1,2] This chapter describes the use of a mushroom-lectin column (derived from *Psathyrella velutina* lectin, PVL) that recognizes and interacts with terminal *N*-acetylglucosamine residue.[3,4]

Methods

Purification of Psathyrella velutina Lectin

Fruiting bodies of *P. velutina* (100 g) collected in the region of Gunma Prefecture (Japan) are homogenized with 1 liter of 145 m*M* NaCl in 5 m*M* sodium phosphate buffer, pH 7.2 (PBS), containing 5 m*M* ethylenediaminetetraacetic acid (EDTA) and 0.5 m*M* phenylmethylsulfonyl fluoride. The homogenate is centrifuged at 20,000 *g* for 20 min at 4°, and the pellet is extracted with 500 ml of fresh extraction buffer. To the combined extracts is added dropwise 20% (v/v) acetic acid with stirring, to adjust the solution to pH 4.0. After standing for several hours, the insoluble material is removed by centrifugation at 20,000 *g* for 20 min at 4°, and the supernatant is neutralized with 1 *N* NaOH. The acid-treated extract is passed through a chitin column (3.5 × 30 cm) at a flow rate of 60 ml/hr, and the column is washed with extraction buffer thoroughly. The bound material is then eluted with 50 m*M* *N*-acetylglucosamine in PBS containing 10% (v/v) glycerol. The eluate is concentrated by ultrafiltration through an Advantec Q0100 membranes (Toyo Roshi, Tokyo, Japan).

[1] R. K. Merkle and R. D. Cummings, this series, Vol. 138, p. 232.
[2] A. Kobata and T. Endo, *J. Chromatogr.* **597,** 111 (1992).
[3] N. Kochibe and K. L. Matta, *J. Biol. Chem.* **264,** 173 (1989).
[4] T. Endo, H. Ohbayashi, K. Kanazawa, N. Kochibe, and A. Kobata, *J. Biol. Chem.* **267,** 707 (1992).

The concentrate is sequentially applied to serially connected columns of DEAE-cellulofine (Chisso, Tokyo, Japan, 1.7 × 25 cm) and CM-Sepharose CL-6B (Pharmacia LKB Biotechnology, Uppsala, Sweden, 1.7 × 25 cm) equilibrated with 10 m*M* sodium phosphate buffer, pH 7.0, containing 10% (v/v) glycerol. After washing thoroughly with the same solution the columns are disconnected, and the material bound to the CM-Sepharose CL-6B column is eluted with 0.4 *M* NaCl. Finally, the lectin is rechromatographed on the chitin column (1.5 × 20 cm) which has been equilibrated with PBS–glycerol. The bound lectin is eluted with 50 m*M* *N*-acetylglucosamine.

Preparation of Lectin Column

Purified PVL thus obtained is coupled to Affi-Gel 10 (Bio-Rad Laboratories, Richmond, CA) as follows. The activated gel, which has been washed with 0.1 *M* $NaHCO_3$, is added to a coupling mixture containing the purified PVL (5 mg/ml), 0.15 *M* NaCl, 10% (v/v) glycerol, and 20 m*M* *N*-acetylglucosamine. The suspension is stirred gently overnight at 4° and then filtered. The gel is resuspended in 1 *M* ethanolamine, allowed to stand for 2 hr at room temperature, and then washed with 10 m*M* Tris-HCl buffer, pH 7.4, containing 140 m*M* NaCl. The amount of PVL bound to 1 ml of Affi-Gel 10 is estimated to be approximately 4.4 mg.

Affinity Chromatography on Lectin Column

Tritium-labeled oligosaccharides [5–10 × 10^2 counts/min (cpm), 25–50 pmol] dissolved in 100 μl of water are applied to a column (8 mm inner diameter) containing 1 ml of PVL–Affi-Gel 10, previously equilibrated with 10 m*M* Tris-HCl buffer, pH 7.4, containing 0.1 *M* NaCl, 1 m*M* $CaCl_2$, 1 m*M* $MgCl_2$, and 1 m*M* $MnCl_2$, and allowed to stand at room temperature for 30 min. Oligosaccharides bound to the column are then eluted with 10 ml of the same buffer, followed by buffer containing 1 m*M* *N*-acetylglucosamine. Fractions of 1.0 ml are collected at a flow rate of 15 ml/hr, and the radioactivity in each fraction is determined by a liquid scintillation method. Recoveries of oligosaccharides based on radioactivities ranges from 98 to 100%.

Behavior of Human Milk Oligosaccharides

Although oligosaccharide **I** in Table I is passed through the column (Fig. 1A), oligosaccharide **II** remains bound to the column and is eluted with the buffer containing 1 m*M* *N*-acetylglucosamine (Fig. 1B). In contrast, isomeric oligosaccharide **III** is only slowed in the column (Fig. 1C).

TABLE I
STRUCTURES OF RADIOACTIVE OLIGOSACCHARIDES

Oligosaccharide	Structure[a]
I	Galβ1→4GlcNAcβ1→3Galβ1→4Glc$_{OT}$
II	GlcNAcβ1→3Galβ1→4Glc$_{OT}$
III	GlcNAcβ1→6Galβ1→4Glc$_{OT}$
IV	GlcNAcβ1 ↘ $^{6}_{3}$Galβ1→4Glc$_{OT}$ ↗ GlcNAcβ1
V	Galβ1→4GlcNAcβ1→2Manα1 ↘ $^{6}_{3}$Manβ1→4GlcNAcβ1→4GlcNAc$_{OT}$ ↗ Galβ1→4GlcNAcβ1→2Manα1
VI	GlcNAcβ1→2Manα1 ↘ $^{6}_{3}$Manβ1→4GlcNAcβ1→4GlcNAc$_{OT}$ ↗ GlcNAcβ1→2Manα1
VII	Manα1 ↘ $^{6}_{3}$Manβ1→4GlcNAcβ1→4GlcNAc$_{OT}$ ↗ Manα1
VIII	Manβ1→4GlcNAcβ1→4GlcNAc$_{OT}$
IX	GlcNAcβ1→4GlcNAc$_{OT}$
X	Galβ1→4GlcNAcβ1→2Manα1 ↘ $^{6}_{3}$Manβ1→4GlcNAcβ1→4GlcNAc$_{OT}$ ↗ GlcNAcβ1→2Manα1
XI	GlcNAcβ1→2Manα1 ↘ $^{6}_{3}$Manβ1→4GlcNAcβ1→4GlcNAc$_{OT}$ ↗ Galβ1→4GlcNAcβ1→2Manα1
XII	GlcNAcβ1 ↓ GlcNAcβ1→2Manα1 4 ↘ $^{6}_{3}$Manβ1→4GlcNAcβ1→4GlcNAc$_{OT}$ ↗ GlcNAcβ1→2Manα1

TABLE I (*continued*)

Oligosaccharide	Structure[a]
XIII	GlcNAcβ1 ↓ Galβ1→4GlcNAcβ1→2Manα1 4 ↘ $^{6}_{3}$Manβ1→4GlcNAcβ1→4GlcNAc$_{OT}$ ↗ GlcNAcβ1→2Manα1
XIV	GlcNAcβ1 ↓ GlcNAcβ1→2Manα1 4 ↘ $^{6}_{3}$Manβ1→4GlcNAcβ1→4GlcNAc$_{OT}$ ↗ Galβ1→4GlcNAcβ1→2Manα1
XV	GlcNAcβ1→2Manα1 ↘ $^{6}_{3}$Manβ1→4GlcNAcβ1→4GlcNAc$_{OT}$ ↗ Manα1
XVI	Manα1 ↘ $^{6}_{3}$Manβ1→4GlcNAcβ1→4GlcNAc$_{OT}$ ↗ GlcNAcβ1→2Manα1
XVII	Manα1 ↘ $^{6}_{3}$Manβ1→4GlcNAcβ1→4GlcNAc$_{OT}$ ↗ GlcNAcβ1→4Manα1
XVIII	GlcNAcβ1→6Manα1 ↘ $^{6}_{3}$Manβ1→4GlcNAcβ1→4GlcNAc$_{OT}$ ↗ Manα1
XIX	Manα1 ↘ ^{6}Manα1 ↘^{6} Manα1 ↗^{3} $_{3}$Manβ1→4GlcNAcβ1→4GlcNAc$_{OT}$ ↗ GlcNAcβ1→2Manaα1

(*continued*)

TABLE I (*continued*)

Oligosaccharide	Structure[a]
XX	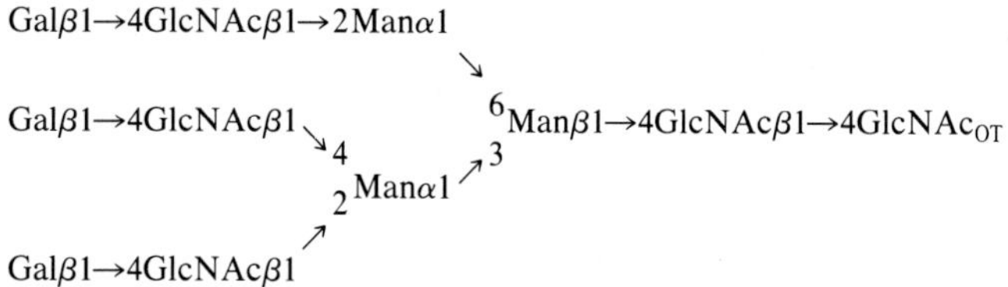 Galβ1→4GlcNAcβ1→2Manα1 ↘ Galβ1→4GlcNAcβ1 ↘4 6Manβ1→4GlcNAcβ1→4GlcNAc$_{OT}$ 2Manα1 ↗3 ↗ Galβ1→4GlcNAcβ1
XXI	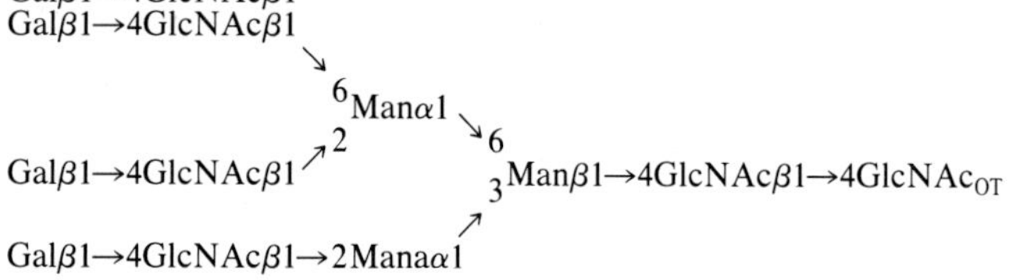 Galβ1→4GlcNAcβ1 ↘ 6Manα1 ↘6 Galβ1→4GlcNAcβ1 ↗2 3Manβ1→4GlcNAcβ1→4GlcNAc$_{OT}$ ↗ Galβ1→4GlcNAcβ1→2Manaα1
XXII	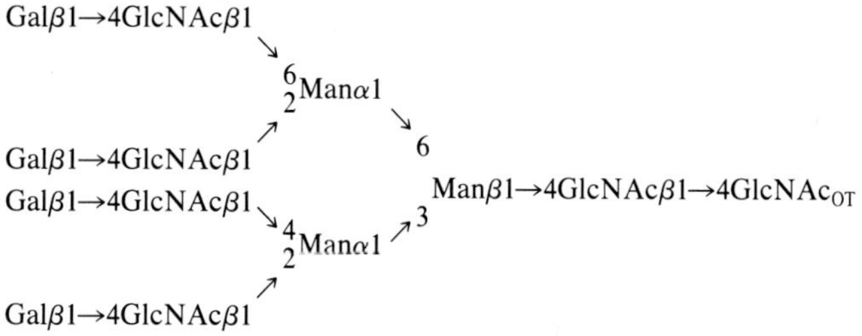 Galβ1→4GlcNAcβ1 ↘ 6/2Manα1 ↗ ↘ Galβ1→4GlcNAcβ1 6 Galβ1→4GlcNAcβ1 Manβ1→4GlcNAcβ1→4GlcNAc$_{OT}$ ↘4/2Manα1 ↗3 ↗ Galβ1→4GlcNAcβ1
XXIII	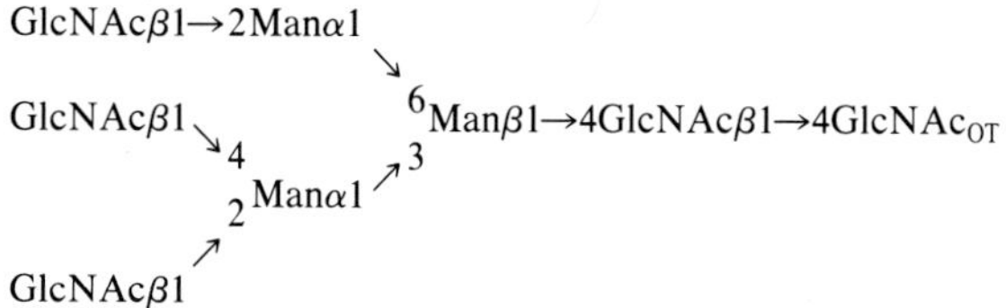 GlcNAcβ1→2Manα1 ↘ GlcNAcβ1 ↘4 6Manβ1→4GlcNAcβ1→4GlcNAc$_{OT}$ 2Manα1 ↗3 ↗ GlcNAcβ1
XXIV	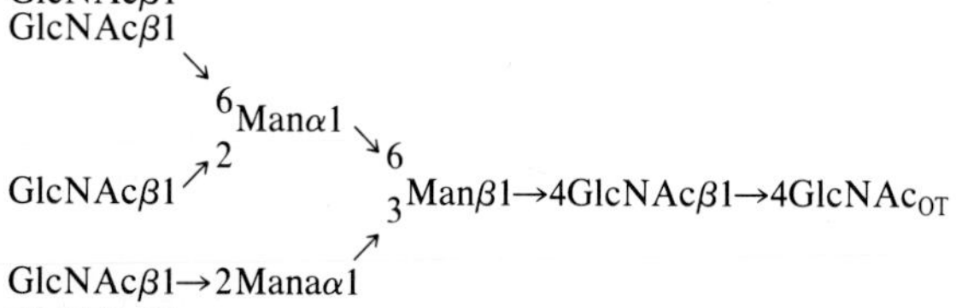 GlcNAcβ1 ↘ 6Manα1 ↘6 GlcNAcβ1 ↗2 3Manβ1→4GlcNAcβ1→4GlcNAc$_{OT}$ ↗ GlcNAcβ1→2Manaα1
XXV	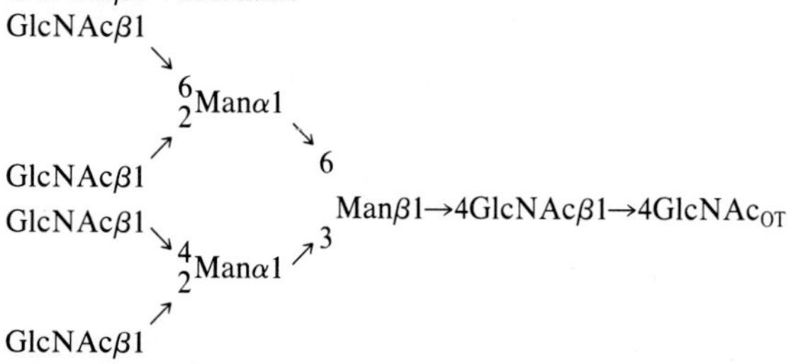 GlcNAcβ1 ↘ 6/2Manα1 ↗ ↘ GlcNAcβ1 6 GlcNAcβ1 Manβ1→4GlcNAcβ1→4GlcNAc$_{OT}$ ↘4/2Manα1 ↗3 ↗ GlcNAcβ1

[a] The subscript OT indicates NaB^3H_4-reduced oligosaccharides.

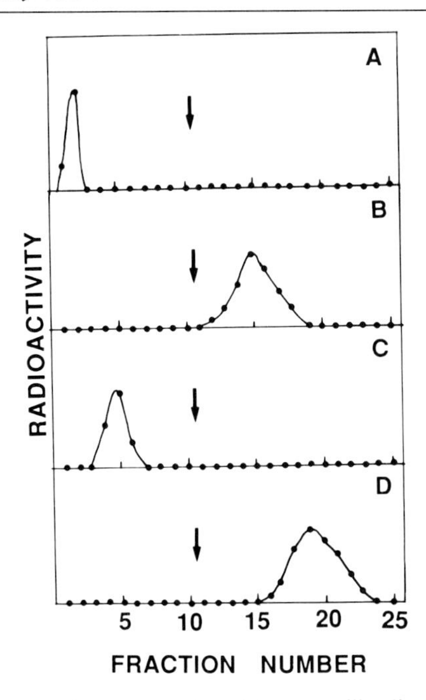

FIG. 1. Affinity column chromatography of human milk oligosaccharides and partial degradation products on the PVL column. (A) Oligosaccharide **I**; (B) oligosaccharide **II**; (C) oligosaccharide **III**; (D) oligosaccharide **IV**. Arrows indicate the positions where the elution buffer was switched to buffer containing 1 m*M* *N*-acetylglucosamine.

These results indicate that the GlcNAcβ1→3Gal group interacts more strongly with PVL than the GlcNAcβ1→6Gal group. This unique specificity is useful for discrimnation of the two isomers produced by endo-β-galactosidase digestion of poly-*N*-acetyllactosaminoglycans found in some *O*-linked oligosaccharides.[5] On the other hand, the oligosaccharide with two *N*-acetylglucosamine residues interacts far more strongly as evidenced by the fact that oligosaccharide **IV** binds to the column and elutes more slowly with the buffer containing 1 m*M* *N*-acetylglucosamine than oligosaccharide **II** (Fig. 1D).

Behavior of N-Linked Oligosaccharides with Different Outer Chain Structures and Partial Degradation Products

Oligosaccharide **VI**, which binds to the column and elutes with the buffer containing 1 m*M* *N*-acetylglucosamine (Fig. 2B), passes through the

[5] J. Amano, P. Strahl, E. G. Berger, N. Kochibe, and A. Kobata, *J. Biol. Chem.* **266,** 11461 (1991).

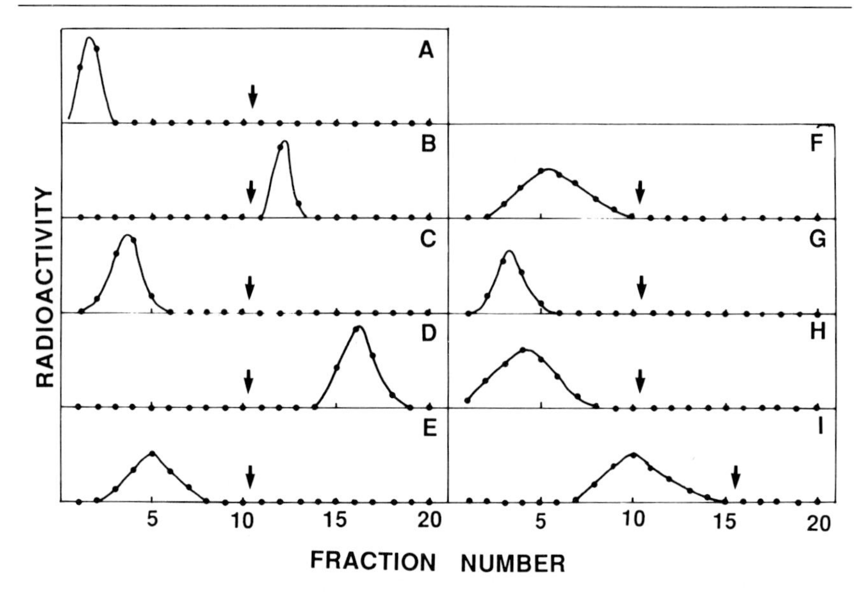

FIG. 2. Affinity column chromatography of N-linked oligosaccharides on the PVL column. (A) Oligosaccharides **V**, **VII**, **VIII**, **XIII**, **XIV**, **XX**, **XXI**, and **XXII**; (B) oligosaccharides **VI**, **XXIII**, and **XXIV**; (C) oligosaccharides **X** and **XI**; (D) oligosaccharide **IX**; (E) oligosaccharide **XII**; (F) oligosaccharides **XV** and **XVI**; (G) oligosaccharides **XVII** and **XVIII**; (H) oligosaccharide **XIX**; (I) oligosaccharide **XXV**. The arrows have the same meaning as in Fig. 1.

column without any interaction after β-*N*-acetylhexosaminidase treatment (oligosaccharide **VII**) (Fig. 2A). Substitution of the two terminal *N*-acetylglucosamine residues of oligosaccharide **VI** by Galβ1→4 residues (oligosaccharide **V** in Fig. 2A) or Galβ1→3 residues (data not shown) completely abolishes its affinity to the column. On the other hand, isomeric biantennary oligosaccharide **X** and oligosaccharide **XI**, which contain one *N*-acetylglucosamine residue as a nonreducing terminal group, are only retarded by the column (Fig. 2C). These results indicate that substitution of one of the two *N*-acetylglucosamine residues of oligosaccharide **VI** with galactose reduces its affinity to the lectin column. The lectin interacts with nonreducing terminal β-linked *N*-acetylglucosamine residues, but does not show any affinity for nonreducing terminal *N*-acetylneuraminic acid and *N*-acetylgalactosamine residues (data not shown).

N,N'-Diacetylchitobiitol (oligosaccharide **IX**) binds to the column and elutes slowly with the buffer containing 1 m*M* *N*-acetylglucosamine (Fig. 2D), but *N*-acetylglucosaminitol passes through the column without any interaction (data not shown). Addition of β-mannosyl residues to the *N,N'*-

diacetylchitobiitol completely abolishes the affinity of oligosaccharide **IX** to the lectin column (oligosaccharide **VIII** in Fig. 2A). These results also confirm that the presence of the terminal β-*N*-acetylglucosamine residue is essential for an oligosaccharide to interact with the column. Addition of an α-fucosyl residue at the C-6 position of the proximal *N*-acetylglucosamine moiety does not affect the behavior of oligosaccharide **VI** at all.

Oligosaccharide **XII**, however, shows much weaker interaction than oligosaccharide **VI** (Fig. 2E), indicating that the affinity of oligosaccharides is reduced by the addition of the bisecting *N*-acetylglucosamine residue. The negative effect of the bisecting *N*-acetylglucosamine residue is also observed in the case of oligosaccharide **XIII** and oligosaccharide **XIV** (Fig. 2A). These results are quite interesting because the addition of a β-linked *N*-acetylglucosamine residue is expected to enhance the binding of oligosaccharides to the lectin column. This unexpected effect probably arises because the bisecting *N*-acetylglucosamine residue cannot bind to the lectin column and sterically interferes with access of the lectin to the GlcNAcβ1→2Man group.

Although all isomeric N-linked oligosaccharides containing a terminal *N*-acetylglucosamine residue are retarded in the column, those with the GlcNAcβ1→2Man group (oligoosaccharides **XV** and **XVI**) (Fig. 2F) interact more strongly than those with the GlcNAcβ1→4Man group (oligosaccharide **XVII**) or the GlcNAcβ1→6Man group (oligosaccharide **XVIII**) (Fig. 2G). These results clearly indicate that the binding affinity to PVL is influenced by the β-*N*-acetylglucosamine linkages and is quite different from that of other *N*-acetylglucosamine-binding lectins from higher plants, which act preferentially at the GlcNAcβ1→4 residue.[6] Although distribution of an GlcNAcβ1→2 residue on the Manα1→6 arm or Manα1→3 arm is not discriminated by the affinity to the lectin column, addition of two α-mannosyl residues on the Manα1→6 arm reduces the interaction of an oligosaccharide with the lectin (compare oligosaccharide **XIX** in Fig. 2H to oligosaccharide **XVI** in Fig. 2F), probably owing to steric hindrance by the mannosyl residues.

All galactosylated tri- and tetraantennary oligosaccharides (**XX**, **XXI**, and **XXII**) do not show any affinity to the lectin column, similar to the case of biantennary ones (Fig. 2A). Both nongalactosylated triantennary oligosaccharides **XXIII** and **XXIV** bind to the column and elute with buffer containing 1 m*M* *N*-acetylglucosamine (Fig. 2B). However, the nongalactosylated tetraantennary oligosaccharide **XXV** does not bind to

[6] I. J. Goldstein and R. D. Poretz, *in* "The Lectins: Properties, Function, and Applications in Biology and Medicine" (I. E. Liener, N. Sharon, and I. J. Goldstein, eds.), p. 33. Academic Press, Orlando, Florida, 1986.

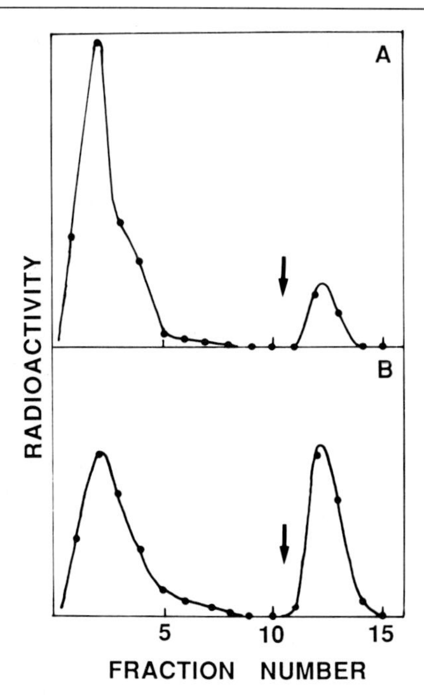

FIG. 3. Elution profiles of immunoglobulin G oligosaccharides by the PVL column. Radioactive oligosaccharide fractions obtained by hydrazinolysis (S. Takasaki, T. Mizuochi, and A. Kobata, this series, Vol. 83, p. 263) of immunoglobulin G from a healthy individual (A) and from a patient with rheumatoid arthritis (B) were subjected to PVL affinity column chromatography. The arrows have the same meaning in Fig. 1.

the column but is retarded in the column (Fig. 2I). These results indicate that the binding affinity to PVL is not simply determined by the number of *N*-acetylglucosamine residues but is influenced by the steric arrangement of the *N*-acetylglucosamine residues.

Application of *Psathyrella velutina* Lectin to Detection of Pathological Conditions

It is well known that manifestation of N-linked oligosaccharides containing terminal *N*-acetylglucosamine in the outer chain moieties can be observed in glycoproteins produced in tissues under diseased states.[7,8]

[7] R. B. Parekh, R. A. Dwek, B. J. Sutton, D. L. Fernandes, A. Leung, D. Stanworth, T. W. Rademacher, T. Mizuochi, T. Taniguchi, K. Matsuta, F. Takeuchi, Y. Nagano, T. Miyamoto, and A. Kobata, *Nature* (*London*) **316,** 452 (1985).

[8] K. Yamashita, A. Hitoi, N. Taniguchi, N. Yokosawa, Y. Tsukada, and A. Kobata, *Cancer Res.* **43,** 5059 (1983).

An immobilized PVL column can be used as an effective tool to study such rare or abnormal N-linked oligosaccharides with β-*N*-acetylglucosamine residues as nonreducing termini. The column is especially useful for detection of oligosaccharide **VI**,[4,9] because no other β-*N*-acetylglucosamine-binding lectin has been reported to bind this oligosaccharide. As an example of such studies, comparative analysis of the radioactive oligosaccharides liberated from immunoglobulin G (IgG) samples purified from sera of healthy individuals and patients with rheumatoid arthritis is shown. It is found that the galactose contents of the N-linked oligosaccharides of serum IgG from patients with rheumatoid arthritis are much lower than those of healthy individuals.[7] In accord with this finding, a clear-cut difference is detected as shown in Fig. 3. This result indicates that an immobilized PVL column can be used as a simple tool to detect the extent of pathological changes in the sugar chains of immunoglobulin G. Quite recently, we have developed a simpler and more sensitive method for detection of the aberrant IgG in the sera of patients with rheumatoid arthritis by using this novel *N*-acetylglucosamine-binding lectin.[10]

[9] M. Tandai, T. Endo, S. Sasaki, Y. Masuho, N. Kochibe, and A. Kobata, *Arch. Biochem. Biophys.* **291,** 339 (1991).
[10] N. Tsuchiya, T. Endo, K. Matsuta, S. Yoshinoya, F. Takeuchi, Y. Nagano, M. Shiota, K. Furukawa, N. Kochibe, K. Ito, and A. Kobata, *J. Immunol.* **151,** 1137 (1993).

[16] Separation of Galβ1,4GlcNAc α-2,6- and Galβ1,3(4)GlcNAc α-2,3-Sialyltransferases by Affinity Chromatography

By SUBRAMANIAM SABESAN, JAMES C. PAULSON, and JASMINDER WEINSTEIN

Introduction

Glycosyltransferases are important biochemical tools in investigations of carbohydrate structure and biological functions.[1,2] With the development of molecular cloning technology, glycosyltransferases have advanced from being rare, expensive reagents to biochemical reagents of choice for large-scale oligosaccharide synthesis.[3] These enzymes

[1] J. C. Paulson and G. N. Rogers, this series, Vol. 138, p. 162.
[2] S. Roth, U.S. Patent 5,180,674 (1992).
[3] Y. Ito, J. J. Gaudino, and J. C. Paulson, *Pure Appl. Chem.* **65,** 753 (1993).

together with methods[4] for the efficient recycling of nucleotide sugars provide a powerful combination for cost-effective, high-yield industrial methods for the rapid assembly of many complex oligosaccharides. Of these, the sialyltransferases occupy a unique role, as knowledge about the involvement of sialic acid in biological functions has grown very rapidly and the need to have available relevant carbohydrate structures is ever increasing. The difficulties associated to date with the chemical synthesis of sialosides[5] and the need to develop carbohydrate-based pharmaceuticals makes the availability of sialyltransferases in pure form essential.

The family of sialyltransferases consists of 10–12 enzymes. These enzymes transfer sialic acid (NeuAc) from CMP-NeuAc to an acceptor oligosaccharide and are classified[6] on the basis of the oligosaccharide sequences to which they transfer sialic acids. It is evident from the vast array of sialoside structures present in glycoproteins and glycolipids that more than 8–10 different sialyltransferases will be required in the biosynthesis. Of these, the Galβ1,4GlcNAc α-2,6- and Galβ1,3(4)GlcNAc α-2,3-sialyltransferases can be obtained either from rat liver or from expression of recombinant glycoproteins.[7,8] These two enzymes bind to the same nucleotide sugar CMP-NeuAc, but differ in their binding to specific oligosaccharide sequences. This chapter describes the affinity purification of Galβ1,3(4)GlcNAc α-2,3-sialyltransferase and separation from the Galβ1,4GlcNAc α-2,6-sialyltransferase based on differential binding to Galβ1,3GlcNAc sequences. The recognition of Galβ1,3GlcNAc sequences by the α-2,3-sialyltransferases and not by the α-2,6-sialyltransferase enables the selective adsorption of α-2,3-sialyltransferases to a synthetic Gal1,3GlcNAc-Sepharose column and thus provides a convenient way to separate the two enzymes from a mixture. The selective adsorption by the α-2,3-sialyltransferase should also be equally valuable for purification of the enzyme from tissue homogenates or culture supernatants containing the recombinant enzyme.

[4] Y. Ichikawa, Y. Lin, D. P. Dumas, G. Shen, E. Garcia-Junceda, M. Williams, R. Bayer, K. Ketcham, L. E. Walker, J. C. Paulson, and C. H. Wong, *J. Am. Chem. Soc.* **114,** 9283 (1992).

[5] K. Okamoto and T. Goto, *Tetrahedron* **46,** 5835 (1990).

[6] T. A. Beyer, J. E. Sadler, J. I. Rearick, J. C. Paulson, and R. Hill, *Adv. Enzymol.* **52,** 23 (1981).

[7] J. Weinstein, U. Souza-e-Silva, and J. C. Paulson, *J. Biol. Chem.* **257,** 13835 (1982).

[8] K. J. Colley, E. U. Lee, B. Adler, J. K. Browne, and J. C. Paulson, *J. Biol. Chem.* **264,** 17619 (1989).

Procedures

General Methods

Sepharose 4B, cyanogen bromide, and ethylenediamine are purchased from Sigma Chemical Co. (St. Louis, MO) and anhydrous methanol from Aldrich Chemical Co. (Milwaukee, WI). BioGel P-2 (200–400 mesh) is purchased from Bio-Rad (Richmond, CA). The ethylenediamine is distilled under atmospheric pressure. The disaccharide Galβ1,3GlcNAc-O-$(CH_2)_5$ $COOCH_3$ (**1**) is prepared according to the procedure described by Lemieux and co-workers[9] except that the 8-methoxycarbonyloctanol is replaced with 5-methoxycarbonylpentanol[10] and the final product is purified by gel-permeation chromatography on a column of BioGel P-2 (200–400 mesh), eluted, and equilibrated with deionized water. The structures of the disaccharides **1** and **2** are established by ^1H nuclear magnetic resonance (NMR) in D_2O, using a GE Omega 300 MHz NMR spectrometer.

The Galβ1,3(4)GlcNAc α-2,3- and Galβ1,4GlcNAc α-2,6-sialyltransferase activities are assayed as follows.[11] The Galβ1,3(4)GlcNAc α-2,3-sialyltransferase (5 μl of the eluant) is incubated for 30–60 min at 37° with an assay mixture (60 μl) containing lacto-*N*-tetraose (10 μg), bovine serum albumin (BSA) (50 μg), cytidine-5′-monophospho-*N*-[^{14}C]acetylneuraminic acid [CMP-sialic acid, 9 nmol, ~6000 counts/min (cpm)/nmol], 50 m*M* sodium cacodylate (pH 6.0), 50 m*M* manganese chloride, and 0.5% Triton CF-54. In the case of Galβ1,4GlcNAc α-2,6-sialyltransferase, the assay mixture contains lactose (6 mg) instead of lacto-*N*-tetraose and 0.6 nmol of CMP[^{14}C]NeuAc (500,000 cpm/nmol). Following incubation, the reaction mixture is diluted with 5 m*M* phosphate buffer (1 ml, pH 6.8) and applied to a Pasteur pipette column containing 1 ml of Dowex 1-X8 (HPO_4^{2-}, 100–200 mesh), and the column is washed with an additional 1 ml of 5 m*M* sodium phosphate, pH 6.8. The combined effluents containing the product are collected in a scintillation vial and assessed for ^{14}C radioactivity by scintillation spectroscopy.

Preparation of Galβ1,3GlcNAcβ-O-$(CH_2)_5CONH(CH_2)_2NH_2$ (Scheme 1)

A solution of compound **1** (260 mg) in methanol–ethylenediamine (3 : 1, v/v; 32 ml) is refluxed for 16 hr (Scheme 1). The reaction mixture is then concentrated to a dry residue, redissolved in deionized water (10 ml), and

[9] R. U. Lemieux, D. R. Bundle, and D. A. Baker, *J. Am. Chem. Soc.* **97,** 4076 (1975).

[10] S. Sabesan and J. C. Paulson, *J. Am. Chem. Soc.* **108,** 2068 (1986).

[11] J. C. Paulson, J. Weinstein, and U. Souza-e-Silva, *J. Biol. Chem.* **257,** 4034 (1982).

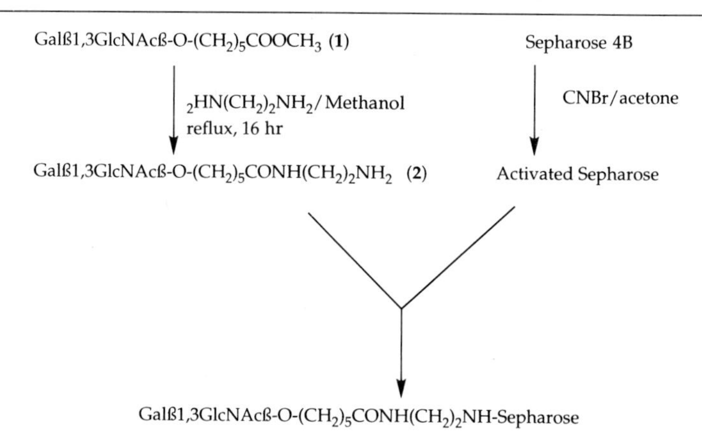

SCHEME 1. Preparation of Galβ1,3GlcNAc-Sepharose affinity matrix.

lyophilized to obtain a colorless material (307 mg). The structure of the product is confirmed by ^{1}H NMR (δ_{HOD} = 4.80 ppm), where the signal at 3.71 ppm for compound **1** disappears in **2** and a new triplet signal appears at 3.25 ppm (CH_2N—CO—). The crude product **2** is directly employed in the next step.

Preparation of Galβ1,3GlcNAcβ-O-$(CH_2)_5$CONH$(CH_2)_2$NH-Sepharose

Sepharose 4B is activated with cyanogen bromide according to the published procedure[12] as described below. Sepharose 4B (40–160 μm, 220 g) is washed successively with deionized water (1.5 liter), 30% aqueous acetone (1 liter), and 60% aqueous acetone (1 liter). The gel is then suspended in 60% aqueous acetone (200 ml) and cooled to $-15°$. A solution of cyanogen bromide in acetone (1 *M*, 67 ml) is added over a period of 3 min to the gently stirred suspension of the gel. After 5 min, 1 *M* triethylamine solution (67 ml) in 60% aqueous acetone is added over 5 min. Subsequently, the reaction mixture is poured into 0.1 *M* hydrochloric acid in acetone (1 liter). The gel is then successively washed with 60% aqueous acetone (1 liter), 30% aqueous acetone (1 liter), deionized water (1 liter), and then with 5 m*M* sodium carbonate solution (700 ml, pH 9.5). The gel is added to a solution of compound **2** (300 mg) in 5 m*M* sodium carbonate buffer (200 ml, pH 9.5) at 4° and gently shaken at that temperature for 48 hr. To this is added 1 *M* methylamine solution (4 ml, pH 8.3), and the mixture is shaken for additional 1 hr. The gel is then filtered on a sintered glass funnel, then washed with 0.2 *M* sodium chloride solution and deion-

[12] J. Kohn and M. Wilchek, *Biochem. Biophys. Res. Commun.* **107,** 878 (1982).

ized water (1 liter). It is finally suspended in a buffer (300 ml) containing 50 m*M* sodium cacodylate (pH 6.1), 0.1 *M* sodium chloride, and 0.02% v/v sodium azide and stored at 4°.

Affinity Chromatographic Separation of Galβ1,4GlcNAc α-2,6- and Galβ1,3(4)GlcNAc α-2,3-Sialyltransferases

A column (1.7 × 30 cm) packed with Galβ1,3GlcNAc-O-$(CH_2)_5$CONH-$(CH_2)_2$NH-Sepharose (25 ml, flow rate 25 ml/hr) at 4° is equilibrated with a buffer (150 ml) containing 2 m*M* cytidine diphosphate (CDP), 50 m*M* sodium cacodylate (pH 6.5), 25% glycerol, 150 m*M* sodium chloride, and 0.1% Triton CF-54. The mixture of Galβ1,4GlcNAc α-2,6- and Galβ1,3(4)-GlcNAc α-2,3-sialyltransferases (obtained from 2 kg of rat liver in step 3 of the published methods)[7,13] in a buffer (505 ml) containing 2 m*M* CDP, 50 m*M* sodium cacodylate (pH 6.5), 25% glycerol, 100 m*M* sodium chloride, and 0.1% Triton CF-54 is passed through the above column over a period of 14 hr (flow rate of 36 ml/hr). Fractions (10 ml) are collected, and these are assayed for the α-2,3- and α-2,6-sialyltransferase activities using lacto-*N*-tetraose and lactose as substrates, respectively (see General Methods). At fraction 60, the column is washed with a buffer (500 ml) containing 2 m*M* CDP, 25% glycerol, 50 m*M* sodium cacodylate (pH 6.5), 250 m*M* sodium chloride, and 0.1% Triton CF-54. As can be seen in Fig. 1, the α-2,6-sialyltransferase activity levels off at fraction 65. No α-2,3-sialyltransferase activity is observed during the loading of the enzyme and as well as during subsequent washing (up to fraction 105, Fig. 1), indicating that the enzyme is tightly bound to the Sepharose column. At this time, the elution is continued with the wash buffer containing 0.2 *M* lactose. The α-2,3-sialyltransferase starts to elute immediately as evidenced by the lacto-*N*-tetraose assay. The fractions containing the α-2,6- and α-2,3-sialyltransferases are pooled separately and concentrated as described in the report of Wen *et al.*[13]

Discussion

Rat liver is a convenient source for obtaining sialyltransferases. The sialyltransferases could be separated from the rest of the protein contaminants by adsorption on a CDP-hexanolamine-Sepharose column followed by elution with a sodium chloride gradient to obtain a partially purified enzyme mixture containing α-2,6- and α-2,3-sialyltransferases.[14] Conven-

[13] D. X. Wen, B. D. Livingston, K. F. Medzihradszky, S. Kelm, A. L. Burlingame, and J. C. Paulson, *J. Biol. Chem.* **267,** 21011 (1992).

[14] J. Weinstein, U. Souza-e-Silva, and J. C. Paulson, *J. Biol. Chem.* **257,** 13845 (1982).

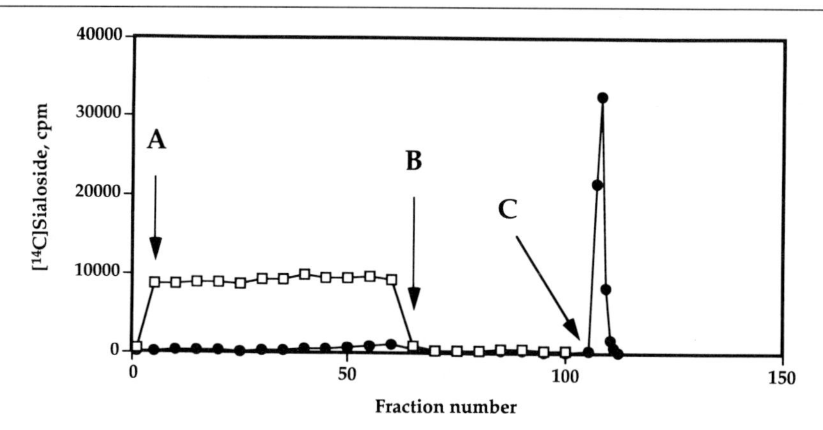

FIG. 1. Separation of (□) Galβ1,4GlcNAc α-2,6- and (●) Galβ1,3(4)GlcNAc α-2,3-sialyltransferases on a column of Galβ1,3GlcNAc-O-Sepharose. The load buffer at arrow A was 2 m*M* CDP–50 m*M* sodium cacodylate–100 m*M* sodium chloride–25% glycerol–0.1% Triton CF-54; at B it was 2 m*M* CDP–50 m*M* sodium cacodylate–250 m*M* sodium chloride–25% glycerol–0.1% Triton CF-54; and at C it was 2 m*M* CDP–50 m*M* sodium cacodylate–100 m*M* sodium chloride–25% glycerol–0.1% Triton CF-54–0.2 *M* lactose; The α-2,3- and α-2,6-sialyltransferase activities were assayed using lacto-*N*-tetraose and lactose, respectively, as substrates.

tionally an additional CDP-hexanolamine-Sepharose step followed by asialoprothrombin-Sepharose affinity chromatography had to be performed in order to separate the two enzymes. In the modified method, the partially purified sialyltransferase mixture was separated in one step by affinity chromatography on a Galβ1,3GlcNAcβ-conjugated Sepharose column. In addition, we have replaced the glycoprotein asialoprothrombin with a relatively inexpensive synthetic disaccharide. The advantages of this new method are several: (1) Galβ1,3GlcNAcβ-O-$(CH_2)_5CONH(CH_2)_2NH_2$ can be made easily in large quantities for the preparation of Sepharose affinity columns on an industrial scale, whereas asialoprothrombin is expensive and available in only limited amounts; (2) whereas the asialoprothrombin-Sepharose is labile and susceptible to protease degradation during storage, the Gal1,3GlcNAc-O-$(CH_2)_5CONH(CH_2)_2NH$-Sepharose column is stable for over 5 years, when stored at 4° in 0.02% sodium azide; and (3) owing to the small size of the synthetic compound **2**, a Sepharose affinity matrix containing a high density of **2** can be prepared. The success of the new methodology in the separation of α-2,6- and α-2,3-sialyltransferases has been described by Wen *et al.*[13]

A monosaccharide, namely, 2-acetamido-2-deoxy-β-D-glucosamine, conjugated to Sepharose has been used earlier to purify galactosyltransferases.[15] Sialyltransferases, whose carbohydrate recognition domain seems to extend beyond a monosaccharide,[16] require a minimum of a disaccharide for efficient adsorption. In this regard it may be of interest to note that Gal1,4GlcNAc-Sepharose affinity columns do not bind α-2,6-sialyltransferases, even though the K_m is in the low millimolar range.

[15] R. Barker, K. W. Olsen, J. H. Shaper, and R. Hill, *J. Biol. Chem.* **247,** 7135 (1972).
[16] K. B. Wlasichuk, M. A. Kashem, P. V. Nikrad, P. Bird, C. Jiang, and A. P. Venot, *J. Biol. Chem.* **268,** 13971 (1993).

[17] Polysaccharide Affinity Columns for Purification of Lipopolysaccharide-Specific Murine Monoclonal Antibodies

By ELEONORA ALTMAN and DAVID R. BUNDLE

Introduction

Immunoabsorption of antibodies is a convenient purification protocol that has been effectively exploited for one-step purification of immunoglobulin G (IgG), Fab, and single-chain antibody.[1,2] Generally the technique has found only limited application in the field of carbohydrate-specific antibodies because intrinsic affinities are weak, with the result that not all the active protein applied to a column is retained. Multipoint attachment of whole antibody to an immobilized antigen composed of polymeric epitopes enhances functional affinity and in the case of bacterial polysaccharides offers a route to recover specific IgM and IgG antibodies. The method described here uses bacterial lipopolysaccharide as a source of polymeric O-antigen, which after derivatization is coupled to commercially available, activated Sepharose gels. The absorbents prepared in this way provide a convenient one-step purification of bacterial O-antigen-specific monoclonal antibodies, three of which have been the subject of successful crystallographic studies.[3-5]

[1] D. R. Bundle, E. Eichler, M. A. J. Gidney, M. Meldal, A. Ragauskas, B. W. Sigurskjold, B. Sinnott, D. C. Watson, M. Yaguchi, and N. M. Young, *Biochemistry* **33,** 5172 (1994).
[2] N. N. Anand, S. Mandal, C. R. MacKenzie, J. Sadowska, B. Sigurskjold, N. M. Young, D. R. Bundle, and S. A. Narang, *J. Biol. Chem.* **266,** 21874 (1991).
[3] M. Cygler, D. R. Rose, and D. R. Bundle, *Science* **253,** 442 (1991).
[4] M. N. Vyas, N. K. Vyas, P. J. Meikle, B. Sinnott, B. M. Pinto, D. R. Bundle, and F. A. Quiocho, *J. Mol. Biol.* **231,** 133 (1993).
[5] D. R. Rose, M. Przybylska, R. J. To, C. S. Kayden, R. P. Oomen, E. Vorberg, N. M. Young, and D. R. Bundle, *Protein Sci.* **2,** 1106 (1993).

Initially an affinity column was prepared by directly coupling the O-antigen of *Salmonella essen* to an epoxy-activated Sepharose gel. The resulting matrix was very effective and exhibited consistent recovery of antibody over 100 cycles. However, under similar conditions, O-antigens from other bacteria did not couple to this activated gel, and a general method for polysaccharide derivatization was subsequently developed.

Preparation of Aminated *O*-Polysaccharides

Lipopolysaccharides (LPS) possess a common, highly conserved, inner core domain that contains heptose sugars.[6] After mild acid cleavage of lipid A, the polysaccharide antigen is terminated by a single 3-keto-D-*manno*-2-octulosonic acid (dOclA) residue. It and the exocyclic diol function of the heptose residues are well suited to the generation of the aldehyde groups by a short, mild periodate oxidation.[7,8] Reductive amination[9,10] of the partially oxidized *O*-polysaccharide in the presence of excess 1,3-diaminopropane then yields a modified polysaccharide bearing several amino groups (Fig. 1A) which may be conveniently coupled to activated Sepharose, either via epoxy (epoxy-activated Sepharose) or carboxyl (CH-Sepharose) groups.

Although there are several examples in the literature of mild periodate conditions preferentially cleaving acyclic diols in the presence of cyclic diol functionalities,[7,8] such differences depend crucially on relative rates. In the examples studied here it was found that 2-O-substituted α-L-rhamnopyranosyl residues of the *Shigella flexneri* antigen could be easily cleaved even under mild periodate conditions (37% of the total rhamnose content of the polysaccharide was destroyed by mild periodate treatment). Consequently the three *O*-polysaccharides (Fig. 2) studied exhibited different susceptibility to periodate oxidation. The *Yersinia enterocolitica* O:9 antigen[11] is resistant to oxidation since it is devoid of diol groupings (Fig. 2C), whereas the *Salmonella essen*[12] and *Shigella flexneri* Y[13] antigens

[6] E. Th. Rietschel, L. Brade, B. Lindner, and U. Zähringer, "Bacterial Endotoxic Lipopolysaccharides" (D. C. Morrison and J. L. Ryan, eds.), Vol. 1, Molecular Biochemistry and Cellular Biology, CRC Press, Boca Raton, Florida, 1992.

[7] W. Dröge, O. Lüderitz, and O. Westphal, *Eur. J. Biochem.* **4,** 126 (1968).

[8] W. Dröge, V. Lehmann, O. Lüderitz, and O. Westphal, *Eur. J. Biochem.* **14,** 175 (1970).

[9] B. A. Schwartz and G. R. Gray, *Arch. Biochem. Biophys.* **181,** 542 (1977).

[10] R. Roy, E. Katzenellenbogen, and H. J. Jennings, *Can. J. Biochem. Cell Biol.* **62,** 270 (1984).

[11] M. Caroff, D. R. Bundle, and M. B. Perry, *Eur. J. Biochem.* **139,** 195 (1984).

[12] D. R. Bundle, "Carbohydrate Chemistry" (J. Theim, ed.), Vol. 154 of Topics in Current Chemistry, p. 1. Springer-Verlag, Berlin, 1990.

[13] L. Kenne, B. Lindberg, K. Petersson, E. Katzenellenbogen, and E. Romanowska, *Eur. J. Biochem.* **91,** 279 (1978).

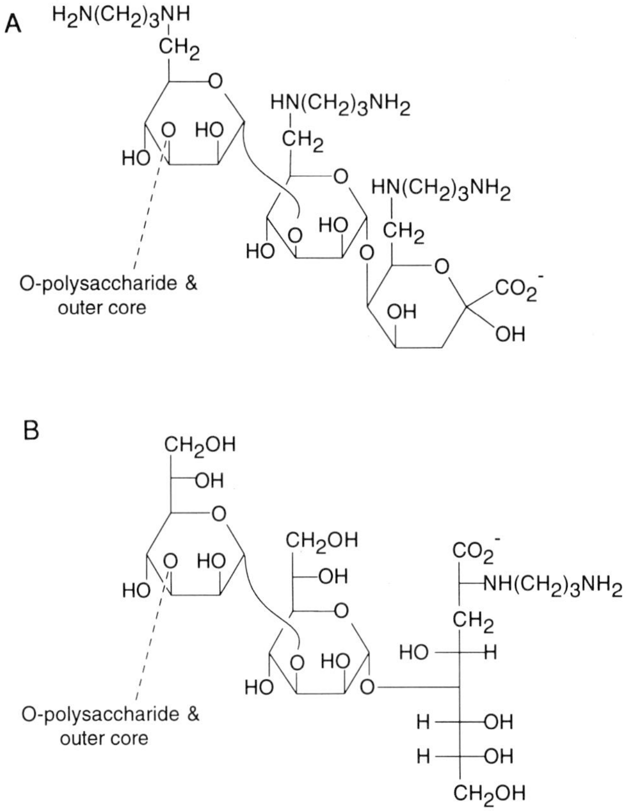

FIG. 1. Proposed structure of the modified polysaccharide containing amino groups introduced into the core oligosaccharide portion of the lipid free antigen. (A) Aldehyde groups generated by mild periodate oxidation and (B) the ketonic carbonyl group of dOclA in the inner core are used for coupling.

contain, respectively, one and two susceptible rhamnopyranosyl residues (Fig. 2A,B). Two of every three rhamnose residues in the *Salmonella* antigen are oxidized by mild periodate treatment, whereas one in three rhamnose residues of the *Shigella O*-polysaccharide is destroyed. The effect is most dramatic for the latter antigen since the susceptible rhamnose residues form part of the minimal recognition epitope, whereas rhamnose does not interact with the binding site of the *Salmonella* antibody. Consequently, affinity gels prepared from periodate-treated *Shigella O*-polysaccharide were inactive, while *Salmonella* counterparts gave active absorbents.

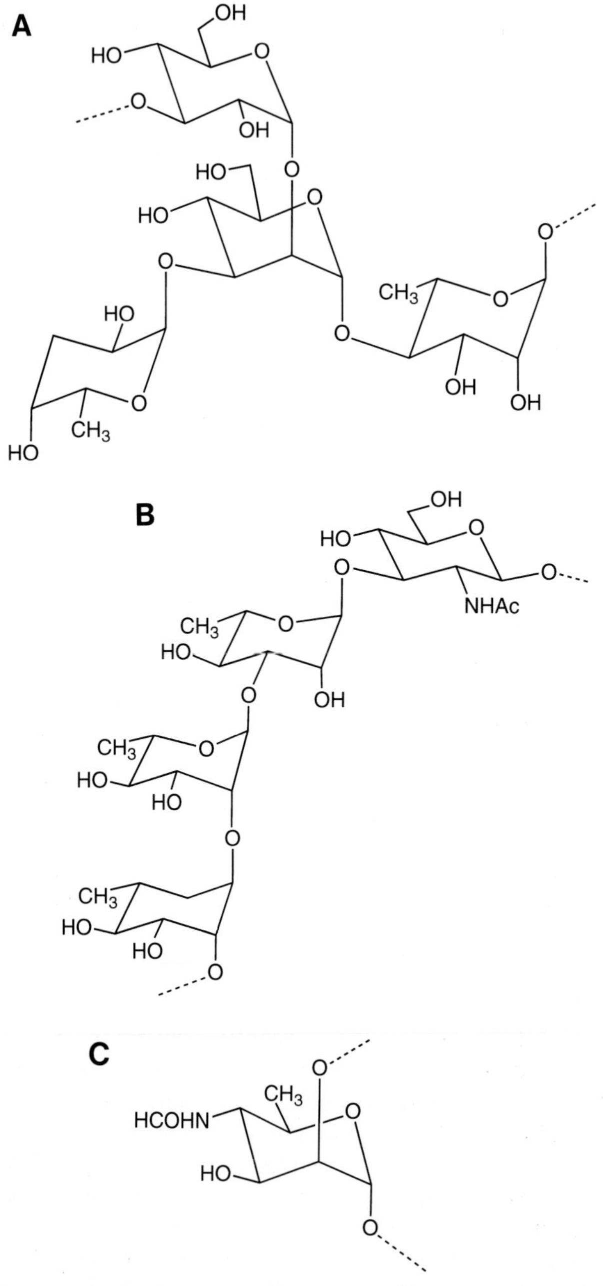

A

B

C

FIG. 2. Structures of bacterial O-antigens; (A) *Salmonella essen* (Ref. 12), (B) *Shigella flexneri* variant Y (Ref. 13), and (C) *Yersinia enterocolitica* O:9 (Ref. 11).

Affinity Absorbents

Because all the LPS-specific antibodies used here bind to the epitopes along the repeating units of the *O*-polysaccharide,[1,14,15] alterations in *O*-polysaccharide chain length may produce inactive affinity gels, attributable to lower functional affinity by destruction of the number of available epitopes along the chain. The preferred approach for coupling polysaccharide to the affinity matrix is direct conjugation of the small tether 1,3-diaminopropane to the terminal dOclA residue[10] of the *O*-polysaccharide (Fig. 1B) followed by covalent attachment of the monoaminated *O*-polysaccharides to succinimide ester-activated Sepharose 4B. This procedure reproducibly affords active affinity absorbents for all three polysaccharide antigens investigated.

The methods section describes the release of *O*-polysaccharide from whole LPS by mild acid cleavage of the acid-sensitive dOclA–dOclA or dOclA to lipid A linkage,[16] followed by introduction of amino groups by reductive amination of native *O*-polysaccharide and mild periodate-treated polysaccharide. Finally, procedures for coupling polysaccharide to the gel are described together with typical conditions for the assay and recovery of purified antibody.

Recovery and Assay of Affinity-Purified Antibody

The performance of the affinity gels prepared from each polysaccharide (Fig. 2) are recorded in Table I. The *Salmonella* antibody Se155.4 is unique among the three antibodies described here since it elutes from the affinity column under unusually mild pH conditions (Fig. 3A), a property that can be attributed to histidine residues in the active site[1] and not to the affinity matrix. Both the *Yersinia* and *Shigella* antibodies YsT9.2 and SYA/J6 require more extreme elution conditions, pH 2.4 (Fig. 3B), denaturing conditions that necessitate careful readjustment to neutral pH by a two-stage dialysis procedure if full antigen binding activity is to be restored. Despite these precautions a certain amount of antibody precipitates owing to a failure to renature all the eluted protein. Assay of recovered antibodies is performed by enzyme immunoassay (EIA) using LPS-coated microtitration plates, and the percentage recovery of active antibody is calculated by comparison with the unit volume signal intensity of crude ascites fluid (Table I).

[14] D. R. Bundle, M. A. J. Gidney, M. B. Perry, J. R. Duncan, and J. W. Cherwonogrodzky, *Infect. Immun.* **46,** 389 (1984).

[15] D. R. Bundle, *Pure Appl. Chem.* **61,** 1171 (1989).

[16] A. Tacken, E. T. Reitschel, and H. Brade, *Carbohydr. Res.* **149,** 279 (1986).

TABLE I
EVALUATION OF AFFINITY COLUMNS AND EXAMINATION OF FACTORS AFFECTING THEIR EFFICIENCY

O-Polysaccharide	Carbonyl groups used in reaction with 1,3-diaminopropane	Column matrix	Antibody recovery (%) based on indirect EIA[a]
Salmonella essen	Generated by mild periodate oxidation	Epoxy-activated Sepharose 6B	95–100
Shigella flexneri Y	Terminal dOclA residue	CH-Sepharose 4B	90
Yersinia enterocolitica O:9	Generated by mild periodate oxidation	CH-Sepharose 4B	70
	Terminal dOclA residue	CH-Sepharose 4B	80

[a] Determined by indirect EIA; $OD_{405\,nm}$ (60 min) of eluted peak ÷ $OD_{405\,nm}$ (60 min) of ascites fluid applied to column.

Methods

Preparation of O-Polysaccharide

Lipopolysaccharides are extracted by the lysozyme–phenol–water method from phenol-killed bacteria[17] and purified by ultracentrifugation. To obtain *O*-polysaccharide, a solution of LPS (1 mg/1 ml) is hydrolyzed in 1.5% (w/v) acetic acid for 2 hr at 100°,[1,13] and lipid A is removed by centrifugation. The liberated *O*-polysaccharide is purified on a Sephadex G-50 (Pharmacia, Uppsala, Sweden) column (2.6 × 95 cm) eluted with 50 m*M* pyridine–acetic acid buffer, pH 4.5.

Mild Periodate Oxidation of O-Polysaccharide

O-Polysaccharide is treated with 50 m*M* sodium metaperiodate (10 mg/ml) in the dark, at 22° for 5 min. Excess periodate is destroyed by reaction with ethylene glycol (1 ml) for 1 hr at 22°. The oxidized *O*-polysaccharide is dialyzed against distilled water until salt-free and then lyophilized.

Coupling of Oxidized O-Polysaccharide to 1,3-Diaminopropane Spacer

A solution of oxidized *O*-polysaccharide (2 μmol, 20 mg) is dissolved in 0.2 *M* sodium borate buffer (2 ml), pH 9.0, and slowly added to a

[17] K. G. Johnson and M. B. Perry, *Can. J. Microbiol.* **22,** 29 (1976).

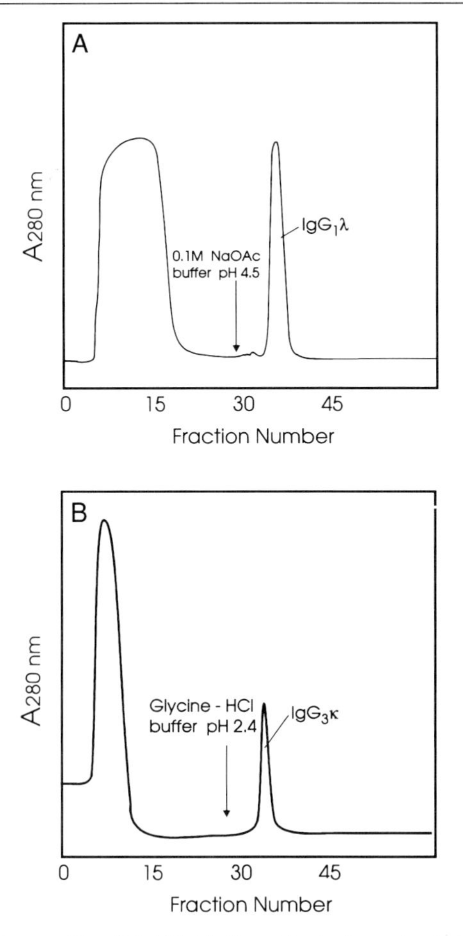

FIG. 3. (A) Elution profile of Se155.4 IgG_1 antibody from the *Salmonella essen* affinity matrix (1.0 × 5.0 cm). Antibody (40 mg in 8 ml of ascites fluid) was applied to the column and eluted with 0.1 *M* sodium acetate buffer, pH 4.5. (B) Elution profile of SYA/J6 IgG_3 antibody from *S. flexneri* Y affinity column (0.7 × 4 cm). Antibody (1.5 mg in 0.5 ml of ascites fluid) was applied to the column and eluted with 0.1 *M* NaCl in 50 m*M* glycine hydrochloride buffer, pH 2.4.

solution of 1,3-diaminopropane dihydrochloride (0.2 mmol, 29.4 mg) in 1 ml of the same buffer containing sodium cyanoborohydride (20 μmol, 1.54 mg recrystallized from tetrahydrofuran in dichloromethane). The final pH is adjusted to 8.0 with sodium hydroxide solution. The reaction is allowed to proceed for 5 days at 45°–50°. The solution is passed through a BioGel

P-2 (Bio-Rad, Richmond, CA) column (1.6 × 95 cm) eluted with 20 m*M* pyridylacetic acid buffer, pH 5.4. Fractions are assayed for the presence of glycose[18] and free amino groups.[16]

Coupling of O-Polysaccharide to 1,3-Diaminopropane Spacer

O-Polysaccharide (2 μmol, 10 mg/ml) is dissolved in 0.2 *M* sodium borate buffer, pH 7.55 (2 ml) (0.2 *M* sodium borate buffer, pH 9.0, in the case of the *Shigella flexneri* serogroup Y *O*-polysaccharide), then added dropwise to a solution of 1,3-diaminopropane dihydrochloride (0.2 mmol, 29.4 mg) in 1 ml of the same buffer containing sodium cyanoborohydride (20 μmol, 1.54 mg). The reaction is carried out for 6 days at 45°, and product is purified on a BioGel P-2 column (1.6 × 95 cm) (Bio-Rad) eluted with 20 m*M* pyridylacetic acid buffer, pH 5.4. Fractions are assayed for the presence of glycose[18] and free amino groups.[19]

Preparation of Sepharose 4B Immunoabsorbent

Coupling to Epoxy-Activated Sepharose 6B. The *O*-polysaccharide-1-(3-amino)propylamino derivative of *S. essen* is covalently coupled to epoxy-activated Sepharose 6B (Pharmacia). Dry epoxy-activated Sepharose 6B gel (4 g) is swollen, washed with water, and filtered, and the moist gel is added to a solution of the aminated polysaccharide (65 mg) in 50 ml of 0.2 *M* NaOH–KCl buffer, pH 12.5. The reaction mixture is shaken at 20° for 18 hr, and then the gel is filtered, washed with water (6 × 50 ml), and suspended in 1 *M* aqueous ethanolamine ssolution for 18 hr at room temperature. The gel is finally washed sequentially with 10 volumes of 0.15 *M* NaCl in 50 m*M* Tris-HCl buffer, pH 8.0, and 10 volumes of 0.5 *M* NaCl in 0.1 *M* sodium acetate buffer, pH 4.5, and finally reequilibrated with the saline Tris buffer for long-term storage at 4°.

Coupling to Succinimide Ester-Activated Sepharose. The *O*-polysaccharide-1-(3-amino)propylamino derivative of *S. flexneri* serogroup Y is covalently bound to succinimide-activated agarose, CH-activated Sepharose 4B (Pharmacia). Dry gel (0.5 g) is suspended in 1 m*M* HCl and washed with the same solution (200 ml/g dry gel). The ligand (10 mg) is dissolved in 1 ml of coupling buffer, namely, 0.1 *M* sodium bicarbonate containing 0.5 *M* sodium chloride (5 ml/g dry gel), and the reaction mixture is rotated end-over-end for 2 hr at room temperature. The gel is washed with coupling buffer (10 ml) and suspended in 1 *M* aqueous ethanolamine solution for

[18] M. Dubois, K. A. Gilles, J. K. Hamilton, P. A. Rebers, and F. Smith, *Anal. Chem.* **28,** 350 (1956).

[19] Y. C. Lee, *Carbohydr. Res.* **67,** 509 (1978).

1 hr at room temperature. It is washed sequentially with 10 volumes of 0.15 *M* NaCl in 50 m*M* Tris-HCl buffer, pH 8.0, and 10 volumes of 0.5 *M* NaCl in 10 m*M* sodium acetate buffer, pH 4.5, and finally reequilibrated with the saline Tris buffer for long-term storage at 4°.

Affinity Purification of Monoclonal Antibodies

Affinity columns are evaluated with ascites fluid containing monoclonal antibodies raised against the respective O-antigens (typical concentration 3.4–5.0 mg antibody protein/ml of ascites). Ascites fluid is centrifuged 5–10 min in an Eppendorf microcentrifuge. The supernatant is filtered through 0.8- and 0.22-μm filters and diluted with 50 m*M* Tris-HCl/0.15 *M* NaCl, pH 8.0 (1 : 1, v/v). The affinity column (1.5 ml gel in a 0.7 × 4 cm column) is equilibrated with the same Tris-HCl buffer, and 0.5 ml of diluted ascites fluid is applied to an affinity column. The column is washed with 50 m*M* Tris-HCl/0.15 *M* NaCl buffer, pH 8.0, until (typically 30 ml) the UV absorbance of the eluate drops to baseline level. A stepwise elution technique is used to determine the optimal elution conditions. The column is first washed with 0.5 *M* NaCl in 0.1 *M* sodium acetate buffer, pH 4.5 (30 ml), followed by 0.1 *M* NaCl in 50 m*M* glycine hydrochloride buffer, pH 3.4 (30 ml), and finally with 0.1 *M* NaCl in 50 m*M* glycine hydrochloride buffer, pH 2.4 (30 ml). Specific *S. essen* Se155.4 IgG[1] is eluted with 0.5 *M* NaCl in 10 m*M* sodium acetate, pH 4.5. The combined fractions of specific antibody are dialyzed against 0.15 *M* NaCl in 50 m*M* Tris-HCl buffer, pH 8.0. Specific *Y. enterocolitica* O:9 YsT9-2 IgG_3 is eluted[14] with 0.1 *M* NaCl dissolved in 50 m*M* glycine hydrochloride buffer, pH 2.4. The combined fractions are dialyzed first against 0.5 *M* NaCl in 0.1 *M* sodium acetate buffer, pH 4.5, followed by 0.15 *M* NaCl in 50 m*M* Tris-HCl buffer, pH 8.0. Alternatively, the eluate is collected in tubes containing 0.5 *M* NaCl in 0.1 *M* sodium acetate buffer, pH 4.5, and the combined fractions are dialyzed against 0.15 *M* NaCl, 50 m*M* Tris-HCl buffer, pH 8.0. Specific *S. flexneri* serotype Y SYA/J6 IgG_3 antibody[15] is eluted with 0.1 *M* NaCl in 50 m*M* glycine hydrochloride buffer, pH 2.4, and the combined fractions are dialyzed using similar conditions.

Indirect Enzyme Immunoassay

The activity of the antibody in eluted fractions is determined by an indirect EIA as previously described[14,20] using goat anti-mouse Ig reagent

[20] N. I. A. Carlin, M. A. J. Gidney, A. A. Lindberg, and D. R. Bundle, *J. Immunol.* **137,** 2361 (1986).

coupled to alkaline phosphatase as the disclosing reagent. The absorbance is measured at 405 nm.

Monoclonal Antibodies

The monoclonal antibodies used in this work are derived from cell fusions between the Sp2/0 cell line and spleen cells from BALB/c mice immunized with phenol-killed cells of the appropriate bacteria.[1,14,20]

Determination of Free Amino Groups

Amino groups introduced into the *O*-polysaccharide are assayed with 2,4,6-trinitrobenzenesulfonic acid (TNBS),[19] and 2-amino-2-deoxy-D-glucose hydrochloride is used as the standard.

Discussion

Of the two strategies used to introduce amino groups into the polysaccharide prior to attachment to the solid matrix, the one based on a monoaminated polysaccharide provided active products independent of the structure of the polysaccharide antigen. The most extensively used affinity column, that involving *Salmonella essen* polysaccharide, has been reused through more than 100 cycles of ascites fluid samples[1] and samples derived from periplasmic extracts of *Escherichia coli*.[2] The capacity of the column remained consistently high, and 1 ml of gel had a capacity of 5.8 mg of antibody.

Intrinsic and functional affinities are of crucial importance to the effective use of carbohydrate immunoabsorbents. Although whole IgG or IgM productively participate in multivalent interaction with single polysaccharide chains or multipoint attachment to the ligands of the gel, this type of binding will not occur for univalent Fab fragments or immobilized polysaccharides that are too short. Two of the antibodies used here, Se155.4 and SYA/J6, have intrinsic affinities for their trisaccharide epitopes of $10^5\ M^{-1}$,[1] whereas YsT9.2 shows only $10^4\ M^{-1}$. Reduced recoveries of the lower affinity antibody occur because not all antibody is absorbed to the affinity matrix. Furthermore, attempts to affinity-purify Fab derived from such low-affinity columns are likely to be unsuccessful. Of the three Fabs prepared and crystallized from the antibodies under study, only Se155.4 has been prepared in this way. In the case of SYA/J6 we have seen that larger pentasaccharide ligands have lower affinity than the smaller trisaccharide epitope that fills the active site. Consequently, binding to the even larger immobilized antigen is likely to carry a penalty owing to unfavorable interactions as the polysaccharide enters and leaves the binding site (D. R. Bundle *et al.*, Auzanneau and Siguskjold). Mild elution conditions are most desirable, and for one case, Se155.4 histidine residues in the binding site render this antibody binding particularly sensitive to pH changes. Elution with an inexpensive, readily available ligand is a preferred strategy when bound antibody may be economically displaced by a small structure.

[18] Streptavidin–Biotinylglycopeptide–Lectin Complex in Detection of Glycopeptides and Determination of Lectin Specificity

By MING-CHUAN SHAO and CHRISTOPHER C. Q. CHIN

Introduction

Streptavidin–biotinylglycopeptide–lectin complexes can be used in the detection of glycopeptides and the determination of lectin specificity at the picomole level. The method is based on the high affinity between streptavidin and biotin derivatives,[1,2] and between lectins and their complementary sugars,[3] and on the ability of streptavidin to adhere to microtiter plate wells[4,5] (Fig. 1). For the detection of glycopeptides, each unknown (glyco)peptide [from a high-performance liquid chromatography (HPLC) run, for example] is first biotinylated and complexed with streptavidin, each conjugate is then coated to the well of a titer plate, and lectin coupled to horseradish peroxidase (HRP) is added. After incubation, the wells are thoroughly washed and probed for the presence of the peroxidase. A positive test shows the presence of a glycan complementary to the lectin used.[4] Based on the same principle, when a series of well-characterized biotinylglycans complexed with streptavidin is precoated in the wells of a titer plate, the sugar-binding specificity of unknown lectins, conjugated to peroxidase, can also be conveniently screened.[5]

[1] N. M. Green, *Adv. Protein Chem.* **29,** 85 (1975).
[2] V. J. Chen and F. Wold, *Biochemistry* **25,** 939 (1986).
[3] I. J. Goldstein and C. E. Hayes, *Adv. Carbohydr. Chem. Biochem.* **35,** 127 (1978).
[4] M. C. Shao and C. C. Q. Chin, *Anal. Biochem.* **207,** 100 (1992).
[5] M. C. Shao, *Anal. Biochem.* **205,** 77 (1992).

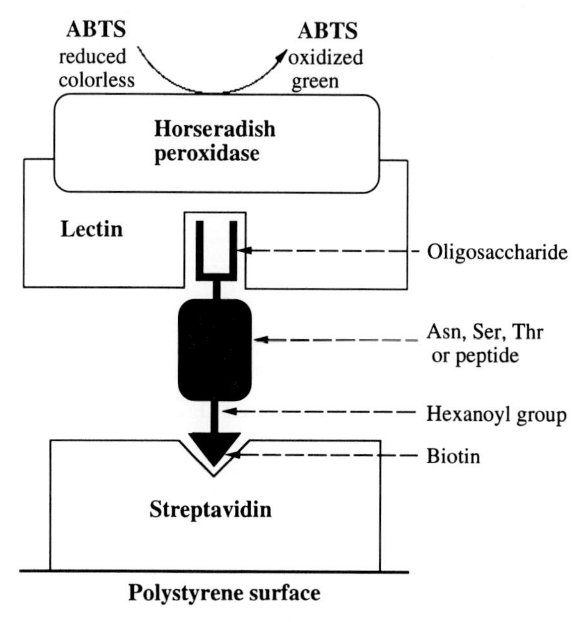

FIG. 1. Principle for the detection of glycopeptides and the determination of lectin specificity with streptavidin–biotinylglycopeptide–lectin complex as a tool. The (glyco)peptides from HPLC peptide maps can be biotinylated, complexed with streptavidin, immobilized on a titer plate, and then detected by the activity of HRP coupled covalently to a lectin, which in turn can bind to specific glycans. Similarly, if a series of well-characterized biotinylglycans complexed with streptavidin are precoated in the wells of a titer plate, the sugar-binding specificity of a given lectin can be screened.

Preparation of the Lectin–Horseradish Peroxidase Conjugate

Reagents

Lectin, 5 nmol by subunit
HRP, 20 nmol
$NaIO_4$, 60 m*M*
Diethylene glycol, 160 m*M*
$NaBH_4$, 0.5%
Saturated ammonium sulfate

Procedure. Briefly, 20 nmol of HRP, dissolved in 75 μl of water, is reacted with 75 μl of 60 m*M* $NaIO_4$ (freshly prepared) at 4° for 30 min; the reaction is terminated by the addition of 75 μl of 160 m*M* diethylene glycol at room temperature and incubated for 30 min. Five nanomoles of lectin by subunit (the molar ratio of the lectin subunit to HRP is about 1:4), dissolved in 150 μl of water, is mixed with the above solution and dialyzed against 50 mM sodium carbonate, pH 9.6, at 4° overnight with gentle stirring. After the dialysis, 30 μl of 0.5% $NaBH_4$ is added to the dialyzed mixture, and the reaction is allowed to proceed at 4° for 2 hr. An equal volume of saturated ammonium sulfate is added to precipitate the lectin–HRP conjugate, leaving unreacted HRP in the supernatant. The lectin–HRP is collected by centrifugation, then suspended in 50% saturated ammonium sulfate or desalted by dialysis against deionized water and lyophilized for storage. The molar ratio of HRP to the lectin in the conjugates prepared usually is 1.5–1.8.

Detection of Glycopeptides in Chromatographic Peptide Maps

Reagents

Buffer A, 50 mM sodium carbonate, pH 9.6, containing 0.02% sodium azide

Buffer B, 50 mM sodium phosphate, pH 7.4, containing 0.05% Tween 20

Buffer C, 50 mM Tris-HCl buffer, pH 7.4, containing 150 mM NaCl, 1 mM $MgCl_2$, 1 mM $MnCl_2$, and 1 mM $CaCl_2$

Buffer D, 50 mM citrate buffer, pH 4.4, containing 0.012% H_2O_2

Glycoprotein (asialofetuin, 3 mg)

Streptavidin (Sigma Chemical Co., St. Louis, MO), 100 μg/ml (1.5 μM) in buffer A

Sulfo-NHSH-biotin, sulfosuccinimidyl 6-(biotinamido)hexanoate (Pierce Chemical Co., Rockford, IL), 3 μM in buffer A

Lectins coupled to HRP: Con A (concanavalin A)–HRP, DSA (*Datura stramonium* agglutinin)–HRP, and PNA (peanut agglutinin)-HRP, 1–2 units of HRP activity in 1 ml of buffer C (the total amount will vary, depending on the HRP activity in each lectin–HRP complex

Bovine serum albumin (BSA), 1% in buffer A

Sodium dodecyl sulfate (SDS), 5% solution

2,2′-Azinobis(3-ethylbenzthiazoline sulfonate) (ABTS) (Sigma), 0.3 mg in 1 ml of buffer D

Procedure

Step 1: Preparation and Assay of Tryptic Peptides of Proteins. The approach is illustrated with the actual peptide mixture from asialofetuin (3 mg).[4] The glycoproteins are reduced with dithiothreitol in 6 M guanidine hydrochloride, reacted with vinylpyridine, dialyzed, and digested with

TPCK-treated trypsin. The digestion is stopped by adjustment of the reaction mixture to pH 3 and heating. An aliquot corresponding to 100 pmol of protein is applied to C_{18} column and eluted. Fractions are collected either manually or automatically. The samples are lyophilized, dissolved in 20% acetonitrile, and stored at −20°.

Step 2: Biotinylation of Proteolytic (Glyco)peptides. Ten-microliter aliquots of the HPLC fractions (containing 1–10 pmol sample) are diluted with buffer A and added to wells of a titer plate, mixed carefully with a 10 μl of sulfo-NHSH-biotin (30 pmol) solution, and incubated for 30 min at 25°. The excess of sulfo-NHSH-biotin is used to maximize the yield of biotinylated (glyco)peptides.

Step 3: Coating of Streptavidin–Biotinylated (Glyco)peptide Complex. After biotinylation, 80 μl of streptavidin (2 μg, 120 pmol of subunits) is added to bind both biotinylated (glyco)peptides and free biotin derivatives, and the resulting conjugates are incubated for 3 hr at 25° to affect adherence to the wall of the wells. A molar ratio of streptavidin subunit to sulfo-NHSH-biotin of 4 : 1 is used to make all the biotin derivatives complex to streptavidin.

Step 4: Blocking with Bovine Serum Albumin. The plate is rinsed 3 times with buffer B, and 350 μl of 1% BSA solution is added to each well. Plates are incubated at 25° for 1.5 hr to coat any remaining sites on the wells.

Step 5: Incubation with Lectin. The plate is again washed 3 times with buffer B; then 100 μl of lectin–HRP solution (1–2 units/ml at an HRP activity) is added to each well, and the plate is incubated at 25° for 1.5 hr.

Step 6: Peroxidase Reaction. The plate is washed 5 times with buffer B, 100 μl of substrate (ABTS) solution is added to each well, and the plate is incubated at 25° C for 10–60 min (longer incubations may be needed to detect low levels of binding). The reaction is stopped by the addition of 100 μl of 5% SDS solution.

Step 7: Measurement of Extent of Binding of Glycopeptides to Lectin–Peroxidase. The absorbance at a wavelength of 405 nm is quantified with the use of a microplate reader.

Comments

A representative assay for the glycopeptides in HPLC maps of the tryptic peptides from asialofetuin is presented in Fig. 2. It is well documented that there are three N-linked glycosylation sites designated N1, N2, and N3 on asialofetuin.[6] All of the N-linked oligosaccharides at the

[6] M. G. Yet, C. C. Q. Chin, and F. Wold, *J. Biol. Chem.* **263,** 111 (1988).

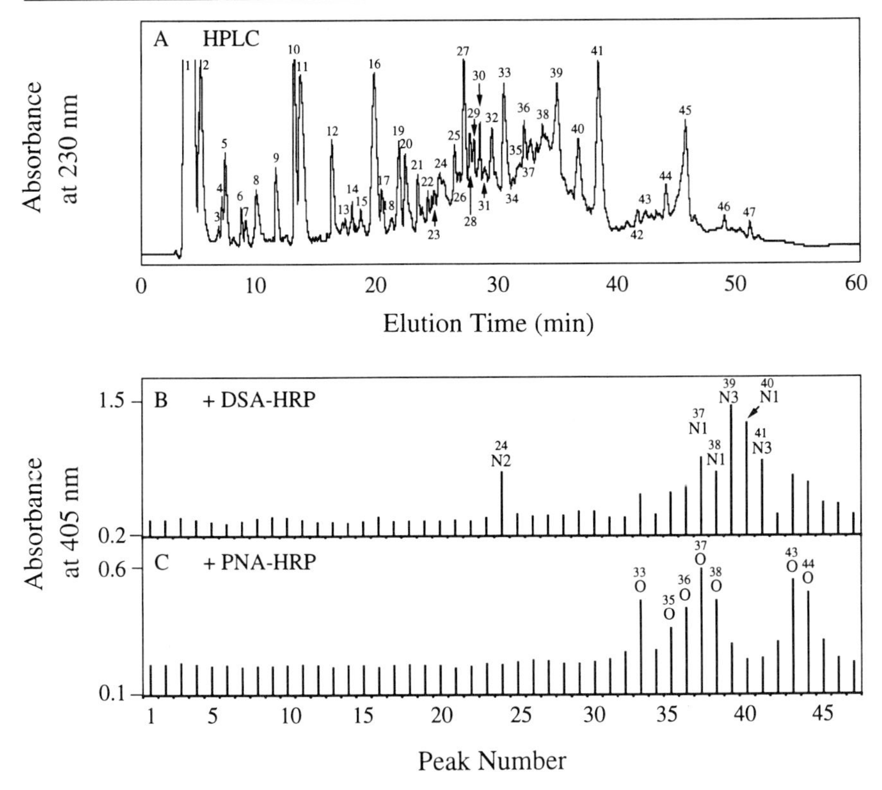

FIG. 2. (A) Reversed-phase HPLC of the tryptic peptides from asialofetuin (peaks 1–47 were collected). (B), (C) Analyses of the HPLC fractions for glycopeptides by the standard procedure described in the text with DSA–HRP (at an HRP activity of about 100 mU/well) (B) or PNA–HRP (at an HRP activity of about 200 mU/well) (C) as probes. Each HPLC fraction was diluted 200-fold with 20% acetonitrile for the assay, and the enzyme reaction was carried out for 15 min in case B and for 60 min in case C. The peaks (24 and 37–41) that reacted positively with DSA–HRP are labeled with an "N," the number to the right of "N" referring to the classification of the glycopeptide identified by amino acid sequence analyses (see text); those peaks (33, 35–38, 43, and 44) that reacted positively with PNA–HRP are labeled with an "O"; the number above "N" and "O" refers to the number of the peak. The other peaks did not react either with DSA–HRP or PNA–HRP.

three sites contain tri- and tetraantennary structures which can react strongly with DSA after removal of sialic acid.[7] It was found in Fig. 2A,B that HPLC peaks 24, 37, 38, 39, 40, and 41 of the tryptic digest of asialofetuin gave positive reactions with DSA–HRP. Based on sequence

[7] K. Yamashita, K. Totani, T. Ohkura, S. Takasaki, I. J. Goldstein, and A. Kobata, *J. Biol. Chem.* **262,** 1602 (1987).

analyses,[4] peaks 37, 38, and 40 correspond to glycopeptide N1, peak 24 to N2, and peaks 39 and 41 to N3. When the tryptic peptides were probed with PNA, which binds tightly to O-linked sugars, containing Gal(β1-3)GalNAc, seven positive peptides were found (Fig. 2C). Sequence analyses showed that all of them share a common peptide fragment, the N-terminal sequence of which was VTCTLFQTQPV, consistent with that of the known tryptic O-linked glycopeptide of asialofetuin on which there are three potential O-linked glycosylation sites.[8,9]

For routine detection of glycopeptides in HPLC peptide maps of an unknown protein, the use of the set of lectins (Con A, DSA, and PNA) should permit the detection in a single experiment of a large number of different glycan structures known to be present in glycoproteins.[4] Con A is quite nonspecific in binding both oligomannose, hybrid, and even biantennary complex glycans. It does not bind tri- and tetraantennary structures, and therefore DSA, which recognizes the Gal(β1-4)GlcNAc structure, is included to cover the latter two types of glycans. PNA, which is specific for Gal(β1-3)GalNAc, a common component of O-linked glycans, is included to be able to screen for both O- and N-linked structures without sialic acid residues. For this reason the test described here should be used only with asialoglycoproteins. Pretreatment of the unknown protein to be assayed for the presence of carbohydrate with neuraminidase[3] and/or acid[10] to remove sialic acid is thus required.

The nature of the reaction of the lectin with the N- or O-linked glycopeptides being detected by this method can be confirmed by inhibiting the reaction with an excess amount of a sugar inhibitor appropriate for the lectin, or by digestion of the biotinylated glycopeptides with N- or O-glycosidases.[4]

The assay is primarily a qualitative one. Because the extent of the biotinylation of the individual peptides may well be quite different and furthermore the lectins may have different affinities for the different oligosaccharide structures on different glycopeptides, quantitative conclusions are quite problematic.

This system can also be used in the detection of glycoprotein and the investigation of glycan structures using several lectins of known specificity and glycosidases. An example is shown in Fig. 3. It is known that the glycoproteins ovalbumin, fetuin, asialofetuin, transferrin, and carboxypeptidase Y contain oligosaccharide structures which can bind to Con A

[8] R. G. Spiro and V. D. Bhoyroo, *J. Biol. Chem.* **249,** 5704 (1974).

[9] K. M. Dziegielewska, W. M. Brown, S.-J. Casey, D. L. Christie, R. C. Foreman, R. M. Hill, and N. R. Saunders, *J. Biol. Chem.* **265,** 4354 (1990).

[10] R. G. Spiro, *J. Biol. Chem.* **235,** 2860 (1960).

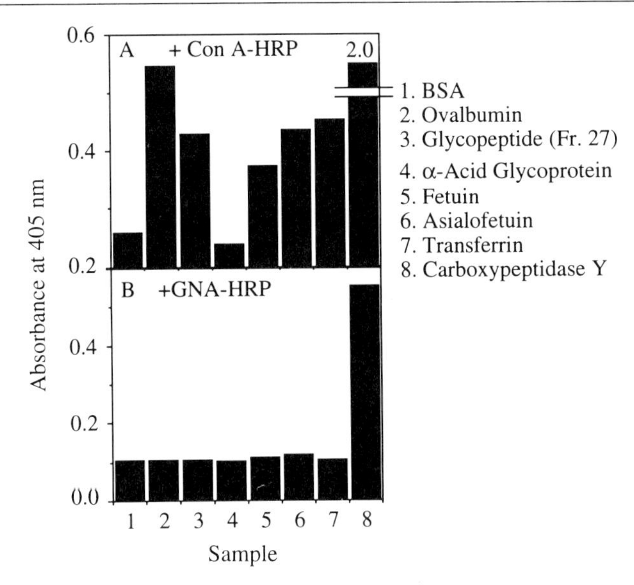

FIG. 3. (A) Detection of glycoproteins by the standard procedure described in the text. Approximately 1 pmol of each sample was used to coat the well of the plate. Con A–HRP at an HRP activity of 100 mU/well was used. The enzyme reaction was carried out for 15 min. It was observed that the glycoproteins ovalbumin, fetuin, asialofetuin, transferrin, and carboxypeptidase Y reacted with Con A–HRP significantly. The glycopeptide (fraction 27) (from HPLC tryptic maps of ovalbumin)[4] was used as a positive control, BSA as a negative control, and α-acid glycoprotein, which has no glycan structures that react with Con A, as another negative control. (B) Determination of the structures of the glycoprotein carbohydrate chain with GNA–HRP (GNA, *Galanthus nivalis* agglutinin) as a probe. The experiments were carried out under the conditions used in (A) with GNA–HRP instead of Con A–HRP. As shown, only carboxypeptidase Y reacted positively with GNA–HRP.

to various extents.[4,10–12] It was found (Fig. 3A) that all of them reacted significantly with Con A–HRP. It was noted that nonbiotinylated glycoproteins can also be coated to the well of a plate, but this nonspecific binding differs greatly for different glycoproteins (data not shown). It has been also established that GNA (*Galanthus nivalis* agglutinin) only recognizes highly branched yeast mannose-type oligosaccharides and does not react with most other glycans.[13] As is shown in Fig. 3B, only carboxypeptidase

[11] K. Yamashita, N. Koide, T. Endo, Y. Iwaki, and A. Kobata, *J. Biol. Chem.* **264,** 2415 (1989).

[12] P. K. Tsai, J. Frever, and C. E. Ballou, *J. Biol. Chem.* **259,** 3805 (1984).

[13] N. Shibuya, I. J. Goldstein, E. M. Van Damme, and W. J. Peumans, *J. Biol. Chem.* **263,** 728 (1988).

High Mannose Type

Hybrid Type

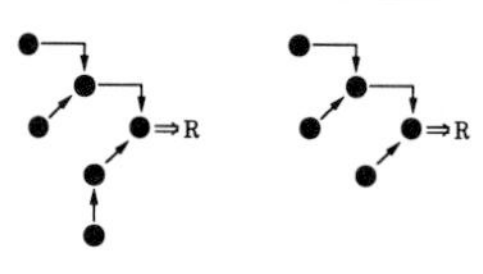
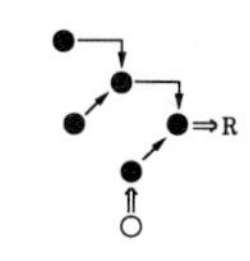
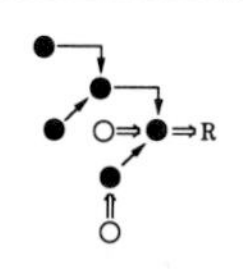
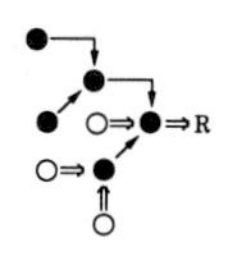
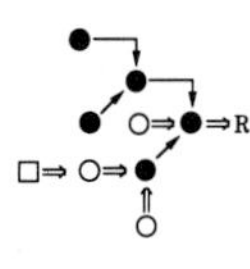

1. Man6-R 2. Man5-R 3. GlcNAcMan5-R 4. GlcNAc2Man5-R 5. GlcNAc3Man5-R 6. GalGlcNAc3-Man5-R

Complex Type

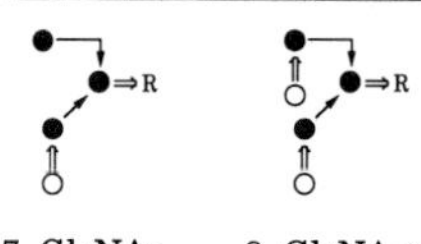
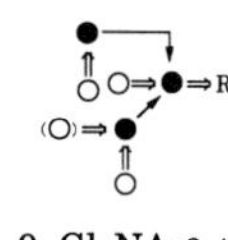
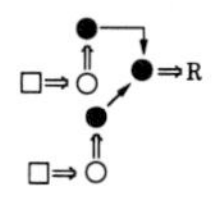
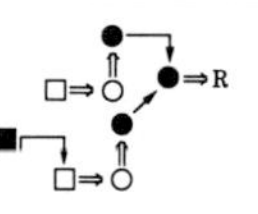
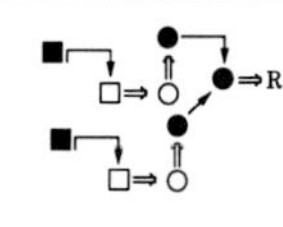

7. GlcNAc-Man3-R 8. GlcNAc2-Man3-R 9. GlcNAc3-4-Man3-R 10. Gal2GlcNAc2-Man3-R 11. SiaGal2-GlcNAc2Man3-R 12. Sia2Gal2 GlcNAc2Man3-R

R=6-(biotinamido)hexanoyl-Asn-GlcNAc2; ●, Mannose; ○, N-Acetylglucosamine; □, Galactose; ■, Sialic acid.

Code: Bond position 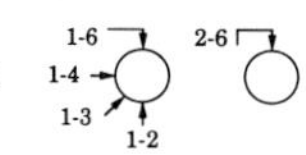 Configuration α → β ⇒

FIG. 4. Structures of the 12 biotinylglycans used in the study of oligosaccharide-binding specificity of Con A. All were bound to streptavidin. The anomeric configuration and the linkage position are encoded as indicated.

Y, which contains highly branched yeast mannose-type oligosaccharides,[12] reacted positively with GNA–HRP.

Assay of Oligosaccharide-Binding Specificity of Lectin

Reagents

Streptavidin, 30 pmol/ml (2 μg/ml) in buffer A

Series of biotinylglycans, each at 150 pmol/ml in buffer A (the structures of 12 biotinylglycans used here are shown in Fig. 4)

The other reagents are the same as those used in the procedure above.

Procedure

Step 1: Preparation and Coating of Streptavidin–Biotinylglycan Complex to Microplates. Apply a 50-μl aliquot of streptavidin (0.1 μg, 1.5 pmol) to each well of a microtiter plate, followed by the addition of 50 μl of a selected biotinylglycan (7.5 pmol) solution, and incubate the mixture at 25° for at least 3 hr; overnight is often convenient. As 1 mol of streptavidin can bind 4 mol of biotinylglycans, a molar ratio of streptavidin to

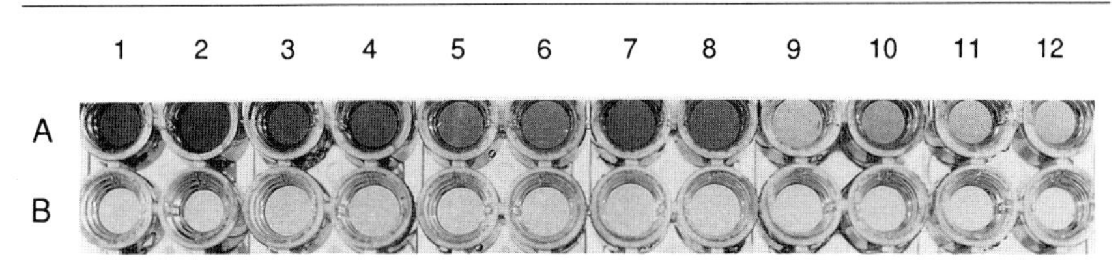

FIG. 5. Typical picture of binding of Con A–HRP (row A) and Con A–HRP plus α-methyl mannoside (row B) to the 12 streptavidin–biotinylglycans with an aminohexanoyl group, coated to wells A1–A12 and B1–B12 of a plate. The structures of the 12 glycan derivatives are shown in Fig. 4. The assays were carried out according to the standard procedure described in the text. The inhibitor for Con A sugar binding, 100 m*M* α-methyl mannoside dissolved in the Con A–HRP solution, was added to each well from B1 to B12. The enzyme reactions were carried out for 15 min.

biotinylglycan of 1 : 5 is used in order to ensure complete saturation of streptavidin by the biotinylglycan. The subsequent steps are the same as those (steps 4–7) involved in the detection of glycopeptides in HPLC peptide maps (see above).

Comments

A representative assay of the oligosaccharide-binding specificity of Con A is presented in Fig. 5. The glycan structures involved are given in Fig. 4. The level of Con A binding to the high-mannose type of oligosaccharides Man_6-R and Man_5-R and hybrid $GlcNAcMan_5$-R is highest. The extent of binding to the other three hybrid structures is decreased, consistent with the previous finding that when $GlcNAcMan_5$-R is bisected with a β1,4-linked *N*-acetylglucosamine residue to the β-linked mannose the binding is weakened.[13,14] The level of Con A binding to the complex type $GlcNAcMan_3$-R and $GlcNAc_2Man_3$-R is similar to that for $GlcNAc_2$ Man_5-R. The compounds $Gal_2GlcNAc_2Man_3$-R, $SiaGal_2GlcNAc_2Man_3$-R, $Sia_2Gal_2GlcNAc_2Man_3$-R, and $GlcNAc_{3-4}Man_3$-R with a bisecting *N*-acetylglucosamine residue gave lower color with Con A, showing that the affinity of the lectin for these glycans is very weak. As predicted, the addition of α-methyl mannoside strongly inhibited the color formation (Fig. 5B). This documents the key role of the interaction between Con A–HRP and oligosaccharides in the assay.

The sugar-binding specificities of a given lectin can be conveniently screened both qualitatively and quantitatively by using streptavidin–

[14] J. U. Baenzinger and D. Fiete, *J. Biol. Chem.* **254,** 2400 (1979).

biotinylglycans as probes. The error of this method usually is less than ±20%.[15]

Acknowledgments

Part of this work was supported by a research Grant (AU-916) from the Robert A. Welch Foundation to Dr. Finn Wold. We thank Dr. Finn Wold for helpful comments and discussions.

[15] C. F. Brewer and L. Bhattacharyya, *J. Biol. Chem.* **261,** 7306 (1986).

Section III

Binding Site Characterization

[19] Spacer-Modified Oligosaccharides as Photoaffinity Probes for Porcine Pancreatic α-Amylase

By JOCHEN LEHMANN and MARKUS SCHMIDT-SCHUCHARDT

Introduction

The probing of receptor binding sites by photoaffinity labeling is a relatively simple method to investigate the structure and functionality of the part of a protein structure that is responsible for the specific attachment of a ligand, and, in the case of an enzyme, also for its chemical conversion. Reviews on the method of photoaffinity labeling have been published[1] and therefore discussion on the general methodology, the different types of photolabile groups, and the scope of general applications is not presented here. Although the use of photolabile, spacer-modified oligosaccharides for the covalent chemical modification of a receptor binding site is described for the specific case of porcine pancreatic α-amylase (PPA), the method is not limited to this enzyme, but can be regarded as a general tool for probing extended binding sites for carbohydrates in general[2,3] and possibly also for other types of ligands such as peptides or oligonucleotides.

"Spacer" Principle

A "spacer-modified oligosaccharide" (SMO) was first used in the synthesis of methyl 4-*O*-(4-α-D-glucopyranosyloxy-4-methoxybutyl)-α-D-glucopyranoside,[4] which, although a simple glucoside linked by an acyclic spacer to the 4-position of another glucoside, is recognized by such typical maltodextrin-binding proteins as α-amylase and the periplasmic maltose-binding protein of *Escherichia coli.* Apparently two glucosyl units, kept at the "right" distance by a spacer, exert a synergistic effect on the affinity of one another to the receptor protein and thereby mimic the "natural" ligand trisaccharide maltotriose. Strong binding enhancement to the hepatic Gal/GalNAc receptor was demonstrated by "clustering" glycosyl

[1] H. Bayley and J. R. Knowles, this series, Vol. 46, p. 69; V. Chowdhry and F. H. Westheimer, *Annu. Rev. Biochem.* **48,** 293 (1979).

[2] J. Lehmann and St. Petry, *Carbohydr. Res.* **204,** 141 (1990); S.-C. Ats, J. Lehmann, and St. Petry, *Carbohydr. Res.* **233,** 125 (1992); S.-C. Ats, J. Lehmann, and St. Petry, *Carbohydr. Res.* **233,** 141 (1992).

[3] J. Lehmann and M. Scheuring, *Carbohydr. Res.* **225,** 67 (1991); C. P. J. Glaudemans, J. Lehmann, and M. Scheuring, *Angew. Chem.* **101,** 1730 (1989).

[4] S. Jegge and J. Lehmann, *Carbohydr. Res.* **133,** 247 (1984).

residues on a tris(hydroxymethyl)aminomethane derivative.[5] The Tris spacer mimics in a way natural multiantennary oligosaccharides, which are galactosylated at their glyconic ends.

The difference between the "cluster glycosides" of Lee and co-workers[5] and the spacer-modified oligosaccharides described here is that the latter are concerned with the length of the spacer linking two or more sugars. Where several monosaccharide units of an oligosaccharide are recognized and bound in fixed subsites, replacement of one or more monosaccharides has to be with a spacer of the same or a similar longitudinal extension. Galactosyltransferase, for instance, binds two GlcNAc residues as acceptor substrates, which are β-glycosidically linked head on to a 10-membered, linear hydrocarbon chain (a nonreducing spacer-modified disaccharide*). The affinity decreases drastically when the chain is shortened by one or two carbon atoms but is affected only in a minor way when it is lengthened.[6] Similar experiences have been gained with spacer-modified trisaccharides, which were modeled after β-(1→6)-galactotetraose.[7]

The concept of spacer-modified oligosaccharides was developed for the covalent chemical modification of receptor binding sites. This applies for the use of spontaneously reacting alkylating groups, such as haloacyl residues, oxiranes, or vinyl oxiranes, but especially for groups that can be converted to highly reactive nitrenes or carbenes on irradiation.

Although spacer-modified oligosaccharides or "cluster glycosides" may be easier to obtain by "shorthand synthesis"[8] than by isolation or synthesis of the original oligosaccharide as the natural ligand of a given receptor, the most direct and also simple route to prepare a reactive ligand would be to attach reactive groups to the native structure by acylation or alkylation. Numerous such reagents are commercially available. With usually very specific oligosaccharide binding receptors, affinity can be greatly reduced by steric hindrance, depending on where the reactive group is placed in the ligand. There are many examples reported in the literature where an apparently small change in the native structure created by introducing a reactive group causes a dramatic drop in affinity and thereby low and unspecific chemical modification of the protein. Replacing

[5] D. T. Connolly, R. R. Townsend, K. Kawaguchi, W. R. Bell, and Y. C. Lee, *J. Biol. Chem.* **257,** 939 (1982).

* In a spacer-modified oligosaccharide, di-, tri-, or tetrasaccharide refers to the actual number of monosaccharide units. If the monosaccharides are in a line, it is reducing; if linked head on, it is nonreducing.

[6] S.-C. Ats, J. Lehmann, and St. Petry, *Carbohydr. Res.* **225,** 325 (1994).

[7] J. Lehmann and M. Scheuring, *Carbohydr. Res.* **225,** 67 (1992).

[8] Y. C. Lee, *Ciba Found. Symp.* **145,** 80 (1989).

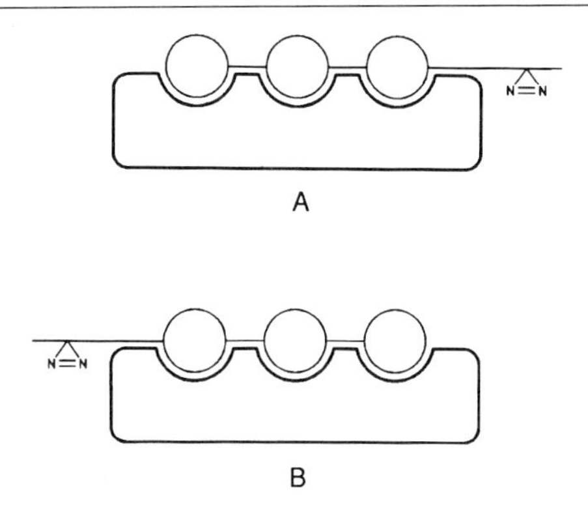

FIG. 1. Schematic presentation of three-centered receptor–ligand complexes. In complex **A** a photolabile group is attached to the glyconic end, and in **B** a photolabile group is attached to the aglyconic end of the ligand.

the amido group of nicotinamide in NAD^+ by azide, a commonly used group for photoaffinity labeling, dramatically reduces the affinity to yeast alcohol dehydrogenase, and of the 7% radiolabel incorporated into the protein, about half was found outside the binding site.[9] More examples are given in Chowdhry and Westheimer,[1] where attention is drawn to "the problems inherent in 'loose' receptor ligand complexes."

One could of course look for and find positions on the native structure of a ligand that are not in contact with the binding site, and where placing of a reactive group would not alter the affinity to the receptor. However, the margin between steric repulsion (formation of a "loose" receptor–ligand complex) and ineffective labeling owing to lack of contact of the reactive group with the binding area is very narrow.

Chemically the simplest approach is the attachment of an aglyconic chain carrying the reactive group to the reducing end of a given oligosaccharide structure as in complex **A** shown in Fig. 1. If the oligosaccharide is linear, chemical modification of the nonreducing end would still be comparatively facile as in complex **B**. Given a three-subsite binding area, each subsite with a comparable binding capacity, and a linear trisaccharide as natural ligand, the reactive groups would actually be placed outside the binding area (Fig. 1); labeling could still occur with acceptable efficiency but not within the binding site of the ligand. This type of labeling could be called "exoaffinity labeling," a term which has been coined for the labeling outside the "active center" of a receptor,[10] which is in our

[9] S. S. Hixson and S. H. Hixson, *Photochem. Photobiol.* **18,** 135 (1973).
[10] M. Cory, J. M. Andrews, and D. H. Bing, this series, Vol. 46, p. 115.

definition not identical with the extended binding area for an oligosaccharide structure. There may still be sufficient affinity to form a "tight" receptor–ligand complex if one monosaccharide from the above trisaccharide is missing, as in the disaccharide derivatives **A-I** or **B-I** (Fig. 2).

If chemo- or photoaffinity labeling is chosen for localizing a restricted area in the binding site, there must ideally be only one preferred binding mode of the receptor–ligand complex. Under the aforementioned assumption that in the three-subsite model all binding contributions of the subsites are equal, there will be the same probability of forming complex **A-I** and complex **A-II** or **B-I** and **B-II** (Fig. 2). In such a case reaction with the protein would cause undesirable random covalent modification and possibly significant lowering of the labeling efficiency. Placing a spacer between two monosaccharide units, as in the spacer-modified disaccharide in complex **C-I**, would reduce the problem of different, energetically equivalent binding modes shown in Fig. 2. The receptor–ligand complex **C-I** would be thermodynamically favored over the complexes **C-II** and **C-III** (Fig. 3). Provided the flanking monosaccharide units in the ligand of **C-I** bind well enough, then the spacer carrying the reactive group is optimally tied down to the binding site for the efficient and regioselective chemical modification of the corresponding peptide segment.

The schematic presentation of the receptor–ligand complex **C-I** is a general model to describe the potential of spacer-modified oligosaccharides. Analogous ligand structures must be adapted to the specific receptor protein under investigation.

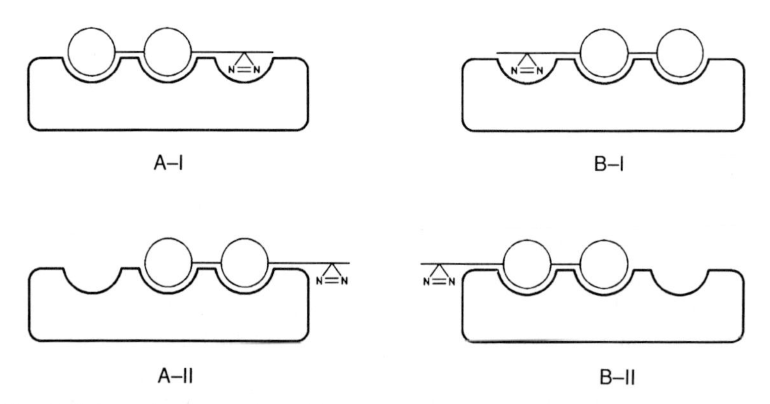

FIG. 2. Schematic presentation of three-centered receptor–ligand complexes. One monosaccharide unit in the ligands, either at the aglyconic (**A-I**, **A-II**) or the glyconic end (**B-I**, **B-II**), is replaced by a photolabile group. All complexes have different binding modes of equal stability. The regioselectivity of photoaffinity labeling is reduced owing to a change in position of the reactive group.

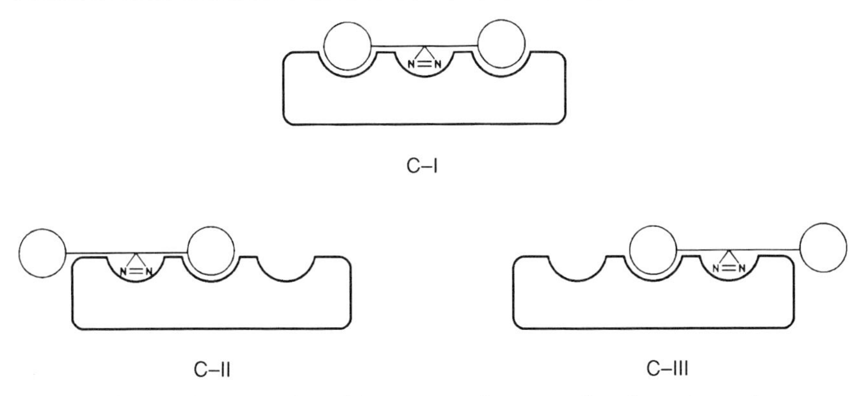

FIG. 3. Schematic presentation of three-centered receptor–ligand complexes. One monosaccharide unit in the ligands is replaced by a photolabile spacer. The binding mode illustrated for complex **C-I** dominates in the equilibrium of complexes **C-I**, **C-II**, and **C-III** owing to superior relative stability. The regioselectivity of photoaffinity labeling is preserved in complex **C-I**.

Choice of Diazirine Group for Photogenerating Carbenes

General affinity labeling needs two reactive species for the formation of a covalent bond. The reactive group attached to the ligand must literally meet an equally reactive group in the receptor binding site. This can be an advantage when a specific group, such as a catalytically active group in the active site of enzymes is to be labeled. Indiscriminate labeling of the adjacent groups, where even totally unreactive alkyl residues become modified, is possible only when the reactive species is a free radical, a nitrene, or a carbene generated within the receptor–ligand complex. Advantages and disadvantages of these reactive species for the purpose of efficient and indiscriminate labeling as well as the availability and stability of the precursors are discussed in detail elsewhere.[1]

For preparation of a photoaffinity reagent for labeling of an oligosaccharide-binding receptor, several criteria must be considered: (1) Introducing a group from which the reactive species is generated must not significantly change the polar properties of the natural ligand. (2) The reagent must not interfere with complex formation yet be close enough to the receptor site. (3) Irradiation, which generates the reactive species, must not damage or alter the native structure of the receptor protein. (4) The lifetime of the reactive species must be significantly shorter than the dissociation rate of the ligand–receptor complex. Among the many different precursor groups, namely, aromatic and aliphatic azides for nitrenes, α-diazoacyl residues, aryldiazomethanes, and diazirines for carbenes, and thiocarbonyl, haloaryl, and other aromatic chromophores for photogeneration of free radicals, only the diazirines meet the aforementioned requirements.

A diazirine can be prepared from any aliphatic ketone or aldehyde.[11] It is not much more space filling than a hydroxymethylene group, but is less polar. Irradiation with UV light (340–380 nm) causes a rapid decay with formation of nitrogen and the reactive carbene. Such low-energy UV light is not absorbed by protein and therefore does not damage the native structure. Diazirines are stable under moderate chemical conditions [bases, acids, oxidizing and reducing agents (except hydrogenolysis) and temperatures up to about 70°]. The only disadvantage of diazirines, when situated next to a carbon carrying a hydroxyl group, is a very rapid intramolecular rearrangement of the generated carbene, whereby the desired intermolecular reaction is totally suppressed[12] (Fig. 4). Not so serious are formations of intramolecular alkene from carbenes when the carbon adjacent to the carbene center carries a hydrogen[13] (Fig. 4). In our experience this hydride shift does not compete to a great extent with intermolecular labeling when the receptor–ligand complex is tight enough and the generated carbene has a "neighboring group" on the binding site with which to react.

Pancreatic α-Amylase as Convenient Receptor Model

Porcine pancreatic α-amylase* (PPA) is one of a multitude of related, starch-degrading enzymes with common structural elements.[14,15] X-Ray structural analysis[16] as well as kinetic studies[17] demonstrate PPA to bind a maltodextrin chain along a 3 nm groove, which can accommodate five consecutive glucosyl units in five subsites with the release of a certain amount of free energy.[18] Cleavage of the chain takes place between subsites **C** and **D** (Fig. 5).

[11] M. T. H. Liu, "Chemistry of Diazirines," Vols. 1 and 2. CRC Press, Boca Raton, Florida, 1987.

[12] J. Lehmann and St. Petry, *Carbohydr. Res.* **239,** 133 (1993).

[13] C.-S. Kuhn, J. Lehmann, and J. Steck, *Tetrahedron* **46,** 3129 (1990).

* Porcine pancreatic α-amylase isoenzyme I was a gift from Prof. G. Marchis-Mouren, laboratoire de Biochimie and Biologie de la Nutrition, Faculté de St. Jerome, Avenue Escadrille Nomandu Niemen, 13397 Marseille, France.

[14] G. Marchis-Mouren and V. Desseaux, *Biochem. Life Sci. Adv.* **8,** 91 (1989).

[15] G. K. Farber and G. A. Petsko *Trends Biochem. Sci.* **15,** 228 (1990).

[16] F. Payan, R. Haser, M. Pierrot, M. Frey, J. P. Astier, B. Abadie, E. Duée, and G. Buisson, *Acta Crystallogr. Sect. A: Cryst. Phys. Diff. Theor. Gen. Crystallogr.* **36,** 416 (1980); G. Buisson, E. Duée, R. Haser, and F. Payan, *EMBO J.* **6,** 3909 (1987).

[17] J. F. Robyt and D. French, *J. Biol. Chem.* **245,** 3917 (1970); C. Seigner, E. Prodanov, and G. Marchis-Mouren, *Eur. J. Biochem.* **148,** 161 (1985).

[18] C. Seigner, E. Prodanov, and G. Marchis-Mouren, *Biochim. Biophys. Acta* **913,** 200 (1987).

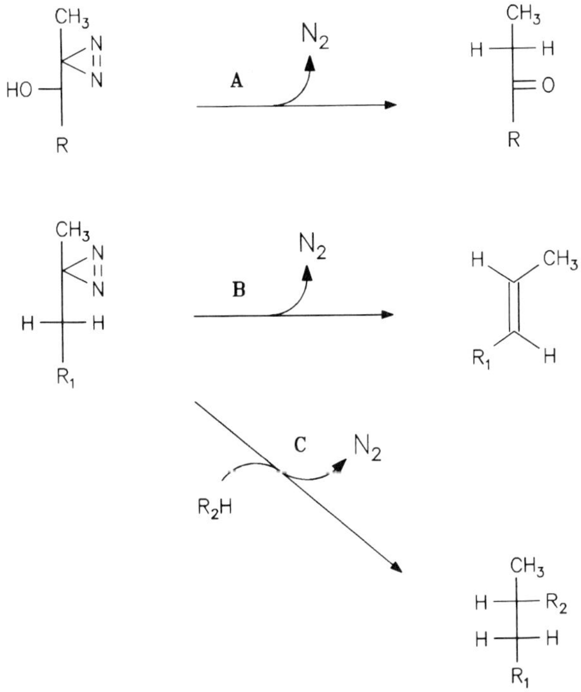

FIG. 4. Different reaction pathways of carbenes generated by photolytic decay of diazirines. In carbenes with α-hydroxyl groups, intramolecular rearrangement to ketones is the dominating reaction (pathway A). In isolated carbenes intramolecular rearrangement yielding olefins (pathway B) competes with intermolecular reactions leading to photoaffinity labeling (pathway C).

For regioselective chemical modification of the binding area by photoaffinity labeling, the ligands ought to be resistant to enzyme-catalyzed cleavage, optimally bound through four of the normally five α-(1→4)-linked glucopyranosyl units, one being replaced by the spacer carrying the photolabile diazirine. Three different spacer-modified tetrasaccharides

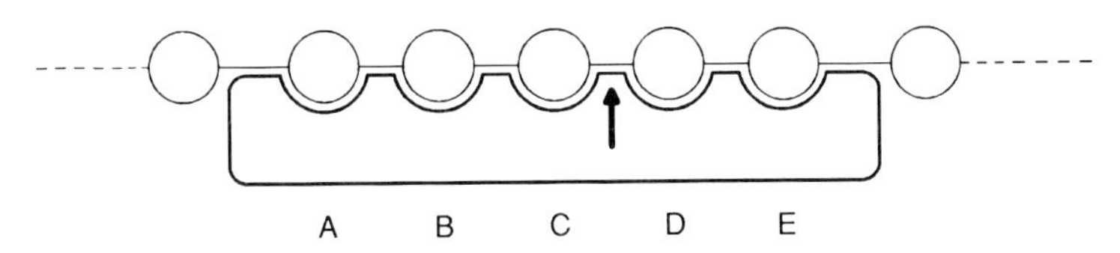

FIG. 5. Schematic presentation of PPA binding area with attached amylose chain. The catalytic site is between subsites **C** and **D**.

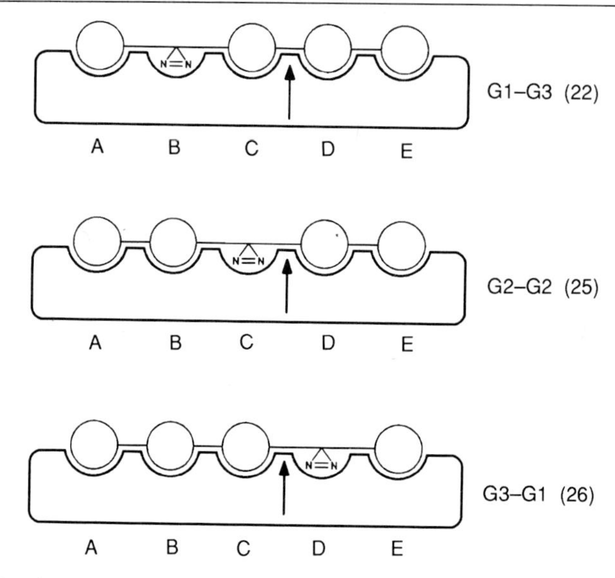

FIG. 6. The three spacer-modified tetrasaccharides are attached in the thermodynamically favored binding mode of PPA.

would theoretically meet these requirements: G1–G3 (**22**), G2–G2 (**25**), and G3–G1 (**26**). The derivatives **22**, **25**, and **26** should be able to modify chemically the area around subsites **B**, **C**, and **D**, respectively, provided there is only one predominant ligand–receptor complex (Fig. 6).

A special problem in the case of an endoglycanase like PPA will be enzyme resistance of the ligand. If the monosaccharide residues occupying the subsites were intact α-D-glucopyranosyl moieties, G1–G3 (**22**, Fig. 6) would immediately be cleaved between $\mathbf{G_C}$ and $\mathbf{G_D}$, thereby losing its affinity and property as a photoaffinity label. For this reason one of the glucosyl units $\mathbf{G_C}$ or $\mathbf{G_D}$ has to be slightly modified to stabilize the reagent against the enzyme. This has to be considered in planning the synthesis.

Synthetic Strategies for Assembly of Spacer-Modified Oligosaccharides, and Their Realization

A spacer-modified oligosaccharide can be assembled essentially in four different ways (Fig. 7): (a) glyconic (**A**) and aglyconic* (**B**) carbohydrate

* The terms aglyconic and glyconic ends or parts refer to the reducing and nonreducing end or ends of an oligosaccharide or derivative. The latter terminus is often misleading and incorrect, because many carbohydrate derivatives are actually not reducing although they have a reducing end. One could equally refer to the different ends as the east and west sides, if formulas are always written in the conventional way, with the reducing end to the right.

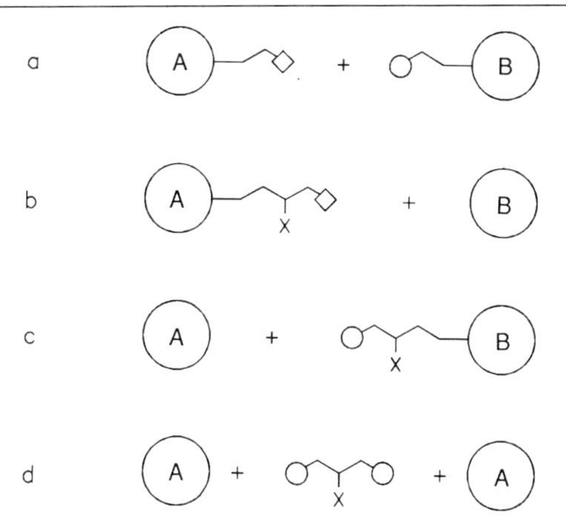

FIG. 7. Strategies of synthesis for assembling spacer-modified oligosaccharides.

ends carry parts of the future spacer, which can be coupled with one another; (b) the aglyconic part (**B**) is used to alkylate the future spacer, which is already attached to the glyconic carbohydrate part (**A**); (c) the glyconic part (**A**) is used to glycosylate the future spacer, which is already attached to the aglyconic carbohydrate part (**B**); (d) if there are two glyconic parts (**A**) which have to be attached head on, the future spacer can be glycosylated in one step at the glyconic and aglyconic ends. Method (a) is only practical if the spacer is longer than six atoms, including the bridgehead atoms, because it is difficult to introduce the diazirine group into a large molecule. Methods (b) and (c) are similar, yet (c) and (d) have the advantage of requiring only relatively simple glycosylations.[3]

For the special case described in this chapter, method (b) is to be preferred for two reasons. First, because the carbohydrate parts have to be maltooligosaccharides, use can be made of the very efficient α-(1→4)-glucosylation of a glucoside by cyclodextrin glucanotransferase (CGTase) and cyclomaltohexaose (α-CD.)[19] A chemically synthesized compound like G1–G2 (**23**) or G1–G1 (**24**) can be enzymically glucosylated at the end of the synthesis to give the desired products G2–G2 (**25**) and G3–G1 (**26**). The yield of the desired products can be increased by trimming higher homologs with β-amylase. Second, as discussed earlier, one of the two glucosyl units ($\mathbf{G_C}$, $\mathbf{G_D}$) in G1–G3 (**22**) has to be slightly modified in order to prevent enzymatic cleavage of the bond in between. This necessitates the use of an activated derivative to produce a modified glucosyl moiety as $\mathbf{G_C}$ in the aglyconic carbohydrate end of G1–G3 (**22**). Because of the

[19] K. Wallenfels, P. Földi, H. Niermann, H. Bender, and D. Linder, *Carbohydr. Res.* **61,** 359 (1978).

reactivity and selectivity of the vinyloxirane grouping in **19**, smaller homologs (**20** and **21**) were equally applicable in the assembly of G1–G2 (**23**) and G1–G1 (**24**).

The glyconic α-D-glucopyranoside carrying the spacer precursor as well as the photolabile diazirine ring must have a nucleophilic aglycon capable of promoting smooth opening of the vinyloxirane at the aglyconic side during the coupling reaction of the glyconic with the aglyconic side. A sulfhydryl group and an amino group[20] can attain almost quantitative coupling. To avoid ionogenic ligands, the sulfhydryl group is chosen.

Experimental Methods

Procedure for Preparation of Glyconic Coupling Component

To prepare the glyconic coupling component **10** (Fig. 8),[21] per-O-benzylated allyl α-D-glucopyranoside[22] (**1**) is treated with ozone followed by dimethyl sulfide,[23] and the isolated 2-oxoethyl 2,3,4,6-tetra-*O*-benzyl-α-D-glucopyranoside (**2**) is reacted with allyl magnesium chloride[24] for chain elongation to yield 2-hydroxy-4-pentenyl 2,3,4,6-tetra-*O*-benzyl-α-D-glucopyranoside (**3**). The diastereomeric alkene **3** is again treated with ozone, and the product reduced with sodium borohydride[25] to yield the diastereomeric mixture of 2,4-dihydroxybutyl 2,3,4,6-tetra-*O*-benzyl-α-D-glucopyranoside (**4**). Blocking of the primary hydroxyl group by tritylation[26] allows oxidation of the remaining hydroxy group by pyridinium chlorochromate (PCC)[27] to yield 2-oxo-4-*O*-(triphenylmethyl)butyl 2,3,4,6-tetra-*O*-benzyl-α-D-glucopyranoside (**5**). The trityl blocking group in **5** is removed[28] to give 4-hydroxy-2-oxobutyl 2,3,4,6-tetra-*O*-benzyl-α-D-glucopyranoside (**6**).

Conversion of the ketone (**6**) to the diazirine[29] 2-azi-4-hydroxybutyl 2,3,4,6-tetra-*O*-benzyl-α-D-glucopyranoside (**7**) is followed by tosylation of the primary hydroxyl group to yield 2-azi-4-*O*-(4-toluenesulfonyl)butyl

[20] J. Lehmann and M. Brockhaus, *Methods Carbohydr. Chem.* **8,** 301 (1980).

[21] J. Lehmann and L. Ziser, *Carbohydr. Res.* **225,** 83 (1992).

[22] E. A. Talley, M. D. Vale, and E. Yanowsky, *J. Am. Chem. Soc.* **67,** 2037 (1945); R. E. Wing and J. N. BeMiller, *Carbohydr. Res.* **10,** 441 (1969).

[23] J. J. Pappas, W. P. Keaveney, E. Gancher, and M. Berger, *Tetrahedron Lett.* 4273 (1966).

[24] M. S. Kharasch and C. F. Fuchs, *J. Org. Chem.* **9,** 359 (1944).

[25] J. Thiem, *in* "Methoden der Organischen Chemie," p. 853. Georg Thieme Verlag, Stuttgart and New York, 1980.

[26] B. Helferich, *Adv. Carbohydr. Chem.* **3,** 79 (1948).

[27] G. Piancatelli, A. Scettri, and M. D'Auria, *Synthesis*, 245 (1982).

[28] M. Bessodes, D. Komiotis, and K. Antonakis, *Tetrahedron Lett.* **27,** 579 (1986).

[29] R. F. R. Church and M. J. Weiss, *J. Org. Chem.* **35,** 2465 (1970).

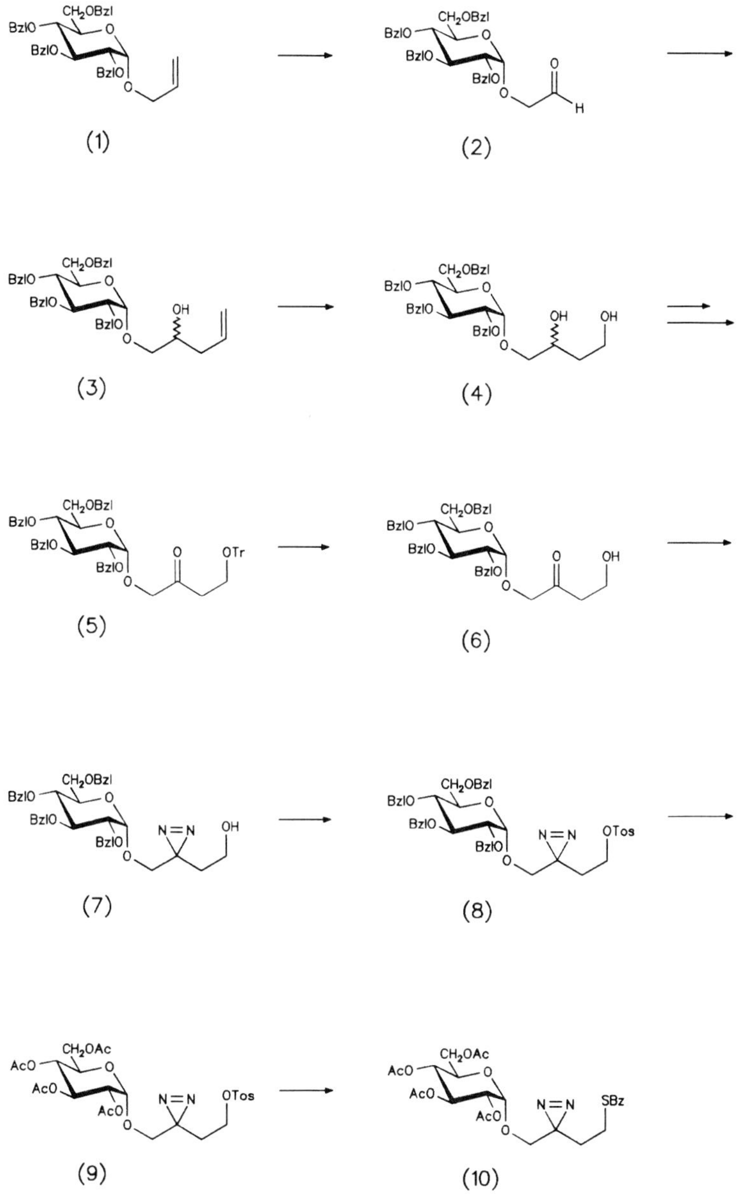

FIG. 8. Synthesis of the glyconic end for assembly of SMO.

2,3,4,6-tetra-*O*-benzyl-α-D-glucopyranoside (**8**). The intermediate alcohol **7** can conveniently be used to introduce a ^3H label via oxidation–reduction. The benzyl groups are exchanged by acetyl groups[30] to give 2-azi-4-*O*-(4-toluenesulfonyl)butyl 2,3,4,6-tetra-*O*-acetyl-α-D-glucopyranoside (**9**). The sulfur nucleophile is introduced by thiobenzoate exchange. 2-Azi-4-*S*-(benzoyl)butyl 2,3,4,6-tetra-*O*-acetyl-α-D-glucopyranoside (**10**) is then ready to be coupled with the aglyconic components. The 12-step synthesis of compound **10** can be achieved in 2.4% yield, starting with glucose (Fig. 8).

Procedure for Preparation of Aglyconic Coupling Components

To prepare the aglyconic coupling components **16–18** (Fig. 9),[21] the starting material in each case is a methyl α-glycoside for reasons of easier isolation and identification by ^1H nuclear magnetic resonance (NMR) spectroscopy (fixed stereochemistry at C-1). Here the preparation of the trisaccharide derivative G3 (**16**) is described. G2 (**17**) and G1 (**18**) are obtained by analogous procedures. Methyl α-maltotrioside (**11**), which has been prepared by enzymatic transglucosylation from methyl α-glucopyranoside, is converted to its 4″,6″-*O*-benzylidene derivative,[31] which is per-O-acetylated to yield methyl 2,2′,2″,3,-3′,3″,6,6′,-octa-*O*-acetyl-4″,6″-*O*-benzylidene-α-maltotrioside (**12**). Removal of the benzylidene group by hydrogenolysis[32] to give **13** is followed by tosylation of the hydroxyl groups to yield methyl 2,2′,2″,3,3′,3″,6,6′-octa-*O*-acetyl-4″,6″-di-*O*-(4-toluenesulfonyl)-α-maltotrioside (**14**). The primary tosyloxy group is selectively exchanged against iodine[33] to give methyl 2,2′,2″,3,3′,3″,6,6′-octa-*O*-acetyl-6″-deoxy-6″-iodo-4″-*O*-(4-toluenesulfonyl)-α-maltotrioside (**15**). Dehydrohalogenation using technical AgF in pyridine[34] gives the unsaturated trisaccharide methyl 2,2′,3,3′,6,6′-hexa-*O*-acetyl-4′-*O*-[2,3-di-*O*-acetyl-6-deoxy-4-*O*-(toluenesulfonyl)-α-D-*xylo*-hex-5-enopyranosyl]-α-maltoside (**16**). Compound **16** cannot be stored for more than 7 days at room temperature.

Coupling of Glyconic Component with Vinyloxiranes

To couple the glyconic component **10** with vinyloxiranes **19–21** (Fig. 10),[21] the glyconic part and aglyconic parts both have to be activated.

[30] K. P. R. Kartha, F. Dasgupta, P. P. Singh, and H. C Srivastava, *J. Carbohydr. Chem.* **5,** 437 (1986).

[31] M. E. Evans, *Methods Carbohydr. Chem.* **2,** 347 (1980).

[32] S. Peat and L. F. Wiggins, *J. Chem. Soc.*, 1088 (1938); H. B. Wood, H. W. Diehl, and H. G. Fletcher, *J. Am. Chem. Soc.* **79,** 1986 (1957).

[33] R. S. Tipson, *Methods Carbohydr. Chem.* **2,** 252 (1963).

[34] B. Helferich and E. Himmen, *Chem. Ber.* **61,** 1825 (1928).

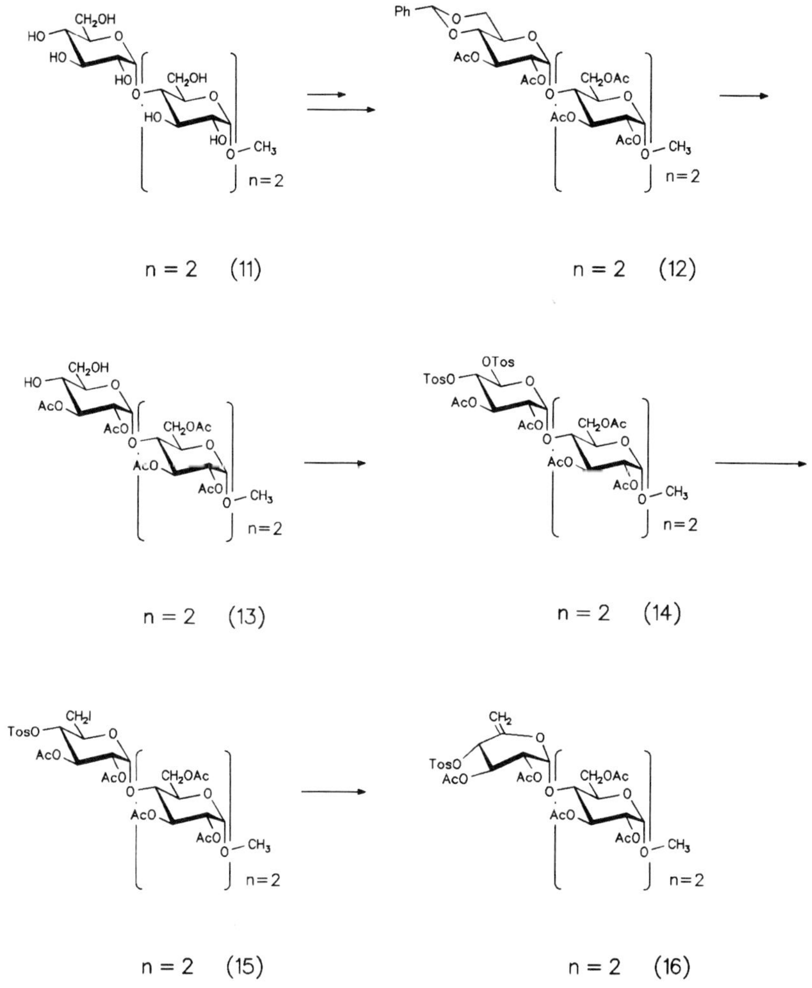

n = 2 (11)

n = 2 (12)

n = 2 (13)

n = 2 (14)

n = 2 (15)

n = 2 (16)

FIG. 9. Synthesis of the aglyconic end for assembly of SMO.

This occurs under the same basic conditions. Compound **16**, **17**, or **18** is dissolved in methanol and a solution of sodium methoxide (1 *M*) in methanol is added. The vinyl oxirane **19**, **20**, or **21** is formed by intramolecular nucleophilic displacement of the tosyloxy group. Precipitated sodium *p*-toluenesulfonate is removed by centrifugation. To the supernatant a solution of compound **10** in methanol is added to complete the conjugation.

The coupling time required to yield G1–G3 (**22**), G1–G2 (**23**) or G1–G1 (**24**) increases with the chain length of the vinyloxiranes. The course of

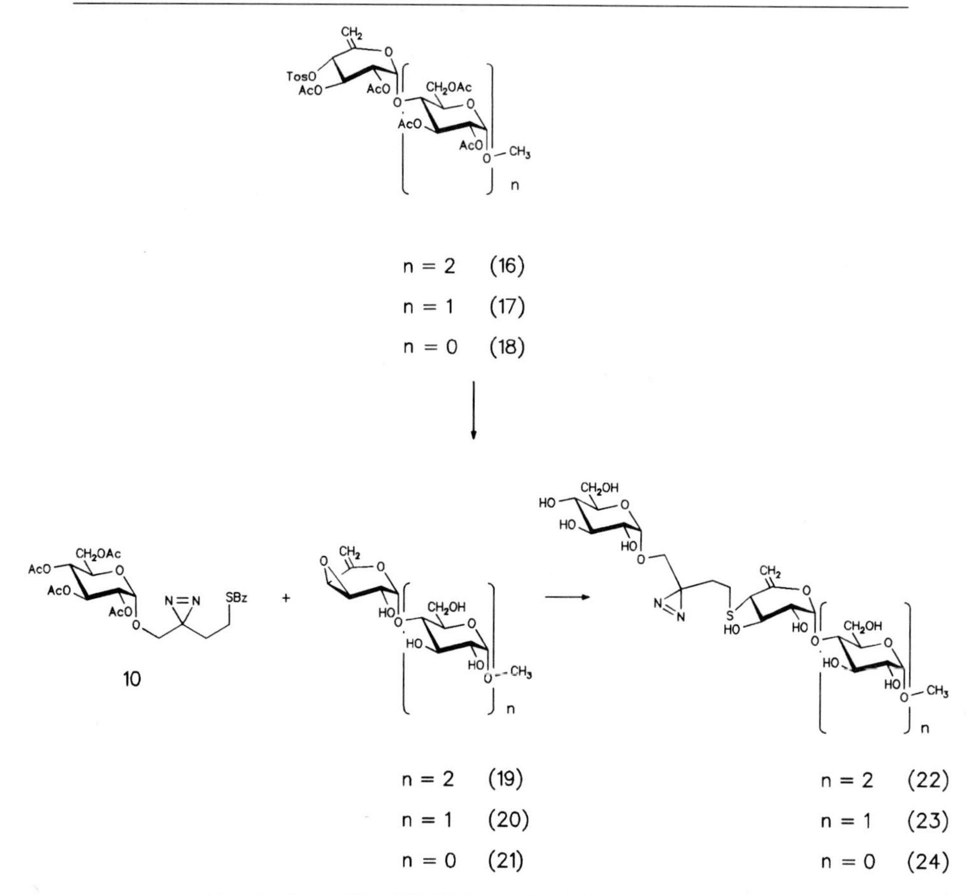

FIG. 10. Assembly of SMO by nucleophilic epoxide opening.

the reaction is conveniently followed by thin-layer chromatography (TLC). Purification and separation of the products G1–G3 (**22**), G1–G2 (**23**), and G1–G1 (**24**) are carried out by high-performance liquid chromatography (HPLC) [Hypersil ODS, 5 μm; 8 × 250 mm column; with 35 : 65 (v/v) methanol–water as eluent for compounds **22** and **23** and 45 : 55 (v/v) methanol–water for compound **24**; flow rate 1.5 ml/min; UV detection at 340 nm]. G1–G3 (**22**) is ready to be used as a ligand for α-amylase.

Preparation of G2–G2 and G3–G1 by Enzymatic Glucosylation and Enzymatic Trimming

To prepare G2–G2 (**25**) and G3–G1 (**26**) (Fig. 11),[21] each of the acceptor molecules, G1–G2 (**23**) and G1–G1 (**24**), in water is incubated with CGTase

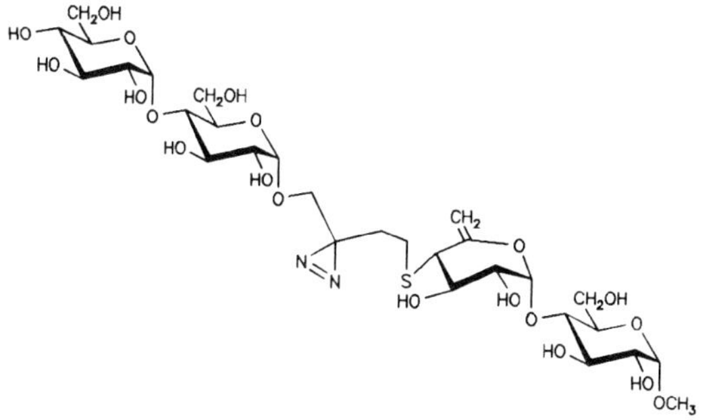

(25)

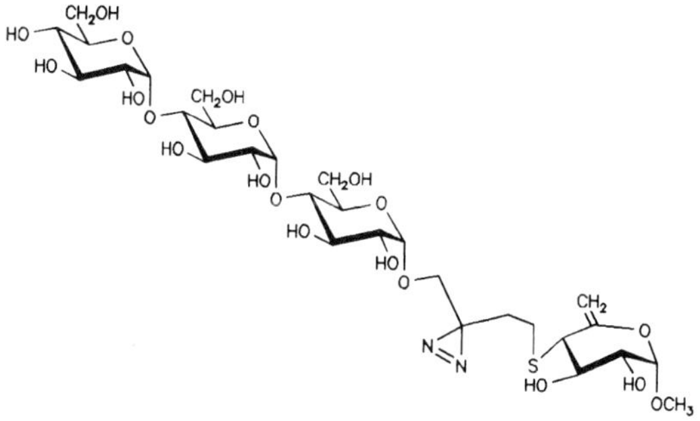

(26)

FIG. 11. Spacer-modified tetrasaccharides **25** and **26** are prepared by enzymatic transglucosylation using **23** and **24**, respectively, as acceptors.

(100 U/1 mmol α-CD) and α-CD (1 mmol α-CD/1.2 mmol acceptor) for 2 hr. The CGTase is denatured by heating the mixture at 95° for 5 min. Then the solution is brought to pH 5 by adding acetic acid (1 *M*) and incubated with β-amylase (70 U/mmol acceptor) for 2 hr. The enzyme is denatured at 95°, and the precipitated protein is separated by centrifugation. The products G2–G2 (**25**) and G3–G1 (**26**) are separated by HPLC.

Preparation of Radiolabeled Ligands

Compound **7** is oxidized with pyridinium dichromate[35] (PDC) to yield the aldehyde (**7a**). Reduction of compound **7a** (2.5 molar excess) with sodium [^{3}H]borohydride (specific activity $\geq$10 Ci/mmol) in dioxane, followed by tosylation of the hydroxyl group with *p*-toluenesulfonyl chloride in pyridine, yields compound **8***. Debenzylation is performed by treating a solution of **8*** in dichloromethane with ozone[36] for 1 hr, followed by

[35] E. J. Corey and G. Schmidt, *Tetrahedron Lett.*, 399 (1979).

[36] P. Angibeaud, J. Defaye, A. Gadelle, and J.-P. Utille, *Synthesis*, 1123 (1985).

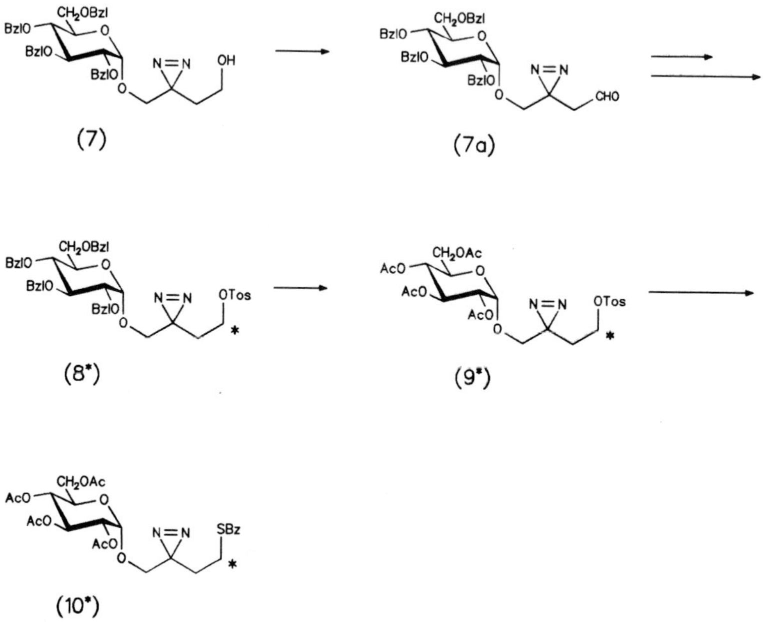

(7) (7a) (8*) (9*) (10*)

FIG. 12. Synthesis of ^{3}H-labeled glyconic end for assembly of SMO.

debenzoylation[37] in methanol with sodium methoxide (1 *M*) and acetylation (acetic anhydride/triethylamine/4-dimethylaminopyridine in dichloromethane) to yield compound **9***. The sulfur nucleophil is introduced by treating a solution of compound **9*** in acetone with a 5 molar excess of sodium thiobenzoate for 2 days at room temperature, to yield compound **10*** (Fig. 12). The coupling reactions, separations of products, enzymatic glucosylations, and HPLC separations are performed as described for the nonradiolabeled compounds.

Kinetic Evaluation of Receptor–Ligand Complex Formation

Binding of a certain ligand to its acceptor is especially easy to determine when the receptor is an enzyme and a convenient and accurate assay is at hand. Very sensitive assays are available for determining low concentrations of α-amylase activity in body fluids.[38] The methods, however, comprise several consecutive reactions and are therefore rather complicated.

[37] G. Zemplén and E. Pacsu, *Chem. Ber.* **62,** 1613 (1929).

[38] H. U. Bergmeyer, J. Bergmeyer, and M. Grassl (eds.), *in* "Methods of Enzymatic Analysis," 3rd Ed., Vol. 4, p. 146. Verlag Chemie, Weinheim, 1984.

TABLE I
INHIBITION CONSTANTS FOR SPACER-MODIFIED OLIGOSACCHARIDES CARRYING DIAZIRINO AND AZIDO GROUPS

Group	Compound	K_i (mM)
Diazirino	G1–G3 **(22)**	0.15
Azido	G1–G3	0.15
Diazirino	G2–G2 **(25)**	2.1
Azido	G2–G2	2.4
Diazirino	G3–G1 **(26)**	2.5
Azido	G1–G3	1.75

Although 4-nitrophenyl α-maltotrioside is a very "slow" and inefficient substrate[39] (with the optical indicator *p*-nitrophenolate being released to less than 20% of all possible fragments), there is a linear dependence of the change of extinction on enzyme activity. As the enzyme concentration can be made sufficiently high, the low sensitivity of the assay is no serious problem. All inhibition constants are determined by Dixon plots,[40] and the results for the spacer-modified tetrasaccharides are listed in Table I.[21]

Procedure. For determination of the inhibition constant, concentrations from 0.2 to 5.3 mM (K_m 2.2 mM) of the substrate are used at 30° in 50 mM triethanolamine hydrochloride buffer (pH 7.0) containing 10 mM. $CaCl_2$. Inhibitor concentrations are 0–1 mM for **22**, 0–4.4 mM for **25**, and 0–4.6 mM for **26**. Each assay is started by adding 22 U/ml of α-amylase. Results are shown in Table I.[21]

Considering the chemical alteration in the native structure of the standard maltopentaoside, the inhibition constants of the analogous spacer-modified tetrasaccharides are remarkably low and similar. In the case of **22**, it was significantly lower than the K_m for maltopentaose (1.08 mM).[16] Replacing the diazirine rings by azido groups does not cause any change of affinity. The K_i values of the azides are also shown in Table I.[41]

Photolysis of Standards, Photoaffinity Labeling Procedure, and Evaluation of Photodeactivation

All diazirines so far investigated decay in aqueous solution at a half-life rate of approximately 2–6 min when irradiated at 350 nm. A typical

[39] C. Seigner, E. Prodanov, and G. Marchis-Mouren, *Biochim. Biophys. Acta* **913,** 200 (1987).
[40] M. Dixon, *Biochem. J.* **55,** 170 (1953).
[41] J. Lehmann and L. Ziser, *Carbohydr. Res.* **205,** 93 (1990).

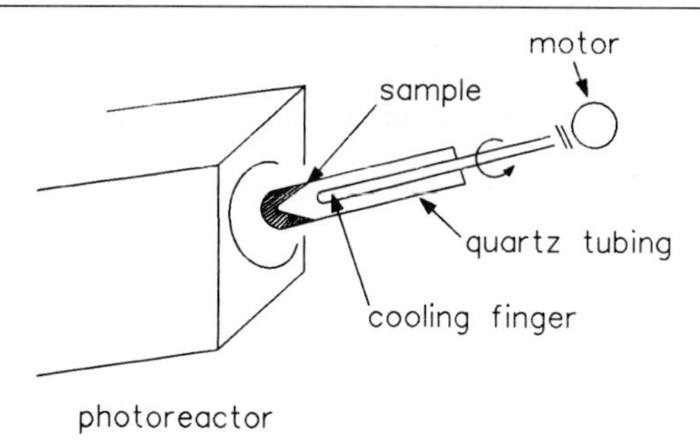

FIG. 13. Apparatus for photolyzing diazirines.

rate determination is carried out by placing the sample in a reactor (Fig. 13), taking aliquots, and measuring the disappearance of the UV absorption at the previously determined λ_{max}, which in the case of the test diazirine 3-azi-1-[β-D-glucopyranosyl(thio)]butane (**27**) is 348 nm.[42] A semilogarithmic plot of the photolysis of **27** is shown in Fig. 14.[42]

Usually irradiation under cooling by ventilation is carried out for $5t/2$. In the test case this would be about 30 min. With a simple compound such as **27** it was shown that 54% of the products are the diastereomeric alcohols (*R*/*S*)-3-hydroxy-[β-D-glucopyranosyl(thio)]butane formed by the desired insertion reaction; the rest are the olefins β-D-glucopyranosyl (thio)-3-butene and β-D-glucopyranosyl(thio)-*cis,trans*-2-butene formed by intramolecular hydride shift.[13] Insertion reactions with the protein can be higher than in the medium. This depends on proximity and the kind of functionality encountered by the carbene.

α-Amylase (1.5 mg) and the ligand (4 K_i) dissolved in 50 m*M* triethanolamine hydrochloride buffer (pH 7.0, 10 m*M* $CaCl_2$) giving a final volume of 1.8 ml are placed in a quartz tubing, which can be rotated at 30 rpm in order to spread the solution as a film over the lower part of the inside wall. The tube is held by a cooling finger connected by a ground glass joint. Before irradiation is started, the solution and the tube are flushed with an inert gas. Irradiation is carried out for 20 min, when almost 100% of the diazirine has been photolyzed. A blank experiment without any ligand is carried out as well as one in the presence of maltotriose (750 m*M*) as a competing ligand. The solutions are dialyzed against 50 m*M* triethanolamine hydrochloride buffer (pH 7.0, 10 m*M* $CaCl_2$) and then brought to equal protein concentrations. The enzyme activity is determined as described for the determination of the inhibition constants.

[42] J. Steck, Dissertation, Universität Freiburg i. Br. (1991).

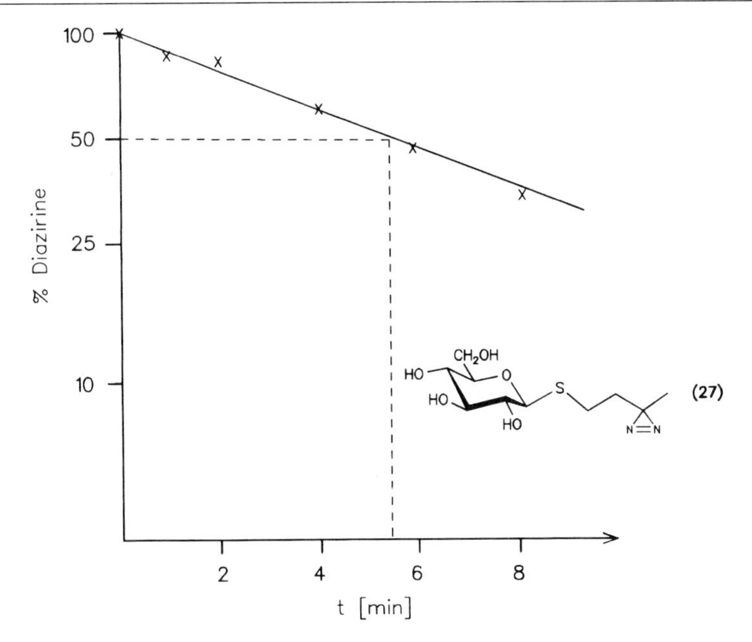

FIG. 14. The rate of photolytic decay is determined by semilogarithmic plotting of diazirine concentration versus time of radiation.

In Fig. 15, the results of photoaffinity labeling of α-amylase with the spacer-modified tetrasaccharides G1–G3 (**22**), G2–G2 (**25**), and G3–G1 (**26**) are shown. In two cases, the chemical modification of the binding site is almost quantitative; the third compound **26** is still very effective. The specificity of labeling is clearly demonstrated by the protective effect of high maltotriose concentrations. All three chemically modified protein samples gave unchanged K_m values with *p*-nitrophenyl α-maltotrioside as substrate. This is proof that unmodified α-amylase molecules remain structurally unchanged by the labeling procedure.

Labeling with Radioactive Ligands and Detection of Labeled Peptides

Chemical modification by photoaffinity labeling can be used to block or deactivate receptors specifically. The purpose of using radioactive spacer-modified oligosaccharides as ligands of an endoglycanase like α-amylase is to label regioselectively the binding area for structural analysis.

As already described, compounds **22**, **25**, and **26** can be prepared with a ^3H label in position 1 of the spacer. Because **22** and **25** showed the highest alkylating efficiency, these two radiolabeled ligands were used to

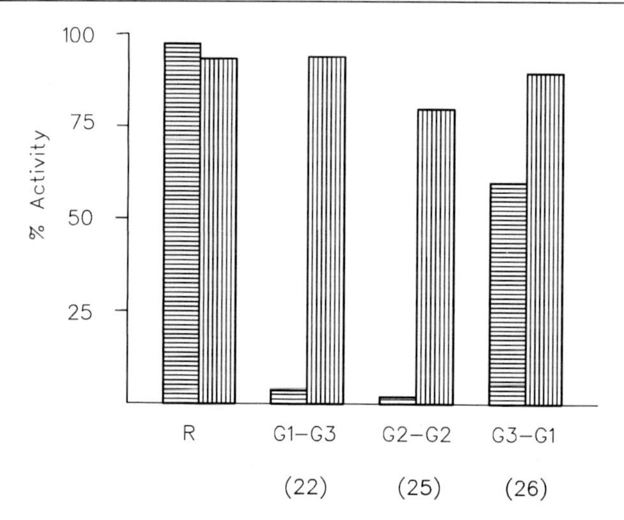

FIG. 15. Catalytic activity of PPA. Irradiation in the presence of compound **22**, **25**, or **26** without (horizontal hatching) and with added maltotriose (vertical hatching) leads to modification of the binding site. R: blank experiment.

introduce a marker into the binding area of α-amylase. We can assume that the photogenerated carbene in a fairly stable receptor–ligand complex modifies exclusively its immediate neighboring group, leading to a high degree of regioselectivity.[43] Refering to Fig. 6, the two ligands in question must be labeling subsite **B** (G1–G3, **22**) and subsite **C** (by G2–G2 **25**). Subsite **C** is adjacent to the catalytic site.

The procedure already described for labeling with nonradioactive ligands is applied with reduced quantities (1/3). After photolysis an aliquot containing 30 μg protein is removed from the original solution, and after dialysis the content of radioactivity is determined. The rest (15 μg) is subjected to sodium dodecyl sulfate–polyacrylamide gel electrophoresis (SDS–PAGE)[44] (10.5% acrylamide, 0.1% SDS, 2% mercaptoethanol). The gel is stained with Serva Blue (Serva, Heidelberg, Germany) to show [^{3}H]PPA as a single band between the 43 and 67-kDa band of the calibration kit. The gel is cut into slices (2 mm), Biolute-S (Zinsser, Frankfurt/Main, Germany, 500 μl) is added, and after standing overnight at room temperature Quickszint 501 (Zinsser, 4.5 ml) is added. After storage for 2 hr at 8°, the radioactivity is determined in a liquid scintillation counter (Table

[43] J. Lehmann, E. Schiltz, and J. Steck, *Carbohydr. Res.* **232,** 77 (1992).
[44] U. K. Laemmli, *Nature* (*London*) **227,** 680 (1970).

TABLE II
RADIOACTIVE LABELING OF PORCINE PANCREATIC α-AMYLASE BY G1–G3 AND G2–G2

Label	Radioactivity covalently attached to protein of SDS gel[a] (%)	Protein (μCi/mg)
G1–G3 **(22)**	35	14
G2–G2 **(25)**	38	15.5

[a] A level of 100% radioactivity would be stoichiometric modification: 1 ligand/molecule PPA.

II). The protein band is radioactive, proving the covalent attachment of the label (Fig. 16).

To the original solution is added 0.5 *M* Tris-HCl buffer (0.5 ml, pH 7.5; 6 *M* guanidinium hydrochloride, 5 m*M* EDTA, and 7 m*M* dithioerythritol), and the mixture is kept for 10 hr at room temperature. Sodium iodoacetate (10 mg) is then added in the dark, followed, after 1 hr, by mercaptoethanol (100 μl). The denatured protein precipitates when dialyzed against water (4 times, 400 ml), and dialysis is carried out until the radioactivity measured in the dialyzate reaches a plateau. The protein is collected by centrifugation.

After suspension in 0.2 *M* $NaHCO_3$ buffer (0.5 ml), the protein is treated with trypsin (1%) for 5 hr at 37°, at which time the same amount of trypsin is added and the digestion continued for another 18 hr. The digestion is monitored by reversed-phase HPLC on a C_{18} column (4.6 × 250 mm) (Vydac 218 TP) with CH_3CN–water containing 5 m*M* trifluoroacetic acid at 1 ml/min. The CH_3CN gradient is as follows: 0–10 min, 10%; 10–80 min, 10–40%; 80–90 min, 40–100%. The effluent is monitored at 220 nm for peptides, and the radioactivity is detected simultaneously.

Analysis of the crude peptide mixtures derived from the two different labeled proteins (Fig. 17) reveal a slightly different labeling pattern, with radioactivity centered in one block in the "hydrophilic part" of the elution profile. By conducting the HPLC separations under a slow gradient, labeled peptides could be eventually isolated, namely, the peak at retention time 15.5 min for G1–G3-labeled PPA and the peaks at retention times 15.5 and 19.1 min for G2–G2-labeled PPA. The total yield after the first purification is 26% for G1–G3I, 15% for G2–G2I, and 50% for G2–G2II, based on the covalently attached radioactivity in the original protein separated by SDS–PAGE (Table II, Fig. 17). (The meaning of I or II is the specification of labeled peptide, obtained with ligands G1-G3 and G2-G2.)

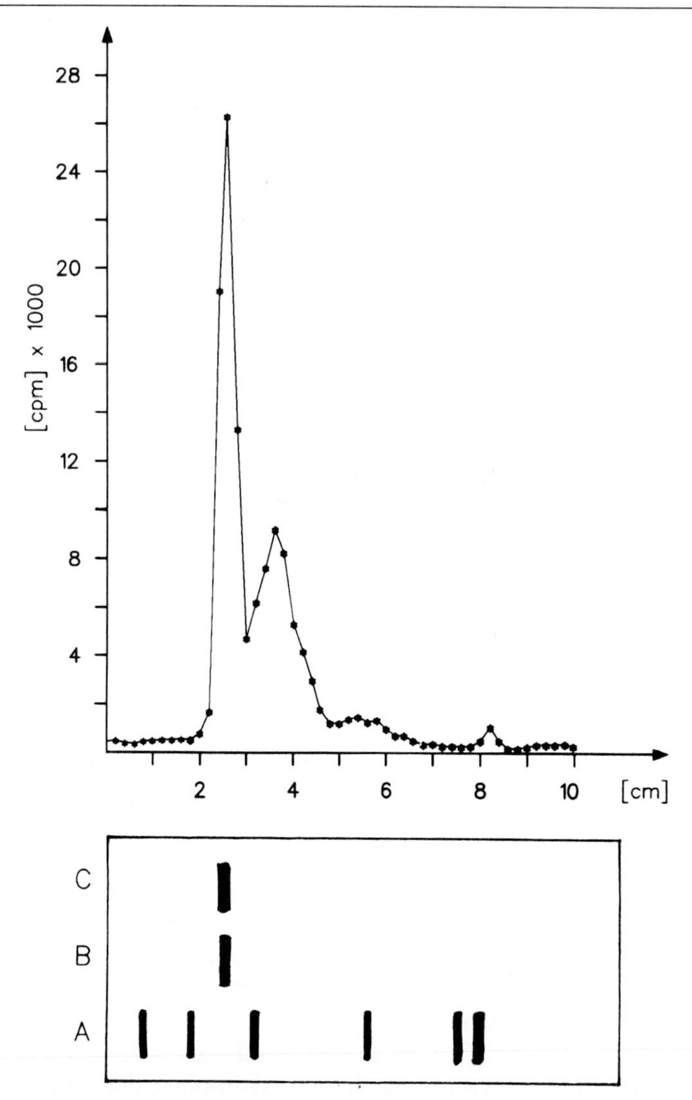

FIG. 16. Electrophoresis of G2–G2-labeled PPA. *Bottom*: Lane A, calibration proteins; lane B, PPA; lane C, [^{3}H]PPA. *Top*: Lane C was cut into strips and counted for radioactivity. The stained gel displays a diffuse zone of glycoforms moving in front of the main protein band (not shown).

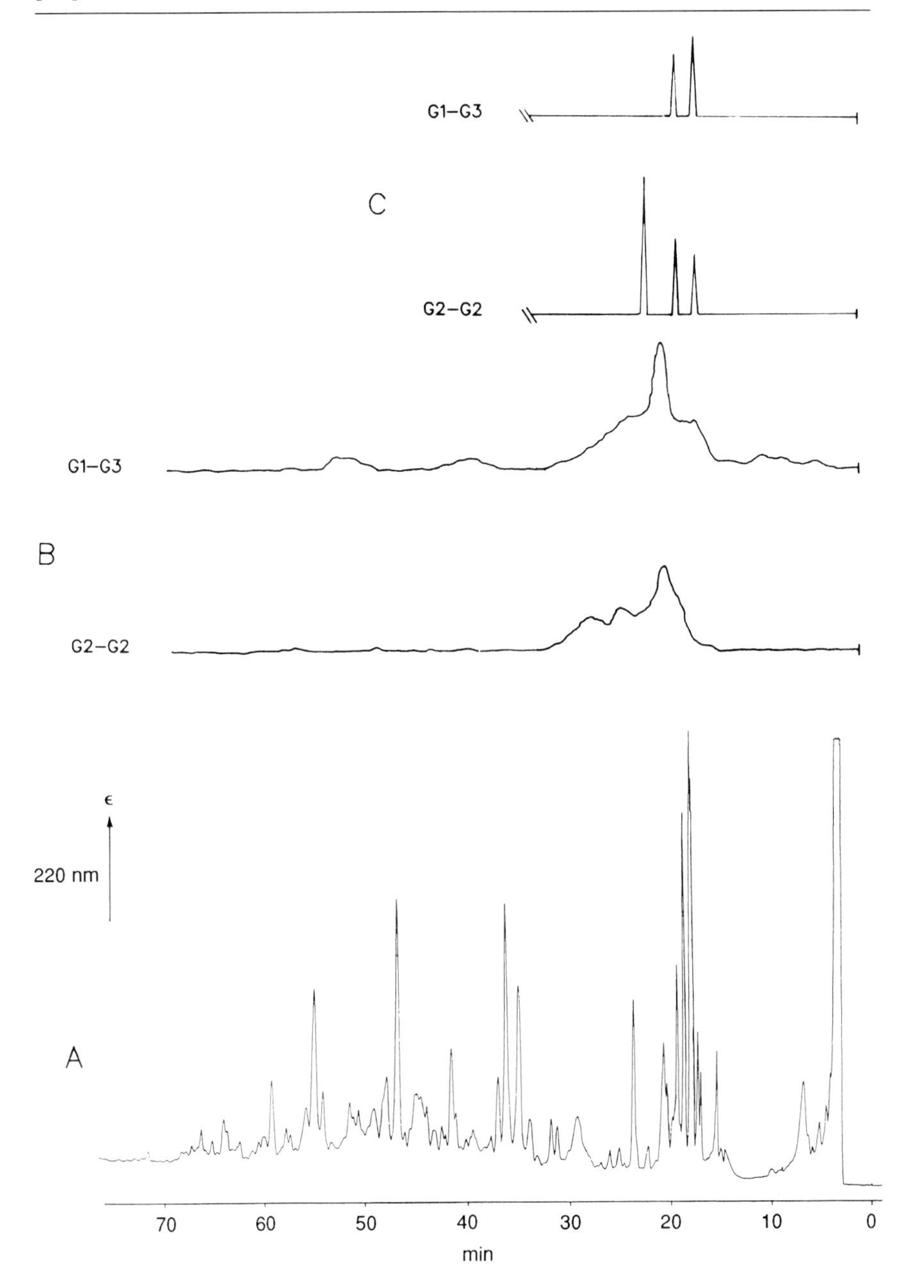

FIG. 17. (A) Elution profile from HPLC separation of tryptic peptide mixtures derived from PPA labeled with G2–G2 or G1–G3. (B) Distribution of radioactivity in the elution mixture monitored by a flow counter cell. (C) Major radioactive peptides purified by fractional HPLC.

To prove the uniformity of all three peaks, the samples are subjected to a second round of reversed-phase HPLC under the same conditions as before, but with CH_3CN–water containing 50 m*M* phosphate buffer (pH 6). G2–G2II remains a single peak, but G1–G3I and G2–G2I each separate into one major and one minor peak (Ia and Ib).

[20] Determination of Accurate Thermodynamics of Binding by Titration Microcalorimetry

By DAVID R. BUNDLE and BENT W. SIGURSKJOLD

Introduction

Virtually all biological processes depend critically on the binding of ligands by specific proteins such as enzymes, receptors, antibodies, lectins, and transport proteins, and the determination of binding affinities for such molecular recognition processes is therefore of central importance in biochemistry. There are relatively few fast, convenient, and accurate methods to measure affinity, and among them titration microcalorimetry is a formidable technique which is, with the limitations in binding constant range and protein concentration outlined below, universally applicable to the task of determining not only the heat of reaction, but also the equilibrium constant, stoichiometry, and entropy in a single experiment. In the present chapter we describe the application of titration microcalorimetry to studies of the binding interactions between monoclonal antibodies and carbohydrate antigens derived from pathogenic bacteria.

Binding interactions are usually quantified in terms of the association constant or the directly related free energy of binding, and numerous methods for determining these measures have been developed.[1] Most methods involve quantification of free ligand either in the presence of the protein or after separation from it. Unless these approaches are able to utilize an intrinsic reporter signal that permits *in situ* determination of the free and bound ligand concentration, they are generally laborious and time-consuming. To obtain a more in-depth thermodynamic description of the interactions and forces driving complex formation, it is desirable to know the enthalpy and entropy. Such information can be obtained by batch or flow calorimetry or van't Hoff analysis of the temperature dependence of binding constants. The latter, even when applied to a

[1] K. A. Connors, "Binding Constants," Wiley, New York, 1987.

convenient spectroscopic technique for measurement of association constants, will involve extensive research time and consume significant quantities of ligand and protein.

Titration microcalorimetry is a more practical approach to the routine determination of thermodynamic parameters. Because nearly all binding interactions are accompanied by a change in enthalpy, virtually all reactions of interest will produce a calorimetric signal, and titration calorimetry offers the possibility of directly determining not only the association constant, and thereby free energy, but also the stoichiometry, enthalpy, and entropy in a single experiment.[2-7] This is a particularly useful technique for studying oligosaccharide–protein interactions, since sugars are almost always devoid of intrinsic spectroscopic reporter groups. Previously the domain of the specialist, calorimetry has become readily accessible with the emergence of sensitive, commercial titration microcalorimeters operated under computer control with sophisticated software for data acquisition and processing. Now binding interactions in biological systems can be measured conveniently and with high accuracy[4-6] in a cost-effective manner provided the receptor protein is available in approximately 0.1–1.0 μmol quantities and the binding constant is in the range 10^4–10^8 M^{-1}.

Theory

The reversible association between a protein P and a ligand L,

$$P + L \rightleftharpoons PL \tag{1}$$

is characterized by its association constant K defined by

$$\frac{[PL]}{[P][L]} = K \tag{2}$$

The association constant K is related to the standard free energy ΔG° by the well-known relation

$$\Delta G^\circ = -RT \ln K \tag{3}$$

where R is the universal gas constant (8.314 J/mol K) and T is the absolute temperature. The standard free energy is again composed of a heat term

[2] N. Langerman and R. L. Biltonen, this series, Vol. 61, p. 261.

[3] R. L. Biltonen and N. Langerman, this series, Vol. 61, p. 287.

[4] T. Wiseman, S. Williston, J. F. Brandts, and L.-N. Lin, *Anal. Biochem.* **179,** 131 (1989).

[5] J. F. Brandts, L.-N. Lin, T. Wiseman, S. Williston, and C. P. Yang, *Am. Lab.* (*Shelton Conn.*), 30 (1990).

[6] E. Freire, O. L. Mayorga, and M. Straume, *Anal. Chem.* **62,** 950A (1990).

[7] B. W. Sigurskjold, E. Altman, and D. R. Bundle, *Eur. J. Biochem.* **197,** 239 (1991).

(enthalpy, ΔH°) and an entropy term (ΔS°), related by another fundamental equation

$$\Delta G^\circ = \Delta H^\circ - T\Delta S^\circ \tag{4}$$

In a titration calorimetric experiment, small volumes of a ligand solution are added to a solution of protein. Each addition gives rise to the evolution or absorption of a certain amount of heat, the magnitude of which depends on the reaction volume, concentrations, molar enthalpy, binding constant, heat of dilution, stoichiometry, and the amount of previously added ligand; in other words, the concentration of unoccupied binding sites begins to decrease and the magnitude of the heat changes decreases correspondingly as ligand is added. The heat evolved after the ith addition will be

$$\Delta Q_i = V_0 \Delta H^\circ \Delta[\mathrm{PL}] + q_\mathrm{d} \tag{5}$$

where V_0 is the reaction volume (volume of the calorimeter cell), $\Delta[\mathrm{PL}]$ is the change in complex concentration, and q_d is the heat of dilution of the ligand for one addition. The stoichiometric concentrations of protein and ligand in the calorimeter cell after the ith addition are, taking into account dilution and displacement, given by

$$[\mathrm{L}]_i = [\mathrm{L}]_0[1 - \exp(-iV_i/V_0)] \tag{6}$$

and

$$[\mathrm{P}]_i = [\mathrm{P}]_0 \exp(-iV_i/V_0) \tag{7}$$

where V_i is the volume of each addition of ligand, $[\mathrm{P}]_0$ is the initial concentration of protein in the cell, and $[\mathrm{L}]_0$ is the concentration of the ligand in the solution from which it is added. Equation (2) can be written as

$$\frac{[\mathrm{PL}]_i}{([\mathrm{P}]_i - [\mathrm{PL}]_i)([\mathrm{L}]_i - [\mathrm{PL}]_i)} = K \tag{8}$$

and solving the quadratic equation yields

$$[\mathrm{PL}]_i = \tfrac{1}{2}([\mathrm{P}]_i + [\mathrm{L}]_i + 1/K) - \{\tfrac{1}{4}([\mathrm{P}]_i + [\mathrm{L}]_i + 1/K)^2 - [\mathrm{P}]_i[\mathrm{L}]_i\}^{1/2} \tag{9}$$

The differential heat which is the amount of heat generated after one addition relative to the change in stoichiometric ligand concentration now becomes

$$\frac{\Delta Q_i}{\Delta[\mathrm{L}]_i} = \frac{V_0 \Delta H^\circ([\mathrm{PL}]_i - (1 - V_i/V_0)[\mathrm{PL}]_{i-1}) + q_d}{[\mathrm{L}]_i - [\mathrm{L}]_{i-1}} \tag{10}$$

and the integral heat (the total accumulated heat of all additions) is obtained from

$$Q_i = V_0 \Delta H^\circ([\mathrm{PL}]_i + V_i/V_0 \sum_{j=1}^{i-1} [\mathrm{PL}]_j) + iq_\mathrm{d} \tag{11}$$

As $[\mathrm{PL}]_i$ is given by Eq. (9), fitting a set of experimentally determined heats to either Eq. (10) or Eq. (11) by nonlinear regression yields estimates of $[\mathrm{P}]_0$, ΔH°, and K as regression parameters, and then ΔG° and ΔS° can be calculated from Eqs. (3) and (4), respectively. Figure 1 shows an example of a thermogram and derived binding isotherms in both heat modes for the binding by an antibody of an antigenic determinant.

Experimental Procedures

The titration calorimetry described here is performed with an OMEGA titration microcalorimeter from MicroCal, Inc. (Northampton, MA). The instrument has been described in detail by Brandts and co-workers.[4,5] For efficient and accurate work it requires that the product of the protein concentration and binding constant $K[\mathrm{P}]_0$ should be in the range from 1 to 1000, preferably between 10 and 100. However, weak binding (small K) can be quantified, if it is possible to increase the ligand concentration sufficiently.

The calorimeter has two cells, a reaction cell and a reference cell, each of approximately 1.3 ml volume. The reference cell functions only as a temperature dummy and is filled with water. A typical concentration of protein would be in the range 0.05–2 mM. The ligand must be in a buffer exactly matching the protein solution and is injected automatically into the reaction cell from a precision syringe that at the same time functions as a stirrer. Typically, 10–25 injections of 5–25 μl spaced 2–5 min apart are used. Heats of dilution either can be determined from Eq. (10) or Eq. (11) by regression analysis or can be measured directly in an experiment where ligand is injected into buffer. The appropriate concentration of the ligand depends on the concentration of the protein, the size of the syringe, and the magnitude of the expected binding constant, but generally titrations leading to a final stoichiometric (free + bound) ligand concentration ([L] + [PL]) twice that of the initial protein concentration will suffice in the $K[\mathrm{P}]_0$ range 10–100.

The instrument is conveniently calibrated with standard electrical pulses. It can detect both exothermic and endothermic reactions equally well and operates in the temperature range from 8° to 90°. For the instrument to be able to respond to both exothermic and endothermic heats,

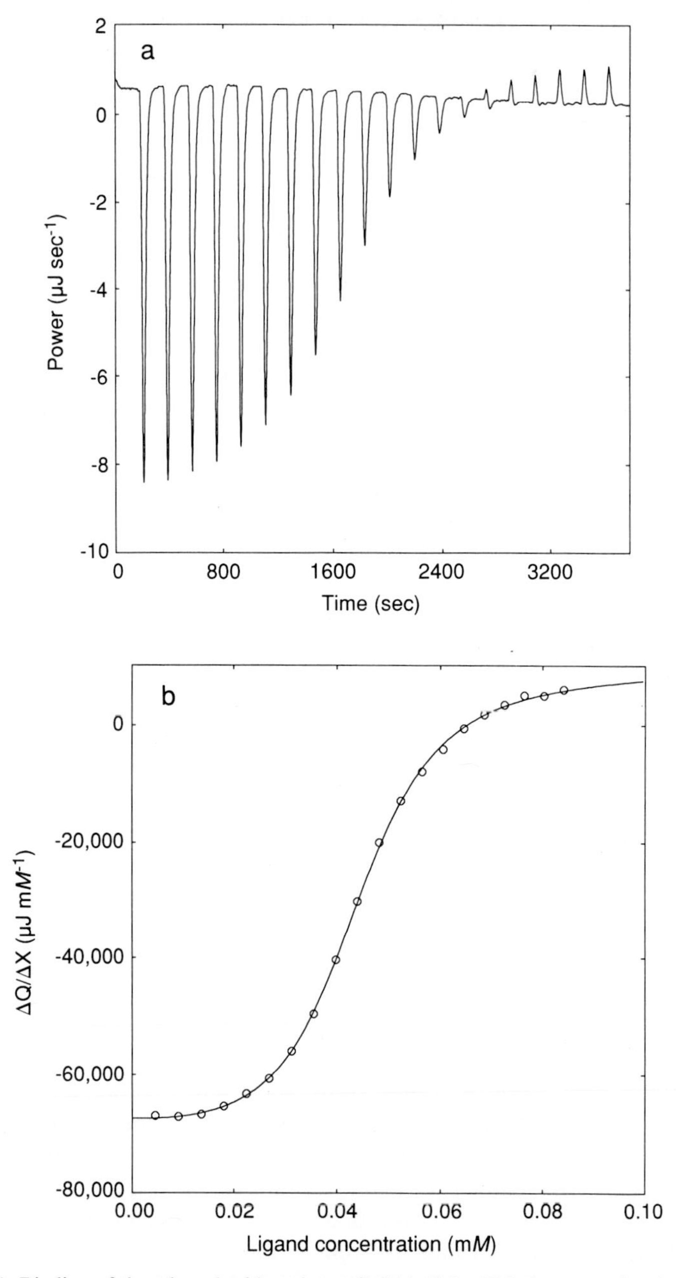

FIG. 1. Binding of the trisaccharide epitope Gal[Abe]ManOMe by monoclonal antibody Se155-4. The thermogram (a) and binding isotherms in the differential heat mode (b) and in the integral heat mode (c) have shown. The lines in (b, c) have been obtained from nonlinear regression according to Eqs. (10) and (11), respectively.

the cells have to be slightly warmer than the insulating jacket. Hence, when experiments are carried out at lower than ambient temperature, a circulating refrigerating bath is used to cool the jacket. The sensitivity is better than 4 μJ, which in a volume of 1.3 ml of pure water corresponds to a temperature change of less than 10^{-6} °.

At least two manufacturers supply sensitive titration microcalorimeters of increasing sophistication. Stability, sensitivity, and ease of use, usually involving full microprocessor or personal computer control of data acquisition and processing, are essential features.

Manufacturer-supplied software (ORIGIN, MicroCal, Inc.) is used for integration of heat signals as well as nonlinear regression analysis and plotting of the results. Slightly modified alternatives to the regression analysis have also been described.[7] Carrying out the regression analysis in either heat mode is in principle equivalent, since the modes merely represent two different mathematical representations of the same experimental data. However, random errors tend to partly cancel one another in the integral heat mode, whereas systematic errors tend to be amplified in this heat mode compared to the differential heat mode. Performing nonlinear regression in the integral heat mode after data analysis in the differential heat mode should therefore lead to relatively smaller standard

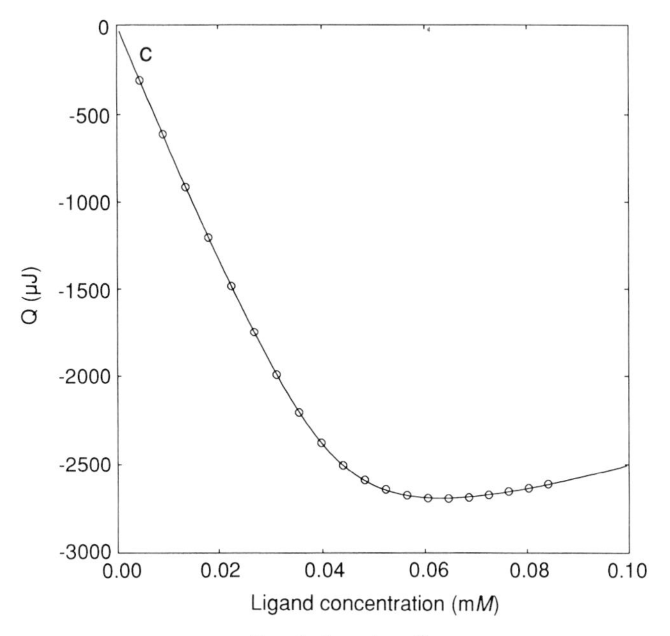

FIG. 1. (*continued*)

TABLE I
CHAIN LENGTH DEPENDENCE OF THERMODYNAMICS OF BINDING OF OLIGOMERS OF *Salmonella* SEROGROUP B ANTIGEN TO MONOCLONAL ANTIBODY Se155-4 AT 25°

Ligand	K ($10^5\ M^{-1}$)	$\Delta G°$ (kJ/mol)	$\Delta H°$ (kJ/mol)	$-T\Delta S°$ (kJ/mol)
Gal[Abe]ManOMe (**1**)	2.1 ± 0.2	−30.5 ± 0.4	−20.5 ± 0.5	−10.0 ± 0.6
$(Gal[Abe]ManRha)_2$	4.9 ± 1.3	−32.6 ± 2.0	−34.1 ± 4.6	1.7 ± 2.3
$(Gal[Abe]ManRha)_3$	4.7 ± 1.0	−32.2 ± 1.6	−44.1 ± 7.1	11.7 ± 2.3
$(Gal[Abe]ManRha)_4$	12 ± 4	−34.7 ± 2.3	−45.5 ± 7.9	10.9 ± 2.9
$(Gal[Abe]ManRha)_5$	5.3 ± 0.9	−32.6 ± 1.3	−71.0 ± 5.4	38.5 ± 1.8

deviations and estimated parameters that lie within the standard deviations obtained in the differential heat mode. If this is not the case, it may be taken as a strong indication of the presence of a systematic error, either during the data collection or in the regression analysis.

Antibody–Antigen Interactions

Dependence on Hapten Chain Length

Titration calorimetry has been used to study the binding of synthetic bacterial antigens by specific monoclonal antibodies. One system of interest is the *Salmonella* serogroup B antigen, a branched polysaccharide with a repeating unit structure of four sugar residues[8]: →3)-α-D-Gal*p*-(1→2)-[α-D-Abe*p*-(1→3)]-α-D-Man*p*-(1→4)-α-L-Rha*p*-(1→, where Gal stands for galactose, Man for mannose, Rha for rhamnose (6-deoxymannose), and Abe for abequose (3,6-dideoxy-D-*xylo*-hexose). A monoclonal antibody, designated Se155-4, has been generated in BALB/c mice and binds specifically to this antigen,[8a] and the crystal structure of the antibody Fab complexed with antigen has been solved.[8b,9] The hapten chain length dependence of the binding free energy, enthalpy, and entropy ($-T\Delta S°$) has been investigated by titration microcalorimetry, and the results are summarized in Table I.[7] Clearly, the molar free energy is virtually constant, whereas the molar enthalpy and entropy show strong but mutually compensating dependencies on the chain length. Because only the first three sugar residues of the repeating unit (Gal[Abe]Man) constitute the binding epitope,[8a,11] it can be argued that the total concentration of epitopes pre-

[8] D. R. Bundle, *Top. Curr. Chem.* **154,** 1 (1990).
[8a] Bundle, D. R., Eichler, E., Gidney, M. A. J., Meldal, M., Ragauskas, A., Sigurskjold, B. W., Sinnott, B., Watson, D. C., Yaguchi, M., and Young, N. M. *Biochemistry* (1994) 33, 5172–5182.
[8b] Bundle, D. R., Baumann, H., Brisson, J. R., Gagne, S. M., Zdanov, A., and Cygler, M. *Biochemistry* (1994) 33, 5183–5192.
[9] M. Cygler, D. R. Rose, and D. R. Bundle, *Science* **253,** 442 (1991).

TABLE II
TEMPERATURE DEPENDENCE OF THERMODYNAMICS OF BINDING OF TRISACCHARIDE EPITOPE Gal[Abe]-ManOMe OF *Salmonella* SEROGROUP B ANTIGEN TO MONOCLONAL ANTIBODY Se155-4

Temperature (K)	K (10^5 M^{-1})	$\Delta G°$ (kJ/mol)	$\Delta H°$ (kJ/mol)	$-T\Delta S°$ (kJ/mol)	ΔC_p (kJ/mol K)
310.06	0.5 ± 0.1	−27.7 ± 0.5	−57.7 ± 3.2	30.0 ± 3.5	
					−1.5 ± 1.0
304.06	0.7 ± 0.1	−28.1 ± 0.2	−48.9 ± 5.0	20.8 ± 5.0	
					−4.9 ± 0.9
298.31	2.1 ± 0.2	−30.5 ± 0.4	−20.5 ± 0.5	−10.0 ± 0.6	
					0.6 ± 0.4
290.83	5.1 ± 3.2	−31.8 ± 0.9	−25.3 ± 2.6	−6.5 ± 2.7	
					0.3 ± 0.8
286.10	1.6 ± 0.7	−28.5 ± 1.0	−26.9 ± 2.5	−1.6 ± 2.7	
					1.5 ± 1.0
282.77	1.3 ± 0.5	−27.7 ± 0.8	−31.8 ± 2.2	4.0 ± 2.6	

sented to the antibody increases with increasing chain length, and a stronger apparent binding might be expected. The enthalpic interactions do in fact become stronger, but this is completely lost in increasingly unfavorable entropy contributions. The calorimetry data indicate that a trisaccharide epitope should fill the antibody site,[8a] a conclusion substantiated by the crystal structure of the Fab–oligosaccharide complex.[8b,9]

Titration Calorimetry to Measure Total Enthalpy and Intrinsic Free Energy

In temperature-dependent binding studies, binding of the trisaccharide epitope Gal[Abe]ManOMe to Se155-4 exhibited an unusually complex behavior, as shown in Table II.[10] The maximum binding both in terms of free energy and binding constant occurred just below room temperature, and both enthalpy and entropy were strongly temperature dependent, again in a mutually compensating manner. The estimated heat capacity changes (ΔC_p) also showed an unusually strong temperature dependence, being large and negative above room temperature and positive below. A van't Hoff plot (Fig. 2) shows a biphasic behavior with two apparent intrinsic enthalpies of approximately −100 kJ/mol above 18° and approximately +100 kJ/mol below. These are both very different from the calorimetrically determined enthalpies ranging from −60 to −20 kJ/mol. Because van't Hoff analysis only detects reactions directly influencing K as this is defined in Eq. (2) and the calorimetrically determined enthalpy contains heats of all reactions taking place in the cell, it was concluded that the discrepancy between these enthalpies was due to concomitant reactions. These would most likely include changes in solvation states and/or temperature-dependent conformational changes in the antibody

[10] B. W. Sigurskjold and D. R. Bundle, *J. Biol. Chem.* **267,** 8371 (1992).

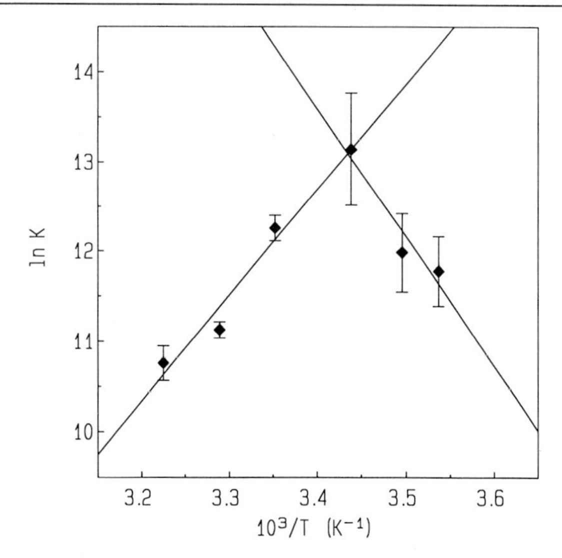

FIG. 2. Van't Hoff plot of the binding of Gal[Abe]ManOMe by monoclonal antibody Se155-4. The two apparent van't Hoff enthalpies are -105.2 ± 19.7 kJ/mol below 18° and $+105.8 \pm 3.7$ kJ/mol above.

and/or ligand, since experiments carried out in different buffer systems ruled out the possibility of changes in ionization states. Furthermore, the thermodynamics and especially the large heat capacity changes suggested that hydrophobic interactions are important for the binding.[10] It is thus important to emphasize that in the interpretation of titration calorimetric results one must be aware that this technique measures the total enthalpy, but only the intrinsic free energy, of the processes studied.[11]

Contributions from Pyranose Residues and Ligand Functional Groups

The energetics of the interactions between the antibody and specific parts or functional groups of the ligand can be elucidated by using chemically modified ligands. The structures of such epitope analogs of the *Salmonella* group B antigen are shown in Scheme 1. Synthetic ligands can be prepared that contain only part of the antigenic determinant, and by observing their binding the relative contributions of each pyranose residue may be estimated. This was accomplished by measuring the binding of the methyl α-glycosides of abequose (**4**) and the disaccharide Abe(1→3) Man (**3**) and comparing these with the full epitope Gal[Abe]ManOMe.

[11] S. N. Timasheff, *Biochemistry* **31,** 9857 (1992).

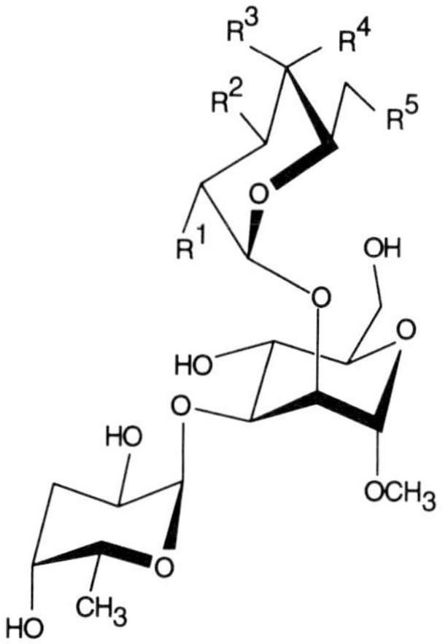

1 $R^1 = R^2 = R^3 = R^5 = OH, R^4 = H$

2 $R^1 = R^2 = R^4 = R^5 = OH, R^3 = H$

5 $R^1 = R^4 = R^5 = OH, R^2 = R^4 = H$

6 $R^1 = R^2 = R^3 = R^5 = OH, R^4 = H$

9 $R^2 = R^3 = R^5 = OH, R^1 = NH_2, R^4 = H$

10 $R^1 = OH, R^2 = OCH_3, R^3 = R^5 = Cl, R^4 = H$

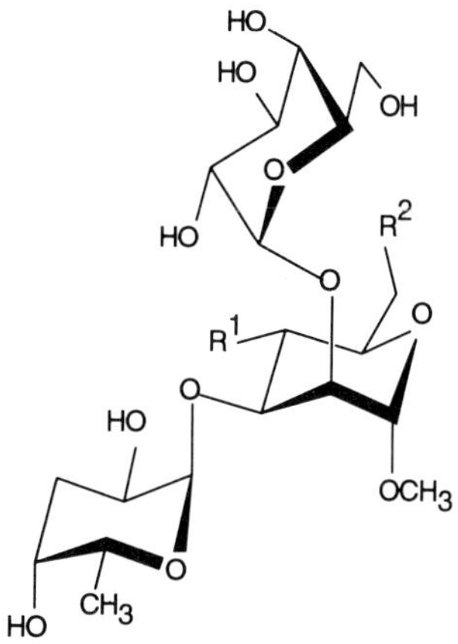

7 $R^1 = H, R^2 = OH$

8 $R^1 = OH, R^2 = H$

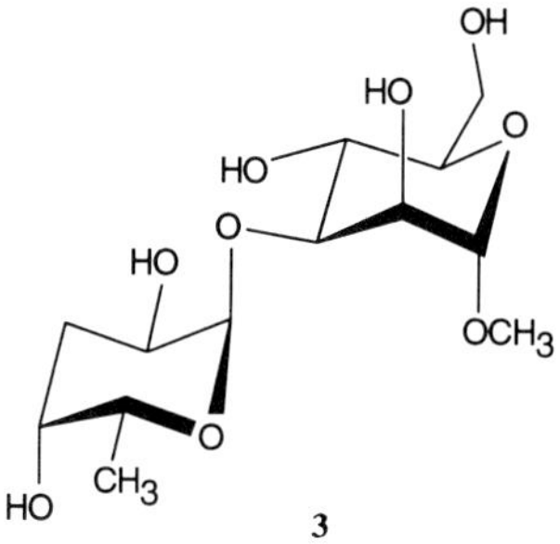

3

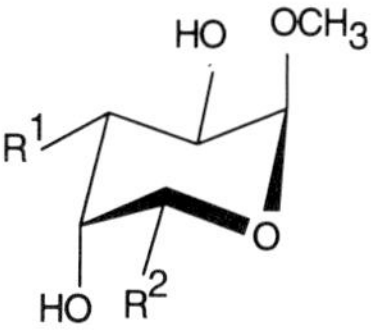

4 $R^1 = H, R^2 = CH_3$

11 $R^1 = R^2 = H$

SCHEME 1

TABLE III
THERMODYNAMICS OF BINDING OF SYNTHETIC *Salmonella* SEROGROUP B EPITOPE DERIVATIVES BY MONOCLONAL ANTIBODY Se155-4 AT 25°[a]

Ligand	K (M^{-1})	$\Delta G°$ (kJ/mol)	$\Delta H°$ (kJ/mol)	$-T\Delta S°$ (kJ/mol)
αGal[Abe]Man (**1**)	$2.1 \pm 0.2 \times 10^5$	-30.5 ± 0.4	-20.5 ± 0.5	-10.0 ± 0.6
αGal[Abe]Man (**2**)	$1.3 \pm 0.2 \times 10^5$	-29.2 ± 0.3	-37.6 ± 1.1	8.4 ± 1.1
AbeMan (**3**)	$2.4 \pm 0.2 \times 10^4$	-25.0 ± 0.2	-28.8 ± 0.7	3.8 ± 0.9
Abe (**4**)	$1.5 \pm 0.4 \times 10^3$	-18.1 ± 0.6	-33.0 ± 4.5	14.9 ± 4.8
(3-Deoxy-Gal)[Abe]Man (**5**)	$8.8 \pm 1.4 \times 10^4$	-28.3 ± 0.4	-33.0 ± 2.2	4.7 ± 2.2
(6-Deoxy-Gal)[Abe]Man (**6**)	$9.0 \pm 2.3 \times 10^4$	-28.3 ± 0.6	-20.4 ± 1.6	-7.9 ± 1.7
Gal[Abe](4-deoxy-Man) (**7**)	$5.1 \pm 1.7 \times 10^4$	-26.9 ± 0.8	-27.7 ± 3.8	0.8 ± 3.9
Gal[Abe](6-deoxy-Man) (**8**)	$8.6 \pm 1.2 \times 10^4$	-28.2 ± 0.3	-26.5 ± 1.2	-1.7 ± 1.3
(2-NH_2-2-deoxy-Gal)[Abe]Man (**9**)	$1.3 \pm 0.3 \times 10^5$	-29.3 ± 0.5	-26.9 ± 1.2	-2.4 ± 1.3
(3-O-CH_3-4,6-Cl_2-Gal)[Abe]Man (**10**)	$3.0 \pm 0.8 \times 10^5$	-31.2 ± 0.6	-34.0 ± 2.3	2.8 ± 2.4
3-Deoxy-Ara (**11**)	$1.2 \pm 0.5 \times 10^2$	-11.9 ± 2.8	-20.3 ± 4.9	8.9 ± 5.7

[a] All compounds are methyl α-glycosides (Scheme 1), thus avoiding anomeric equilibria.

Further mapping of the antibody binding site by thermodynamic characterization was carried in a series of experiments with synthetic ligands. These possessed functional group replacements at single sites within the epitope. The results are summarized in Table III. These may be used to deduce the respective contributions from the three pyranose residues and to rationalize the hydrogen bonding map (Fig. 3), deduced from crystallographic studies.[8b,9]

The binding of several of the trisaccharide ligands to the antibody display favorable entropy contributions. Because both the protein and ligand lose entropy on association owing to translational, rotational, vibrational, and conformational restrictions,[12] the favorable entropy contribution must come from the third participant in the process, namely, water. The release of structured water molecules from the interacting protein and carbohydrate surfaces into bulk water is associated with a substantial entropy gain. This effect (akin to the hydrophobic effect) is probably a major driving force for the binding and is optimized by the precise and close fit between the epitope and the binding pocket.[13]

From the differences in thermodynamic values between compounds **1** and **3** the binding contributions of the Gal unit in the trisaccharide can be estimated as follows: $\Delta\Delta G° = -5.5 \pm 0.4$, $\Delta\Delta H° = +8.3 \pm 0.9$, and

[12] A. V. Finkelstein and J. Janin, *Protein Eng.* **3,** 1 (1989).

[13] R. U. Lemieux, L. T. J. Delbaere, H. Beierbeck, and U. Spohr, *Ciba Found. Symp.* **158,** 231 (1991).

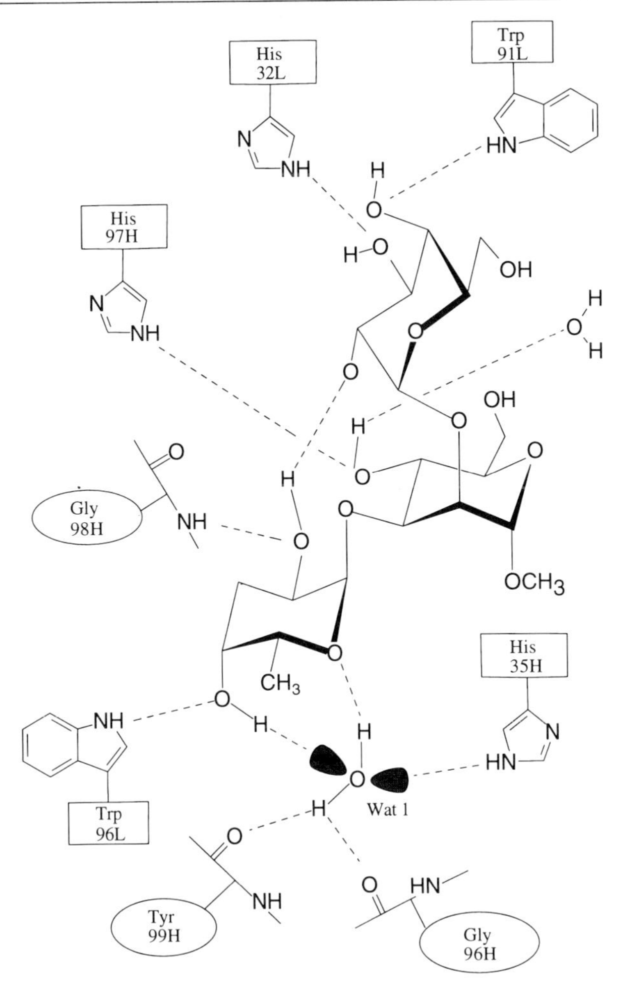

FIG. 3. Hydrogen bonding scheme for the Fab of antibody of Se155-4 complexed with a dodecasaccharide hapten representing the O-antigen of *Salmonella* serogroup B.

$-T\Delta\Delta S^\circ = -13.8 \pm 1.1$ kJ/mol. Likewise, from the difference between compounds **3** and **4** contributions from Man can be estimated: $\Delta\Delta G^\circ = -6.9 \pm 0.6$, $\Delta\Delta H^\circ = +4.2 \pm 4.6$, and $-T\Delta\Delta S^\circ = -11.1 \pm 4.9$ kJ/mol. All of the enthalpy of binding Gal[Abe]ManOMe stems from interactions with the abequose residues, whereas Gal and Man bind relatively weakly and mainly entropically, the major driving force for Gal and Man presumably being solvent displacement. This clearly demonstrates that abequose is the dominant moiety of the epitope for binding to the antibody, and this may be appreciated from the high enthalpic contribution to binding, $\Delta H° = -33.0 \pm 4.5$ kJ/mol, for the simple monosaccharide glycoside **4**. Binding of 3-deoxyarabinose (**11**) (a pentose corresponding to Abe without the 6-CH_3 group) relative to **4** provides an estimate of the contribution made by the methyl group. The lost van der Waals interactions result in a significant reduction in the enthalpy of binding, whereas the entropy is slightly less unfavorable, indicating that the peptide chain of the binding pocket has gained some motional freedom.

The monodeoxy derivatives (**5–8**) all display slightly weaker binding. Most significant is the 3.6 kJ/mol lower activity of the 4-deoxy-Man compound (**7**). The magnitude of this effect suggests that this OH group forms a neutral acceptor–neutral donor hydrogen bond, which is known to approximate 4 kJ/mol.[14] Compound **9** in which the 2-OH of Gal has been replaced by an amino group retains its activity, and it may, since the amino group is uncharged at pH 8.0, still participate in the postulated hydrogen bonds. These involve donation of a hydrogen bond from this hydroxyl group to either a glycosidic oxygen atom or possibly a water molecule. The activity of the deoxy compounds **5**, **6**, and **8** suggests that none of the OH groups at these positions form hydrogen bonds with the antibody. This conclusion as well as that predicting no hydrogen bond to the Gal 4-OH are supported by the activity of compound **10**. A methyl group at O-3 of Gal mimics the anomeric carbon of the preceding rhamnose unit in the polysaccharide antigen, and the two chlorine atoms which are largely isosteric with hydroxyl groups replace the 4-OH and 6-OH of Gal. This is the only compound that binds as strongly, if not slightly more strongly, than the natural epitope, αGal[Abe]ManOMe **1**. Thus it seems that the hydroxyls at positions 4 and 6 of Gal mainly are involved in van der Waals interactions with the antibody. The increase in enthalpic interactions for compound **10** is consistent with this conclusion, since Cl produces stronger van der Waals interactions than OH. Consistent with these inferences, epimerization of the 4-OH of Gal produces a slightly weaker binding (compound **2**), although the changes in enthalpy and entropy are considerably larger than the change in free energy. This suggests that the postulated hydrogen bond involving this OH group is very weak in solution, since it is exposed to bulk water.

Significance of Enthalpy–Entropy Compensation

An enthalpy–entropy compensation plot ($-\Delta H°$ versus $-T\Delta S°$) of the data in Table III is shown in Fig. 4. Nine of the eleven points show a

[14] I. P. Street, C. R. Armstrong, and S. G. Withers, *Biochemistry* **25**, 6021 (1986).

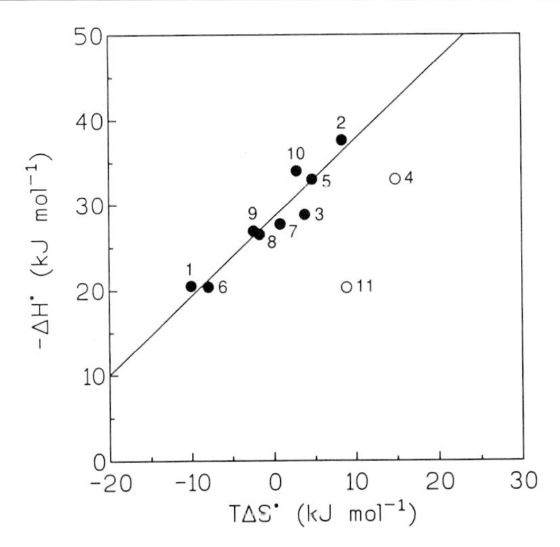

FIG. 4. Isothermal enthalpy–entropy compensation plot of the binding of synthetic ligands with various functional group replacements (Scheme 1) to monoclonal antibody Se155-4 at 25°. The straight line obtained by linear regression has a slope of 0.92 ± 0.11 and an intercept ($-\Delta H^\circ$ when $\Delta S^\circ = 0$) of 28.5 ± 0.6 kJ/mol ($r = 0.951$). The outliers **4** and **11** were not included in the regression analysis.

strong enthalpy–entropy compensation with a slope near unity (0.92 ± 0.11) and an intercept of 28.5 ± 0.6 kJ/mol (the two monosaccharides **4** and **11** constitute outliers and were not included in the regression analysis). The enthalpy–entropy compensation signifies that changes in free energy are much smaller in magnitude than the corresponding changes in enthalpy and entropy. A completely similar enthalpy–entropy compensation plot was obtained for a series of engineered mutants of the Fab of the antibody and their binding of the epitope (compound **1**).[15] Enthalpy–entropy compensation is a ubiquitous phenomenon and in aqueous solution is usually characterized by slopes near unity.[13,16–18] This is probably directly related to the role of solvent water molecules in the association process. Release of structured water molecules (desolvation) from the interacting surfaces (hydrophilic, hydrophobic, or amphiphilic) will result in an entropy gain

[15] D. A. Brummell, V. P. Sharma, N. N. Anand, D. Bilous, G. Dubuc, J. Michniewicz, C. R. MacKenzie, J. Sadowska, B. W. Sigurskjold, B. Sinnott, N. M. Young, D. R. Bundle, and S. A. Narang, *Biochemistry* **32,** 1180 (1993).

[16] J. E. Leffler and E. Grunwald, "Rates and Equilibria in Organic Reactions," Wiley, New York, 1963.

[17] R. Lumry and S. Rajender, *Biopolymers* **9,** 1125 (1970).

[18] W. Linert and R. F. Jameson, *Chem. Soc. Rev.* **18,** 477 (1989).

compensated by loss in enthalpy owing to enthalpically weaker hydrogen bonds in bulk water. The magnitudes of the enthalpy and entropy changes for this process tend to be roughly equal, and this is the origin of the compensation slopes near unity.

It is important to note that enthalpy–entropy compensation plots usually contain experimental errors on both axes, and these errors are statistically correlated since the entropy values have been calculated from the enthalpy values. It is therefore important to use a proper regression method that take these circumstances into account.[19]

Extent of Antigen Obscured by Antibody Binding

Titration calorimetry determines the stoichiometry of a reaction, and this approach therefore allows an estimate of the number of monosaccharide residues that become inaccessible to other antibody molecules when the Fab component binds to its antigenic determinant. A monoclonal antibody, MAb735, specifically binds the capsular antigen of group B *Meningococcus*, a homopolymer of α-(2→8)-linked sialic acid (*N*-acetylneuraminic acid) residues.[20] This interaction has a peculiar chain length dependence in that at least 10 sugar residues are required for binding to be detected.[21] An estimate of the extent to which an individual Fab fragment renders adjacent segments of the antigen chain unavailable for binding was made in an experiment where the Fab fragment of the antibody was titrated with a preparation of colominic acid [α-(2→8)-linked polymer of sialic acid] of approximately 12 kDa corresponding to 41 sialic acid residues. The binding isotherm revealed only one apparent binding constant (Fig. 5). This isotherm was fitted four times under the assumptions that one ligand molecule could contain a number of epitopes ($N = 1, 2, 3$, or 4, respectively). The results are shown in Table IV, and only for $N = 2$ did the apparent concentration of Fab ($73 \pm 8\ \mu M$) obtained from the nonlinear regression analysis correspond to the actual Fab concentration of 76 μM determined spectrophotometrically. Hence, if the epitope size is 10 residues, at least another 10 flanking sialic acid residues are rendered inaccessible to binding with other antibody molecules. The association between this antibody and the polymer is enthalpy driven, with a relatively large unfavorable contribution from entropy. This is typical for binding of long chains since large linear and flexible molecules lose a great amount of motional freedom on complexation.

[19] M. L. Johnson, *Anal. Biochem.* **148,** 471 (1985).

[20] M. Frosch, I. Görgen, G. J. Boulnois, K. N. Timmis, and D. Bitter-Suermann, *Proc. Natl. Acad. Sci. U.S.A.* **82,** 1194 (1985).

[21] J. Häyrinen, D. Bitter-Suermann, and J. Finne, *Mol. Immunol.* **26,** 523 (1989).

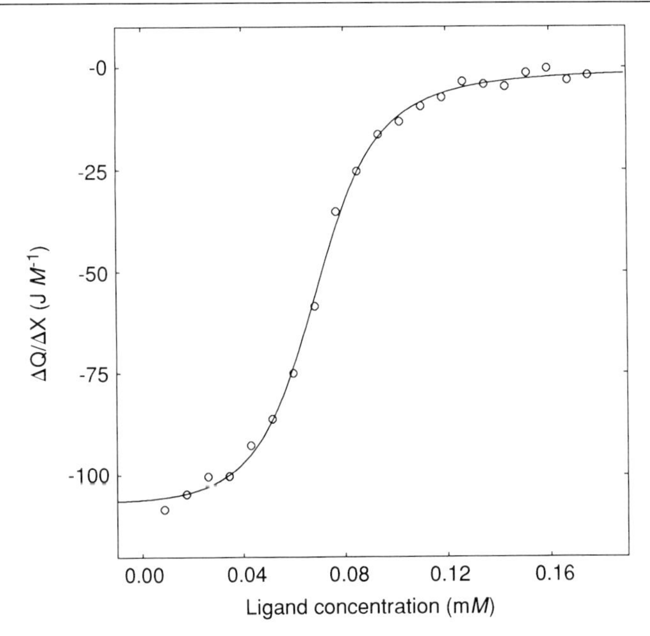

FIG. 5. Binding isotherm for the binding of a 12-kDa preparation of colominic acid [α-(2→8)-linked sialic acid] to a Fab fragment of monoclonal antibody MAb735. Only one apparent binding constant can be determined from the plot, and the four different fits summarized in Table IV all give the same sigmoid curve shown.

When linear oligosaccharides are bound by an antibody, the upper limit to the size of a binding site has been experimentally determined as 6–7 sugar residues. This is also consistent with docking model determinants into antibody sites.[22] It is therefore remarkable that this antibody needs at least 10 residues in order to bind and takes up the space of 15–20 residues of the polysaccharide.

The thermodynamics of the binding of the 12-kDa sialic acid polymer by the Fab of MAb735 is compared to the binding of oligomers containing 9 to 15 sialic acid residues by the whole antibody (Table V). The binding of these ligands is about one order of magnitude weaker. There seems to be very little chain length dependence for any of the thermodynamic parameters in this range of chain lengths, which is in contradiction to what has been reported previously.[21] However, differences in chain length dependence between systems in which both components are in solution (calorimetry) and systems in which one species is immobilized (e.g., enzyme-linked immunosorbent assays) have been observed before.[7] These

[22] R. P. Oomen, N. M. Young, and D. R. Bundle, *Protein Eng.* **4,** 427 (1991).

TABLE IV
THERMODYNAMICS AND DETERMINATION OF EPITOPE SIZE OF THE BINDING OF A 12 kDa POLYMER OF α-(2→8)-LINKED SIALIC ACID BY MAb735[a]

N	K ($10^5\ M^{-1}$)	$\Delta G°$ (kJ/mol)	$\Delta H°$ (kJ/mol)	$-T\Delta S°$ (kJ/mol)	$[\text{Fab}]_{\text{app}}$ (μM)
1	7.3 ± 4.0	−33.8 ± 1.4	−162 ± 15	128 ± 16	37 ± 4
2	3.4 ± 2.0	−31.9 ± 1.5	−81.6 ± 8.5	49.8 ± 9.3	73 ± 8
3	2.4 ± 1.4	−31.0 ± 1.4	−54.1 ± 5.3	23.1 ± 6.1	110 ± 13
4	1.7 ± 0.9	−30.1 ± 1.4	−41.0 ± 4.1	10.9 ± 4.9	146 ± 17

[a] The 12-kDa polymer contained approximately 41 residues. The concentration of MAb735 Fab in the experiment was 76 μM, which corresponds to $N = 2$.

differences presumably arise from changes in entropy occurring as a consequence of the immobilization. There is a relatively strong chain length dependence between binding of the polymer and the shorter oligomers. The enthalpy of binding for the polymer is twice that of the smaller ligands. It thus seems as if the antibody is able to take advantage to a fuller extent of the enthalpic interactions for the whole epitope of about 20 residues. The entropy change, on the other hand, has become increasingly more unfavorable, being three times more unfavorable for the polymer than for the oligomers and thus compensating a substantial amount of the enthalpy gain. This reflects the fact that there is a large entropy penalty on binding longer chains, offsetting the binding constant in this case by approximately 3–4 orders of magnitude.

Conclusion

Titration microcalorimetry can in a single experiment yield the binding constant (free energy), enthalpy, entropy, and stoichiometry (concentra-

TABLE V
CHAIN LENGTH DEPENDENCE OF THERMODYNAMICS OF BINDING OF α-(2→8)-LINKED SIALIC ACID OLIGOMERS TO MAb735

N	K ($10^5\ M^{-1}$)	$\Delta G°$ (kJ/mol)	$\Delta J°$ (kJ/mol)	$-T\Delta S°$ (kJ/mol)
9	3.9 ± 0.9	−26.2 ± 0.6	−45.4 ± 6.7	19.1 ± 6.8
11	4.8 ± 0.9	−26.7 ± 0.5	−42.6 ± 2.9	15.8 ± 2.9
14	5.9 ± 2.9	−27.2 ± 1.2	−38.7 ± 2.8	11.5 ± 3.1
15	4.4 ± 1.9	−26.5 ± 1.1	−41.6 ± 6.8	15.9 ± 7.3
Average	4.4 ± 0.9	−26.6 ± 0.5	−41.0 ± 2.6	14.4 ± 2.7

tion of binding sites or concentration of ligand) for binding of a ligand by a protein. Thermodynamic descriptions provide the most valuable information if parameters such as temperature or the chemical structure of the ligands are varied, or if they are combined with other information such as three-dimensional structures of the complexes. The technique usually does not require large amounts of ligand, but it is not a microanalytical technique in terms of the amount of protein. The measurements described in this chapter with antibodies of a molecular mass per binding site of approximately 75 kDa required 3–7 mg of protein per experiment. This material can be recovered, if affinity purification procedures exist for the particular system. Because the stoichiometry can be determined in the nonlinear least-squares analysis, the precise concentration of the protein need not be known; however, the overall accuracy is improved if the binding site concentration is determined independently.

A titration experiment can be carried out in a couple of hours including equilibration and cleaning, and the data analysis and plotting routines are not very time-consuming either. The biggest limitation of the technique is that it is generally difficult to determine the binding constant if the binding is too strong ($K > 10^8\ M^{-1}$) or too weak ($K < 10^4\ M^{-1}$). However, the enthalpy component is virtually always obtainable, unless, of course, the process is completely or nearly completely entropy driven ($\Delta H^\circ \sim 0$). Independent enthalpy measurements can be carried out by single injections of ligand into excess protein or protein into excess ligand.

[21] Mapping of Hydrogen Bonding between Saccharides and Proteins in Solution

By CORNELIS P. J. GLAUDEMANS, PAVOL KOVÁČ, and EUGENIA M. NASHED

Introduction

Noncovalent binding interactions between saccharides and proteins can be mediated by a variety of charged, hydrophobic, and hydrogen bond interactions. Hydrogen bonds are weak interactions between a hydrogen atom of a proton donor group (such as a hydroxyl or amino group) with a localized site of high electron density. We wish to address the phenomenon that a hydrogen atom becomes shared between the saccharide and the protein in such an interaction. The hydrogen atom involved can originate from either the saccharide or the protein. These bonds can be probed by binding studies employing the use of synthetic saccharide ligands in which a selected hydroxyl group(s) (C–OH) has been replaced by fluorine (to give C–F) or by hydrogen (to give C–H). In the former case the electronegative fluorine atom can still accept hydrogen bonding[1–3] (because fluorine has a radius that is smaller than that of a hydroxyl group, no spatial considerations are involved), but in the latter case the methylene group can neither donate nor accept hydrogen bonding. Thus, if the latter replacement results in a significant reduction of binding of the ligand to the protein, it reveals the involvement of that particular hydroxyl group. Observations on the binding of the corresponding fluoro-substituted ligand can then discriminate between proton donation by the protein or by the ligand.

Below we give a few examples of the preparation of carbohydrate ligands and some deoxyfluoro derivatives known to bind to monoclonal antibodies and their use in binding studies. Two cases are discussed, namely, the synthesis and binding of certain D-galactose derivatives to a monoclonal antibody capable of binding $\beta(1\rightarrow6)$-linked D-galactans and the synthesis and subsequent binding of some derivatives of D-glucose to monoclonal antibodies specific for $\alpha(1\rightarrow6)$-dextrans. Two reviews are available for further information on the intricacies of the glycosidation reaction.[4,5]

Syntheses of Some Galactosyl Ligands and Studies on Galactan-Binding Monoclonal Antibodies

Methyl O-(β-D-Galactopyranosyl)-(1→6)-β-D-galactopyranoside

The disaccharide methyl *O*-(β-D-galactopyranosyl-(1→6)-β-D-galactopyranoside (Gal→Gal→Me) can be prepared in several ways.[6–8] Here, its preparation (Scheme 1) as well as that of the corresponding (by-product) α-linked disaccharide are described. Synthesis proceeds via the condensation of 2,3,4,6-tetra-*O*-acetyl-α-D-galactopyranosyl bromide with methyl 2,3,4-tri-*O*-acetyl-β-D-galactopyranoside (whose preparation is first described below), followed by deprotection of the resulting product.[6]

[1] P. Murray-Rust, W. C. Stallings, C. T. Monti, R. K. Preston, and J. P. Glusker, *J. Am. Chem. Soc.* **105,** 3206 (1983).
[2] F. A. Quicho and N. K. Vyas, *Nature (London)* **310,** 381 (1984).
[3] C. P. J. Glaudemans, P. Kováč, and A. S. Rao, *Carbohydr. Res.* **190,** 267 (1989).
[4] H. Paulsen, *Angew. Chem., Int. Ed. Engl.* **21,** 155 (1982).
[5] K. Toshima and K. Tatsuta, *Chem. Rev.* **93,** 1503 (1993).
[6] P. Kováč, E. A. Sokoloski, and C. P. J. Glaudemans, *Carbohydr. Res.* **128,** 101 (1984).
[7] E. M. Nashed and C. P. J. Glaudemans, *J. Org. Chem.* **52,** 5255 (1987).
[8] E. M. Nashed and C. P. J. Glaudemans, *J. Org. Chem.* **54,** 6116 (1989).

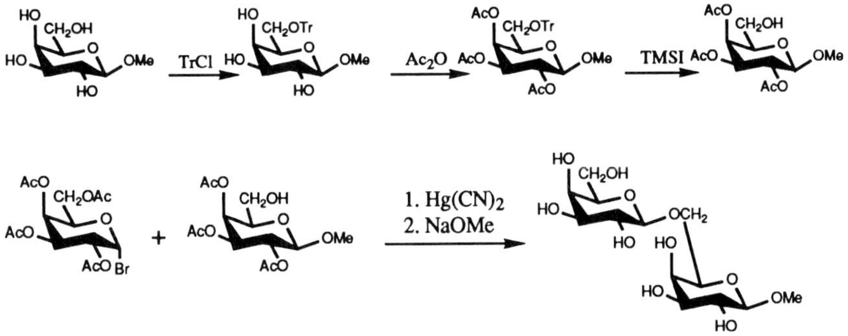

SCHEME 1. Preparation of methyl O-(β-D-galactopyranosyl)-(1→6)-β-D-galactopyranoside. Tr Cl, Trityl chloride; Ac_2O, acetic anhydride; TMSI, iodotrimethylsilane; NaOMe, sodium methoxide.

Methyl β-D-galactopyranoside (Sigma Chemical Co., St. Louis, MO, 5 g, 25.7 mmol) is dissolved in dry pyridine (25 ml). Trityl chloride (7.9 g, 28.3 mmol) is added to the stirred solution over a period of 30 min, after which stirring is continued for 2 hr at room temperature and for 3 hr at 50°. Thin-layer chromatography using dichloromethane–methanol* shows the reaction to be complete. Acetic anhydride is added (10 ml) in portions over 1 hr, and the mixture is kept overnight at room temperature. Methanol (10 ml) is added in portions with cooling. After 1 hr, the mixture is concentrated to half the volume and poured into a vigorously stirred aqueous solution of sodium bicarbonate and ice. The solid precipitate is collected by filtration, washed with water, and dissolved in dichloromethane. The solution is washed with water, dried (anhydrous sodium sulfate), and concentrated. The residue is purified on a column of silica gel (75 g) using carbon tetrachloride–acetone. Crystallization from ethanol gives 11.3 g (80%) methyl 2,3,4-tri-O-acetyl-6-O-trityl-β-D-galactopyranoside. An analytical sample has mp 143°–145°, and $[\alpha]_D$ −52.7° (c 1.6, chloroform).

This trityl derivative (4.7 g) is dissolved in acetic acid (25 ml), cooled to 0°, and detritylated by treatment with 33% hydrobromic acid (Fluka, Ronkonkoma, NY) in acetic acid (2.5 ml) for 1 min. The precipitated trityl bromide is removed by filtration, and the filtrate is poured onto ice and water (100 ml). The solid formed from the filtrate on contact with the

* Particular thin-layer chromatography plates used will exert an influence on the mobility of compounds in any given solvent system. Therefore, we have indicated only the components of the solvent systems used; however, relative amounts always decrease in the order given.

ice is collected by filtration, washed with ice water, and dissolved in dichloromethane. The solution is combined with the dichloromethane extract of the filtrate and washings. The combined solution and extracts are washed with water, then dried (anhydrous sodium sulfate), and the organic layer is concentrated under reduced pressure. Traces of acetic acid are removed by repeated evaporations of toluene, and the residue is separated on a column of silica gel (carbon tetrachloride–acetone) to first give the product of acetyl migration, methyl 2,3,6-tri-*O*-acetyl-β-D-galactopyranoside (0.31 g, 11%). It has mp 108°–109°, $[\alpha]_D$ +1° (*c* 1.2, chloroform) after recrystallization from ether containing a little dichloromethane. Subsequently eluted is the desired methyl 2,3,4-tri-*O*-acetyl-β-D-galactopyranoside (1.3 g, 48%). It has mp 125°–126°, $[\alpha]_D$ +5.2° (*c* 1.1, chloroform) after recrystallization from dichloromethane–isopropyl ether.

2,3,4,6-Tetra-*O*-acetyl-α-D-galactopyranosyl bromide (Fluka, 1.86 g, 4.5 mmol) is added to a mixture of methyl 2,3,4-tri-*O*-acetyl-β-D-galactopyranoside (0.96 gm, 3 mmol), mercuric cyanide (0.57 g, 2.25 mmol), mercuric bromide (0.25 g), and Drierite (3 g) in dry benzene (5 ml) that has been stirred for 2 hr. The suspension, protected from atmospheric moisture, is stirred for 16 hr at room temperature, when silica gel thin-layer chromatography (toluene–acetone) shows the presence of one major and some minor products. The mixture is filtered, the solids are washed with dichloromethane, and the combined filtrate and washings are washed with an aqueous solution of potassium bromide to remove mercuric salts. The organic layer is dried over anhydrous sodium sulfate, filtered, and concentrated. The residue is chromatographed on silica gel to first give the α-linked disaccharide methyl *O*-(2,3,4,6-tetra-*O*-acetyl-α-D-galactopyranosyl)-(1→6)-2,3,4-tri-*O*-acetyl-β-D-galactopyranoside (150 mg, 7%), mp 165° (after two recrystallizations from ethanol), $[\alpha]_D^{25}$ +86° (*c* 0.94, chloroform). The corresponding desired β-linked disaccharide methyl *O*-(2,3,4,6-tetra-*O*-acetyl-β-D-galactopyranosyl)-(1→6)-2,3,4-tri-*O*-acetyl-β-D-galactopyranoside (1.49 g, 76%), is obtained next as an amorphous substance, $[\alpha]_D^{25}$ −13.5° (*c* 1.2, chloroform).

Deacetylation of the protected β-linked disaccharide so obtained is achieved as follows. A solution of methyl *O*-(2,3,4,6-tetra-*O*-acetyl-β-D-galactopyranosyl)-(1→6)-2,3,4-tri-*O*-acetyl-β-D-galactopyranoside (80 mg) in methanol (5 ml) is treated with sodium methoxide in methanol (1 *M*, 1 ml) for 24 hr. The solution is neutralized with Amberlite IR 120 H^+ resin (Fluka Chem. Co., Ronkonkoma, NY), filtered, and concentrated. The desired methyl *O*-(β-D-galactopyranosyl)-(1→6)-β-D-galactopyranoside is obtained in nearly theoretical yield. After recrystallization from methanol, and drying for 3 hr at 100° (vacuum), it has mp 218–219°, $[\alpha]_D^{25}$ −9.5° (*c* 1.25, water).

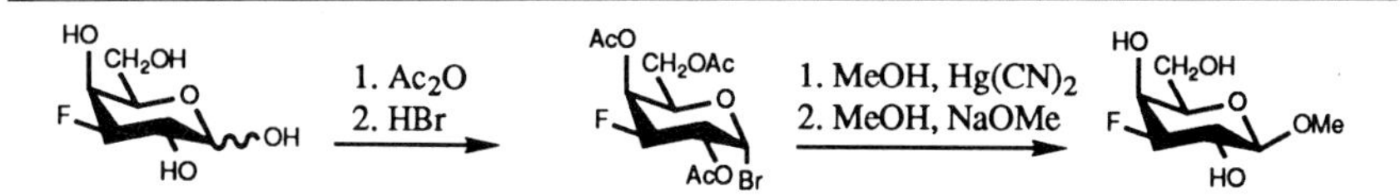

SCHEME 2. Preparation of methyl 3-deoxy-3-fluoro-β-D-galactopyranoside. MeOH, Methanol.

Preparation of Methyl 3-Deoxy-3-fluoro-β-D-galactopyranoside

The glycoside methyl 3-deoxy-3-fluoro-β-D-galactopyranoside (3FGal→Me) is prepared by acetylation of the deoxyfluorogalactose under conditions favoring formation of the pyranose form, followed by conversion to the corresponding deoxyfluoro-galactopyranosyl bromide, followed in turn by treatment with methanol to obtain the methyl glycoside (Scheme 2). A solution of 3-deoxy-3-fluoro β-D-galactopyranose[9] (0.55 g) in dry pyridine (6 ml) is frozen, acetic anhydride (30 ml) is added, and, with occasional shaking, the mixture is allowed to warm to room temperature. After 16 hr thin-layer chromatography shows the reaction to be complete, and the solution is concentrated to dryness. The residue is taken up in dichloromethane (6 ml) and treated with hydrobromic acid in glacial acetic acid (33%, 9 ml) for 1 hr. The solution is diluted with dichloromethane, extracted with ice–water, then with cold, dilute, aqueous sodium bicarbonate, and dried over anhydrous sodium sulfate. After filtration, the filtrate is concentrated, and the residue (85% yield) is crystallized from ether–petroleum ether to give chromatographically pure 2,4,6-tri-*O*-acetyl-3-deoxy-3-fluoro-α-D-galactopyranosyl bromide, mp 101°–102°, $[\alpha]_D^{25}$ +251.6° (*c* 1.25, chloroform).

The above bromide (212 mg) is dissolved in a minimum volume of dichloromethane and added to a stirred (2 hr) mixture of methanol (7 ml), mercuric cyanide (150 mg), mercuric bromide (50 mg), and Drierite (1.5 g), with the solution being stirred for 30 min longer. At that time all of the bromide has reacted, as shown by thin-layer chromatography (toluene–acetone). The mixture is filtered, the filtrate is concentrated, the residue is extracted with dichloromethane, and the extract is successively washed with an aqueous solution of potassium bromide and water. The organic layer is dried (anhydrous sodium sulfate), concentrated, and the residue chromatographed over silica gel (slurry-packed, using toluene–acetone) to give methyl 2,3,4-tri-*O*-acetyl-3-deoxy-3-fluoro-β-D-galactopyranoside (137 mg, 73%), which is crystallized from ethanol–isopropyl ether, mp 81.5°–82.5°, $[\alpha]_D^{25}$ −3.3° (*c* 0.87 chloroform).

[9] P. Kováč and C. P. J. Glaudemans, *Carbohydr. Res.* **123,** 326 (1983).

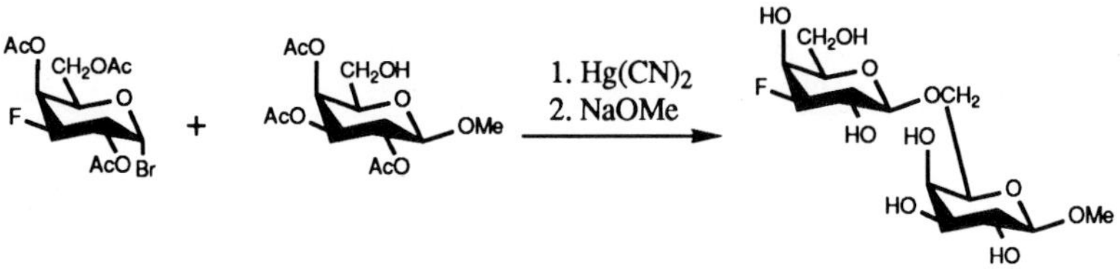

SCHEME 3. Preparation of methyl *O*-(3-deoxy-3-fluoro-β-D-galactopyranosyl)-(1→6)-β-D-galactopyranoside.

A solution of sodium methoxide in methanol (1 *M*, 0.5 ml) is added to a solution of the above methyl 2,3,4-tri-*O*-acetyl-3-deoxy-3-fluoro-β-D-galactopyranoside (100 mg) in methanol (5 ml), and the mixture is kept overnight at room temperature. After neutralization with Dowex 50W (H^+) ion-exchange resin (Fluka Chem. Co., Ronkonkoma, NY), filtration, and concentration of the filtrate, methyl 3-deoxy-3-fluoro-β-D-galactopyranoside is obtained in nearly theoretical yield, mp 161°–162° (crystallized from ethanol–ethyl acetate), $[\alpha]_D^{25}$ −11° (*c* 0.85, water).

Methyl O-(3-Deoxy-3-fluoro-β-D-galactopyranosyl)-(1→6)-β-D-galactopyranoside

The disaccharide methyl *O*-(3-deoxy-3-fluoro-β-D-galactopyranosyl)-(1→6)-β-D-galactopyranoside (3FGal→Gal→Me) is prepared by the condensation of 2,4,6-tri-*O*-acetyl-3-deoxy-3-fluoro-α-D-galactopyranosyl bromide with methyl 2,3,4-tri-*O*-acetyl-β-D-galactopyranoside followed by deprotection of the resulting product[10] (Scheme 3). 2,4,6-Tri-*O*-acetyl-3-deoxy-3-fluoro-α-D-galactopyranosyl bromide (see the preparation above, 556 mg, 1.5 mmol) is added to a suspension of methyl 2,3,4-tri-*O*-acetyl-β-D-galactopyranoside (see above, 320 mg, 1 mmol), mercuric cyanide (190 mg, 0.75 mmol), mercuric bromide (100 mg), and Drierite (1 g) in benzene (5 ml) that has been stirred first for 2 hr, and the mixture is stirred overnight with the exclusion of moisture. The reaction mixture is processed as described for the methyl glycoside above. Silica gel chromatography (slurry-packed, using toluene–acetone) gives first a small amount of the α-linked isomer, followed by the β-linked isomer methyl *O*-(2,3,4-tri-*O*-acetyl-3-deoxy-3-fluoro-β-D-galactopyranosyl)-(1→6)-2,3,4-tri-*O*-acetyl-β-D-galactopyranoside (435 mg, 71%), mp 181°–181.5° (crystallization from ethanol), $[\alpha]_D^{25}$ −0.8° (*c* 0.63, chloroform).

Sodium methoxide in methanol (1 *M*) is added to the β-linked disaccharide in methanol until the solution is strongly alkaline. After 2 hr at room

[10] P. Kováč, H. J. C. Yeh, and C. P. J. Glaudemans, *Carbohydr. Res.* **140,** 277 (1985).

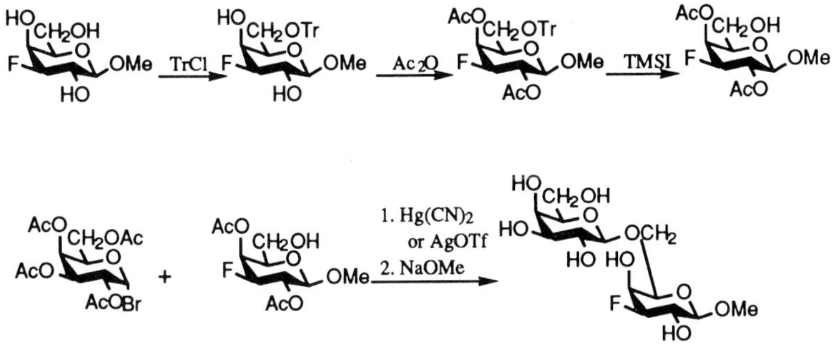

SCHEME 4. Preparation of methyl *O*-(β-D-galactopyranosyl)-(1→6)-3-deoxy-3-fluoro-β-D-galactopyranoside. AgOTf, Silver trifluoromethane sulfonate.

temperature thin-layer chromatography shows the reaction to be complete. The mixture is neutralized with ion-exchange resin (Dowex 50W, H^+), then filtered, and the filtrate is concentrated. Crystallization from methanol–acetone gives methyl *O*-(3-deoxy-3-fluoro-β-D-galactopyranosyl)-(1→6)-β-D-galactopyranoside in 90% yield, mp 200°–201°, $[\alpha]_D^{25}$ −18° (*c* 0.8, water).

Methyl O-(β-D-Galactopyranosyl)-(1→6)-3-deoxy-3-fluoro-β-D-galactopyranoside

The disaccharide methyl *O*-(β-D-galactopyranosyl)-(1→6)-3-deoxy-3-fluoro-β-D-galactopyranoside (Gal→3FGal→Me) is obtained by condensation of methyl 2,4-di-*O*-acetyl-3-deoxy-3-fluoro-β-D-galactopyranoside with 2,3,4,6-tetra-*O*-acetyl-α-D-galactopyranosyl bromide, followed by deblocking of the product[11] (Scheme 4). Chlorotriphenylmethane (3 g, 11 mmol) is added to a solution of methyl 3-deoxy-3-fluoro-β-D-galactopyranoside, whose preparation is described above (2 g, 10 mmol), in dry pyridine (10 ml). The solution is stirred at room temperature for 5 hr, and then at 50° for 16 hr. Thin-layer chromatography (dichloromethane–acetone) then shows the reaction to be complete. After concentration the residue is partitioned between dichloromethane and aqueous sodium bicarbonate, and the organic layer is concentrated. Remnants of pyridine are removed by repeated evaporations of water under reduced pressure. Water is removed by azeotropic distillation with toluene. Crystallization of the residue from dichloromethane–isopropyl ether gives methyl 6-*O*-trityl-3-deoxy-3-fluoro-β-D-galactopyranoside (3.4 g, 76%), mp 105°–109°, $[\alpha]_D^{25}$ −35.5° (*c* 1.1, chloroform).

[11] P. Kováč and C. P. J. Glaudemans, *Carbohydr. Res.* **140,** 289 (1985).

Methyl 6-*O*-trityl-3-deoxy-3-fluoro-β-D-galactopyranoside (1.7 g, 3.87 mmol) in pyridine (5 ml) is treated with acetic anhydride (2.3 ml, 22.5 mmol). After 2 hr at room temperature, thin-layer chromatography (toluene–acetone) shows that no starting material is left and that two products have formed. More acetic anhydride (~5 ml) is added, and after a further 4 hr at 50° all starting material has been converted to a single product. The mixture is concentrated under reduced pressure, and remnants of pyridine are removed as described above. Crystallization from dichloromethane–ethanol gives methyl 6-*O*-trityl-2,4-di-*O*-acetyl-3-deoxy-3-fluoro-β-D-galactopyranoside (1.85 g, 91%), mp 234°, $[\alpha]_D^{25}$ −58.3° (*c* 1.9, chloroform).

A mixture of methyl 6-*O*-trityl-2,4-di-*O*-acetyl-3-deoxy-3-fluoro-β-D-galactopyranoside (2.1 g, 4 mmol) and sodium iodide (1.8 g, 12 mmol) in dry acetonitrile (40 ml), protected from atmospheric moisture, is stirred until a clear solution is formed. The solution is cooled in ice, and chlorotrimethylsilane (1.53 ml, 12 mmol) is added while stirring under anhydrous conditions. Evolved iodine (dark) is evident. After 2 min, ice–water (100 ml) is added, and the mixture is stirred for 15 min. The precipitated triphenylmethanol is removed by filtration, then washed with cold water, and the filtrate and washings are stirred with an aqueous sodium thiosulfate solution. The mixture is partitioned between dichloromethane and water, the organic layer is dried (anhydrous sodium sulfate) and filtered, and the filtrate is concentrated to a small volume. Addition of ether gives crystalline methyl 2,4-di-*O*-acetyl-3-deoxy-3-fluoro-β-D-galactopyranoside (0.94 g). After two recrystallizations from ether the material has mp 118°–119°, $[\alpha]_D^{25}$ +27° (c_1, chloroform).

The above nucleophile, methyl 2,4-di-*O*-acetyl-3-deoxy-3-fluoro-β-D-galactopyranoside (280 mg, 1 mmol), mercuric cyanide (190 mg, 0.75 mmol), mercuric bromide (10 mg), and Drierite (1 g) are stirred for 2 hr in dry benzene (5 ml). 2,3,4,6-Tetra-*O*-acetyl-α-D-galactopyranosyl bromide (Fluka, 0.61 g, 1.5 mmol) is added, and the mixture is stirred for 20 hr at room temperature, with the exclusion of atmospheric moisture. Thin-layer chromatography (toluene–acetone and carbon tetrachloride–acetone) shows that one major product is formed. Dichloromethane (30 ml) is added, the mixture is filtered, solids are washed with the same solvent, and the filtrate and washings are shaken with aqueous potassium bromide solution. The organic phase is dried, filtered, and concentrated. The residue is treated with acetic anhydride and pyridine (to convert any product of hydrolysis of the galactosyl bromide used to the faster moving acetate). Excess acetic anhydride is destroyed by the addition of methanol while cooling, and the solution is concentrated under reduced pressure. The residue is separated by silica gel chromatography. First eluted is the α-linked disaccharide methyl 2,3,4,6-tetra-*O*-acetyl-α-D-galactopyranosyl-(1→6)-2,4-di-*O*-acetyl-3-deoxy-3-fluoro-β-D-galactopyranoside (25 mg, 4%), mp 184°–185° (twice recrystallized from ethanol), $[\alpha]_D^{25}$ +94° (*c* 0.7, chloroform). Eluted next is the amorphous, desired β-linked disaccharide methyl 2,3,4,6-tetra-*O*-acetyl-β-D-galactopyranosyl-(1→6)-2,4-di-*O*-acetyl-3-deoxy-3-fluoro-β-D-galactopyranoside (518 mg, 85%), $[\alpha]_D^{25}$ −1.3° (*c* 0.9, chloroform).

Formation of the undesired α anomer can be avoided by different condensation conditions: An anhydrous solution of silver trifluoromethane sulfonate (282 mg, 1.1 mmol) and 2,4,6-trimethylpyridine (112 μl, 0.85 mmol) in 1 : 1 (v/v) toluene–nitromethane (2 ml) is added dropwise to a cold (−25°) stirred anhydrous solution of 2,3,4,6-tetra-*O*-acetyl-α-D-galactopyranosyl bromide (452 mg, 1.1 mmol) and 2,4-di-*O*-acetyl-3-deoxy-3-fluoro-β-D galactopyranoside (280 mg, 1 mmol) in the same solvent mixture (3 ml). Silver bromide precipitates immediately, and after 15 min at −25° thin-layer chromatography (carbon tetrachloride-acetone) shows that the reaction is essentially complete. Besides the major product, three minor products are also present, but no methyl 2,3,4,6-tetra-*O*-acetyl-α-D-galactopyranosyl-(1→6)-2,4-di-*O*-acetyl-3-deoxy-3-fluoro-β-D-galactopyranoside is detected. 2,4,6-Trimethylpyridine (100 μl) is added to neutralize the excess triflic acid, followed by dichloromethane (30 ml), and the mixture is filtered. The filtrate is washed with aqueous sodium thiosulfate solution, dried, and concentrated. Elution (carbon tetrachloride–acetone) of the residue from a column of silica gel first gives a small amount (45 mg) of methyl 2,4,6-tri-*O*-acetyl-3-deoxy-3-fluoro-β-D-galactopyranoside, mp 81°–82°. Eluted next is the desired disaccharide, methyl 2,3,4,6-tetra-*O*-acetyl-β-D-galactopyranosyl-(1→6)-2,4-di-*O*-acetyl-3-deoxy-3-fluoro-β-D-galactopyranoside (396 mg, 65%), identical [^{13}C nuclear magnetic resonance (NMR) spectrum and optical rotation] to the material isolated above.

To obtain the deprotected disaccharide, the above methyl 2,3,4,6-tetra-*O*-acetyl-β-D-galactopyranosyl-(1→6)-2,4-di-*O*-acetyl-3-deoxy-3-fluoro-β-D-galactopyranoside is dissolved in methanol to give an approximately 2% solution. A 1 *M* solution of sodium methoxide in methanol is added, until the solution is strongly alkaline (litmus). After 16 hr the solution is neutralized (Dowex 50 H^+ ion-exchange resin), and the solvent is evaporated to give methyl β-D-galactopyranosyl-(1→6)-3-deoxy-3-fluoro-β-D-galactopyranoside in nearly theoretical yield. After two recrystallizations from methanol–ethanol it has mp 175°–176°, $[\alpha]_D^{25}$ −13.3° (*c* 0.7, water).

General Considerations in Antibody Studies

In measuring the affinity between carbohydrate ligands and antibody, the method[12] uses the incremental change in the tryptophanyl fluorescence of the antibody, as a function of ligand aliquots added. The carbohydrate ligands themselves are transparent to UV radiation and do not show any fluorescence. This method has been verified by comparison of the derived affinity constant with those obtained by other methods.[13,14] It is far more accurate than equilibrium dialysis, especially when the values for affinity are low. The maximal ligand-induced change in protein fluorescence is an intrinsic property of each particular immunoglobulin, and it appears to be related to the number of perturbable tryptophanyl residues in the general immunoglobulin combining area.

Briefly, measurements are performed at 25° in 10 mM phosphate-buffered saline (PBS), pH 7.4, using an antibody site concentration of approximately 5×10^{-7} M. Excitation and emission wavelengths are 295 and 340 nm, respectively (280/340 can be used if tyrosine residues are involved). The instrument used is a Perkin-Elmer (Norwalk, CT) LS-50 spectrofluorimeter. The symbol ΔF stands for the incremental changes in protein fluorescence due to ligand additions. This is divided by the maximally attained fluorescence change at infinite ligand concentration, ΔF_{max} [when all antibody (Ab) sites carry ligand], to give the fraction ν $(=\Delta F/\Delta F_{max})$ of antibody sites occupied by ligand at any given concentration of ligand (C_{ligand}). The value of ΔF_{max} can be determined by plotting $\Delta F/C_{ligand}$ (ordinate) versus ΔF (abscissa). At infinite ligand concentration $\Delta F/C_{ligand}$ becomes 0, so the intercept on the abscissa is ΔF_{max}. For

$$\text{Ab} + \text{ligand} \rightleftharpoons \text{Ab..ligand}$$

the equilibrium constant is

$$K_a = C_{\text{Ab..ligand}}/C_{\text{Ab}} \times C_{\text{ligand}}$$

It follows that

$$K_a = \nu/(1 - \nu)C_{\text{ligand}} \qquad \text{or} \qquad \nu/C_{\text{lig}} = K_a(1 - \nu)$$

Thus, a Scatchard plot of ν/C_{ligand} (ordinate) versus ν (abscissa) readily yields the K_a as the intercept on the ordinate.

Titration of the antibody can be done using, for example, 1.0 ml of immunoglobulin A (IgA) (or Fab solution), which has an OD_{280} of 0.04 in PBS ($\sim 1.8 \times 10^{-7}$ M for whole antibody), by adding incremental amounts

[12] M. E. Jolley and C. P. J. Glaudemans, *Carbohydr. Res.* **33,** 377 (1974).
[13] D. G. Streefkerk and C. P. J. Glaudemans, *Biochemistry* **16,** 3760 (1977).
[14] L. G. Bennet and C. P. J. Glaudemans, *Carbohydr. Res.* **72,** 315 (1979).

(1 μl or more) of ligand stock solution. For ligands with a K_a of approximately 10^3, the ligand stock solution should be around 0.1 *M*, and a total amount of about 30 μl should be sufficient to approach the end point of the titration. For ligands having a K_a of approximately 10^4, a ligand (stock) solution of approximately 10–15 m*M* should be correct. All ligands are to be checked for the absence of fluorescent or fluorescence-quenching impurities.

Studies on Antigalactan Monoclonal Antibodies

Monoclonal antibody specific for β(1→6)-linked polygalactopyranan is obtained by the hybridoma technique following intravenous injection into BALB/c mice of a β-galactosyl–bovine serum albumin conjugate and gum Ghatti, a predominantly β(1→6)-linked galactopyranan.[15] Antibodies are purified by affinity chromatography using a column of *p*-aminophenyl β-D-galactopyranoside insolubilized on agarose (Sigma).[16] The antibody is eluted with 0.2 *M* methyl β-D-galactopyranoside (Sigma), from which any trace of polymeric galactan has previously been removed by dialyzing it through a membrane, in PBS, pH 7.4. Eluted fractions are monitored for A_{280}. The ligand is then removed from the column eluate containing the antibody by exhaustive dialysis, or by membrane filtration (using an Amicon, Danvers, MA, YM30 membrane) and washing of the membrane-retained antibody. In the latter case, that is followed by removal of the last traces of ligand by submitting the membrane-retained antibody dissolved in a small amount of buffer solution to gel-sizing chromatography. Small disposable desalting columns (Pierce, Rockford, IL, or Pharmacia, Piscataway, NJ) are suitable. The reader is cautioned that gel-sizing columns can be either polyacrylamide or dextran. For the latter type, use for antidextran antibodies could possibly lead to retention of antibody. Affinity constants between antibody and synthetic ligands can be measured either on whole antibody or on the Fab fragment. The preparation of the Fab has been described.[17] The measurements here discussed were done on the anti-β(1→6)-galactan hybridoma HyGal 10, an IgM (μ,κ).[15]

Discussion of Results

Amino acid sequencing[18,19] of all but one (X44) of the anti-β(1→6)-D-galactan immunoglobulins of this group showed the presence of two

[15] M. Pawlita, M. Potter, and S. Rudikoff, *J. Immunol.* **129,** 615 (1982).

[16] M. Potter and C. P. J. Glaudemans, this series, Vol. 28, p. 388.

[17] M. E. Jolley, S. Rudikoff, M. Potter, and C. P. J. Glaudemans, *Biochemistry* **12,** 3039 (1973).

[18] C. P. J. Glaudemans and P. Kováč, *ACS Symp. Ser.* **374,** 78 (1988).

[19] S. Rudikoff, N. D. Rao, C. P. J. Glaudemans, and M. Potter, *Proc. Natl. Acad. Sci. U.S.A.* **77,** 4270 (1980).

TABLE I
AFFINITY CONSTANTS FOR HYGAL 10 WITH CERTAIN LIGANDS

Ligand	K_a	ΔF_{max} (%)
Gal→Me	0.7×10^3	+26
Gal→Gal→Me	3.1×10^4	+41
Gal→Gal→Gal→Me	2.1×10^5	+41
Gal→Gal→Gal→Gal→Me	3.2×10^5	+43
Gal→Gal→Gal→Gal→Gal→Me	3.9×10^5	+42
3FGal→Me	0	0
3FGal→Gal→Me	3.0×10^3	+24
Gal→3FGal→Me	3.1×10^4	+46
3FGal→Gal→Gal→Me	1.5×10^5	+42

tryptophanyl residues: one at position 33 of the heavy chain (33H) and one at position 91 of the light chain (91L). The crystal structure of a related antigalactan (IgA J539), shows[20] these residues to occur at the interface of the H and L chain in the general area that in antibodies is known to be the antigen-combining region. It had been shown that all the monoclonal antibodies of this family could maximally bind four sequential saccharides, that is, the β(1→6)-linked galactotetraoside. Thus, there are four subsites, each binding a single saccharide in the antigen chain. Also, from observing the maximal ligand-induced fluorescence change, it had been suggested that the monosaccharide methyl β-D-galactopyranoside perturbed one tryptophanyl residue, whereas the disaccharide methyl β-D-galactopyranosyl-(1→6)-β-D-galactopyranoside perturbed two tryptophanyl residues. From this it appeared that these residues might be spaced apart a distance similar to that of the two saccharide residues in the disaccharide. A pyranoside ring is about 4–5 Å in size, and, indeed, the crystal structure of IgA Fab J539, a representative antigalactan for this family of immunoglobulins, revealed the above two tryptophanyl residues to be around 10 Å apart.[20] The monosaccharide must of necessity first fill the subsite having the highest affinity. From the above observation it therefore appears that the neighboring subsite must be near the second perturbable tryptophanyl residue. Optimal affinity occurred for the methyl glycoside of the (1→6)-linked tetrasaccharide, but no further increases in protein fluorescence (perturbation) were noticed for the tri- or tetrasaccharide when compared with the disaccharide.

Table I lists the affinity constants, as well as the maximal ligand-induced change in the tryptophanyl fluorescence of the antibody obtained

[20] S. W. Suh, T. N. Bhat, M. A. Navia, G. H. Cohen, D. N. Rao, S. Rudikoff, and D. R. Davies, *Proteins: Struct. Funct. Genet.* **1,** 74 (1986).

with HyGal 10 and the ligands whose preparation has been described. In addition, methyl β-D-galactopyranoside, or Gal→Me, is obtained from Sigma. The affinities of the methyl β-glycosides of higher, β(1→6)-linked D-galactosyl oligosaccharides (denoted as Gal→Gal→Gal→Me, etc.) are also listed.

If the four antibody subsites, each capable of binding a sequential galactosyl residue in the antigenic chain, are labeled **A**, **B**, **C**, and **D** in the order of their affinity, the measurements listed in Table I reveal that the order on the antibody surface of the first three must be **C–A–B**. Consider: Gal→Me must fill the subsite possessing the highest affinity. We label this subsite **A** (because even the monosaccharide shows ligand-induced tryptophanyl fluorescence change, this subsite has a perturbable tryptophan residue associated with itself). As Gal→Gal→Me binds some 40 times stronger, and shows enhanced ligand-induced protein fluorescence change (from 26 to 41%), it must involve the second perturbable tryptophanyl residue, which is associated with subsite **B**. The position of the two subsites **A** and **B** and this ligand could be either

Gal→Gal→Me or Gal→Gal→Me
A **B** **B** **A**

Because Gal→Gal→Me and Gal→3FGal→Me show the same affinity constant, as well as produce essentially the same (41, 46%) maximal change in antibody fluorescence, their binding mode is taken to be the same. The subsite binding the monosaccharide, that is, possessing the highest subsite affinity, **A**, apparently requires a hydrogen bond donation involving the 3-OH group of the galactosyl residue capable of filling this subsite, since 3FGal→Me does not bind to the **A** subsite. Thus the correct mode must be

Gal→Gal→Me or Gal→3FGal→Me
A **B** **A** **B**

thereby avoiding the possibility that the 3FGal moiety would bind in the **A** subsite. The increased fluorescence induced by binding of these disaccharides indicates the involvement of a tryptophanyl residue near the **B** subsite. It then follows that disaccharide 3FGal→Gal→Me must bind as

3FGal→Gal→Me
A **B**

which agrees with the observation that it perturbs only one tryptophanyl residue (ΔF_{max} 24%).

The increased affinity of 3FGal→Gal→Me, when compared to that of Gal→Me, must be essentially the affinity of another subsite (**C**), located on the other side of **A**, capable of accommodating another galactosyl residue (and apparently a 3-deoxy-3-fluoro group would not greatly affect the binding in that subsite **C**). Thus, the arrangement becomes

3FGal→Gal→Me
C **A** **B**

The additional affinity for subsite **C**, so measured, is found to be close to the additional affinity found for Gal→Gal→Gal→Me compared to that of Gal→Gal→Me. Thus the arrangement of the subsites is indeed

Gal→Gal→Gal→Me
C **A** **B**

This is entirely consistent with the fact that 3FGal→Gal→Gal→Me shows the same maximal ligand-induced fluorescence change, and essentially the same affinity, as does Gal→Gal→Gal→Me. As it is clear from Table I that the entire maximally binding determinant is the tetrasaccharide, there is to be a fourth subsite. It was subsequently found that subsite **D** was situated next to **B** (data not shown here). All observations were confirmed by affinity determinations with oligosaccharides containing deoxygalactosyl residues[21] (data not shown). For further reports on studies of many antibodies of this family using this approach, the reader is referred to a comprehensive review.[22]

Syntheses of Some Glucosyl Ligands and Studies on Antidextran Monoclonal Antibodies

Methyl O-(6-Deoxy-6-fluoro-α-D-glucopyranosyl)-(1→6)-α-D-glucopyranoside

The disaccharide methyl *O*-(6-deoxy-6-fluoro-α-D-glucopyranosyl)-(1→6)-α-D-glucopyranoside (6FGlc→Glc→Me) is prepared by the condensation of 2,3,4-tri-*O*-benzyl-6-deoxy-6-fluoro-β-D-glucopyranosyl fluoride with methyl 2,3,4-tri-*O*-benzyl-α-D-glucopyranoside, followed by deblocking of the product[23] (Scheme 5). 2,3,4,6-Tetra-*O*-benzyl-D-glucose (Pfanstiehl Chemical Co., Waukeegan, IL, 50 g) is added to a solution of acetic acid (100 ml), acetic anhydride (300 ml), and concentrated sulfuric acid (1.5 ml), and the mixture is swirled until all reagents have dissolved (~1 min), after which it is left for 20 min at room temperature. Sodium acetate

[21] T. Ziegler, V. Pavliak, T.-H. Lin, P. Kováč, and C. P. J. Glaudemans, *Carbohydr. Res.* **204,** 167 (1990).

[22] C. P. J. Glaudemans, *Chem. Rev.* **91,** 25 (1991).

[23] P. Kováč, V. Sklenář, and C. P. J. Glaudemans, *Carbohydr. Res.* **175,** 201 (1988).

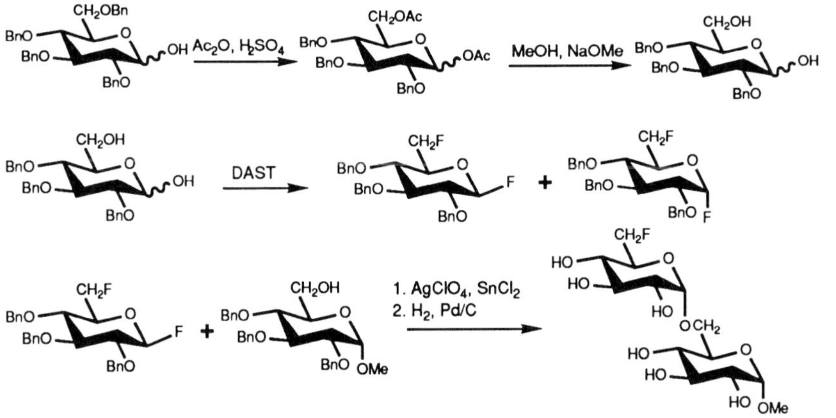

SCHEME 5. Preparation of methyl O-(6-deoxy-6-fluoro-α-D-glucopyranosyl)-(1→6)-α-D-glucopyranoside. Bn, Benzyl; DAST, diethylaminosulfur trifluoride.

trihydrate (20 g) is added, and the mixture is concentrated to dryness by azeotropic distillation with toluene. The residue is taken up in dichloromethane, dried (anhydrous sodium sulfate), filtered and concentrated. Column chromatography on silica gel (carbon tetrachloride–acetone) gives, as a first fraction, 29 g of a syrupy material which partially crystallizes on standing for 1 week. Recrystallization from methanol gives 1,6-di-O-acetyl-2,3,4-tri-O-benzyl-α-D-glucopyranose (18 g), mp 68°–69°, $[\alpha]_D^{25}$ +61° (c 1, chloroform). The second fraction (1.2 g) consists mostly of the corresponding β anomer, 1,6-di-O-acetyl-2,3,4-tri-O-benzyl-β-D-glucopyranose, mp 63°, $[\alpha]_D^{25}$ +17.4° (after recrystallization from ethanol). Deacetylation of either anomer yields 2,3,4-tri-O-benzyl-D-glucose in quantitative yield as a mixture of α and β anomers, mp 98°–100°, with the melting point varying depending on the ratio of anomers in the mixture.

Either anomer or a mixture of anomers of the above, 2,3,4-tri-O-benzyl-D-glucose (0.9 g, 2 mmol) in 1,2-dimethoxyethane (10 ml) containing triethyl amine (1 ml) is treated dropwise at 0° with diethylaminosulfur trifluoride (DAST, 1.2 ml, Aldrich Chemical Co., Milwaukee, WI). The mixture is warmed and held at 70° for 1 hr. It is then cooled to −10°, and a few milliliters of methanol is added, followed by the addition of solid sodium bicarbonate. The mixture is concentrated, and the residue is partitioned between dichloromethane and aqueous, saturated sodium chloride solution. The organic phase is dried and concentrated. The residue is chromatographed on silica gel using toluene–ethyl acetate containing 0.1% pyridine. Two major products are obtained in a combined yield of around 60%: 2,3,4-tri-O-benzyl-6-deoxy-6-fluoro-α-D-glucopyranosyl fluoride,

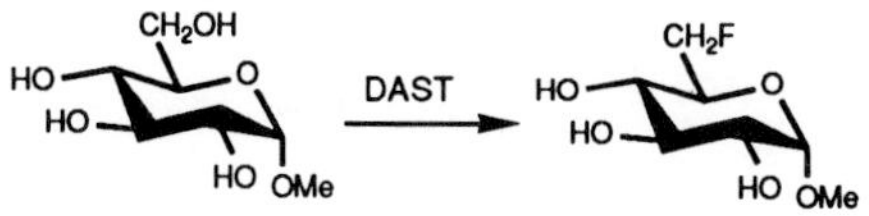

SCHEME 6. Preparation of methyl 6-deoxy-6-fluoro-α-D-glucopyranoside.

mp 88°–89° (from ethanol), $[\alpha]_D^{25}$ −3.5° (*c* 0.5, chloroform), and 2,3,4-tri-*O*-benzyl-6-deoxy-6-fluoro-β-D-glucopyranosyl fluoride, mp 101°, $[\alpha]_D^{25}$ +23° (*c* 0.8, chloroform).

A solution of silver perchlorate in ether (80 m*M*, 15 ml, 1.2 mmol) is added at −20° to a stirred solution of methyl 2,3,4-tri-*O*-benzyl-α-D-glucopyranoside (0.46 g, 1 mmol),[24,25] 2,3,4-tri-*O*-benzyl-6-deoxy-6-fluoro-β-D-glucopyranosyl fluoride (0.45 g, 1 mmol), and stannous chloride (190 mg, 1 mmol) in ether (5 ml) containing molecular sieves 4 Å (2 g). The mixture is stirred for 30 min while it is allowed to warm to room temperature. Thin-layer chromatography (carbon tetrachloride–ethyl acetate) shows that two main products are formed. The reaction is neutralized (2,4,6-trimethylpyridine), then filtered through Celite, and the filtrate is concentrated and chromatographed, to first give the desired α-linked disaccharide methyl *O*-(2,3,4-tri-*O*-benzyl-6-deoxy-6-fluoro-α-D-glucopyranosyl)-(1→6)-2,3,4-tri-*O*-benzyl-α-D-glucopyranoside (560 mg, 62%), mp 129°–130° (from isopropyl ether–ethanol), $[\alpha]_D^{25}$ +70.3°. Continued elution then gives the β-linked disaccharide methyl *O*-(2,3,4-tri-*O*-benzyl-6-deoxy-6-fluoro-β-D-glucopyranosyl)-(1→6)-2,3,4-tri-*O*-benzyl-α-D-glucopyranoside (145 mg, 16%), mp 162°–164°, $[\alpha]_D^{25}$ +18°.

A solution of the above methyl *O*-(2,3,4-tri-*O*-benzyl-6-deoxy-6-fluoro-α-D-glucopyranosyl)-(1→6)-2,3,4-tri-*O*-benzyl-α-D-glucopyranoside (0.85 g) in 2-methoxyethanol (50 ml) is stirred in a hydrogen atmosphere at room temperature and atmospheric pressure in the presence of a 5% palladium-on-charcoal catalyst (0.3 g) until the uptake of hydrogen ceases (~2 hr). Thin-layer chromatography (ethyl acetate–propanol–water) shows that a single product is formed. The mixture is filtered, the catalyst is washed with the solvent, and the filtrate is concentrated to dryness. Elution of the residue from a small column of silica gel with ethyl acetate–propanol–water gives, in nearly theoretical yield, methyl *O*-(6-deoxy-6-fluoro-α-D-glucopyranosyl)-(1→6)-α-D-glucopyranoside, $[\alpha]_D^{25}$ +166°.

Methyl 6-deoxy-6-fluoro-α-D-glucopyranoside (6FGlc→Me) is prepared (Scheme 6) by the method of Card.[26] Methyl 6-deoxy-α-D-glucopyranoside (6dGlc→Me) is commercially available (Sigma).

[24] R. Eby and C. Schuerch, *Carbohydr. Res.* **34,** 79 (1974).
[25] P. Kováč, J. Alfödi, and M. Košík, *Chem. Zvesti* **28,** 820 (1974).
[26] P. Card, *J. Org. Chem.* **48,** 393 (1983).

TABLE II
AFFINITY CONSTANTS FOR 16.4.12E WITH CERTAIN LIGANDS

Ligand	K_a	ΔF_{max} (%)
Glc→Me	4.5×10^3	+51
Glc_2→Me	5.2×10^4	+45
Glc_3→Me	8.7×10^4	+41
Glc_4→Me	3.9×10^5	+47
Glc_5→Me	2.4×10^5	+47
6dGlc→Me	0	0
6FGlc→Me	30	+47
6FGlc→Glc→Me	3.6×10^2	+40

Studies on Antidextran Antibodies

The anti-α-(1→6)-dextran IgA antibody[27] 16.4.12E is purified by affinity chromatography using Sephadex G-150 as the immunoadsorbent and (predialyzed) methyl α-D-glucopyranoside (0.2 *M*) as the eluant. The ligand is removed from the purified immunoglobulin in the same way as for the antigalactan antibody.[3]

Table II lists the affinity constants and the maximal ligand-induced antibody fluorescence change for the glucosyl ligands whose synthesis is described above. In addition, these values are also given for the series of α-(1→6)-linked glucosyl oligosaccharides up to and including the pentasaccharide (Glc_5→Me). It can be seen that the antibody maximally binds the isomaltotetrasaccharide, and in that series all saccharides, from mono- to oligosaccharides, show the same ligand-induced fluorescence change. That means that the monosaccharide already perturbs every perturbable tryptophanyl residue in the antibody-combining area, that is, they most probably are near the highest affinity subsite. The structure of IgA 16.4.12E shows three such residues, at positions 33H, 100^AH, and 96L, located in space near one another in a deep cavity at the H/L interface. This may account for the high ligand-inducible fluorescence change of that antibody.[28] Notice that 6dGlc→Me shows no binding and 6FGlc→Me binds only marginally, approximately 100 times less than the regular monosaccharide Glc→Me. It must be kept in mind that the nonreducing terminal glucosyl residue in a (1→6)-linked dextran is the only residue bearing a free 6-OH group. Thus, the fact that the removal of hydrogen bonding potential at the 6-OH position of the residue binding in the highest binding subsite leads to

[27] T. Matsuda and E. A. Kabat, *J. Immunol.* **142,** 863 (1989).

cessation of its binding shows that residue to be critically involved in hydrogen bonding, apparently hydrogen bond donation.

The free energies of binding of the four subsites **A**, **B**, **C**, and **D** were shown to be additive to give the binding free energy of the entire determinant.[3,28] Thus, subsite **B** has a free energy of binding its glucosyl residue given by $\Delta G_{\text{subsite } \mathbf{B}} = \Delta G_{\text{Glc2}\rightarrow\text{Me}} - \Delta G_{\text{Glc}\rightarrow\text{Me}}$, and so on. It can be seen that the free energy of binding of the subsite to the 6FGlc residue in 6FGlc→Glc→Me is given by $\Delta G_{\text{6FGlc}\rightarrow\text{Me}} = \Delta G_{\text{6FGlc}\rightarrow\text{Glc}\rightarrow\text{Me}} - \Delta G_{\text{subsite } \mathbf{B}}$. Thus the derived partial affinity of the 6FGlc residue in that disaccharide is computed as 31 M^{-1}, which is remarkably close to the value found directly for 6FGlc→Me (30 M^{-1}). For more detailed discussions of the binding of dextrans to this and other antibodies, the reader is referred to a more comprehensive report.[28]

[28] E. M. Nashed, G. R. Perdomo, E. A. Padlan, P. Kováč, T. Matsuda, E. A. Kabat, and C. P. J. Glaudemans, *J. Biol. Chem.* **265,** 20699 (1990).

Section IV

Biomedical Applications

[22] Carbohydrate–Lysyllysine Conjugates as Cell Antiadhesion Agents

By TATSUSHI TOYOKUNI and SEN-ITIROH HAKOMORI

Introduction

The molecular basis of cell–cell adhesion provides a fundamental mechanism for the formation of multicellular organisms. This mechanism is also essential for understanding a wide variety of biological processes, including fertilization, differentiation, cellular immune response, cell migration, and pathogen infectivity. Improper functioning of cell recognition underlies oncogenesis, tumor metastasis, and inflammatory diseases. The accumulating evidence suggests that cell surface carbohydrates are the specific determinants in intercellular interactions.[1] Carbohydrate-binding proteins known as lectins or selectins have been described as cell surface mediators for cell–cell interactions,[2] and findings identifying oligosaccharides as ligands for selectins, a family of cell adhesion molecules involved in leukocyte trafficking and recruitment to sites of inflammation, have substantiated the potential use of carbohydrates as cell adhesion inhibitors.[3–5] We have shown that carbohydrate–carbohydrate interactions can also serve as a basis for cell recognition.[6,7]

Cell–cell interactions are composed of a number of cooperative interactions. Thus, inhibitors must be designed to possess multivalency so that they are capable of multisite binding to the receptors on cells. A variety of methods is available for constructing multivalent structures. Some of the frequently used methods include conjugation of carbohydrate ligands with proteins (i.e., neoglycoproteins) or other polymers,[8,9] polymerization of glycosylated monomers (i.e., glycopolymers),[10,11] and formation of lipo-

[1] S. Hakomori, *Annu. Rev. Biochem.* **50,** 733 (1981).
[2] N. Sharon and H. Lis, *Science* **246,** 227 (1989).
[3] J. L. Winkelhake, *Glycoconjugate J.* **8,** 381 (1991).
[4] S. Hakomori, *Cancer Cells* **3,** 461 (1991).
[5] K.-A. Karlsson, *Trends Pharmacol. Sci.* **12,** 265 (1991).
[6] I. Eggens, B. Fenderson, T. Toyokuni, B. Dean, M. Stroud, and S. Hakomori, *J. Biol. Chem.* **264,** 9476 (1989).
[7] S. Hakomori, *Pure Appl. Chem.* **63,** 473 (1991).
[8] C. P. Stowell and Y. C. Lee, *Adv. Carbohydr. Chem. Biochem.* **37,** 225 (1980).
[9] J. H. Pazur, *Adv. Carbohydr. Chem. Biochem.* **39,** 405 (1981).
[10] V. Horejsi, P. Smolek, and J. Kocourek, *Biochim. Biophys. Acta* **538,** 293 (1978).
[11] R. Roy, F. D. Tropper, and A. Romanowska, *Bioconjugate Chem.* **3,** 256 (1992).

somes using glycolipids or neoglycolipids.[6,12] Although these approaches have been successful, the products are ambiguous in composition and structure. In the following section we describe the use of commercially available L-lysyl-L-lysine (**1**) (Fig. 1) to construct chemically well-defined trivalent conjugates and their application as cell antiadhesion agents. The three amino groups of **1** (one α- and two ε-amino groups) are utilized to introduce three carbohydrate residues.

Trivalent Lacto-*N*-Fucopentaose III–Lysyllysine Conjugate as Anticompaction Agent

Carbohydrate structures of glycolipids and glycoproteins change during embryogenesis.[13] The Lewis X trisaccharide [Le^x, Galβ1→4(Fucα1→3)GlcNAcβ1→R], also known as stage-specific embryonic antigen 1, is highly expressed at the morula stage of the mouse embryo, the beginning of embryo compaction. The trivalent conjugate of the Le^x oligosaccharide [lacto-*N*-fucopentaose III (LNF III), Galβ1→4(Fucα1→3)GlcNAcβ1→3Gal→4Glc] with lysyllysine (**2**) effectively inhibits the compaction process and disrupts embryogenesis (Fig. 2).[14] On the other hand, free LNF III and the trivalent Le^a oligosaccharide [LNF II, Galβ1→3(Fucα1→4)GlcNAcβ1→3Galβ1→4Glc] **3**, a positional isomer of LNF III, have no effect.

The conjugation method is based on reductive amination of reducing oligosaccharides by sodium cyanoborohydride ($NaBH_3CN$),[15,16] which at pH 6–8 reduces the iminium ion faster than a carbonyl group.[17] Although this method destroys a ring structure of the reducing end, prior modification of oligosaccharides is not required.

Materials. LNF II and LNF III can be purified from human milk by pretreatment with ethanol, followed by repeated BioGel P-2 (Bio-Rad, Richmond, CA) column chromatography with water as the eluent[18] and reversed-phase (C_{18}) high-performance liquid chromatography (HPLC) with water.[19] They are also available from Accurate Chemical and Scientific Corp. (Westbury, NY). L-Lysyl-L-lysine dihydrochloride is obtained

[12] M. S. Stoll, T. Mizuochi, R. A. Childs, and T. Feizi, *Biochem. J.* **256,** 661 (1988).

[13] B. A. Fenderson, E. M. Eddy, and S. Hakomori, *BioEssays* **12,** 173 (1990).

[14] B. A. Fenderson, U. Zehavi, and S. Hakomori, *J. Exp. Med.* **160,** 1591 (1984).

[15] B. A. Schwartz and G. R. Gray, *Arch. Biochem. Biophys.* **181,** 542 (1977).

[16] R. Roy, E. Katzenellenbogen, and H. J. Jennings, *Can. J. Biochem. Cell Biol.* **62,** 270 (1983).

[17] R. F. Borch, M. D. Bernstein, and H. D. Durst, *J. Am. Chem. Soc.* **93,** 2897 (1971).

[18] A. Kobata, this series, Vol. 28, p. 262.

[19] V. K. Dua and C. A. Bush, *Anal. Biochem.* **133,** 1 (1983).

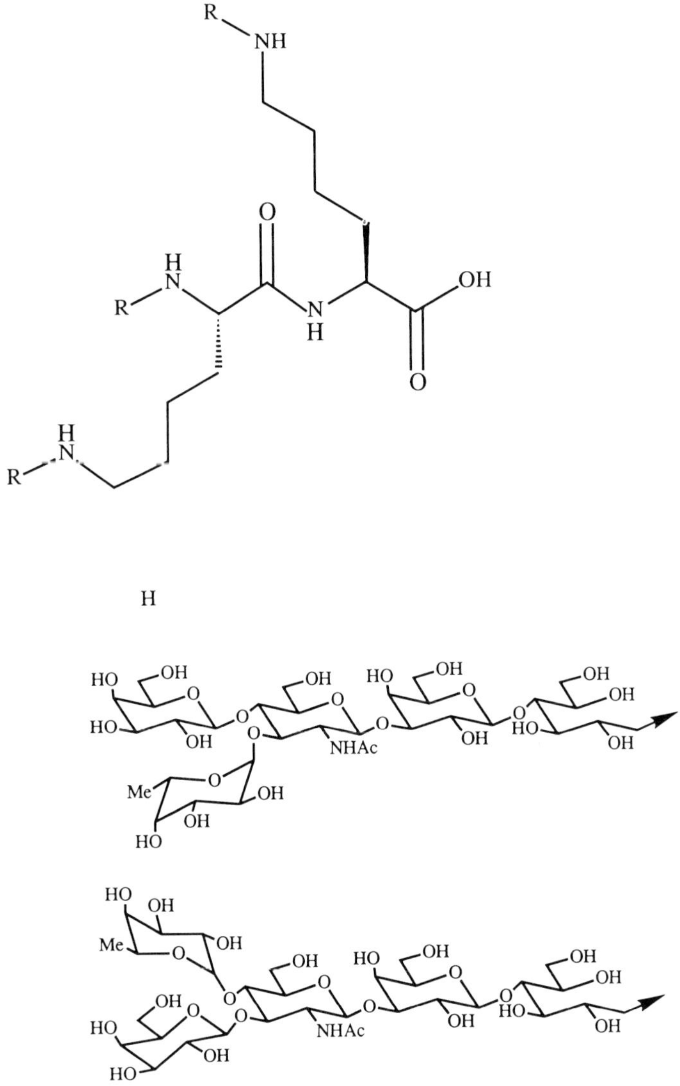

FIG. 1. Structure of L-lysyl-L-lysine (**1**) and conjugates (**2** and **3**).

from Bachem Bioscience (Philadelphia, PA). All other reagents and solvents can be purchased from common commercial suppliers and used without further purification. The structures of the conjugates are characterized by ^1H nuclear magnetic resonance (NMR) spectroscopy and fast

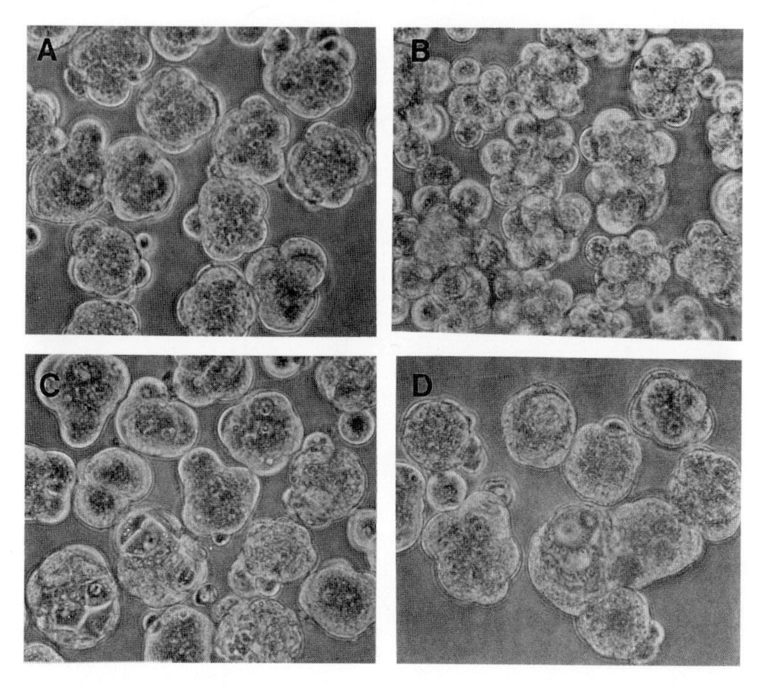

FIG. 2. Compacted 8- to 16-cell mouse embryos (A) were incubated with 1 mM of trivalent LNF III–lysyllysine conjugate (**2**) (B), 1 mM of trivalent LNF II–lysyllysine conjugate (**3**) (C), or 5 mM of free LNF III plus 1 mM of lysyllysine (**1**) (D) for 12 hr × 150. [Adapted, with permission, from Fenderson *et al.*[14] Copyright (1984) The Rockefeller University Press].

atom bombardment-mass spectrometry (FAB-MS). For collection of embryos, Swiss Webster mice are purchased from Tyler Labs (Seattle, WA).

*Lacto-N-fucopentaose III–Lysyllysine Conjugate (**2**).* A mixture of LNF III, $NaBH_3CN$, and **1** in a molar ratio of 0.35 : 0.033 : 1.00 in 0.2 M potassium phosphate (pH 8) is stirred at 37° for 72 hr. The reaction mixture is acidified (pH ~2) by addition of 0.1 M HCl. The acidification must be performed in a well-ventilated hood owing to the evolution of HCN. The mixture is then placed on a BioGel P-2 column and eluted with water. The conjugate **2**, which is not reactive with fluorescamine or ninhydrin, is eluted from the column immediately after the void volume and identified in a spot test using an orcinol spray. The LNF II–lysyllysine conjugate (**3**) is prepared in a similar way.

Preimplantation Embryo Culture. Four- to eight-cell embryos are collected by flushing from the oviducts of Swiss Webster mice on day 3 postcoitum. The precompaction embryos are incubated overnight in Whitten's medium under mineral oil on 50-mm petri dishes at 37° in a 5% CO_2-in-air humidified incubator. Compacted 8- to 16-cell embryos, thus obtained, are transferred briefly into Tyrode's salt solution containing 0.4% polyvinylpyrrolidone (pH 2.5) to remove zonae pellucidae and used for the compaction assay.

Compaction Assay. The zona-free embryos are transferred to 60-μl droplets of Whitten's medium containing free LNF III (5 m*M*), **2** (1 m*M*), or **3** (1 m*M*). After incubation at 37° for 12 hr, effects on compaction are photographed using a Polaroid camera attached to a Nikon inverted-phase microscope (Nikon, Japan).

Trivalent Oligosaccharide–[^{3}H]Lysyllysinol Conjugate as Probe of Carbohydrate–Carbohydrate Interaction

A search for an Le^x-binding molecule involved in mouse embryonic cell adhesion described above has led us to hypothesize carbohydrate–carbohydrate (CH_2O–CH_2O) interactions,[6] that is, the Le^x-recognition molecule per se is Le^x. Since then a number of CH_2O–CH_2O interactions have been identified[4] based on the aggregation of glycosphingolipid (GSL) liposomes, the adherence of radiolabeled liposomes containing GSLs to a solid phase coated with other GSLs, or affinity chromatography of trivalent oligosaccharide–[^{3}H]lysyllysinol[20] conjugates on a GSL-bound C_{18} column (Fig. 3).

Indeed, an Le^x-bound column shows specific affinity with the trivalent LNF III–[^{3}H]lysyllysinol conjugate (**5**) in the presence of Ca^{2+} but not with trivalent lactose conjugate (**7**) (Fig. 4A).[7] Likewise the trivalent α(2→3)-sialosyllactose (G_{M3} trisaccharide) conjugate (**6**) is adsorbed specifically on a G_{g3} (GalNAcβ1→4Galβ1→4Glcβ1→1Cer)-bound column (Fig. 4B).[21] In either case the adsorbed conjugate is eluted with an EDTA-containing medium. Interaction between G_{g3} and G_{M3} (NeuAcα2→3Galβ1→4Glcβ1→1Cer) has been suggested to be the basis of the specific interaction of the mouse lymphoma L5178 AA12 cell line (high expresser of G_{g3}) and the mouse B16 melanoma cell line (high expresser of G_{M3}).[22]

Materials. Gangliosides G_{M3} and G_{g3} are prepared from dog and guinea pig, respectively, as previously described.[23] α(2→3)-Sialosyllacatose is isolated from human milk by paper chromatography.[18] Di-*tert*-butyl dicar-

[20] Lysyllysinol is (*S*)-N^2-L-lysyl-2,6-diamino-1-hexanol.

[21] N. Kojima and S. Hakomori, *J. Biol. Chem.* **266,** 17552 (1991).

[22] N. Kojima and S. Hakomori, *J. Biol. Chem.* **264,** 20159 (1989).

[23] S. Hakomori, *in* "Sphingolipid Biochemistry" (J. Kanfer and S. Hakomori, eds.), p. 1. Plenum, New York, 1983.

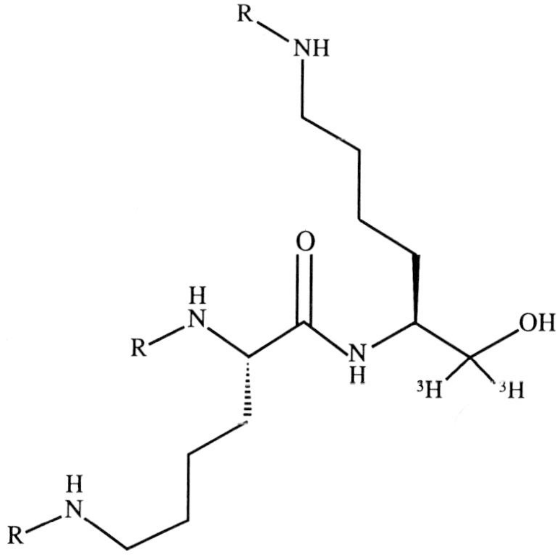

4 R= H

5 R=

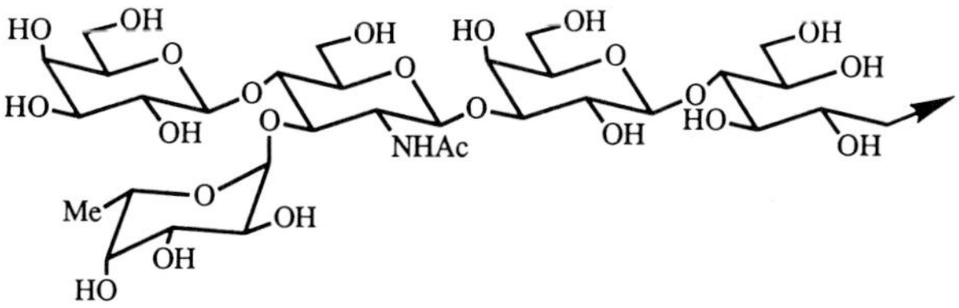

6 R=

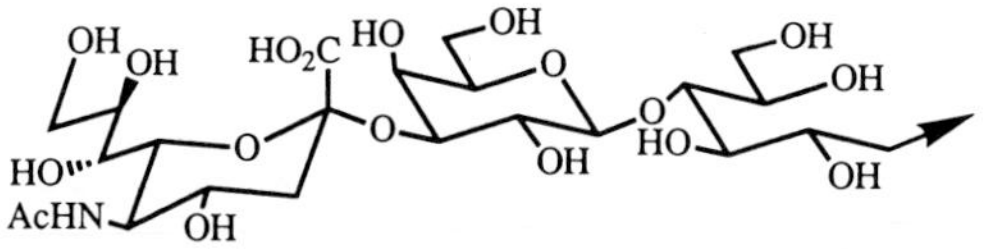

7 R=

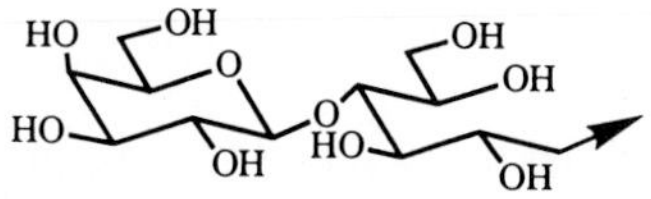

FIG. 3. Structures of [^{3}H]lysyllysinol (**4**) and oligosaccharide–lysyllysinol conjugates (**5–7**).

bonate, *N*-hydroxysuccinimide (NHS), and *N*-ethyl-*N'*-(3-dimethylamino-propyl)carbodiimide (EDC) are purchased from Aldrich (Milwaukee, WI). The products are characterized by ^{1}H NMR spectroscopy and FAB-MS.

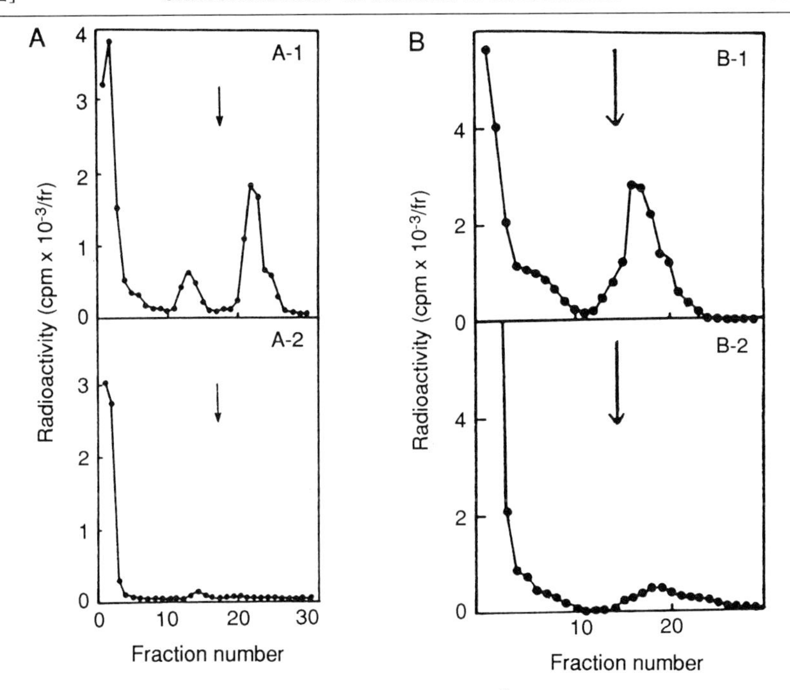

FIG. 4. (A) Affinity chromatography of trivalent Le^x–[^{3}H]lysyllysinol conjugate (**5**) (A-1) or trivalent lactose–[^{3}H]lysyllysinol conjugate (**7**) (A-2) on a Le^x glycolipid-bound column. The column was eluted with Tris-buffered saline containing 1 m*M* $CaCl_2$ and 150 m*M* NaCl. (B) Affinity chromatography of trivalent $\alpha(2\rightarrow3)$-sialosyllactose–[^{3}H]lysyllysinol (**6**) (B-1) or **7** (B-2) on a G_{g3} glycolipid-bound column. The column was eluted with Tris-buffered saline containing 1 m*M* $CaCl_2$ and 0.5 m*M* $MgCl_2$. Arrows indicate points where eluents were changed to Tris-buffered saline containing 5 m*M* EDTA. [(A) Adapted, with permission, from Hakomori.[7] Copyright (1991) International Union of Pure and Applied Chemistry. (B) Adapted, with permission, from Kojima and Hakomori.[21] Copyright (1991) The American Society for Biochemistry and Molecular Biology, Inc.]

Preparation of Trivalent Oligosaccharide–[^{3}H]Lysyllysinol Conjugates

Tritium is readily introduced to the lysyllysine backbone by converting the C-terminal carboxyl group to a primary hydroxyl group through a succinimidyl ester[24] with NaB^3H_4 (Scheme 1).

*N,N′,N″-Tris(tert-butoxycarbonyl)-L-lysyl-L-lysine (**11**).* A solution of di-*tert*-butyl dicarbonate (3.7 g, 17 mmol) in 1,4-dioxane (5 ml) is added dropwise to a solution of **1** dihydrochloride (1.7 g, 4.9 mmol) in water (20 ml) containing $NaHCO_3$ (1.2 g, 14.6 mmol) stirred at room temperature. After stirring overnight, the mixture is acidified (pH ~2) by the addition of 0.1 *M* HCl and extracted with ethyl acetate (3 times, 50 ml). The ethyl

[24] J. Nikawa and T. Shiba, *Chem. Lett.*, 981 (1979).

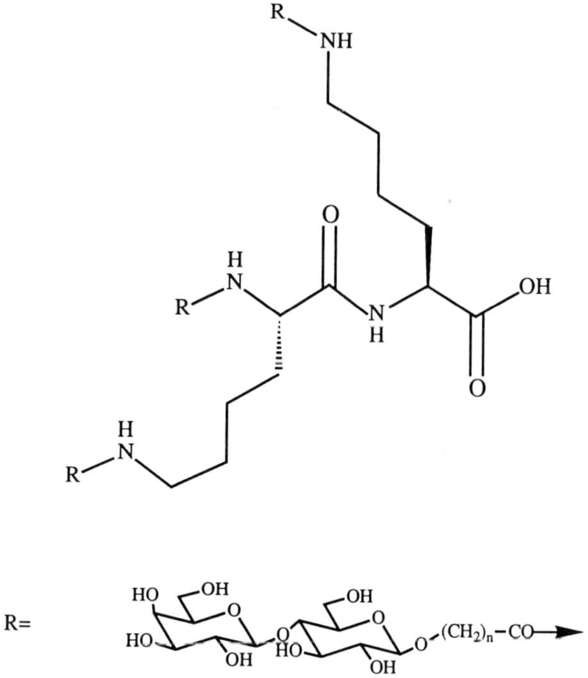

R=

8	n=1
9	n=3
10	n=8

FIG. 5. Structure of β-lactoside–lysyllysine conjugates (**8–10**).

acetate layer is dried (Na_2SO_4) and concentrated to dryness. The residue is dissolved in CH_2Cl_2 (10 ml), and the product is precipitated by addition of hexane. After decantation, **11** is obtained as a colorless gummy syrup (2.1 g, 77%); R_f 0.3 [5 : 1 (v/v) $CHCl_3$–methanol]; $[\alpha]_D$ −3.7° (*c* 1.5, methanol).

*[^{3}H]Lysyllysinol (**4**).* The reagent EDC (25 mg, 130 μmol) is added portionwise to a cloudy solution of **11** (50 mg, 87 μmol) and NHS (13 mg, 113 μmol) in tetrahydrofuran (THF) (3 ml). The mixture is stirred at room temperature for 4 hr and filtered to remove a gummy precipitate. The filtrate is evaporated (<30°) to dryness, and the residue is dissolved in CH_2Cl_2 (5 ml). The CH_2Cl_2 layer is washed with water (3 times, 5 ml), dried (Na_2SO_4), and concentrated to give the succinimidyl ester **12** as a colorless syrup [>95% purity on thin-layer chromatography (TLC)]: R_f 0.7 [5 : 1 (v/v) $CHCl_3$–methanol]; UV-positive.

Next, NaB^3H_4 (50 mCi) is added to a solution of **12** in THF (3 ml) with vigorous stirring at room temperature. After 1 hr, $NaBH_4$ (2 mg) is added, and the mixture is stirred for an additional 1 hr. The mixture is poured into 10% aqueous citric acid (5 ml) and extracted with ethyl acetate (4 times, 3 ml). The ethyl acetate layers are combined, washed with saturated aqueous $NaHCO_3$ (5 ml) and water (5 ml), dried (Na_2SO_4), and concentrated to give the alcohol **13**. Treatment of **13** with trifluoroacetic acid (TFA) (3 ml) at room temperature for 10 min, followed by precipitation by the addition of diethyl ether, yields the TFA salt of **4** [10 mg, 1×10^7 counts/min (cpm)].

Trivalent oligosaccharide–[^{3}H]lysyllysinol conjugates **5**, **6**, and **7** are prepared in a similar procedure as described for the preparation of **2**.

Affinity Chromatography

Preparation of Glycosphingolipid-Bound C_{18} Column. The GSL (Le^x or G_{g3}) (15 mg) is dissolved in warm water (5 ml) and applied to a C_{18} column (0.6 × 0.8 cm). After washing with water (2 ml), the column is equilibrated with Tris-buffered saline (pH 7.4) containing 1 m*M* $CaCl_2$ and 150 m*M* NaCl for Le^x, or with Tris-buffered saline (pH 7.4) containing 1 m*M* $CaCl_2$ and 0.5 m*M* $MgCl_2$ for G_{g3}. Based on the recovery of GSL, approximately 5 mg of GSL is adsorbed on the column.

*Affinity Chromatography of **5** on Le^x-Bound Column.* A solution of **5** (15,000 cpm/100 nmol) in Tris-buffered saline (pH 7.4) containing 1 m*M* $CaCl_2$ and 150 m*M* NaCl is applied to the Le^x-bound column. The column is eluted with the same buffer (7.5 ml), followed by Tris-buffered saline containing 5 m*M* EDTA (7.5 ml). Fractions (0.5 ml) are collected, and aliquots (100 μl) are counted in a scintillation counter.

*Affinity Chromatography of **6** on G_{g3}-Bound Column.* The experiment is carried out in a similar manner except that Tris-buffered saline containing 1 m*M* $CaCl_2$ and 0.5 m*M* $MgCl_2$ is used.

Trivalent β-Lactoside–Lysyllysine Conjugates as Antimetastasis Agents

Tumor cell adhesion mediated by cell surface carbohydrates has been suggested to play the key steps in tumor cell metastasis. The adhesion could be based on carbohydrate–lectin (including selectins)[25,26] or carbohydrate–carbohydrate interactions.[21] In either case carbohydrates are po-

[25] H. J. Gabius and S. Gabius, *Naturwissenschaften* **77,** 505 (1990).

[26] S. D. Hoff, T. Irimura, Y. Matsushita, D. M. Ota, K. R. Cleary, and S. Hakomori, *Arch. Surg.* **125,** 206 (1990).

tential candidates[4,27–30] for prevention of metastatic spread by disrupting the requisite carbohydrate-initiated interactions between tumor cells and other cells such as endothelial cells and platelets.

We have shown that methyl β-lactoside significantly reduces the formation of lung tumor colonies in mice injected with mouse B16 melanoma cells.[30] Follow-up studies have suggested that the inhibitory effect of methyl β-lactoside might be due to competitive blocking of the interaction between LacCer (Galβ1→4Glcβ1→1Cer), expressed on lung capillary endothelial cells, and G_{M3}, expressed on B16 melanoma cells.[21,31] To increase the efficacy of β-lactoside, trivalent β-lactoside conjugates with lysyllysine (**8–10**) have been synthesized (Fig. 5).[27] Interestingly, the inhibitory activity of the conjugates decreases with lengthening of the spacer arm, indicating the importance of the spatial arrangement of the lactoside for binding affinities (Fig. 6). The conjugation method has two key steps: introduction of a spacer arm to lactose via a β-linkage and conjugation with **1** by the NHS ester method. A similar methodology has been employed to synthesize tumor-associated Tn antigen (GalNAcα1→OSer/Thr) clusters.[32]

Materials. The following reagents are obtained from Aldrich: azelaic acid monomethyl ester, methyl glycolate, 1,4-butyrolactone, trimethylsilyl trifluoromethanesulfonate (TMSOTf), hydrazine acetate, tricholoacetonitrile, and a 1.0 *M* solution of borane (BH_3)–THF complex in THF. Silica gel used for flash column chromatography[33] (FCC) is purchased from EM Science (Gibbstown, NJ). The identity of the products is verified by ^1H NMR spectroscopy and FAB-MS.

Preparation of Spacer Arms

The longest spacer arm **16** (Fig. 7), developed by Lemieux *et al.*,[34] has been frequently used to prepare artificial carbohydrate antigens via covalent attachment to carrier proteins.

[27] B. Dean, H. Oguchi, S. Cai, E. Otsuji, K. Tashiro, S. Hakomori, and T. Toyokuni, *Carbohydr. Res.* **245,** 175 (1993).

[28] J. Beuth, H. L. Ko, V. Schirrmacher, G. Uhlenbruck, and G. Pulverer, *Clin. Exp. Metastasis* **6,** 115 (1988).

[29] D. Platt and A. Raz, *J. Natl. Cancer Inst.* **84,** 438 (1992).

[30] H. Oguchi, T. Toyokuni, B. Dean, H. Ito, E. Otsuji, V. L. Jones, K. K. Sadozai, and S. Hakomori, *Cancer Commun.* **2,** 311 (1990).

[31] N. Kojima, M. Shiota, Y. Sadahira, K. Handa, and S. Hakomori, *J. Biol. Chem.* **267,** 17264 (1992).

[32] T. Toyokuni, B. Dean, and S. Hakomori, *Tetrahedron Lett.* **31,** 2673 (1990).

[33] W. C. Still, M. Kahn, and A. Mitra, *J. Org. Chem.* **43,** 2923 (1978).

[34] R. U. Lemieux, D. R. Bundle, and D. A. Baker, *J. Am. Chem. Soc.* **97,** 4076 (1975).

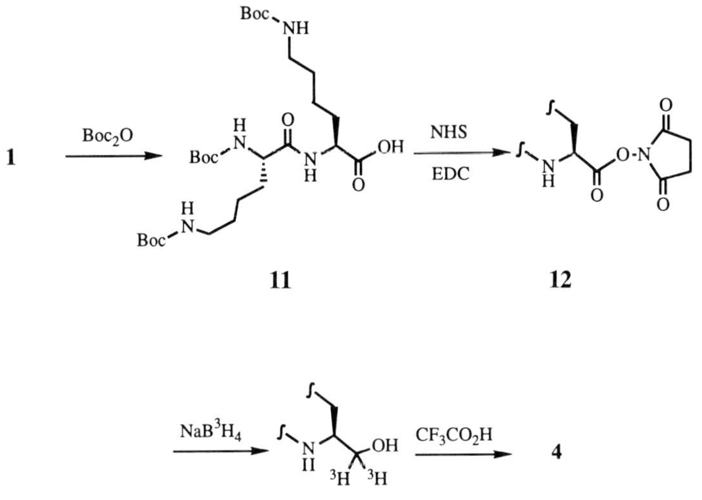

Boc = *tert*-BuOCO

SCHEME 1

*Benzyl Hydroxyethanoate (**14**).* Methyl glycolate (1.0 g, 10.9 mmol) is dissolved in 0.1 *M* sodium phenylmethoxide in anhydrous benzyl alcohol (50 ml), and the mixture is stirred under dry N_2 at room temperature overnight. The base is neutralized with Amberlite IR-120 (H^+) resin, and a large portion of benzyl alcohol is removed by evaporation at 70°. The oily residue is purified by FCC [2 : 1 (v/v) hexane–ethyl acetate] to give **14** (1.0 g, 55%) as a colorless oil: R_f 0.2 [2 : 1 (v/v) hexane–ethyl acetate].

*Benzyl 4-Hydroxybutanoate (**15**).* 1,4-Butyrolactone (1.0 g, 11.5 mmol) is treated with 0.1 *M* sodium phenylmethoxide in ahydrous benzyl alcohol (50 ml) as described above. FCC [1 : 1 (v/v) hexane–ethyl acetate] yields **15** (1.18 g, 53%) as a colorless oil: R_f 0.1 [2 : 1 (v/v) hexane–ethyl acetate].

*Methyl 9-Hydroxynonanoate (**16**).* A 1.0 *M* solution of BH_3–THF complex in THF (80 ml) is added dropwise to a stirred solution of azelaic acid monomethyl ester (30 g, 0.15 mol) in anhydrous THF (75 ml) at −15° (ice–NaCl) under dry N_2. The addition is completed in about 20 min. The mixture is maintained at −15° for an additional 2 hr and at room temperature overnight. The mixture is diluted with ethanol (30 ml), and 80% aqueous acetic acid (30 ml) is added. After 30 min, the mixture is

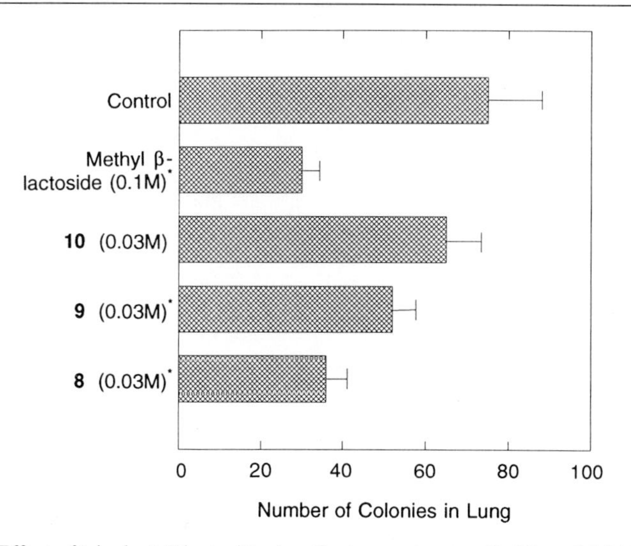

FIG. 6. Effect of trivalent β-lactoside–lysyllysine conjugates (**8–10**) on inhibition of lung tumor colonization. Error bars represent standard errors of the mean ($n = 6$), and asterisks indicate a significant difference ($p < 0.05$) as compared to the control. [Adapted with permission from Dean *et al.*[27] Copyright (1993) Elsevier Science Publishers B.V.]

concentrated, and the residue is suspended in water (150 ml). The solution is brought to approximately pH 7.5 by the addition of solid $NaHCO_3$. The mixture is extracted with diethyl ether (3 times, 150 ml). The combined extracts are washed with water (150 ml) and dried (Na_2SO_4). Filtration and concentration yield **16** (22 g, 79%) as a colorless oil.[34] The product sometimes contains a trace amount of the starting material, which can be removed by extensive washing with saturated aqueous $NaHCO_3$.

Introduction of Spacer Arms to Lactose

The trichloroacetimidate method, introduced by Schmidt and Michel,[35] is employed for introducing spacer arms to lactose (Scheme 2). The acidic character of this glycosylation medium reduces the formation of an orthoester, thus leading to a better yield of glycosides.

O-(2,3,4,6-Tetra-O-acetyl-β-D-Galactopyranosyl)-(1→4)-2,3,6-tri-O-Acetyl-D-Glucose (***17***). Hydrazine acetate (950 mg, 10.3 mmol) is added portionwise to a solution of lactose octaacetate (7.0 g, 10.3 mmol) in anhydrous *N,N*-dimethylformamide (100 ml) stirred at room temperature under dry N_2. After 4 hr, the mixture is diluted with ethyl acetate (1 liter)

[35] R. R. Schmidt and J. Michel, *Angew. Chem., Int. Ed. Engl.* **19,** 731 (1980).

HO—$(CH_2)_n$—CO—OR	**14**	n=1, R=Bn	Bn=benzyl
	15	n=3, R=Bn	
	16	n=8, R=Me	

FIG. 7. Structures of spacer arms, (**14–16**).

and washed with water (3 times, 300 ml). The ethyl acetate layer is dried (Na_2SO_4) and concentrated to dryness. Purification by FCC [3 : 1 (v/v) $CHCl_3$–acetone] yields **17** (5.5 g, 84%) as a colorless syrup: R_f 0.3 [3 : 1 (v/v) $CHCl_3$–acetone].

*O-[O-(2,3,4,6-Tetra-O-acetyl-β-D-galactopyranosyl)-(1→4)-2,3,6-tri-O-acetyl-α-D-glucopyranosyl] Trichloroacetimidate (**18**).* Trichloroacetonitrile (2.5 ml, 25 mmol) is added to a mixture of **17** (5.3 g, 8.3 mmol) in anhydrous CH_2Cl_2 (100 ml) containing K_2CO_3 (1.2 g, 8.3 mmol) stirred at room temperature under dry N_2. The suspension is stirred vigorously at room temperature in the dark overnight. After filtration, the filtrate is concentrated to dryness. Purification by FCC [3 : 1 (v/v) $CHCl_3$–acetone] gives **18** (5.1 g, 78%) as a colorless amorphous solid: R_f 0.6 [3 : 1 (v/v) $CHCl_3$–acetone]; $[\alpha]_D$ +48° (*c* 4.2, $CHCl_3$).

*Benzyloxycarbonylmethyl O-(2,3,4,6-Tetra-O-acetyl-β-D-galactopyranosyl)-(1→4)-2,3,6-tri-O-acetyl-β-D-glucopyranoside (**19**).* A 0.22 *M* solution of TMSOTf in anhydrous CH_2Cl_2 (1.6 ml) is added dropwise to a mixture of **14** (2.4 g, 14.5 mmol) and **18** (5.6 g, 7.2 mmol) in anhydrous CH_2Cl_2 (200 ml) vigorously stirred at −15° under dry N_2. After 30 min, the mixture is washed with ice-cold saturated $NaHCO_3$ (3 times, 100 ml) and water (100 ml), then dried (Na_2SO_4). Evaporation followed by FCC [1 : 1 (v/v) hexane–ethyl acetate] yields **19** (4.0 g, 71%) as a colorless syrup: R_f 0.3 [1 : 1 (v/v) hexane–ethyl acetate]; $[\alpha]_D$ −17° (*c* 4.4, $CHCl_3$).

*3-Benzyloxycarbonylpropyl O-(2,3,4,6-Tetra-O-acetyl-β-D-galactopyranosyl)-(1→4)-2,3,6-tri-O-acetyl-β-D-glucopyranoside (**20**).* Glycosylation of **15** (0.5 g, 2.6 mmol) with **18** (1.0 g, 1.3 mmol) is carried out as described

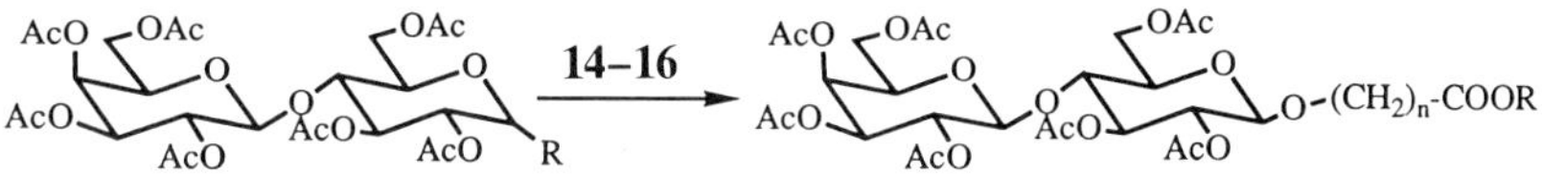

AcO OAc O AcO O AcO AcO OAc O AcO R —**14–16**→ AcO OAc O AcO O AcO AcO OAc O AcO O-$(CH_2)_n$-COOR

17 $R_{\alpha,\beta}$=OH

18 R_α=-O-C(NH)CCl_3

19 n=1, R=Bn

20 n=3, R=Bn

21 n=8, R=Me

SCHEME 2

above to give **20** (0.7 g, 67%) as a colorless syrup: R_f 0.5 [3 : 1 (v/v) $CHCl_3$–acetone]; $[\alpha]_D$ −13° (*c* 2.0 $CHCl_3$).

*8-Methoxycarbonyloctyl O-(2,3,4,6-Tetra-O-acetyl-β-D-galactopyranosyl)-(1→4)-2,3,6-tri-O-acetyl-β-D-glucopyranoside (**21**).* A similar glycosylation of **16** (0.58 g, 3.1 mmol) with **18** (1.2 g, 1.5 mmol) gives **21** (0.89 g, 72%) as a colorless syrup: R_f 0.4 [5 : 1 (v/v) $CHCl_3$–acetone]; $[\alpha]_D$ −11° (*c* 5.0, $CHCl_3$).

*Coupling of β-Lactosides to **1***

Owing to the hydrophobic nature of the spacer **16**, the unprotected lactoside **26** is soluble in nonaqueous solvent, suitable for the preparation of NHS esters, whereas the corresponding lactosides with C_2 and C_4 spacer arms are not. Accordingly, for the latter cases, the acetylated compounds **22** and **23** are used for coupling (Scheme 3).

*Carboxymethyl O-(2,3,4,6-Tetra-O-acetyl-β-D-galactopyranosyl)-(1→4)-2,3,6-tri-O-acetyl-β-D-glucopyranoside (**22**).* A solution of **19** (1.4 g, 1.8 mmol) in ethanol (20 ml) is stirred vigorously with 10% palladium on carbon (0.1 g) under H_2 (1 atm) at room temperature for 2 hr. The mixture is filtered and concentrated to give **22** (1.2 g, 97%) as a colorless solid: R_f 0.3 [10 : 1 (v/v) $CHCl_3$–methanol]; $[\alpha]_D$ −2.5° (*c* 3.2, $CHCl_3$).

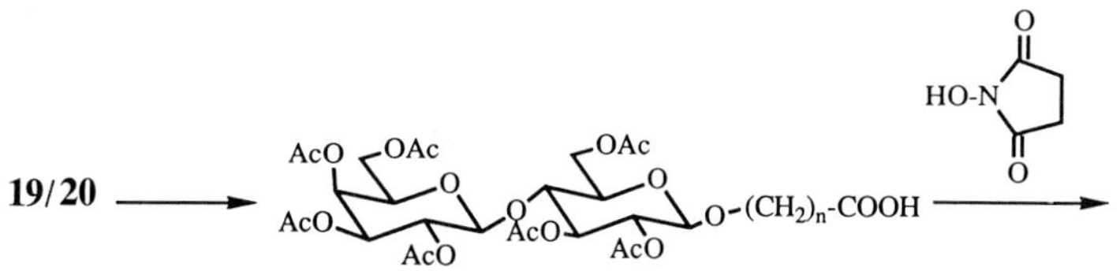

19/**20** → O-$(CH_2)_n$-COOH (HO-N succinimide) →

22 n=1

23 n=3

O-$(CH_2)_n$-COO-N (succinimide) — 1) **1**; 2) MeOH-H_2O-Et_3N → **8**/**9**

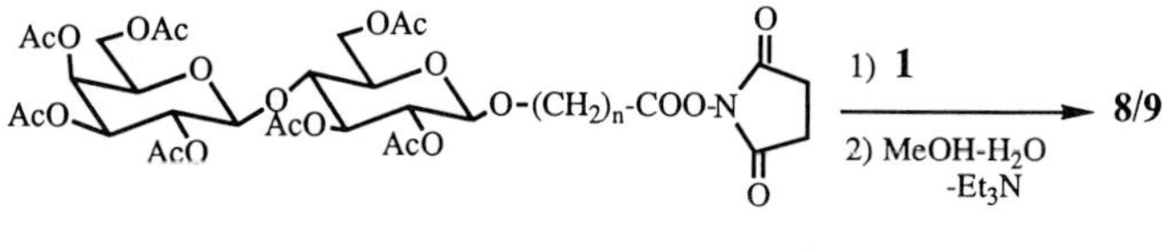

24 n=1

25 n=3

SCHEME 3

*Carboxypropyl O-(2,3,4,6-Tetra-O-acetyl-β-D-galactopyranosyl)-(1→4)-2,3,6-tri-O-acetyl-β-D-glucopyranoside (**23**).* Hydrogenolysis of **20** (1.3 g, 1.6 mmol) with 10% palladium on carbon (0.1 mg) in ethanol (20 ml) as described above yields, after FCC [12 : 1 (v/v) $CHCl_3$–methanol], compound **23** (0.9 g, 78%) as a colorless syrup: R_f 0.3 [10 : 1 (v/v) $CHCl_3$–methanol].

*Succinimidoxycarbonylmethyl O-(2,3,4,6-Tetra-O-acetyl-β-D-galactopyranosyl)-(1→4)-2,3,6-tri-O-acetyl-β-D-glucopyranoside (**24**).* EDC (0.45 g, 2.3 mmol) is added portionwise to a mixture of **22** (1.2 g, 1.7 mmol) and NHS (0.29 g, 2.5 mmol) in anhydrous CH_2Cl_2 (100 ml) stirred in an ice bath. The mixture is stirred at room temperature for 8 hr and washed with water (3 times, 50 ml). The organic layer is dried (Na_2SO_4) and concentrated to give **24** (1.3 g, 95%) as a colorless solid: R_f 0.6 [20 : 1 (v/v) $CHCl_3$–ethanol].

*3-Succinimidoxycarbonylpropyl O-(2,3,4,6-Tetra-O-acetyl-β-D-galactopyranosyl)-(1→4)-2,3,6-tri-O-acetyl-β-D-glucopyranoside (**25**).* Treatment of **23** (0.8 g, 1.1 mmol) and NHS (0.15 g, 1.3 mmol) in anhydrous CH_2Cl_2 (80 ml) with EDC (0.23 g, 1.2 mmol) as described above gives **25** (0.83 g, 91%) as a homogeneous solid: R_f 0.9 [10 : 1 (v/v) $CHCl_3$–methanol]; $[\alpha]_D$ −10° (*c* 1.4, $CHCl_3$).

*8-Carboxyoctyl O-β-D-Galactopyranosyl-(1→4)-β-D-glucopyranoside (**26**).* Compound **21** (2.5 g, 3.1 mmol) is dissolved in 9 : 1 (v/v) methanol–water (100 ml), and the mixture is adjusted to approximately pH 11 by the addition of 1 *M* aqueous LiOH (Scheme 4). After overnight, the

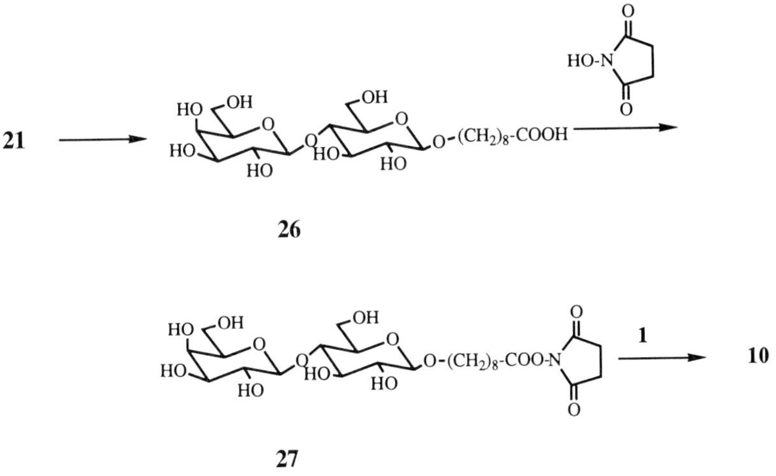

SCHEME 4

mixture is neutralized with Amberlite IR-120 (H^+) resin. The filtrate is concentrated to give **26** (1.4 g, 91%) as a colorless solid: R_f 0.5 [2 : 1 : 1 (v/v/v) butanol–acetic acid–water]; $[\alpha]_D$ −6.5° (*c* 0.9, water).

*8-Succinimidoxycarbonyloctyl O-β-D-Galactopyranosyl-(1→4)-β-D-Glucopyranoside (**27**).* *N,N'*-Dicyclohexylcarbodiimide (3.2 g, 15.5 mmol) is added portionwise to a mixture of **26** (5.2 g, 10.4 mmol) and NHS (1.5 g, 13 mmol) in anhydrous DMF (50 ml) stirred in an ice bath. The mixture is stirred at room temperature for 12 hr. After removal of the precipitated urea by filtration, the filtrate is concentrated (<35°) to one-third of its original volume, poured into precooled ethyl acetate (200 ml), and kept in a cold room overnight. The precipitated crystals are collected by filtration to give **27** (4.54 g, 73%) as colorless crystals: mp 154°–156°; R_f 0.5 [5 : 4 : 1 (v/v/v) $CHCl_3$–methanol–water]; $[\alpha]_D$ −7° (*c* 5, DMF).

*Trivalent Lactose–Lysyllysine Conjugate through C_2 Spacer Arm **8**.* A solution of **1** dihydrochloride (0.102 g, 0.29 mmol) in water (8 ml) containing triethylamine (0.13 ml, 0.93 mmol) is added dropwise to a solution of **24** (0.9 g, 1.0 mmol) in DMF (45 ml) stirred in an ice bath. The mixture is stirred at room temperature for 1 hr and treated with Amberlite IR-120 (H^+) resin to remove triethylamine and free amino compounds. After filtration, the filtrate is concentrated to dryness, and the residue is purified by Sephadex LH-20 column chromatography with acetone as the eluent. The orcinol-positive fractions [R_f 0.8, 2 : 1 : 1 (v/v/v) butanol–acetic acid–water] are combined and concentrated to give the protected conjugate (0.47 g, 69% based on **1**) as a colorless solid: $[\alpha]_D$ −13.5° (*c* 2.4, $CHCl_3$). De-O-acetylation with 10 : 6 : 1 (v/v/v) methanol–water–triethylamine (100 ml) at room temperature for 3 hr, followed by neutralization with Amberlite IR-120 (H^+) resin, yields a crude product. Purification by BioGel P-2 column chromatography with water as the eluent, followed by lyophilization, gives **8** (0.28 g, 97%) as an amorphous solid: R_f 0.2 [2 : 1 : 1 (v/v/v) propanol–acetic acid–water]; $[\alpha]_D$ −3.5° (*c* 2.5, water).

*Trivalent Lactose–Lysyllysine Conjugate through C_4 Spacer Arm **9**.* A similar coupling reaction between **25** and **1**, followed by de-O-acetylation, yields **9**. Protected conjugate: 73% yield based on **1**; R_f 0.7 [2 : 1 : 2 (v/v/v) butanol–acetic acid–water]; $[\alpha]_D$ −11.5° (*c* 1.2, acetone). Conjugate **9**: 95% yield; R_f 0.2 [2 : 1 : 1 (v/v/v) propanol–acetic acid–water]; $[\alpha]_D$ −5.5° (*c* 1.3, water).

*Trivalent Lactose–Lysyllysine Conjugate through C_9 Spacer Arm **10**.* A solution of **27** (2.4 g, 4.0 mmol) in DMF (37 ml) is added dropwise to a solution of **1** dihydrochloride (0.35 g, 1.0 mmol) in water (15 ml) containing $NaHCO_3$ (0.51 g, 6.1 mmol) stirred in an ice bath. The mixture is stirred at room temperature for 2 hr and concentrated to dryness. The residue is purified by BioGel P-2 column chromatography with water as the eluent to give, after lyophilization, **10** (1.6 g, 93% based on **1**): R_f 0.3 [2 : 1 : 1 (v/v/v) butanol–acetic acid–water]; $[\alpha]_D$ −5.5° (*c* 3.0, water).

In Vivo Assay

Murine BL6 cells (2×10^4), highly metastatic clones of the B16 melanoma cell line, are preincubated with 30 m*M* of trivalent β-lactoside conjugates (**8**, **9**, or **10**) in RPMI 1640 medium (0.15 ml) at 37° for 10 min. The mixture is then injected into the tail vein of syngeneicC57/BL female mice at 8 weeks of age. Control mice are given the same amount of cells and RPMI 1640 without the lactoside conjugates. After 18 days, the mice are sacrificed, the lungs are fixed in 10% formaldehyde in PBS (pH 7.4), and tumor cell colonies are counted under a disecting microscope. Six mice are used for each treatment and control group.

Acknowledgment

The authors thank Ms. Jennifer Stoeck for preparation of figures and schemes.

[23] Ligand-Based Carrier Systems for Delivery of DNA to Hepatocytes

By MARK A. FINDEIS, CATHERINE H. WU, and GEORGE Y. WU

Introduction

Although many techniques have been developed for transferring foreign genes into cells, only a few of these techniques appear to be practical for therapeutic applications with respect to efficiency of gene transfer and viability of transfected cells. The potential of therapeutic gene transfer has resulted in a diverse and growing range of research efforts to explore and develop gene therapy strategies.[1–3] Examples of these gene transfer

[1] R. C. Mulligan, *Science* **260,** 926 (1993)
[2] W. F. Anderson, *Science* **256,** 808 (1992).
[3] A. D. Miller, *Nature (London)* **357,** 455 (1992).

techniques include the uses of retroviruses,[4,5] adenoviruses,[6,7] receptor-mediated endocytosis,[8,9] direct injection,[10] and liposomes.[11,12] Receptor-mediated targeted delivery of DNA has been successfully applied using protein ligands to the hepatic asialoglycoprotein receptor (ASGr),[8,13,14] and, subsequently, the transferrin receptor.[9] A major advantage to the use of the ASGr is that it is a cell surface receptor that is highly selective for hepatocytes. Thus, genes targeted to this receptor can be delivered specifically to the liver.

The liver has been recognized as a useful target for receptor-mediated delivery via the ASGr.[15] Early experiments focused on the delivery of agents covalently bound to carriers such as asialofetuin[15–17] and asialoorosomucoid[18] or galactose-terminated neoglycoproteins.[19,20] The ASGr-mediated endocytosis of the glycoproteins carried the bound agent into cells. This strategy has been extended to the use of asialoorosomucoid–polylysine (ASOR–PL) conjugates to carry DNA into hepatocytes.[13] The DNA is bound in a tight electrostatic complex by the ASOR–PL. Receptors on cell surfaces bind and internalize the ASOR in the complex, bring-

[4] R. C. Mulligan, *in* "Etiology of Human Disease at the DNA Level" (J. Lindsten and U. Pettersson, eds.), p. 143. Raven, New York, 1991.

[5] D. J. Jolly, J.-K. Yee, and T. Friedmann, this series, Vol. 149, p. 10.

[6] M. A. Rosenfeld, K. Yoshimura, B. C. Trapnell, K. Yoneyama, E. R. Rosenthal, W. Dalemans, M. Fukayama, J. Bargon, L. E. Stier, L. Stratford-Perricaudet, M. Perricaudet, W. B. Guggino, A. Pavirani, J.-P. Lecocq, and R. G. Crystal, *Cell (Cambridge, Mass.)* **68,** 143 (1992).

[7] J. Herz and R. D. Gerard, *Proc. Natl. Acad. Sci. U.S.A.* **90,** 2812 (1993).

[8] G. Y. Wu and C. H. Wu, *Biochemistry* **27,** 887 (1988).

[9] E. Wagner, M. Zenke, M. Cotten, H. Beug, and M. L. Birnstiel, *Proc. Natl. Acad. Sci. U.S.A.* **87,** 3410 (1990).

[10] G. J. Nabel and P. L. Felgner, *Trends Biotechnol.* **11,** 211 (1993).

[11] N. Zhu, D. Liggitt, and R. Debs, *Science* **261,** 209 (1993).

[12] R. Stribling, E. Brunette, D. Liggitt, K. Gaensler, and R. Debs, *Proc. Natl. Acad. Sci. U.S.A.* **89,** 11277 (1992).

[13] M. A. Findeis, J. R. Merwin. G. L. Spitalny, and H. C. Chiou, *Trends Biotechnol.* **11,** 202 (1993).

[14] G. Y. Wu and C. H. Wu, *in* "Liver Diseases, Targeted Diagnosis and Therapy Using Specific Receptors and Ligands" (G. Y. Wu and C. H. Wu, eds.), p. 127. Dekker, New York, 1991.

[15] G. Y. Wu, C. H. Wu, and R. J. Cristiano, *Proc. Natl. Acad. Sci. U.S.A.* **80,** 3078 (1983).

[16] G. Y. Wu, C. H. Wu, and M. I. Rubin, *Hepatology (Baltimore)* **5,** 709 (1985).

[17] V. Keegan-Rogers and G. Y. Wu, *Cancer Chemother. Pharmacol.* **26,** 93 (1990).

[18] G. Y. Wu, V. Keegan-Rogers, S. Franklin, S. Midford, and C. H. Wu, *J. Biol. Chem.* **263,** 4719 (1988).

[19] A. Ponzetto, L. Fiume, B. Forzani, S. H. Song, C. Busi, A. Mattioli, C. Spinelli, M. Marinelli, A. Smedile, E. Chiaberge, F. Bonino, G. B. Gervasi, M. Rapicetta, and G. Verme, *Hepatology (Baltimore)* **14,** 16 (1991).

[20] L. Fiume, C. Busi, A. Mattioli, P. G. Balboni, and G. Barbanti-Brodano, *FEBS Lett.* **129,** 261 (1981).

ing the DNA into the cell at the same time. This technique has been used successfully both *in vitro*[8,21–23] and *in vivo*[24–28] for the delivery of DNAs coding for a variety of genes including the gene for the low-density lipoprotein receptor (LDLr) in Watanabe rabbits. Expression of the LDLr gene resulted in lowered serum cholesterol levels.[27]

Asialoorosomucoid is readily available by isolation of orosomucoid (also referred to as α_1-acid glycoprotein) from human plasma[29] followed by desialylation to expose penultimate galactose groups. Although orosomucoid is commercially available, it is readily isolable from plasma in good yield.[29] Orosomucoid is quite a robust protein and can be desialylated by acid treatment with heating, thus avoiding the expense of using neuraminidase to cleave sialic acid residues.[30] Poly(L-lysine) (PL) is available commercially as a polymer with varying molecular weight ranges. Conjugates formed by cross-linking PL with proteins have been reported using carbodiimide-mediated amide bond formation[22,23] and thiol reagents.[8,9,21] The product mixtures obtained in the formation of conjugates of this type are heterogeneous and highly charged. Previously described methods for the preparation of ASOR–PL complexes targeted to the ASGr have used dialysis,[18] gel filtration,[18,21] and ion-exchange high-performance liquid chromatography (HPLC)[9,22] for purification of the conjugate from reaction by-products and residual starting materials. We describe here methods for the synthesis and purification of ASOR–PL conjugates which effectively bind DNA in an electrostatic complex for targeted delivery to the ASGr-bearing cells.

Experimental Methods

Preparation of Human Orosomucoid: α_1-Acid Glycoprotein

Buffers used in the isolation are prepared at the stated concentration of sodium acetate. Glacial acetic acid is added to obtain the desired pH.

[21] G. Y. Wu and C. H. Wu, *J. Biol. Chem.* **262,** 4429 (1987).

[22] G. Y. Wu and C. H. Wu, *J. Biol. Chem.* **267,** 12436 (1992).

[23] R. J. Cristiano, L. C. Smith, and S. L. C. Woo, *Proc. Natl. Acad. Sci. U.S.A.* **90,** 2122 (1993).

[24] G. Y. Wu and C. H. Wu, *J. Biol. Chem.* **263,** 14621 (1988).

[25] G. Y. Wu, J. M. Wilson, and C. H. Wu, *J. Biol. Chem.* **264,** 16985 (1989).

[26] G. Y. Wu, J. M. Wilson, F. Shalaby, M. Grossman, D. A. Shafritz, and C. H. Wu, *J. Biol. Chem.* **266,** 14338 (1991).

[27] J. M. Wilson, M. Grossman, J. A. Cabrera, C. H. Wu, N. R. Chowdhury, G. Y. Wu, and J. R. Chowdhury, *J. Biol. Chem.* **267,** 963 (1992).

[28] J. M. Wilson, M. Grossman, J. A. Cabrera, C. H. Wu, and G. Y. Wu, *J. Biol. Chem.* **267,** 11483 (1992).

[29] P. H. Whitehead, and H. G. Sammons, *Biochim. Biophys. Acta* **124,** 209 (1966).

[30] K. Schmid, A. Polis, K. Hunziker, R. Fricke, and M. Yayoshi, *Biochem. J.* **104,** 361 (1967).

Buffer 1 contains 50 m*M* sodium acetate, pH 4.5; buffer 2, 100 m*M* sodium acetate, pH 4.0; buffer 3, 50 m*M* sodium acetate, pH 3.0. Diethylaminoethyl-cellulose (DEAE-cellulose, sufficient to pack a 500-ml bed volume column once swollen) is suspended in water, allowed to swell for at least 2 hr, and then washed successively with 0.5 *N* HCl, 0.5 *N* NaOH, and 10 m*M* ethylenediaminetetraacetic acid (EDTA).[31] The DEAE-cellulose is packed in a 5 cm diameter by 30 cm height Waters (Milford, MA) AP-5 column. Using a peristaltic pump, the column is equilibrated at a flow rate of 10 ml/min with buffer 1 until the column eluate reaches pH 4.5.

Outdated pooled human plasma obtained from the American Red Cross (4 units, ~1.1 liter) is transferred to dialysis tubing (12,000–14,000 molecular weight cutoff) and dialyzed overnight at 4° against 20 liters of buffer 1. The dialyzed plasma is then centrifuged at 10,000 rpm in a Beckman (Palo Alto, CA) JA-14 rotor (15,000 *g*) for 10 min at 4°. The supernatant is then filtered through Whatman (Clifton, NJ) No. 1 paper, and the precipitate is discarded. The dialyzed and filtered plasma is applied to the DEAE-cellulose column that is then washed with buffer 1 until the eluate has an absorbance at 280 nm of less than 0.10 cm^{-1}. The column is then eluted with buffer 2. The eluate is collected starting when the A_{280} of the eluate begins to increase and ending after the A_{280} has peaked and drops below 0.10 cm^{-1}. After the orosomucoid-rich fraction is collected, the column is washed with buffer 3 (1 liter) and reequilibrated with buffer 1.

The orosomucoid-enriched eluate is brought to 50% saturation with ammonium sulfate (313 g/liter of eluate) and stirred overnight at 4°. The solution is then centrifuged at 14,000 rpm in the JA-14 rotor for 15 min at 4° and the supernatant retained. Ammonium sulfate (320 g/liter of 50% saturation supernatant) is slowly added to bring the solution to 92% saturation. The solution is stirred for at least 4 hr at 4° and then centrifuged (10,000 rpm for 30 min, JA-14 rotor, 4°). The pellet is retained, dissolved in a minimal volume of water, transferred to dialysis tubing (leaving a 3-fold volume for expansion of the dialysate), and dialyzed for 2 days at 4° against 20 liters of water that is replaced after 1 day. The resulting dialysate is lyophilized and stored at 20°. The orosomucoid is analyzed by sodium dodecyl sulfate–polyacrylamide gel electrophoresis (SDS–PAGE)[32,33] and shows a single band at an apparent molecular weight of 44,000 (the compound has a molecular weight of 36,800 but runs on

[31] T. G. Cooper, "The Tools of Biochemistry," Chap. 6. Wiley, New York, 1977.

[32] J. Sambrook, E. F. Fritsch, and T. Maniatis, "Molecular Cloning: A Laboratory Manual," 2nd Ed., Chap. 6. Cold Spring Harbor Laboratory, Cold Spring Harbor, New York, 1989.

[33] B. D. Hames, *in* "Gel Electrophoresis of Proteins" (B. D. Hames and D. Rickwood, eds.), Chap. 1. IRL Press, Oxford and Washington, D.C., 1981.

SDS–PAGE with a higher apparent molecular weight)[34,35] after staining with Coomassie blue. The typical yield of lyophilized salt-free orosomucoid using this procedure is 350–400 mg/liter of plasma.

Preparation of Asialoorosomucoid

Asialoorosomucoid (ASOR) is prepared by cleavage of terminal sialic acid residues from the branched carbohydrates on the surface of orosomucoid. Sialic acid cleavage can be performed by treatment of orosomucoid with neuraminidase or by acid hydrolysis.[30] Although less mild than enzymatic treatment, acid hydrolysis affords ASOR that is no less effective of a ligand for the ASGr. This procedure is also more practical and economical on a larger scale.

Orosomucoid (10 mg/ml) isolated as above is dissolved in water. An equal volume of 0.1 *N* sulfuric acid (2.78 ml/liter of concentrated H_2SO_4 in water) is added to the solution, and the resulting mixture is heated at 80° for 1 hr in a water bath to hydrolyze sialic acids from the protein. The acidolysis mixture is removed from water bath, neutralized with NaOH solution, dialyzed against water for 2 days, and then lyophilized. The thiobarbituric acid assay of Warren[36] or Uchida *et al.*[37] is then used to verify desialylation of the orosomucoid. Targetability of ASOR samples is verified by labeling with ^{125}I and measuring liver uptake in rats or mice.[24]

Synthesis of Asialoorosomucoid–Polylysine Conjugates

The ASOR (200 mg) is dissolved in water (10 ml) and the solution filtered through a 0.2-μm syringe filter. The filter is washed with water (5 ml), the filtrates are combined, and the solution is adjusted to pH 7.4. Poly(L-lysine) (PL, as the HBr salt, 160 mg) of average molecular weight 41,000 is dissolved in water (10 ml), and the solution is adjusted to pH 7.4 with 0.1 *N* NaOH. 1-(3-Dimethylaminopropyl)-3-ethylcarbodiimide (EDC, 92 mg) is dissolved in water (1 ml) and added directly to the ASOR solution. The PL solution is added to the ASOR–EDC solution, and the resulting mixture is readjusted to pH 7.4 with 0.1 *N* NaOH. The reaction mixture is covered and maintained at 37° for 24 hr. The reaction mixture is then transferred to 12,000–14,000 molecular weight cutoff dialysis tubing

[34] B. T. Chait and S. B. H. Kent, *Science* **257,** 1885 (1992).

[35] K. Schmid, *in* "Alpha$_1$-Acid Glycoprotein: Genetics, Biochemistry, Physiological Functions, and Pharmacology" (P. Baumann, C. B. Eap, W. E. Müller, and J.-P. Tillement, eds.), p. 7. Alan R. Liss, New York, 1989.

[36] L. Warren, *J. Biol. Chem.* **234,** 1971 (1959).

[37] Y. Uchida, Y. Tsukuda, and T. Sugimori, *J. Biochem.* (*Tokyo*) **82,** 1425 (1977).

and dialyzed at 4° against water (20 liters) for 2 days with one change of water. The dialysate can be lyophilized and weighed prior to further purification at this point.

Analytical Acid–Urea Gel Electrophoresis

Analytical gels are prepared using the recipe in Table I adapted from the method of Panyim and Chalkley.[38] Prior to the addition of TEMED, acrylamide solutions are degassed under reduced pressure (~20 mm Hg) for 10 min. After samples are loaded on cast gels, electrophoresis is run from anode to cathode (electrodes connected in *reverse* polarity from the usual) at 90–150 V. Methyl green can be used as a tracking dye in these gels although we usually omit it. Gels are stained with 0.1% Coomassie Brilliant Blue R250 in ethanol–acetic acid–water (40 : 10 : 50, v/v/v).

Preparative Acid–Urea Gel Elution Electrophoresis

Preparative gels are cast and run in the Bio-Rad (Richmond, CA) Model 491 Prep Cell apparatus (for recipe, see Table I) in a manner similar to that used for analytical gels and according to the instructions for the use of the apparatus. Running gel solutions are degassed under reduced pressure, poured into the gel holder, and allowed to stand overnight to ensure full polymerization. Stacking gel solutions are degassed with bubbling of nitrogen gas prior to the addition of TEMED and allowed to polymerize for 30 min. After assembly of the elution electrophoresis apparatus with the electrodes connected (in *reverse* polarity as above), the elution chamber is eluted with a peristaltic pump at 1 ml/min, sample is loaded, and electrophoresis conducted at 11 W for small gels and 22 W for large gels. The eluate is passed through a UV detector operating at 280 nm. Fractions (5 ml) are collected starting after the ion front (first sharp peak) is eluted from the gel (Fig. 1).

Cation-Exchange Chromatographic Fractionation of Asialoorosomucoid–Polylysine Complexes

Cation-exchange HPLC is performed using a 1 × 10 cm Waters AP-1 column packed with CM 15HR packing material. The gradient described in Table II is used to chromatograph ASOR–PL mixtures that had been freed of unreacted PL by preparative electrophoresis. Fractions of interest elute after the buffer pH is lowered to pH 2.3 and the salt concentration begins to increase. Typically, the first major peak observed under these

[38] S. Panyim and R. Chalkley, *Arch. Biochem. Biophys.* **130,** 337 (1969).

TABLE I
PREPARATION OF GELS

Solutions for casting acid–urea gels[a]	Components and concentration	Gel recipe[b]	Concentrations	
			10% Running gel	4% Stacking gel
8 M Urea	48 g urea plus water to 100 ml	30% Bis/acrylamide	13.3 ml	1.3 ml
Running gel buffer (RGB)	48 ml 1 M Potassium hydroxide, 17.2 ml glacial acetic acid, water to 100 ml	Buffer	5 ml RGB	2.5 ml 4×SB
4× Sample buffer (4×SB)	1.34 g Potassium hydroxide, 2 ml glacial acetic acid, water to 100 ml	8 M urea	20 ml	5 ml
10× Running buffer (10×RB)	311.8 g β-Alanine, 80.4 ml acetic acid, water to 1 liter	Water	1.2 ml	1.3 ml
30% Bis/acrylamide	30 g Acrylamide, 0.8 g bisacrylamide, 100 ml water	10% Ammonium persulfate	0.3 ml	0.1 ml
10% Ammonium persulfate	100 mg in 1 ml water	TEMED[c]	0.2 ml	0.012 ml
Methyl green stock	0.1% Methyl green in 1 : 3 (4×SB : 8 M urea)			
Coomassie stain	1 g/liter Coomassie blue in 40% ethanol, 10% acetic acid in water			
Destaining solution	10% Ethanol, 7.5% acetic acid in water			

[a] B. D. Hames, *in* "Gel Electrophoresis of Proteins" (B. D. Hames and D. Rickwood, eds.), p. 68. IRL Press, Oxford and Washington, D.C., 1981.
[b] Proportions of solutions to use for casting running gels and stacking gels, respectively.
[c] *N,N,N′,N′*-Tetramethylethylenediamine.

conditions will have the best properties for DNA binding, resulting in ASOR–PL–DNA complexes with increased solubility (Fig. 2).

Gel Retardation Assay

Concentrations of DNA are based on an extinction coefficient of 20 $(mg/ml)^{-1}$ cm^{-1} at 260 nm, and agarose gel electrophoresis of DNA is

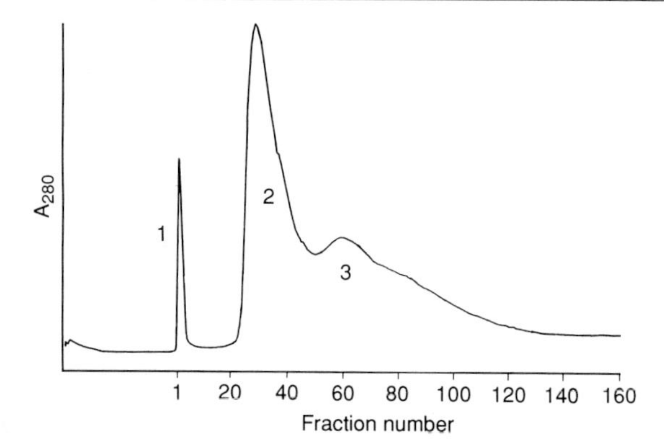

FIG. 1. Elution profile from preparative acid–urea elution electrophoresis. The sample was a crude product mixture (170 mg) from the cross-linking of ASOR with 41,000 average molecular weight PL as described in the text. Fractions (5 ml) were collected starting with peak 1 (the ion front and unbound PL). Peaks 2 (fractions 20–45, 41 mg) and 3 (fractions 50–90, 34 mg) were collected and pooled for further fractionation by cation-exchange HPLC.

TABLE II
HPLC GRADIENT[a]

t (min)	%A	%B	%C	%D
0	25	0	75	0
8	25	0	75	0
9	0	25	75	0
20	0	25	75	0
21	0	25	70	5
35	0	25	70	5
36	0	25	65	10
50	0	25	65	10
51	0	25	60	15
65	0	25	60	15
66	0	25	55	20
80	0	25	55	20
81	0	25	25	50
95	0	25	25	50

[a] For cation-exchange chromatography on Waters CM 15HR packing in a 10 mm diameter by 100 mm length column, flow rate, 1.8 ml/min. Buffer A is 0.4 *M* sodium acetate, pH 5.0; buffer B, 0.4 *M* sodium acetate, pH 2.3; buffer C, water; buffer D, 2 *M* NaCl. Buffers are made up to the stated concentration in sodium acetate. The pH is adjusted by adding HCl.

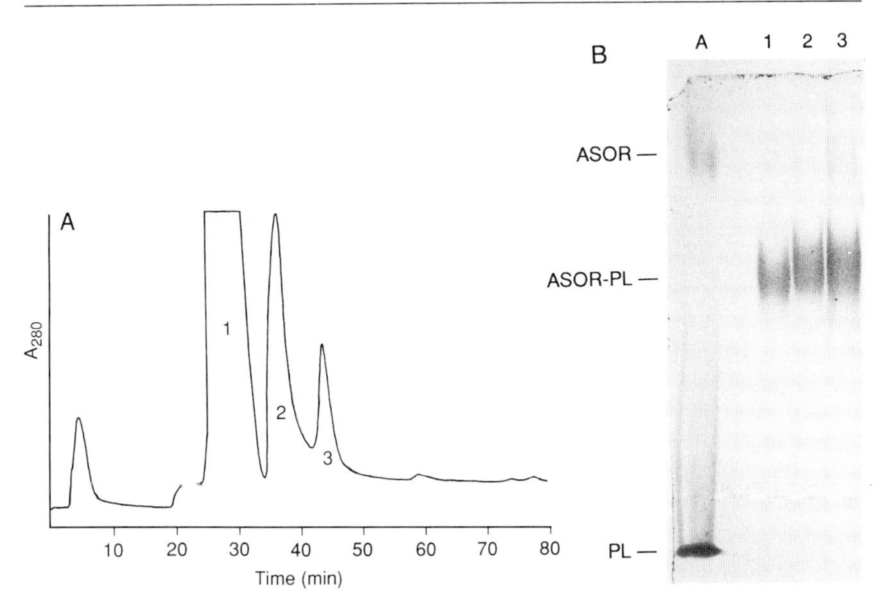

FIG. 2. (A) High-performance liquid chromatography ion-exchange chromatogram of ASOR–PL (peak 2 from Fig. 1) and (B) associated acid–urea electrophoresis gel (see Table II for HPLC gradient conditions). The acid–urea gel shows that after both preparative electrophoresis and cation-exchange HPLC the fractions isolated are free of unbound polylysine (see lane A, a mixture of ASOR and PL). The ASOR–PL isolated from peak 1 in A (lane 1 in B) produces DNA complexes with greater solubility than the ASOR–PL isolated from peaks 2 and 3.

performed as described by Sambrook *et al.*[32] A series of samples are prepared by adding 20-μl aliquots of ASOR–PL solutions in 0.15 *M* NaCl to 20-μl samples of plasmid DNA at 200 ng/μl and 0.15 *M* NaCl to give ASOR–PL to DNA mass ratios of 4.0 : 1, 3.5 : 1, 3.0 : 1, 2.5 : 1, 2.0 : 1, 1.5 : 1, 1.0 : 1. After a 15-min incubation at room temperature, the test samples and a sample of DNA alone are electrophoresed on a 1% w/v agarose gel eluted with 1 × TPE[32] at 50 V for 45 min. The optimum ASOR–PL to DNA ratio for complex formation is determined from the migration of the samples as the lowest ratio that gives 80–90% complete DNA retardation (Fig. 3A).

DNA–Conjugate Complex Formation

To form targetable complexes, two methods may be used. In the first, plasmid DNA, 0.5 mg in 1 ml of 2 *M* NaCl, is added to ASOR–PL conjugate, 0.15 mg (with respect to ASOR content), in 600 μl of 2 *M* NaCl at

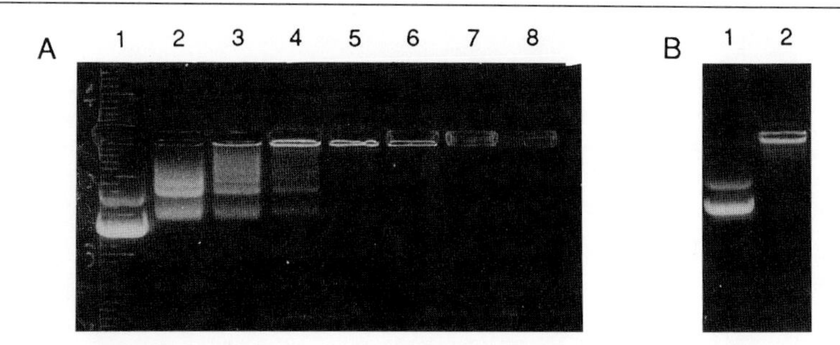

FIG. 3. (A) Gel retardation assay of the binding of ASOR–PL synthesized and purified as described in the text in complex with plasmid DNA (plasmid psvHBV, provided by Dr. H. C. Chiou, TargeTech, Inc.). Lane 1 is plasmid in the absence of conjugate. Lanes 2 through 8 contain weight ratios of ASOR-PL to DNA of 1.0, 1.5, 2.0, 2.5, 3.0, 3.5, and 4.0, respectively. (B) Gel assay of an ASOR–PL–DNA complex prepared at a ratio of 2 : 1 as described in the text. Lane 1 is DNA before complexation, and lane 2 shows retarded complexed DNA.

25°. The mixture is then placed in 1.0-cm (flat width) dialysis tubing with an exclusion limit of 12,000–14,000 and stepwise dialyzed successively at 4° for 0.5 hr against 1.0 liter of NaCl in each of the following concentrations: 1.5, 1.0, 0.5, 0.25, and 0.15 *M*. If the size of the plasmid is larger than 8 kilobases (kb), it is advisable to dialyze more slowly, taking 24 hr for each change. After the final dialysis, the complex is filtered through 0.45-μm membranes. Complete complexation is determined by testing samples for retardation by agarose gel electrophoresis. No free DNA should be present.

In the second method, an ASOR–PL–DNA complex is formed at a 2 : 1 ASOR–PL to DNA ratio by adding 2.5 ml of an ASOR–PL solution at 1.5 mg/ml and 0.15 *M* NaCl to 2.5 ml of a stirred solution of DNA at 1.0 mg/ml and 0.15 *M* NaCl at 25 μl/min via a peristaltic pump. After complete addition the complex is filtered through a 0.45-μm filter, and the concentration of soluble complexed DNA is measured by the UV absorbance at 260 nm (final concentration of complexed DNA: 374 μg/ml, 75% recovery). Complexation of the DNA is then confirmed by agarose gel electrophoresis (Fig. 3B). Complexes in 0.15 *M* NaCl are usually stable at 4° for at least 2–3 days.

Comments

The procedures described here represent general guidelines for the preparation and use of ASOR–PL conjugates to complex DNA for targeted delivery. Considerable flexibility is available within these guidelines. In our laboratories we have prepared conjugates using PLs over a wide range of average molecular weights (4000–60,000) with similar results in DNA binding and delivery. Proper purification of the ASOR–PL is important to allow the formation of a soluble DNA complex. Especially for applications *in vivo*, the ASOR–PL should be free of both unbound PL and the highly cross-linked ASOR–PL species that migrate more slowly in the acid–urea gels. The ASOR–PL purified in this manner allows the formation of DNA complexes at a concentration suitable for intravenous injection into experimental animals (up to 1 mg/ml DNA). For *in vitro* experimentation, for which complexes can be made up at lower concentrations, it is possible (though not preferable) to use less rigorously purified ASOR–PL to prepare soluble targetable complexes. Complexes of oligonucleotides[22] are more soluble than those of plasmid DNAs and also may be prepared with less highly purified ASOR–PL.

Acknowledgments

Research was supported in part by grants from the National Institutes of Health (DK42182 to G.Y.W.), March of Dimes Birth Defects Foundation (No. 1-0786 to G.Y.W.), and TargeTech, Inc. (to C.H.W.). G. Y. Wu and C. H. Wu hold equity in Immune Response Corp. The technical assistance of Pei-li Zhan is gratefully acknowledged. The data presented in Figs 1–3 were collected by T. D. McKee and M. E. DeRome (TargeTech, Inc.).

[24] Induction of Rabbit Immunoglobulin G Antibodies against Synthetic Sialylated Neoglycoproteins

By René Roy, Craig A. Laferrière, Robert A. Pon, and Andrzej Gamian

Introduction

Sialic acids constitute a group of some 30 derivatives of neuraminic acid among which *N*-acetylneuraminic acid (NeuAc) appears to be the most ubiquitous representative. This unusual carbohydrate is most often found at the terminal nonreducing end of the oligosaccharide sequences of glycolipids, in glycoproteins, and in capsular polysaccharides. As such, they constitute the first antigens encountered in several biological and immunological interactions. Recognition and adherence phenomenon in some viral[1] and bacterial infections[2] have therefore been attributed to this family of carbohydrates. Sialic acids also play crucial roles in masking antigenic determinants.[3] In a few circumstances, however, they have demonstrated antigenic properties.[4]

Perhaps the most important are the findings in which sialic acids and NeuAc in particular have been implicated as immunodominant epitopes in certain forms of cancer, in Waldenström macroglobulinemia, in acute inflammatory and Salla diseases, in sialuria, and in chronic lymphocytic leukemia.[5] As a consequence, the total and free NeuAc levels in serum or urine of the patients afflicted with the above diseases are dramatically elevated in comparison to those of healthy individuals.

The prime event in influenza virus infections has been demonstrated to depend on the recognition and binding of the viral glycoprotein hemagglutinin to the cell surface sialic acids.[6] As multivalencies have been recognized as a key factor in the tight binding, a number of NeuAc conjugates have been synthesized as potential inhibitors of adherence.[7]

To address the critical functions of NeuAc as a single immunochemical entity, the synthesis of artificial immunogenic conjugates has been undertaken. Such neoglycoproteins could possibly be used in two different ways. In the first case, they could act as high-avidity inhibitors against the viral hemagglutinin. In the second case, they could be used as immunogens to raise anti-NeuAc antibodies which would be useful in serodiagnosis or in specific cell targeting. This chapter demonstrates the immunogenicities of sialic acid-containing neoglycoproteins in rabbits and raises a warning in the implications of using protein carriers for multivalent sialosides. Previous work has been done to evaluate the immunogenicities of synthetic sialic acid conjugates.[8]

Materials and Methods

Sodium cyanoborohydride ($NaBH_3CN$), solvents, and other reagent-grade chemicals are obtained from Aldrich Chemical Co. (Milwaukee,

[1] J. C. Paulson, *in* "The Receptors" (M. Conn, ed.), Vol. 2, p. 131. Academic Press, Orlando, Florida, 1985.

[2] H. J. Jennings, E. Katzenellenbogen, C. Lugowski, F. Michon, R. Roy, and D. L. Kasper, *Pure Appl. Chem.* **56,** 893 (1984).

[3] R. Schauer *in* "The Molecular Immunology of Complex Carbohydrates" (A. M. Wu and L. G. Adams, eds.), Vol. 228, p. 47. Plenum, New York, 1988.

[4] R. Schauer, this series, Vol. 138, p. 132.

[5] W. Reutter, E. Köttgen, C. Bauer, and W. Gerok, *Cell Biol. Monogr.* **10,** 263 (1982).

[6] N. K. Sauter, M. D. Bednarski, B. A. Wurzburg, J. E. Hanson, G. M. Whitesides, J. J. Skehel, and D. C. Wiley, *Biochemistry* **28,** 8388 (1989).

[7] R. Roy, F. O. Andersson, G. Harms, S. Kelm, and R. Schauer, *Angew. Chem., Int. Ed. Engl.* **31,** 1478 (1992), and references therein.

[8] D. F. Smith and V. Ginsburg, *J. Biol. Chem.* **255,** 55 (1980).

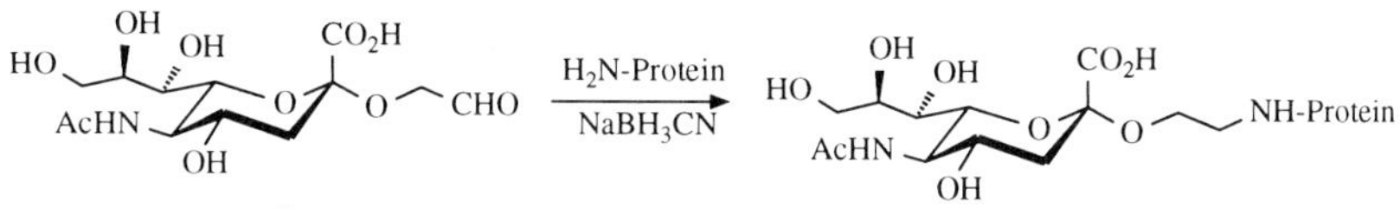

SCHEME 1

WI) and are used without purification. Bovine (No. A-7638) and chicken serum albumin (Fraction V, No. A-3014) are from Sigma Chemical Co. (St. Louis, MO). Tetanus toxoid is a gift from Dr. P. Rousseau (Institut Armand-Frappier, Laval, Québec, Canada). The horseradish peroxidase-labeled goat anti-rabbit immunoglobulin G (IgG) (H+L) is obtained from Kirkegaard and Perry Laboratories, Inc. (Gaithersburg, MD). Linbro microtiter plates are from Flow Laboratories (Mississauga, ON, Canada). Freund's complete and incomplete adjuvants are from Difco Laboratories (Detroit, MI). Allyl (**3**), methyl (**2**), and 2-oxoethyl (**1**) α-glycosides of *N*-acetylneuraminic acid are prepared as previously described (Schemes 1 and 2).[9] The *N*-acetylneuraminic acid glycopolymer **4** is obtained as described.[10]

Preparations of Neoglycoconjugates

The covalent coupling of the 2-oxoethyl α-sialoside derivative of NeuAc (**1**) to the ε-amino groups of the lysine residues on bovine (BSA) and chicken (CSA) serum albumin (M_r 67,000) and on tetanus toxoid (TT, M_r 150,000) is accomplished by direct reductive amination with sodium cyanoborohydride.[9] A time course study for the coupling is depicted in Fig. 1. To a solution of TT (30 mg, 0.2 μmol, ~13 μmol NH_2 groups), CSA (30 mg, 0.44 μmol), or BSA (30 mg, 0.44 μmol, 26 μmol, NH_2 groups) in 0.2 *M* sodium phosphate (or borate) buffer,[9] pH 8.0 (10 ml), are added 2-oxoethyl α-sialoside (**1**) (52 mg, 0.15 mmol) and $NaBH_3CN$ (83 mg, 1.32 mmol) (Scheme 1). These conditions represent a molar ratio for NH_2–sugar–$NaBH_3CN$ of 1 : 9 : 51 (0.5 : 9 : 51 for TT). The solutions are heated at 37° with aliquots (1 ml) removed over a period of 6 days. The pH of all fractions is lowered to pH 5.4 with acetic acid in order to destroy the excess reagent. After exhaustive dialysis against running water and lyophilization, the sialylated neoglycoprotein conjugates are analyzed for sialic acid content. The results agree with amino acid analyses (within ±2 residues).[11] The residues from the individually lyophilized fractions are

[9] R. Roy and C. A. Laferrière, *Can. J. Chem.* **68,** 2045 (1990).

[10] R. Roy and C. A. Lafferière, *Carbohydr. Res.* **177,** C1 (1988).

[11] R. Roy, F. D. Tropper, A. Romanowska, M. Letellier, L. Cousineau, S. J. Meunier, and J. Boratynski, *Glycoconjugate J.* **8,** 75 (1991).

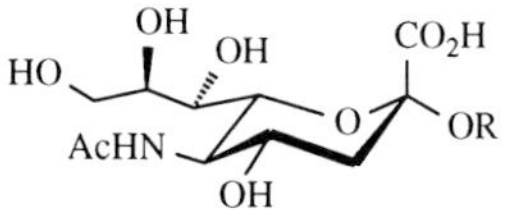

2 R = Me
3 R = CH_2-CH=CH_2

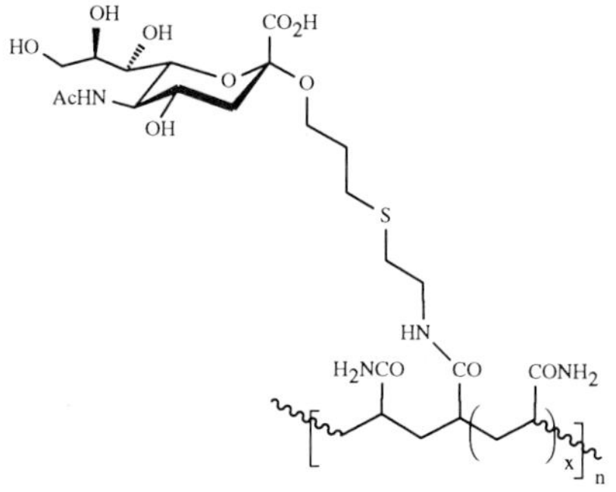

4

SCHEME 2

diluted to a known volume and analyzed by the resorcinol method.[12] The BSA and CSA conjugates with an incorporation of 20 sialic acid residues or more give precipitin bands with wheat germ agglutinin (WGA, not shown) in double radial diffusion experiments. The TT conjugates do not precipitate WGA in all cases.

Preparation of Rabbit Antisera

Californian white rabbits (2–3 kg) are immunized with the NeuAc–BSA (22 NeuAc/BSA) or NeuAc–TT (10 NeuAc/TT) conjugates prepared in phosphate-buffered saline (PBS, 0.5 mg/ml) suspended in Freund's complete adjuvant (1 : 1, v/v). Injections are done intramuscularly and subcutaneously at few sites (50 μg conjugates/200 μl). The serum samples are collected on day 24 following booster treatments on days 7 and 14 with Freund's incomplete adjuvant.

Absorption of Rabbit Antisera

The BSA (1 mg) or TT (1 mg) is added to the homologous rabbit sera (2 ml). The solutions are incubated for 4 days at 4°. The immune precipitates are centrifuged for 5 min at 14,000 *g* at 25° in a Fisher (Ottawa, Ontario, Canada) microcentrifuge, Model 235B. The sera are then col-

[12] L. Svennerholm, *Biochim. Biophys. Acta* **24,** 604 (1957).

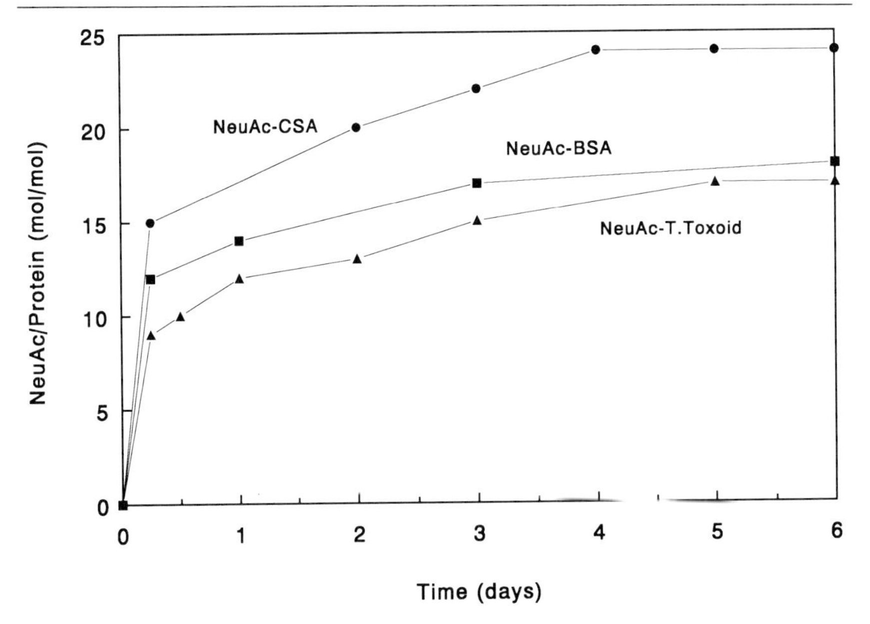

FIG. 1. Time course of conjugation of **1** to tetanus toxoid (▲) and bovine (■) and chicken (●) albumin by reductive amination with $NaBH_3CN$.

lected and tested for residual antiprotein carrier antibodies by an enzyme-linked immunosorbent assay (ELISA) (see below). When necessary, the procedure is repeated in the same manner until the absorbances at a serum dilution no less than 10^3 become part of the background readings (OD at 414 nm $\leq$ 0.1).

Quantitative Precipitation Analyses

Quantitative precipitation experiments are carried out using the method of Kabat and Mayer.[13] The sera (1 ml) are diluted with PBS (4 ml). Aliquots (100 μl) are added to microcentrifuge tubes (1.5 ml, Eppendorf), and increasing concentrations of the glycoconjugates or protein carriers (5–100 μg) in a total volume of 200 μl (adjusted with PBS) are added. The tubes are incubated at 37° for 1 hr and 3 days at 4°. Each tube is then centrifuged at 15,000 *g* for 10 min at 25°. The precipitates are washed with cold PBS (3 times), and the protein contents are determined by the method

[13] E. A. Kabat and M. M. Mayer, *in* "Experimental Immunochemistry" (E. A. Kabat, ed.), 2nd Ed. Thomas, Springfield, Illinois, 1961.

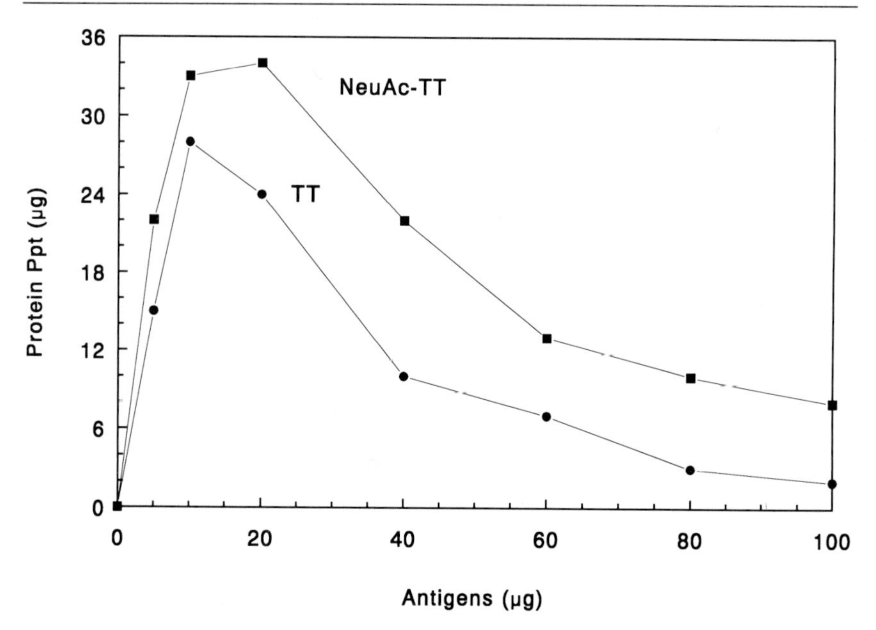

FIG. 2. Quantitative precipitin curves of rabbit anti-NeuAc–TT serum against tetanus toxoid (●) and NeuAc–TT (■) antigens.

of Lowry *et al.*[14] The results of these experiments are illustrated in Figs. 2 and 3.

Quantitative Precipitation of Antibodies with Sialylated Glycopolymer

Undiluted NeuAc–TT or NeuAc–BSA antisera (100 μl) are placed in microcentrifuge tubes. Aliquots (5 to 100 μl) of glycopolymer **4** [14% (w/w) NeuAc, 1 mg/ml PBS] are added and the final volumes adjusted to 300 μl (PBS). The tubes are incubated and centrifuged as above. The precipitin complexes are diluted with 0.1 *M* NaOH (100 μl), and the protein contents are determined by the method of Lowry *et al.*[14] using human γ-globulin as standard (Sigma). The quantitative precipitin curves are depicted in Fig. 4.

Enzyme-Linked Immunosorbent Assay with Protein Conjugates

The wells of Linbro enzyme immunoassay microtitration plates are coated at 100 μl/well using glycoconjugate or protein carrier (TT and

[14] O. H. Lowry, N. J. Rosebrough, A. L. Farr, and R. J. Randall, *J. Biol. Chem.* **193,** 264 (1951).

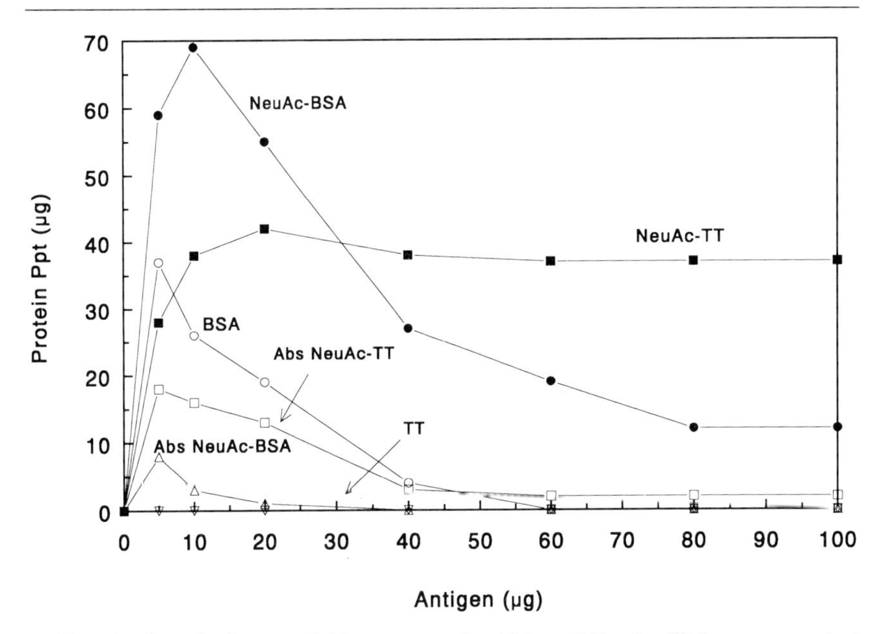

FIG. 3. Quantitative precipitin curves of rabbit anti-NeuAc–BSA serum against NeuAc–BSA (●), NeuAc–TT (■), BSA (○), and TT (▽) and BSA preabsorbed serum against NeuAc–BSA (△) and NeuAc–TT (□), showing residual antisialic acid activities in both sera.

BSA) solutions at 100 μg/ml in PBS. The coatings are done at 4° overnight following a pretreatment at room temperature for 3 hr. The plates are washed with 1% BSA in PBS for 10 min at room temperature. The wells are then filled with 100 μl of 10-fold serial dilutions (PBS) of native or preabsorbed antisera and left for 2 hr at room temperature. After washing, the IgG levels in the plates are determined with horseradish peroxidase-labeled goat anti-rabbit IgG (H+L) (100 μl of a 2000-fold diluted 1-mg sample) using ammonium 2,2′-azinobis(3-ethylbenzothiazoline sulfonate) (ABTS) and H_2O_2 as substrates in cacodylic acid buffer. The absorbance at 414 nm is measured after 10 min. The results of the ELISA with the absorbed sera and the carrier proteins alone (BSA or TT) and the corresponding NeuAc conjugates are shown in Fig. 5.

Inhibition of Assay with Sialylated Glycopolymer Used as Coating Antigen

Microtitration plates are rinsed with 95% (v/v) aqueous ethanol, dried, and coated overnight at 4° with the sialylated polymer **4** [32% (w/w)

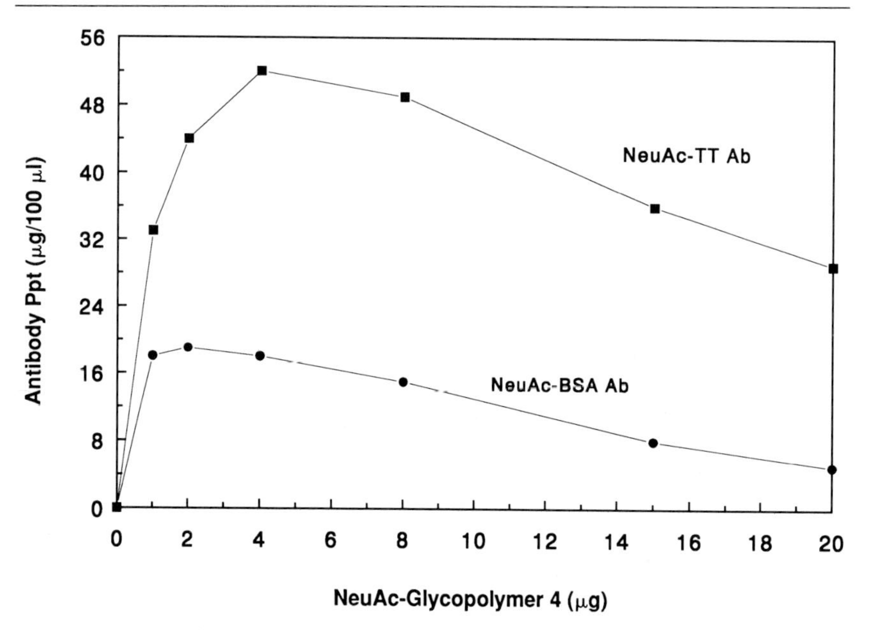

FIG. 4. Quantitative precipitin curves of rabbit anti-NeuAc–BSA (●) and anti-NeuAc–TT (■) sera against the NeuAc–glycopolymer **4**.

NeuAc, 10 μg/ml PBS, 100 μl/well]. The plates are blocked with CSA (0.1% (w/v) in PBS, 100 μl/well) for 20 min at room temperature. The inhibitors **2** and **3** are added (50 μl/well) in serial 2-fold dilutions from a stock solution at 2 mg/ml in PBS. Rabbit antisera (50 μl/well) against the NeuAc–TT conjugate (diluted 5000 times) or against the NeuAc–BSA conjugate (diluted 1250 times) are then added. The final immune sera dilutions are at 1/10,000 and at 1/5000, respectively. The solutions are allowed to equilibrate for 6 hr at room temperature. The plates are washed with water 5 times, and 100 μl of 1/2000 diluted affinity purified horseradish peroxidase-labeled goat anti-rabbit IgG (H+L) antibody is added to each well. The incubation is allowed to proceed for 3 hr at room temperature. The plates are washed again as above, and ABTS solution (50 μl, 1 mg/ml) in 0.2 *M* phosphate–citrate buffer, pH 4.0, containing 0.01% (v/v) H_2O_2 is added. After 30 min, 5 μl KCN (0.27 μ*M*) is added to stop the enzymatic reaction, and the absorbance is read at 410 nm on a Dynatech (Alexandria, VA) MR 600 microtiter plate reader.

The calculations are done as follows. Samples in edge wells are excluded. The values of three samples are averaged. Control wells are filled

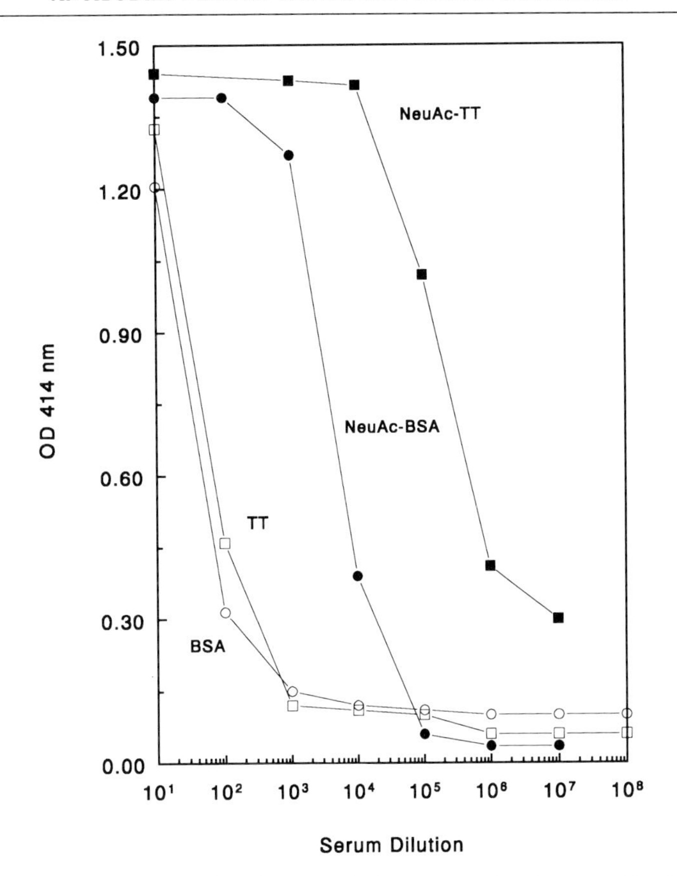

FIG. 5. Microtiter plate ELISA of absorbed rabbit sera against NeuAc–TT (■), TT (□), NeuAc–BSA (●), and BSA (○) with affinity-purified horseradish peroxidase-labeled goat anti-rabbit IgG using ABTS/H_2O_2 as enzyme substrates.

with an unrelated glycopolymer, *N*-acetylglucosamine-*co*-acrylamide, instead of sialylated polymer **4**, and the wells are blocked with CSA and processed with the rabbit antisera as above. The absorbance of control wells is subtracted from the absorbance of the samples after corrections for the absorbances at 570 nm, following the manufacturer's recommendation:

$$\% \text{ Inhibition} = \frac{A_{(\text{no inhibitor})} - A_{(\text{with inhibitor})}}{A_{(\text{no inhibitor})}} \times 100$$

where $A_{(\text{no inhibitor})}$ is the absorbance of the sample after correction for the control and $A_{(\text{with inhibitor})}$ is the absorbance of the samples in the presence of inhibitor after correction for the control. The inhibition curves for both the anti-NeuAc–TT and anti-NeuAc–BSA sera are illustrated in Fig. 6.

Results and Discussion

Neoglycoproteins were prepared by reductive amination of 2-oxoethyl α-sialoside (**1**) with the ε-amino groups of the lysine residues of bovine (BSA) or chicken serum albumin (CSA) or tetanus toxoid (TT) using sodium cyanoborohydride ($NaBH_3CN$) as the reducing reagent (Scheme 1). The time course for the couplings is illustrated in Fig. 1. The lower

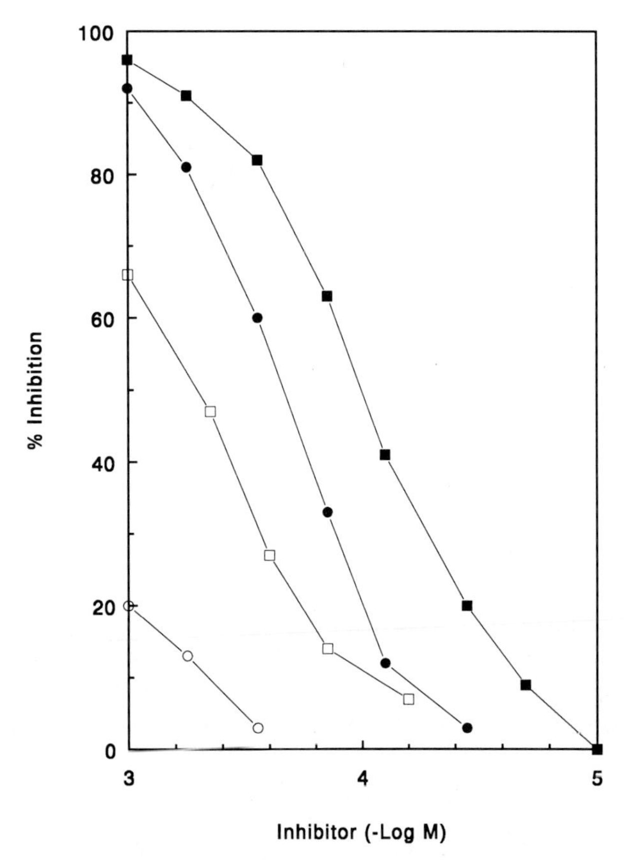

FIG. 6. Inhibition of microtiter plate ELISA using NeuAc–polymer **4** as coating antigen and rabbit anti-NeuAc–TT (■, □) and anti-NeuAc–BSA (●, ○) sera with allyl (■, ●) (**3**) and methyl (□, ●) (**2**) α-sialoside inhibitors. Detection of rabbit IgG was achieved with horseradish peroxidase-labeled goat anti-rabbit IgG.

level of incorporation of the α-sialoside relative to that of neutral oligosaccharides has been attributed to the anionic nature of α-sialosides. The buildup of negative charges on the newly formed neoglycoproteins prevents further incorporation beyond a certain level. This can be avoided by coupling the α-sialoside as its methyl ester. In this manner, up to 45 α-sialoside residues per molecule of BSA (76% yield, based on 59 L-lysine residues) have been coupled.[9]

The immunogenicities of the NeuAc–BSA and NeuAc–TT conjugates were evaluated in rabbits using Freund's complete adjuvant. The antibody titers and specificities were measured using quantitative precipitation, enzyme-linked immunosorbent assays (ELISA), and inhibition of ELISA assays. The ELISA using affinity-purified goat anti-rabbit IgG (H+L) supports the fact that the antisialic acid antibodies of both sera belong to the IgG isotypes.

The quantitative precipitation curves of the two rabbit sera against their homologous neoglycoprotein antigens NeuAc–BSA and NeuAc–TT are illustrated in Figs. 2 and 3, respectively. The two rabbit antisera exhibited the same pattern in that both showed higher titers of antibodies against their conjugates than against the respective protein carriers alone. This indicated that antibodies were also formed against the NeuAc hapten. This behavior was also observed in the ELISA experiments (Fig. 5).

The most striking evidence for the presence of specific antisialic acid antibodies was obtained using a NeuAc-containing water-soluble polymer (**4**)[10] as coating antigen, since the polymer **4** shares only the sialic acid residues in common with the neoglycoproteins used as immunogens. Figure 4 shows the quantitative precipitin curves of the sera titrated with polymer **4**. The amount of protein precipitated in the assays should be directly correlated with the titer of the sialic acid-specific antibody since the polymer **4** gave no color in the colorimetric protein analysis. Figure 6 shows that the glycopolymer **4** coated on the microtiter plates could capture the antibodies and that the inhibition of binding of rabbit sera to the polymer **4** by methyl (**2**) and allyl (**3**) α-sialosides is quite different using α-sialosides containing two different aglycons (**2** and **3**). Using the concentrations of inhibitors required for 50% inhibition (Fig. 6), the anti-NeuAc–BSA antibodies showed a marked preference for **3** (240 μM) over **2** (no inhibition at 3 mM), thus indicating some linker effect. This effect is less pronounced in the case of the anti-NeuAc–TT antibodies (100 μM for **3** and 630 μM for **2**).

In conclusion, the synthetic NeuAc–protein conjugates were immunogenic in rabbits. Both conjugates elicited immunoglobulins of the IgG class. The antibody titer against the NeuAc–TT conjugate was significantly higher than that against the NeuAc–BSA conjugate in spite of the fact that the latter appeared to contain more linker-specific antibodies.

[25] Syntheses and Functions of Neoproteoglycans: Lipid-Derivatized Chondroitin Sulfate with Antiadhesion Activity

By NOBUO SUGIURA and KOJI KIMATA

Introduction

Proteoglycans are macromolecules composed of core proteins with covalently bound glycosaminoglycans (GAGs). The GAGs are linear polymers of repeating uronosylhexosamine disaccharide units with sulfate in various positions. Numcrous acidic groups easily adsorb a number of water molecules to occupy a large volume and also strongly interact with various types of cationic molecules. Aggrecan in cartilage is a typical example for the involvement of such characteristics of GAGs in the function of proteoglycans.[1] However, studies[2,3] have revealed that some specific structural domains in GAGs can interact with other functional molecules. Heparan sulfates have the activity to bind basic fibroblast growth factor (bFGF) through specific units in the chains and greatly affect their activities and metabolism.[3]

The core protein moiety generally has the ability to bind to other extracellular matrix molecules. Aggrecan, for example, is bound at the amino-terminal globular domains to hyaluronan and link proteins and also at the carboxyl-terminal domains with C-type lectinlike activity to some ligand molecules in the matrix.[4] Proteoglycans are considered, therefore, as molecules which enable GAGs to be immobilized in extracellular and/or cell surface proteins by such core protein-mediated interactions.

Adhesion of cells to extracellular substrate molecules is one of the fundamental processes for formation and maintenance of animal tissues and bodies. We have previously shown that chondroitin sulfate proteoglycans had an inhibitory effect on all types of cell adhesion to substrates and suggested a general role of chondroitin sulfate proteoglycans in modulating a variety of cell–substrate interactions.[5] Not only treatment of the

[1] K. Kimata, H.-J. Barrach, K. S. Brown, and J. P. Pennypacker, *J. Biol. Chem.* **256,** 6961 (1981).

[2] H. Habuchi, S. Suzuki, T. Saito, T. Tamura, T. Harada, K. Yoshida, and K. Kimata, *Biochem. J.* **285,** 805 (1992).

[3] L. Kjellen and U. Lindahl, *Annu. Rev. Biochem.* **60,** 443 (1991).

[4] K. Doege, M. Sasaki, E. Horigan, J. R. Hassell, and Y. Yamada, *J. Biol. Chem.* **262,** 17757 (1987).

[5] M. Yamagata, S. Suzuki, S. K. Akiyama, K. M. Yamada, and K. Kimata, *J. Biol. Chem.* **264,** 8012 (1989).

proteoglycans with various proteases or chondroitin sulfate-lyases (e.g., chondroitinase AC II) but also blocking the proteoglycan immobilization to substrate-coated dishes abolished the inhibitory effects on the cell adhesion.[5] Results suggested that chondroitin sulfate chains in the proteoglycans are active sites but the core protein-mediated immobilization of the chains is essential for the activity.

Based on these results, we have come to the idea that chondroitin sulfate chains alone would show the inhibitory effects on cell adhesion if they are immobilized onto substrate dishes, which natural chondroitin sulfate proteoglycans do. We synthesized new compounds by conjugating glycosaminoglycans (GAGs) to dipalmitoyl-L-α-phosphatidylethanolamine (PE), whose lipophilic nature would favor immobilization. Chondroitin sulfate-derivatized PE (CS–PE) was completely substitutive for chondroitin sulfate proteoglycans, preserving the inhibitory activity of cell adhesion. When compared to other types of GAG-derivatived PE in the activity, CS–PE was more inhibitory than any other GAG-derivatized PEs. In addition, a significant amount of GAG-derivatized PE was found to be immobilized onto the cell surface as we expected.[6] The compounds can totally mimic the function of natural chondroitin sulfate proteoglycans, which has encouraged us to name them "neoproteoglycans." As a neoproteoglycan, CS–PE may be useful as a modulator of a variety of cellular events such as cell growth, cell migration, and cell morphology which involve cell–substrate interactions.

In this chapter we describe methods for the conjugation and purification of these new compounds, neoproteoglycans, and also demonstrate their proteoglycan-like natures and functions.

Materials

Chondroitin sulfate (CS) from shark cartilage (M_r 45,000; 4-sulfate to 6-sulfate ratio ~1 : 5), CS from whale cartilage (M_r 20,000; 4-sulfate to 6-sulfate ratio ~4 : 1), CS from bovine trachea cartilage (M_r 15,000; 4-sulfate to 6-sulfate ratio ~1 : 1.4), dermatan sulfate (DS) from porcine skin (M_r 15,000), heparin (HP) from porcine intestinal mucosa (M_r 15,000), chondroitinase ABC and Δ^4-unsaturated disaccharide standards are obtained from Seikagaku Corp. (Tokyo, Japan). Chondroitin sulfates of various sizes, hyaluronan (HA) from cock's comb (M_r 10,000), and heparan sulfate (HS) from mouse EHS tumor (M_r 40,000) are gifts of Seikagaku Corp. Fluorescein-labeled GAGs, in which fluorescein groups are linked

[6] N. Sugiura, K. Sakurai, Y. Hori, K. Karasawa, S. Suzuki, and K. Kimata, *J. Biol. Chem.* **268,** 15779 (1993).

to 1–2% of the hexuronic acid carboxyl residues, are prepared by the carbodiimide method using fluoresceinamine.[7] Dipalmitoyl-L-α-phosphatidylethanolamine (PE), dipalmitoyl-L-α-phosphatidic acid, trypsin (Type XIII, TPCK treated), soybean trypsin inhibitor (type 1-S), and bovine serum albumin are obtained from Sigma Chemical Co. (St. Louis, MO).

Superose 6HR 10/30 and Sepharose CL-4B are obtained from Pharmacia LKB Biotechnology (Tokyo, Japan); TSK-gel G4000 PWXL, G3000 PWXL, G2500 PWXL, TSK-gel Phenyl 5PW, and Phenyl Toyopearl 650M are from Tosoh (Tokyo, Japan); Ultrafree C3 (0.22 μm) filters are from Millipore Japan (Tokyo, Japan). Monoclonal antibody MO-225 directed to chondroitin sulfate is obtained from Seikagaku Corp. Peroxidase-conjugated goat anti-mouse immunoglobulin M (IgM) is from Organon Teknika Corp. (Durham, NC); *Streptomyces chromofucus* phospholipase D is from Asahi Chemical Ind. Co. (Shizuoka, Japan). The molecular weight of each GAG is determined with a light-scattering spectrophotometer (Model DLS-700, Otsuka Electronics Co., Tokyo, Japan)[8] and/or by gel filtration on three columns of TSK-gel G4000 PWXL, G3000 PWXL, and G2500 PWXL using CS molecular weight standards.

Laminin, bovine plasma vitronectin, and pig skin Type I collagen are obtained from Nitta Gelatin Inc. (Osaka, Japan). Bovine plasma fibronectin is prepared by the method of Yamada.[9] Dulbecco's modified Eagle's medium, Eagle's minimal essential medium, Hanks' balanced salt solution, and Dulbecco's phosphate-balanced salt solution are obtained from Nissui Seiyaku Co. (Tokyo, Japan); fetal calf serum is from Flow Laboratories (McLean, VA); 96-well polystyrene microtitration plates (MS-3496F) and plate seals (MS-30010) are from Sumitomo Bakelite Co. (Tokyo, Japan); and tissue culture plates are from Becton Dickinson Labware (Falcon) (Lincoln, Park, NJ).

BHK-21 (baby hamster kidney) cells, CHO-K1 (Chinese hamster ovary) cells, bovine aorta endothelial cells, bovine smooth muscle cells, and chick embryonic fibroblasts are cultured in Eagle's minimal essential medium containing 10% v/v fetal calf serum or in Dulbecco's modified Eagle's medium containing 10% v/v fetal calf serum as described previously.[5]

[7] A. Ogamo, K. Matsuzaki, H. Uchiyama, and K. Nagasawa, *Carbohydr. Res.* **105,** 69 (1982).

[8] Y. Ueno, Y. Tanaka, K. Horie, and K. Tokuyasu, *Chem. Pharm. Bull.* **36,** 4971 (1988).

[9] K. M. Yamada, *in* "Immunochemistry of the Extracellular Matrix" (H. Futhmayer, ed.), p. 111. CRC Press, Boca Raton, Florida, 1981.

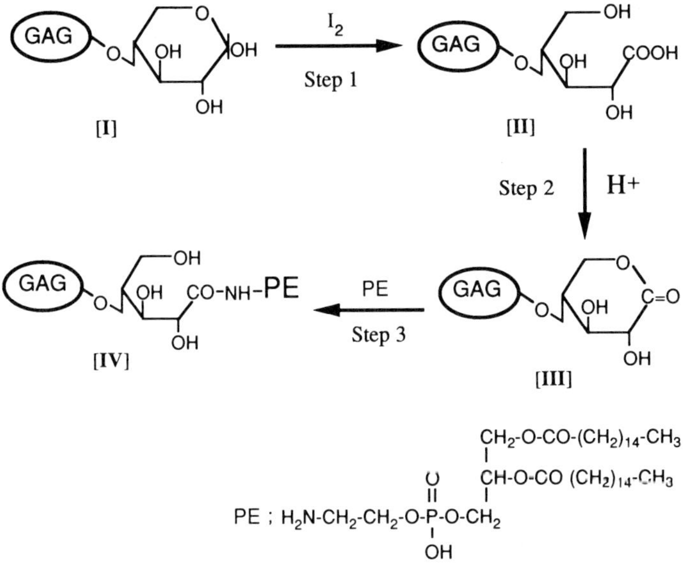

FIG. 1. Chemical synthesis of neoproteoglycans involving GAG-derivatized dipalmitoyl-L-α-phosphatidylethanolamine (GAG–PEs).

Methods and Results

Preparations of Neoproteoglycans

Chemical Syntheses of Glycosaminoglycan–Dipalmitoyl-L-α-phosphatidylethanolamine Conjugates. An outline of the principle of the methods[10] is presented in Fig. 1. Intact GAG chains (Fig. 1, **I**) are oxidized selectively at the reducing terminal sugar (**II**) to generate the lactone group (**III**) which can easily couple to amines. The lactone-carrying GAGs can be dissolved in the amphiphilic solvent dimethylformamide (DMF) for the subsequent N-acylation of PE to form GAG–PEs (**IV**).

1. The reducing ends of CS sodium salt from shark cartilage (M_r 20,000) (10 g, 0.5 mmol) are oxidized with iodine (1.3 g, 5 mmol) in 500 ml of methanol–water (1 : 10, v/v) at room temperature for 6 hr. The solution is then concentrated to about 100 ml on a rotary evaporator, and polysaccharides are precipitated with 3 volumes of 95% (v/v) ethanol containing 1.3% (w/v) potassium acetate.

[10] K. Kobayashi, H. Sumitomo, and Y. Ina, *Polymer J.* **15,** 667 (1983).

2. The precipitate (**II**) is dissolved in 100 ml water and passed through a 90-ml column of Dowex 50W-X8 (H^+). The aqueous solution is concentrated at 40° *in vacuo,* and the remaining water is replaced by the addition of DMF followed by evaporation of water at 40° *in vacuo* (several cycles). The DMF solution (500 ml) is allowed to stand at 4° for 72 hr.

3. Dipalmitoyl-L-α-phosphatidylethanolamine (PE) (1.7 g, 2.5 mmol) in 50 ml of chloroform is added to the above solution, and the mixture is stirred at 60° for 6 hr. The reaction mixture is concentrated *in vacuo*. Fifty milliliters of 0.2 *M* NaCl is added to the residue, and undissolved material (excess PE) is removed by centrifugation. The reaction product (**IV**) is obtained by ethanol precipitation, followed by further purification as described below.

The overall yield through the above steps is about 10%. Other PE-derivatized GAGs are synthesized by an essentially similar method. There is a variation in yields between preparations, presumably owing to differences in the efficiency of the coupling reaction.

Other coupling trials are also performed in which periodate-oxidized GAGs are reacted with PE to form Schiff bases, which are then reduced with cyanoborohydride.[11,12] However, very low overall yields are obtained (<3%). The following reasons might be suggested.[6] The small extent of coupling between GAGs and PE can be due to partial cleavage of internal hexuronic acid residues (which contain periodate-sensitive vicinal hydroxyl groups) during the oxidation step. Moreover, long GAG chains are far less soluble in the hydrophobic solvents used for the reductive amination with PE. Heterogeneity of the reducing terminal saccharides of GAGs may also have some effects on reactivity with PE. However, we have not determined what causes the problem.

Purification of Glycosaminoglycan Derivatives. The reaction product (**IV**) is easily purified by two alternative ways as described in the following with similar yields, taking advantage of the characteristics of these novel molecules. (1) The crude product (2 g) is dissolved in 10 ml of 0.2 *M* NaCl and applied to a column of Sepharose CL-4B (2.5 × 120 cm) equilibrated with the same solution. Because CS–PE aggregates owing to its hydrophobic property (see below), CS–PE is eluted at the void volume (V_0) region and separated from free CS retarded in the gel. (2) The crude product in 0.2 *M* NaCl is mixed gently with a hydrophobic resin, TSK-gel Phenyl Toyopearl 650M (200 ml), at 4° overnight. Unadsorbed free CS is washed

[11] P. W. Tang, H. C. Gool, M. Hardy, Y. C. Lee, and T. Feizi, *Biochem. Biophys. Res. Commun.* **132,** 474 (1985).

[12] M. S. Stoll, T. Mizouchi, R. A. Childs, and T. Feizi, *Biochem. J.* **256,** 661 (1988).

out with 600 ml of the same salt solution. Adsorbed CS–PE is eluted with 30% (v/v) methanol–water and ethanol-precipitated as described above.

Properties of Glycosaminoglycan Derivatives

The purified preparation of CS–PE is subjected to analyses for molecular properties by molecular sieving high-performance liquid chromatography (HPLC), hydrophobic HPLC, light scattering, and enzymatic cleavage.

Self-Aggregation. The elution profiles of CS–PE are studied on a Superose 6HR 10/30 column equilibrated with phosphate-buffered saline, pH 7.4, in the absence and presence of 0.1% (w/v) Triton X-100. In the absence of Triton X-100, the majority of the material is eluted in the V_0 fraction. The presence of Triton X-100 results in a shift of all materials to the eluting position of free CS. The results suggest that amphipathic CS–PE molecules in phosphate-buffered saline tend to form detergent-sensitive aggregates. All other GAG–PE samples tested are identical with regard to this property. Measurement of the size of the CS–PE in the above solution by light scattering indicates the occurrence of M_r 450,000–600,000 aggregates that are about 15–20 times larger on average than the free CS.

Hydrophobic Interaction. The CS–PE is adsorbed onto an analytical hydrophobic column of TSK-gel Phenyl 5PW and eluted with reversed linear gradient elution (0.2 *M* NaCl to distilled water). Free CS appears exclusively in the pass-through fraction.[6] Hydrophobic chromatography of other PE-derivatized GAGs shows elution profiles similar to that for CS–PE. The results suggest that GAG–PEs can bind via hydrophobic interactions to other hydrophobic molecules. The immobilization property of GAG–PE onto plastic plates should also be due to the same hydrophobic interaction.

Phospholipase D Digestion. Digestion of CS–PE with phospholipase D, which is capable of hydrolyzing phosphatidylethanolamine to yield ethanolamine and phosphatidic acid, completely destroys the hydrophobic properties described above. The behavior on thin-layer chromatography of CS–PE before and after phospholipase D digestion and the biochemical properties of the digestion products are consistent with those expected from the proposed structure of CS–PE as shown in Fig. 1.[6]

Functions of Neoproteoglycans: Cell Antiadhesion Activity of Chondroitin Sulfate Derivative

Immobilization onto Polystyrene Plates. The stock solutions of GAG–PEs (2–10 mg/ml) should be prepared in Hanks' solution/20 m*M*

N-(2-hydroxyethyl)piperazine-*N'*-2-ethanesulfonic acid (HEPES) overnight at 4°, and insoluble materials, if any, are removed by filtration through a 0.22-μm Millipore filter. Dilutions of the stock solutions are made with Hanks' balanced salt solution/20 m*M* HEPES just before use. Each well in a polystyrene plate, which has been coated with various adhesive proteins such as fibronectin in 0.1 *M* $NaHCO_3$ (0.5 μg/100 μl/0.32 cm^2 well) overnight at 4° followed by three washes with Dulbecco's phosphate-buffered saline, is then filled with 100 μl of Hanks' salt solution/20 m*M* HEPES containing 0.01–10 μg/100 μl of GAG–PEs or related substances. The plates are usually allowed to stand at room temperature for 2 hr, then rinsed three times with Hanks' solution.

The GAG–PEs are immobilized onto the plates in a dose-dependent manner (Fig. 2). The amounts of GAG–PEs immobilized on plastic plates are determined using PE-derivatized fluorescein-labeled GAGs. Various concentrations of the fluorescent derivatives (from 0.005 to 10 μg/well) are used to coat the wells under the same conditions as described above for unlabeled GAG–PEs. After washing, each well is treated with 0.2 *M* NaOH/0.5% (w/v) Triton X-100 (200 μl/well) for 2 hr at room temperature with gentle shaking, and the solubilized fluorescence is measured (excitation at 490 nm, emission at 515 nm) in an F-3010 fluorescence spectropho-

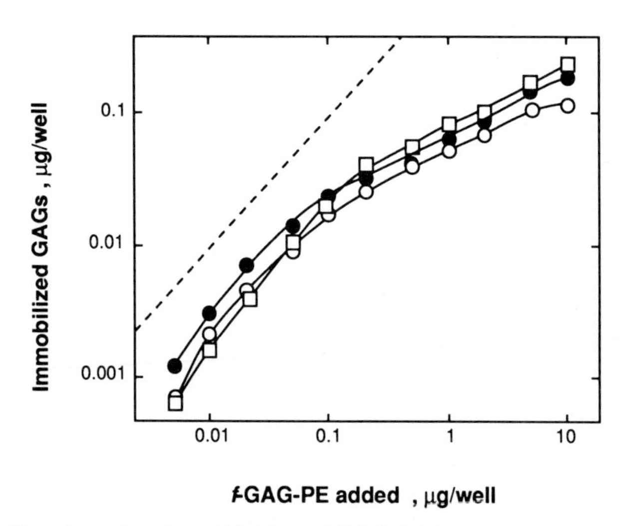

FIG. 2. Dose-dependent immobilization of GAG–PEs onto 96-well microtiter plates. Wells in fibronectin-precoated polystyrene plates were further coated with the indicated concentrations of fluorescein-labeled CS–PE (●), HA–PE (○), and HP–PE (□) at room temperature for 2 hr. The amounts of immobilized GAG–PEs were determined by measuring the fluorescence intensity of each well after treatments described in the text.

tometer (Hitachi Co., Tokyo, Japan). Enzyme-linked immunosorbent assay (ELISA) using MO-225, the monoclonal antibody directed to the chondroitin sulfate chain,[13] can also be applied to measure the immobilization of CS–PE onto the plates, although only relative values are obtained.[6]

Figure 2 also indicates that immobilizations of GAG–PEs approach saturation as the concentration of added GAG–PEs is increased (e.g., at 2 hr, the proportion of immobilized CS–PE decreases from ~20 to ~8% of the input GAG–PE, as the concentration of CS–PE added to each well increases from 0.05 to 0.5 μg/100 μl). The GAG–PE molecules may interact preferentially with polystyrene plates in such a way that the molecules are oriented vertically to the plane of the surface, with the hydrophobic PE moiety associating with the solid polystyrene phase and the hydrophilic GAG chains protruding into the aqueous phase. Consistent with this notion, CS–PE immobilized on the plates is sensitive to chondroitinase ABC added to the medium.[6] The level of CS–PE immobilization (2 hr) onto fibronectin-coated plates is compared with that on uncoated plates. The former is about 1.5–2 times higher than the latter. Because the fibronectin-dependent increase in immobilization is not affected by excess heparin (1 mg/ml) added to the medium, this increase appears not to be due to a specific binding of the chondroitin sulfate moiety of CS–PE to GAG-binding sites on fibronectin.[6]

Effects of Immobilized Glycosaminoglycan Derivatives on Cell–Substrate Adhesion. Adhesion of BHK cells is examined using coated plates described above. An aliquot (1×10^4 cells/100 μl) of a single-cell suspension in Hanks' solution is added to each well, and the plates are incubated for 1 hr at 37°. After cells are fixed at 4° for 5 min by adding 100 μl of 6% (w/v) paraformaldehyde in phosphate-buffered saline to each well, unattached cells are removed by two washes with Hanks' solution. The relative number of adherent cells is determined by the bicinchoninic acid (BCA) protein assay method.[14] Briefly, each well is filled with 200 μl of BCA working solution (Pierce Chemical Co., Rockford, IL). The plate is covered with an adhesive plate seal (MS-30010, Sumitomo Bakelite Co.) and incubated at 60° for 30 min. After the plate is allowed to cool to room temperature, the cover sheet is removed, and the absorbance of each well is determined at 550 nm with a microplate reader (MTP-100, Corona Electric Co., Ibaragi, Japan).

Fibronectin-mediated BHK cell adhesion is progressively inhibited by increasing the concentration of CS–PE used to coat the plates. The initial

[13] M. Yamagata, K. Kimata, Y. Oike, K. Tani, N. Maeda, K. Yoshida, T. Shinomura, M. Yoneda, and S. Suzuki, *J. Biol. Chem.* **262,** 4146 (1987).

[14] G. P. Tuszynski and A. Murphy, *Anal. Biochem.* **184,** 189 (1990).

M_r 20,000 CS–PE concentration required for 50% inhibition (IC_{50}) is 0.06 μg/100 μl/well (Table I). This effect is actually brought about by 0.024 μg of CS–PE immobilized onto the well. The CS–PE derivative is inhibitory not only for the adhesion of other types of cells such as B16F10 melanoma cells to fibronectin-coated plates but also for the cell adhesion to plates coated with other substrate molecules such as laminin, although quantitative comparisons have not been made among different combinations of cell and substrate types. In all the experiments, the ability of CS–PE to inhibit cell–substrate adhesion is completely abolished by treatment of the CS–PE-coated plates with phospholipase D or chondroitinase ABC, prior to the seeding.[6]

Interestingly, at concentrations of GAG–PE lower than 0.1 μg/100 μl/well, none of other GAG–PEs tested significantly inhibit the adhesion of BHK cells to fibronectin-coated plates. These compounds have inhibitory effects only at much higher concentrations (Table I). The results are consistent with the previous observations[5] that the inhibition by proteoglycans is specific with respect to the structure of GAG chain.

TABLE I
INHIBITORY EFFECTS OF NEOPROTEOGLYCANS ON BHK CELL–FIBRONECTIN ADHESION

Sample	Source of GAG moiety	Molecular weight[a]	IC_{50} (μg/well)[b]
	Chondroitin sulfate		
CS–PE	Shark cartilage	45,000	0.060
CS–PE	Shark cartilage	20,000	0.063
CS–PE	Shark cartilage	10,000[c]	0.223
CS–PE	Shark cartilage	6500[c]	1.72
CS–PE	Shark cartilage	3500[c]	>100
CS–PE	Whale cartilage	20,000	0.075
CS–PE	Bovine trachea	15,000	0.078
	Dermatan sulfate		
DS–PE	Porcine skin	15,000	0.40
	Hyaluronan		
HA–PE	Cock's comb	10,000[c]	2.16
	Heparan sulfate		
HS–PE	Mouse EHS tumor	40,000	5.64
	Heparin		
HP–PE	Porcine intestine	15,000	5.86

[a] Average molecular weight was determined by gel filtration and/or light scattering.
[b] Amount of GAG–PE giving 50% inhibition in cell adhesion assay on fibronectin, expressed as GAG–PE immobilization. Note that any GAG–PEs that failed to bind to plates were removed by washing with Hanks' solution before cell seeding.
[c] Partial degradation products produced by digestion with testicular hyaluronidase.

Discussion and Comments

Effect of Chain Size and Sulfation Pattern of Chondroitin Sulfate Moiety

The PE-derivatized shark cartilage chondroitin sulfate with different chain sizes were tested for their inhibitory effect on cell–substrate adhesion. A decrease of average chain size from M_r 45,000 to 10,000 had little effect on the activity. However, a further decrease to M_r 6500 caused a significant reduction of the activity, and the M_r 3500 sample never showed the inhibition activity. The 4-sulfate to 6-sulfate ratio of the chondroitin sulfate moiety seems not to be critical for the inhibitory effect, since the compounds derived from chondroitin sulfates with different ratios showed inhibition activity almost at the same range (Table I).

Heparin-Insensitive Action of Chondroitin Sulfate Derivatives on Cell Adhesion

Numerous reports have noted that heparin inhibits a variety of cell adhesive functions, possibly by displacing cell surface proteoglycans from GAG-binding sites on adhesive proteins (see examples in Refs. 15 and 16). Interestingly, even at the highest concentration (up to 100 mg/100 μl), heparin did not have any significant effect on the fibronectin-mediated cell adhesion, nor did it interfere with the inhibitory effect of CS–PE on fibronectin-coated plates.[6] We also observed a similar level of inhibitory effect of CS–PE on BHK cell adhesion to the 120-kDa RGD-containing fragment of fibronectin, which does not contain known GAG-binding domains.[17] The results suggest that the inhibitory effect of CS–PE on BHK cell adhesion to fibronectin is not related to the cellular recognition events involved in its GAG-binding domains.

Binding of Neoproteoglycan to BHK Cell Surfaces and Effects on Cell–Substrate Adhesion

In aqueous solution, CS–PE, once exposed to intact cells, can associate with the surface of the cells, owing to its hydrophobic lipid moiety.[6] The incorporated CS–PE molecules on the cell surface caused a significant reduction in the rate of cell adhesion to fibronectin-coated plates. How-

[15] K. Hayashi, J. A. Madri, and P. M. Yurchenco, *J. Cell Biol.* **119,** 945 (1992).

[16] J. D. San Antonio, A. D. Lander, T. C. Wright, and M. J. Karnovsky, *J. Cell. Physiol.* **150,** 8 (1992).

[17] M. D. Pierschbacher, E. G. Hayman, and E. Ruoslahti, *Cell (Cambridge, Mass.)* **26,** 259 (1981).

ever, the amount of cell surface CS–PE required for 40% inhibition is 1.0 $\mu g/cm^2$ surface area of cells. The same degree of inhibition was attained only with 0.044 μg of the CS–PE located on the same surface area of fibronectin-coated plates, indicating that the inhibition activity of CS–PE varies depending on its topological location, being about 20-fold higher on the fibronectin-coated plate than on the cell surface. We do not know the reason at present. Iida *et al.*[18] have postulated a model in which chondroitin sulfate proteoglycan located on the surface of melanoma cells may modify the function of integrin on the same surface. This might explain partly the observed difference in the inhibitory activity owing to the variation of the topological location.

Mechanism for Cell Antiadhesion Activity of Neoproteoglycan

It is possible that the inhibition of fibronectin-mediated cell adhesion may have been caused, at least in part, by an interaction of CS–PE molecules on a solid phase with a cell membrane factor capable of modulating the properties of receptors (e.g., integrins) required for cell adhesion, although we do not have any direct evidence. As has been found in various systems for synthetic neoglycolipids,[11,12,19,20] carbohydrate conjugates to lipids confer novel properties such as ability to bind to natural or synthetic matrices to have multivalency of the carbohydrate chains for more efficient interaction with carbohydrate-binding proteins (e.g., receptors, antibodies, and lectins). This carbohydrate-clustering effect may be a critical factor in considering the inhibitory effect of CS–PE on cell adhesion.

Possible Applications of Neoproteoglycans

The specific and sensitive mode of action of CS–PE suggests that this compound may be useful in ascertaining whether chondroitin sulfate proteoglycan-induced changes in a variety of biological processes (e.g., cell migration, cell proliferation, cell differentiation, cell–cell recognition and adhesion, tissue morphogenesis, wound healing, and metastasis) result from the function of the GAG chains attached. In addition, we have demonstrated that CS–PE, but not free chondroitin sulfate, inhibits not only the invasion of pannus onto rheumatoid articular cartilage[21] but also

[18] J. Iida, A. P. N. Skubitz, L. T. Furcht, E. A. Wayner, and J. B. McCarthy, *J. Cell Biol.* **118,** 431 (1992).

[19] S. Roseman, *Trends Glycosci. Glycotechnol.* **3,** 438 (1991).

[20] Y. C. Lee, *Trends Glycosci. Glycotechnol.* **4,** 251 (1992).

[21] N. Sugiura, S. Iwasaki, S. Aoki, Y. Hori, S. Suzuki, and K. Kimata, *Cell Struct. Funct.* **17,** 510 (1992).

the experimental metastasis of B16F10 mouse melanoma cells[22] in both animal and culture model systems in which the involvement of cell adhesion is supposed to be essential.

Acknowledgments

We thank Dr. H. Sumitomo for helpful discussion on the synthesis of GAG–PE derivatives, other members in our laboratory, M. Otsuji, M. Yamagata, T. Shinomura, N. Yamakawa, and S. Suzuki, in particular, for stimulating discussion and providing some compounds and informations, K. Yoshida and T. Harada, Seikagaku Corp., for providing some compounds and enzymes, and also K. Karasawa, Y. Hori, and K. Sakurai, collaborators of Seikagaku Corp. Tokyo Institute.

[22] K. Karasawa, N. Sugiura, Y. Hori, J. Onoya, K. Sakurai, S. Suzuki, and K. Kimata, *Seikagaku* **63,** 667 (1991).

[26] *In Vivo* Quantification of Asialoglycoprotein Receptor

By MASATOSHI KUDO, DAVID R. VERA, and ROBERT C. STADALNIK

Introduction

Technetium-labeled galactosyl neoglycoalbumin (NGA) is a receptor-binding radiopharmaceutical[1] designed for diagnosis of liver disease via kinetic modeling of liver and heart time–activity data. With this radioligand *in vivo* measurement of asialoglycoprotein receptor (ASGP-R) biochemistry can be employed. The most relevant quantities pertaining to biochemical function are (1) receptor concentration $[R]_o$, which represents the total amount of ASGP-R per liter of hepatic plasma, and (2) the forward binding rate constant k_b, which represents the affinity of ASGP-R for the radiolabeled neoglycoprotein. In this chapter we present a method of *in vivo* quantification of the asialoglycoprotein receptor using radiolabeled neoglycoproteins.

Theory

Assay Principle

The goal of the assay is the noninvasive measurement of a specific biochemical within its native environment. This is in contrast to a destruc-

[1] R. C. Stadalnik, M. Kudo, W. C. Eckelman, and D. R. Vera, *Invest. Radiol.* **28,** 64 (1993).

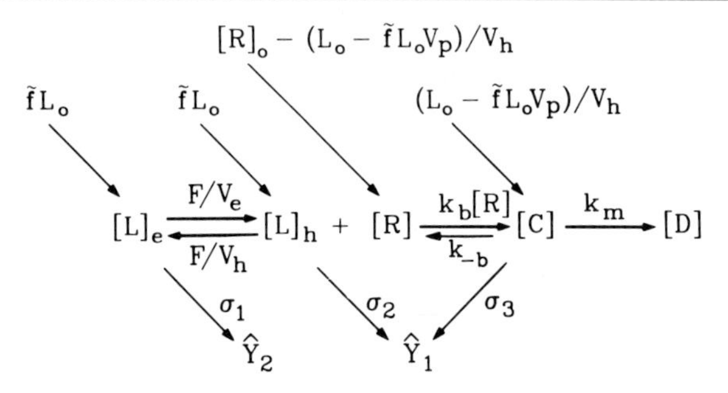

FIG. 1. The kinetic model for technetium-99*m*-labeled galactosyl neoglycoalbumin (Tc-NGA) receptor binding uses a bimolecular chemical reaction. The objective of the kinetic analysis is the simultaneous estimation of parameters $[R]_0$ k_b, V_h, V_e, and F. These parameters represent the ASGP-R concentration, TcNGA–ASGP-R forward binding rate constant, hepatic plasma volume, extrahepatic plasma volume, and hepatic plasma flow, respectively.

tive assay, which is carried out within a test tube. Because the *in vivo* assay is not invasive, we can sample the entire liver rather than a small biopsy specimen. Additionally, because the assay is not carried out in a test tube, we can measure receptor concentration and affinity without disrupting the tissue.

There are three characteristics of the assay which differ from the standard ligand-binding experiment. First, it uses a single injection of the radioligand. Typically, *in vitro* assays require multiple dilutions of the ligand. Second, the *in vivo* assay does not assume steady-state conditions. This assumption, which is typically employed by most *in vitro* assays, prevents the unique measurement of each rate constant for the receptor or enzyme under study.[2] Third, the *in vivo* assay is a transient state experiment. Up to 180 data points are acquired by a gamma camera to monitor the rate at which the liver extracts the radioligand from the blood of the subject. For this reason, we refer to the assay as a functional imaging study.

Kinetic Model

The kinetic model[3] used to describe the ASGP-R-mediated extraction of a labeled neoglycoalbumin is illustrated in Fig. 1. The receptor–ligand reaction is bimolecular and therefore displays second-order kinetics.[4] The

[2] G. G. Hammes and P. R. Schimmel, *in* "The Enzymes" (P. D. Boyer, ed.), 3rd Ed., Vol. 2. Academic Press, New York, 1970.

[3] D. R. Vera, R. C. Stadalnik, W. L. Trudeau, P. O. Scheibe, and K. A. Krohn, *J. Nucl. Med.* **32,** 1169 (1991).

[4] D. R. Vera, E. S. Woodle, and R. C. Stadalnik, *J. Nucl. Med.* **30,** 1519 (1989).

TABLE I
MODEL STATES AND PARAMETERS

Symbol	Definition	Units
$[L]_e$	NGA[a] concentration in the extrahepatic plasma	μM
$[L]_h$	NGA concentration in the hepatic plasma	μM
$[C]$	NGA–ASGP-R[b] complex concentration in the hepatocyte	μM
$[D]$	Metabolic product	μM
F	Hepatic plasma flow	liters
V_e	Extrahepatic plasma volume	liters
V_h	Hepatic plasma volume	liters
k_b	Forward binding rate constant	μM^{-1} $\min^{-1}$
k_{-b}	Reverse binding rate constant	$\min^{-1}$
k_m	Metabolic rate constant	$\min^{-1}$
$[R]_o$	Initial ASGP-R concentration[c]	μM
L_o	Amount of NGA injected	μmol
f	Fraction of injected NGA per liter of plasma	L^{-1}
Y_1	Liver ROI[d] count rate	cpm
Y_2	Heart ROI count rate	cpm
σ	Detector sensitivity coefficient	cpm nM^{-1}

[a] Galactosyl neoglycoalbumin.
[b] Asialoglycoprotein receptor.
[c] Defined as the total amount of asialoglycoprotein receptor divided by V_h.
[d] Region of interest.

model uses four states expressed in units of concentration and eight parameters. The symbols and units are listed in Table I.

The model employs the following assumptions. (1) The receptor concentration $[R]_o$ can be defined as the total quantity of receptor in the liver divided by the ligand–receptor reaction volume.[5] (2) The ligand–receptor reaction volume equals the hepatic plasma volume V_h. This relationship represents the upper limit of the reaction volume.[5] (3) Binding of NGA to the ASGP receptor is operationally irreversible.[6] (4) The hepatic extraction of NGA is sufficiently low such that $[L]_e = [L]_h$.[5] (5) The amount of NGA injected L_o per liter of plasma volume is greater than the concentration of endogenous ASGP ligand.[4] (6) The concentration of NGA within the plasma is sufficiently homogeneous within 3 min after injection. This is based on smooth heart time–activity curves 3 min postinjection.[3] (7) Nonspecific and nonhepatic binding, as well as interstitial diffu-

[5] D. R. Vera, K. A. Krohn, P. O. Scheibe, and R. C. Stadalnik, *IEEE Trans. Biomed. Eng.* **BME-32,** 312 (1985).

[6] D. R. Vera, K. A. Krohn, R. C. Stadalnik, and P. O. Scheibe, *J. Nucl. Med.* **25,** 779 (1984).

sion, of NGA is insignificant.[7] (8) At early time points (<5 min) NGA is conserved between states $[L]_e$, $[L]_h$, and [C] with virtually no metabolized NGA; therefore $L_o = [L]_e V_e + [L]_h V_h + [C] V_h$.[7] (9) Replacement of free receptor by recycling or synthesis is negligible during the 30-min study.[8] (10) Regional variation of hepatic plasma flow, ligand–receptor, forward binding rate constant, and receptor concentration can be ignored.[5] Consequently, the model measures the average value of each parameter. (11) The gamma camera detects Tc-NGA within hepatic plasma and the Tc-NGA–ASGP-R complex with equal efficiency.[5] Therefore, σ_2 and σ_3 are equal.

The kinetic model can be described by the following set of differential equations,[9] which by virtue of the bimolecular ligand–receptor reaction is nonlinear and must be solved by numerical simulation:

$$\frac{d[L]_e}{dt} = \frac{F}{V_e}[L]_h - \frac{F}{V_e}[L]_e$$

$$\frac{d[L]_h}{dt} = \frac{F}{V_h}[L]_e - \frac{F}{V_h}[L]_h - k_b[L]_h([R]_o - [C]) + k_{-b}[C]$$

$$\frac{d[C]}{dt} = k_b[L]_h([R]_o - [C]) - k_{-b}[C] - k_m[C]$$

$$\frac{d[D]}{dt} = k_m[C]$$

Observational equations are used to simulate the detection of radiation emitted from the liver and heart:

$$Y_1 = \frac{1}{\Delta t}\int_t^{t+\Delta t} (\sigma_2[L]_h + \sigma_3[C])dt$$

$$Y_2 = \frac{1}{\Delta t}\int_t^{t+\Delta t} (\sigma_1[L]_e + \sigma_4[L]_h + \sigma_5[C])dt$$

Because liver and heart counts are accumulated during a fixed time interval Δt, the simulated liver (Y_1) and heart (Y_2) data at a given time t are produced by integrating the observed states over t through $t + \Delta t$. Observations of the liver include detection of radiation from the radioligand within the

[7] D. R. Vera, K. A. Krohn, R. C. Stadalnik, and P. O. Scheibe, *Radiology* **151,** 191 (1984).

[8] R. C. Stadalnik, D. R. Vera, E. S. Woodle, W. L. Trudeau, B. A. Porter, R. E. Ward, K. A. Krohn, and L. F. O'Grady, *J. Nucl. Med.* **26,** 1233 (1985).

[9] D. R. Vera, P. O. Scheibe, K. A. Krohn, W. L. Trudeau, and R. C. Stadalnik, *IEEE Trans. Biomed. Eng.* **BME-39,** 356 (1992).

hepatic plasma and the ligand–receptor complex (states $[L]_h$ and [C]); therefore, the detection coefficient σ_2 is set equal to σ_3. Observations of the heart include a sample of the radioligand with the extrahepatic plasma (state $[L]_e$) and scattered radiation from the liver. Because the scattered radiation is a small fraction of the activity detected directly from the heart, σ_1 will be significantly larger than σ_4 and σ_5.

Methods

Radiolabeled Neoglycoproteins

The synthetic ligand is prepared by covalent coupling of 2-imino-2-methoxyethyl-1-β-D-galactopyranoside (IME-thiogalactose) to normal human serum albumin based on the method outlined elsewhere in this volume.[10,11] Two variations of radiolabeled neoglycoalbumin exist. The first is Tc-NGA,[10] which uses an electrolytic method to label the albumin backbone directly with technetium-99*m*. The second, Tc-GSA,[11] employs an indirect technetium label via chelation to diethylenetriaminepentaacetic acid, which is covalently attached to lysine moieties of the albumin. When Tc-NGA is used as the radiopharmaceutical, each patient receives an intravenous injection of 1.8×10^{-9} mol per kilogram of body weight. If Tc-GSA is used, each patient receives a fixed amount of ligand, 4.5×10^{-8} mol. Injected radionuclide activity for each protocol is 5 mCi. A dilution of the injectate is saved as a counting standard.

Receptor Imaging and Data Acquisition

Measurement of the ASGP-R concentration[1] of a subject requires a gamma camera and an image processing computer; both are standard equipment in any modern nuclear medicine clinic. Patients are imaged in the supine position under the gamma camera fitted with a low energy (150 keV) parallel hole collimator. The field-of-view is positioned over the entire liver and heart. Prior to the study the image processing computer is set up to acquire 120 digital images in the form of a 128 × 128 matrix at a rate of four per minute. The study is started by initiation of the computer followed by the intravenous injection of the radiopharmaceutical. Three minutes after injection, approximately 1 ml of blood is drawn from a peripheral vein of the opposing arm. Computer acquisition is halted

[10] D. R. Vera, R. C. Stadalnik, M. Kudo, and K. A. Krohn, this volume [29].

[11] M. Kudo, K. Washino, Y. Yamamichi, and K. Ikekubo, this volume [27].

TABLE II
INITIAL VALUES FOR AUTOMATED CURVE-FITTING OF TcNGA FUNCTIONAL IMAGING DATA

Parameter	Equation
V_h	0.078[a] TPV[b]
V_e	TPV − V_h
F	TBW[c] (1 − hct[d]/100) × 0.023 liter/min/kg[e]
$[R]_o$	Based on curvature of liver data[f]
k_b	NGA galactose density × 0.075 μM^{-1} min^{-1} [g]
k_{-b}	Zero
k_m	Zero
$[L]_e(\tau)$[h]	$\tilde{f}L_o$
$[L]_h(\tau)$	$\tilde{f}L_o$
$[C](\tau)$	$L_o - \tilde{f}L_o(V_e + V_h)/V_h$
$[D](\tau)$	Zero
σ_1[i]	$Y_2(\tau)/\{\tilde{f}L_o\}$
σ_2	$0.1\sigma_1$
σ_3	$Y_1(\tau)/\{L_o - V_h[L]_h(\tau) - [C](\tau)\}/V_h$

[a] Assume 13% of TPV in liver and 60% of liver plasma in hepatic sinusoids [C. V. Greenway and R. D. Stark, *Physiol. Rev.* **51,** 23 (1971)].

[b] Total plasma volume based on total body weight, height, and hematocrit from S. B. Nadler, J. U. Hidalgo, and T. Bloch, *Surgery* **51,** 224 (1962).

[c] Total body weight.

[d] Hematocrit.

[e] J. Caesar, S. Shaldon, J. Chiandussi, L. Guevara, and S. Sherlock, *Clin. Sci.* **21,** 43 (1961).

[f] D. R. Vera, R. C. Stadalnik, W. L. Trudeau, P. O. Scheibe, and K. A. Krohn, *J. Nucl. Med.* **32,** 1169 (1991).

[g] D. R. Vera, K. A. Krohn, R. C. Stadalnik, and P. O. Scheibe, *J. Nucl. Med.* **25,** 779 (1984).

[h] Time postinjection of blood sample.

[i] D. R. Vera, P. O. Scheibe, K. A. Krohn, W. L. Trudeau, and R. C. Stadalnik, *IEEE Trans. Biomed. Eng.* **BME-39,** 356 (1992).

after 30 min, and time–activity curves for the blood and liver are generated from regions of interest (ROI) for the whole liver and heart.

Computer Analysis

A computer program is used to calculate the model parameters for each functional imaging study.[3,9] Each of four steps is completely auto-

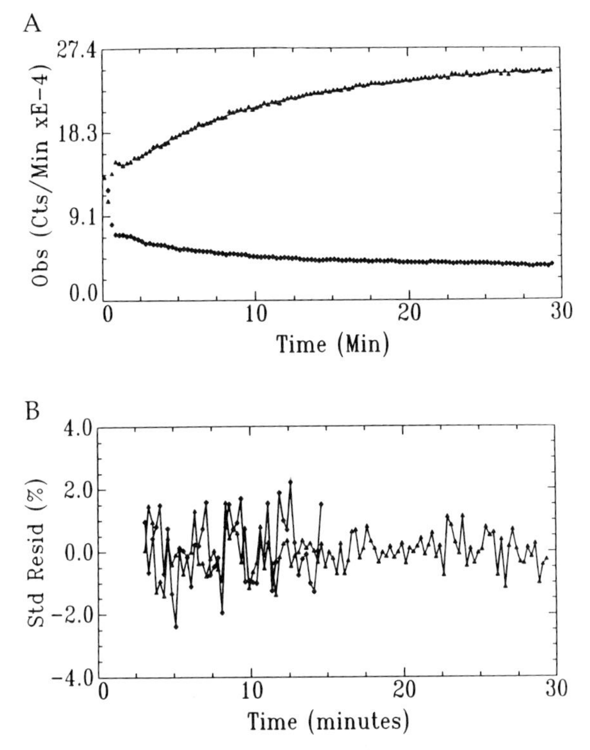

FIG. 2. A Tc-NGA functional imaging study of a healthy subject. (A) The curve-fit produced a receptor concentration of 0.752 ± 0.087 μM and a forward binding rate constant of 2.502 ± 0.908 μM^{-1} $\min^{-1}$. The data points represent decay-corrected count rates for the liver (▲) and heart (◆) regions of interest. The smooth lines represent the kinetic simulations. The reduced χ^2 was 0.738. (B) When plotted against time, the standardized residuals cluster about zero for curve-fits to both liver (▲) and heart (◆) data and did not exceed 3%.

mated. First, the liver and heart time–activity data are transferred from the image processing computer to either a VAX (Digital Equipment Corp., Maynard MA) or Macintosh (Apple Computer, Cupertino, CA) computer. Second, each parameter is assigned an initial value (see Table II) using the following information: the carbohydrate density of the labeled neoglycoalbumin and the sex, height, weight, and hematocrit of the patient. Additionally, the fraction of injected NGA per liter of plasma is calculated from the peripheral hematocrit of the subject, the weight of the 3-min blood sample, and the radioactivity assay (100–200 keV) of the counting standard and the blood sample. Third, using standard least-squares tech-

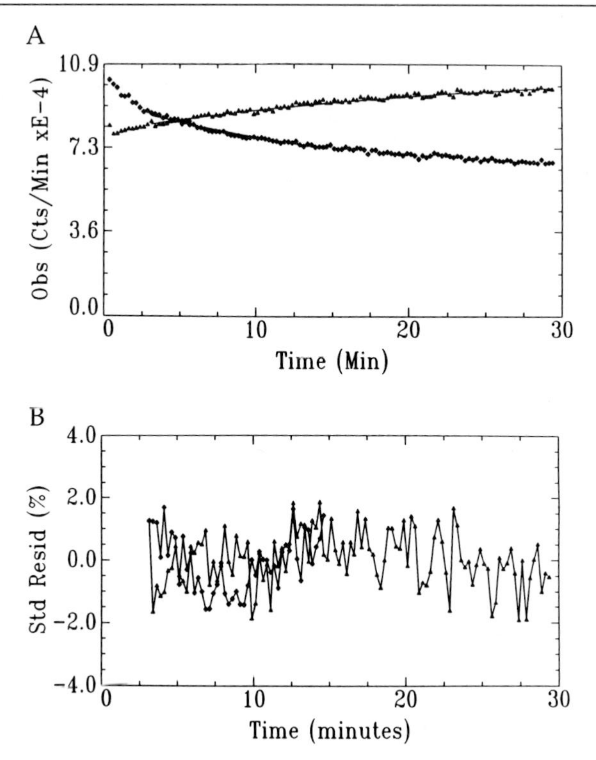

FIG. 3. A Tc-NGA functional imaging study of a patient with liver cirrhosis of moderate severity. (A) The curve-fit produced a receptor concentration of 0.341 ± 0.031 μM and a forward binding rate constant of 2.155 ± 1.076 μM^{-1} min^{-1}. The reduced χ^2 was 1.314. (B) The standardized residuals cluster about zero for curve-fits to both liver (▲) and heart (◆) data and did not exceed 2%.

niques, the model parameters ($[R]_o$, k_b, V_h, V_e, and F) are adjusted[12] until the simulated curves match the heart and liver time–activity data. Finally, standard deviations for each parameter estimate are determined by local identifiability analysis.[9]

Examples from two functional imaging studies are shown in Figs. 2A and 3A. Figure 2A is a curve-fit to time–activity data from a healthy subject; the activity was rapidly extracted from the blood. In Fig. 3A, which is a curve-fit to data from a patient with a cirrhotic liver, the time–activity curve has a very shallow slope, an indication of poor hepatic

[12] J. A. Nelder and R. Mead, *Comput. J.* **7,** 308 (1965).

TABLE III
In Vivo MEASUREMENTS[a] FROM TcNGA FUNCTIONAL IMAGING STUDIES

Parameter	Subject		Units
	Normal	Cirrhosis	
$[R]_o$	0.752 ± 0.087	0.341 ± 0.031	μM
k_b	2.502 ± 0.908	2.155 ± 1.076	μM^{-1} min^{-1}
V_p[b]	2.323 ± 0.021	2.645 ± 0.094	liters
V_h	0.272 ± 0.001	0.232 ± 0.004	liters
F	0.958 ± 0.634	1.287 ± 7.964	liters min^{-1}

[a] Estimates ± one standard error.
[b] Total plasma volume equals $V_e + V_h$.

function. Table III lists the model parameters resulting from each curve-fit. Note that the ASGP-R concentration ($[R]_0$) of the cirrhotic patient was significantly lower than that of the healthy subject.

Quality Control

After termination of the curve-fit, systemic error is measured by the reduced chi-square (χ^2). Visual and statistical analysis of the curve-fit residuals are used as quality control of the curve-fitting process. Standardized residuals are calculated and plotted against time.[9] Curve-fits with a reduced χ^2 greater than 2.75 or standardized residuals that display trends over time or have many points that exceed 5% are indications of a functional imaging study of poor technical quality. This may be the result of an incomplete injection, a diluted blood sample, or patient movement. The standardized residuals displayed in Figs. 2B and 3B indicate that each functional imaging study is of good technical quality.

Results

Validation of in Vivo Asialoglycoprotein Receptor Measurements

To establish the *in vivo* ASGP-R measurement as a reliable procedure, the results obtained by an *in vitro* binding assay of ASGP-R should correlate well with the results obtained by the *in vivo* method. The binding assay was carried out under the conditions outlined by Van Lenten and Ashwell,[13] with modifications to accommodate needle biopsy microsam-

[13] L. Van Lenten and G. Ashwell, *J. Biol. Chem.* **247**, 4633 (1992).

ples. Receptor concentration was calculated via linear regression of bound/free on concentration bound, which forms a standard Scatchard plot. A good correlation was observed between *in vivo* ASGP-R measurements and ASGP-R quantities measured *in vitro*, validating the *in vivo* measurement by Tc-NGA functional imaging.[14]

Normal and Disease

There are several reports of ASGP-R concentration as measured by Tc-NGA or Tc-labeled galactosyl human serum albumin (Tc-GSA) functional imaging. Using the Tc-GSA imaging protocol in Japan, Kudo and co-workers[15] reported 0.792 ± 0.080 μM for normal $[R]_0$ values in healthy subjects. Patients with chronic liver disease, such as cirrhosis, had significantly lower values ($p < 0.01$) depending on the degree of hepatocellular damage; values for mild, moderate, and severe damage were 0.584 ± 0.150, 0.510 ± 0.139, and 0.241 ± 0.094 μM, respectively. Patients with acute disease, such as hepatitis, also had $[R]_0$ values that were significantly lower than normal; values for mild, moderate, and severe disease were 0.528 ± 0.082, 0.460 ± 0.070, and 0.279 ± 0.036 μM, respectively. Using Tc-NGA Kudo *et al.*[16] reported 0.333 ± 0.120 μM in patients with hepatocellular carcinoma superimposed on liver cirrhosis. Using the Tc-NGA imaging protocol Virgolini and co-workers[17,18] determined a normal range of 0.8–1.2 μM and a mean $[R]_0$ of 0.65 ± 0.16 μM for breast cancer patients with liver metastases.

Acknowledgments

This work was supported by grants from the National Institutes of Health (RO1-AM34768 and RO1-AM34706) and the Department of Energy (FG03-87ER60553).

[14] M. Kudo, D. R. Vera, W. L. Trudeau, and R. C. Stadalnik, *J. Nucl. Med.* **32,** 1177 (1991).

[15] M. Kudo, A. Todo, K. Ikekubo, K. Yamamoto, D. R. Vera, and R. C. Stadalnik, *Hepatology (Baltimore)* **17,** 814 (1993).

[16] M. Kudo, D. R. Vera, R. C. Stadalnik, W. L. Trudeau, K. Ikekubo, and A. Todo, *Am. J. Gastroenterol.* **85,** 1142 (1990).

[17] I. Virgolini, C. Müller, J. Höbart, W. Scheithauer, P. Angelberger, H. Bergmann, J. O'Grady, and H. Sinzinger, *Hepatology (Baltimore)* **15,** 593 (1992).

[18] I. Virgolini, G. Kornek, J. Höbart, S. R. Li, M. Raolerer, H. Bergmann, W. Scheithauer, T. Pantev, P. Angelberger, H. Sinzinger, and R. Höfer, *Br. J. Cancer* **68,** 549 (1993).

[27] Synthesis and Radiolabeling of Galactosyl Human Serum Albumin

By MASATOSHI KUDO, KOMEI WASHINO, YOSHIHIRO YAMAMICHI, and KATSUJI IKEKUBO

Introduction

Asialoglycoprotein receptor (ASGP-R) is a hepatic membrane receptor that binds and transports desialylated plasma glycoproteins from hepatic blood to hepatocellular lysosomes. A receptor-binding radiopharmaceutical, which is a potential tool for the investigation of interactions between the radioligand and its receptor *in vivo*, can be obtained by synthesis and radiolabeling of an analog ligand of endogenous asialoglycoprotein. The use of a receptor-binding radioligand is a unique approach for the assessment of the receptor function, and consequently organ function. In addition, the tracer kinetic technique using such a radioligand is particularly suited for the assessment of receptor function, since the receptor quantities observed *in vivo* are too low to be assessed by other techniques.[1] In this chapter, we describe methods of sterile, nonpyrogenic synthesis and radiolabeling of galactosyl human serum albumin (GSA), an analog ligand for asialoglycoprotein, for routine clinical use.

Methods

Synthesis of Galactosyl Human Serum Albumin

Glassware for the synthesis of galactosyl human serum albumin (GSA) and subsequent ^{99m}Tc labeling is soaked in 0.1 *N* HCl solution for 24 hr, then rinsed with purified water from Milli-Q (Millipore, Bedford, MA) and baked at 180° for 3 hr in order to inactivate trace pyrogen. The synthesis of GSA is shown in Fig. 1, and the chemical structure of GSA is shown in Fig. 2.

Galactose is derivatized with cyanomethyl 2,3,4,6-tetra-*O*-acetyl-1-thio-β-D-galactose (CNM-thiogalactose) according to the literature.[2-4] Procedures reported in literature can be followed without substantial modifi-

[1] M. R. Kilbourn and M. R. Zalutsky, *J. Nucl. Med.* **26,** 655 (1985).
[2] M. Barczai-Martos and F. Korosy, *Nature* (*London*) **165,** 369 (1950).
[3] S. Chipowsky and Y. C. Lee, *Carbohydr. Res.* **31,** 339 (1973).
[4] Y. C. Lee, C. P. Stowell, and M. J. Krantz, *Biochemistry* **15,** 3956 (1976).

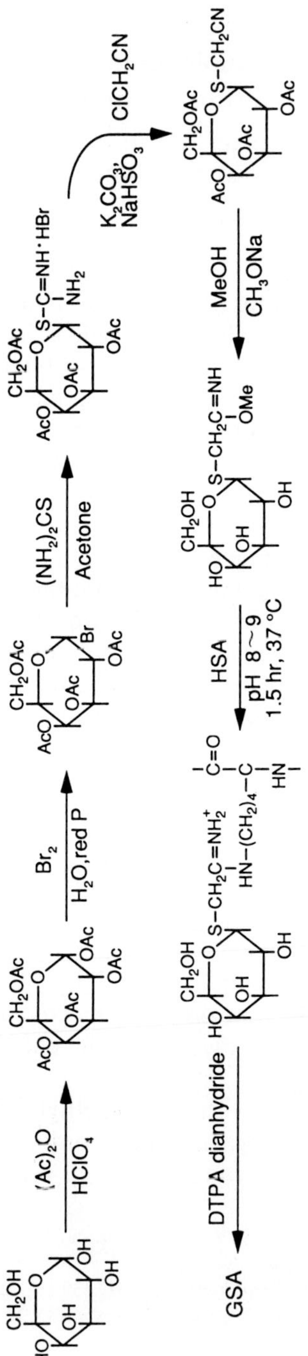

FIG. 1. Synthetic route to galactosyl human serum albumin (GSA).

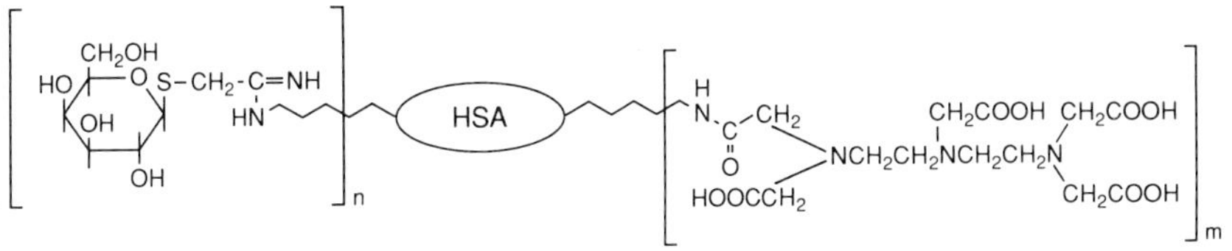

FIG. 2. Estimated chemical structure of GSA. *n*, Number of galactoses bound to HSA; n = 30–40/1 (molar ratio). *m*, Number of DTPA molecules bound to HSA; m = 4–7/1 (molar ratio).

cations until the final step of CNM-thiogalactose synthesis. The CNM-thiogalactose is purified by repeated recrystallization (at least three times) from dry methanol, since any impurity will be carried over to the following reaction steps.

CNM-thiogalactose is the best chemical form for storage at room temperature. However, a dark, cold, dry atmosphere is recommended for long-term storage. Two days before GSA synthesis, CNM-thiogalactose is dissolved in dry methanol to a concentration of 0.1 *M* in a pear-shaped flask, which is used as the reaction vessel throughout the synthesis. The solution can be heated to 40° to dissolve CNM-thiogalactose quickly. Sodium methoxide is then added as a catalyst at 10 m*M*, and the reaction mixture is kept at room temperature for 48 hr. The catalyst promotes the conversion of the galactose derivative to 2-imino-2-methoxyethyl-1-thio-β-D-galactose (IME-thiogalactose) with estimated 55% yield.[4] The solvent is then evaporated *in vacuo* at around 40° using a rotatory evaporator. In some experiments, aliquots of resulting IME-thiogalactose solution are transferred to other glass flasks and evaporated in the same manner.

Human serum albumin (HSA) supplied as injectable solution (200 mg/ml) is diluted with an equal volume of 0.4 *M* borate buffer (pH 8.5). The diluted solution is then added to the pear-shaped flask containing IME-thiogalactose while being stirred with a powerful mechanical stirrer, and stirring is continued for 1.5 hr at 37°. A IME-thiogalactose to HSA molar ratio of 55/1 for yields HSA derivatives with 30–40 mol of Gal coupled per mole of HSA. After the galactose coupling reaction, powdered cyclic dianhydride of diethylenetriaminepentaacetic acid (cDTPA) is added slowly to the reaction mixture, so that a cDTPA to HSA molar ratio of 11/1 is attained. The DTPA coupling is allowed to proceed for 10 min at 37° under continuous stirring. The reaction proceeds very rapidly to yield HSA derivatives containing 4.5–7.0 mol DTPA per mole HSA. The GSA monomer is then isolated with a high-performance liquid chromatography (HPLC) system using a pyrogen-free TSK-G3000SW column (5.5 cm diameter × 60 cm length, Tosoh, Tokyo, Japan; flow rate 25 ml/min) and sterile 0.1 *M* NaCl solution as an eluent. The protein concentration of the final mixture to be injected into the HPLC column should be less than 60 mg/ml for better resolution. Although GSA is relatively stable in neutral aqueous solutions, freeze-drying would be recommended for long-term storage.

Because most HPLC columns are usually heavily contaminated with pyrogens, the columns should be pyrogen-purged before they are used for the purification of products for human consumption. Pyrogen can be removed from the column by extensive washing with pyrogen-free saline as eluent. The volume of eluent necessary for pyrogen purging is usually more than 500 times the column bed volume and may vary with columns.

The amount of galactose bound to HSA to determined by removing galactose from GSA and quantifying free galactose by colorimetry according to the method reported in the literature.[5–7] A GSA solution containing 1–3 mg GSA in 0.5 ml is placed in vial and mixed with 1 ml of 0.2 *M* mercury(II) acetate solution (in 0.1 *M* acetic acid solution) and 3 ml of purified water. After sealing the vial with rubber stopper and aluminum cap, the vial is heated to 95°–100° for 10 min. The sample is cooled by tap water, mixed with 0.5 ml of 30% (w/v) trichloroacetic acid solution, then filtered through a syringe packed with 2.5 ml of cation-exchange resin (AG 50W-X8, Bio-Rad, Richmond, CA) to remove protein precipitates and Hg^{2+} cation. In this procedure, thioglycoside bonds are selectively hydrolyzed to give free galactose which can be determined by phenol–sulfuric acid colorimetry using galactose standard solution. The protein concentration of the sample is determined either by the Lowry method or by the absorption at 278 nm. The degree of galactose conjugation is expressed here as a molar ratio of galactose to HSA.

The degree of DTPA conjugation is determined by measuring the amount of $^{111}In^{3+}$ bound using a concentration of indium [cold indium cation (In^{3+}) with trace radioactive $^{111}In^{3+}$] that will saturate all the chelators on the GSA molecule. One hundred fifty microliters of 1.0 m*M* $InCl_3$ solution in 0.01 *N* HCl is mixed with 0.2 ml of carrier-free $^{111}InCl_3$ solution (0.4 mCi in 0.01 *N* HCl, Nihon Medi-Physics) and 0.7 ml of 0.1 *M* sodium citrate buffer (pH 5.7). The In^{3+} cation does not form colloidal aggregates in the citrate buffer. Three hundred microliters of GSA solution (~3 mg/

[5] M. J. Krantz and Y. C. Lee, *Anal. Biochem.* **71,** 318 (1976).

[6] M. Debois, K. A. Gilles, J. K. Hamilton, P. A. Rebers, and J. Smith, *Anal. Chem.* **28,** 350 (1956).

[7] K. Kawaguchi and Y. C. Lee, *Tanpakusitu Kakusan Koso* **25,** 707 (1980).

ml) in saline is then added to the above In^{3+} solution and kept for 30 min at room temperature to allow complexation of DTPA with In^{3+}. The concentration of In^{3+} present in the reaction mixture should be 2–10 times higher than the estimated DTPA concentration on GSA.

After that, the amount of DTPA equal to $InCl_3$ (0.15 ml of 1.0 m*M* DTPA solution) is added to the reaction mixture and kept for 15 min to scavenge In^{3+} that remains free or weakly bound to GSA. The final mixture is analyzed by cellulose acetate thin-layer electrophoresis using 60 m*M* barbital buffer (pH 8.6) as electrode buffer for 30–40 min in a constant voltage mode (150 V). A few microliters of sample mixture is applied 2 cm from the cathode wig on a thin cellulose acetate film (2 × 11 cm, Fuji film, Tokyo, Japan). In the analysis, unbound In^{3+} is detected as ^{111}In-DTPA, whereas In^{3+} bound to GSA is found as ^{111}In-GSA which is identified by Ponceau 3R staining. The ^{111}In-labeled DTPA and GSA migrate toward the anode approximately 4 cm and 2 cm, respectively, under these conditions. Thus, the fraction of In^{3+} bound to GSA can be easily obtained by a conventional radiochromatogram scanner (Aloka, Tokyo, Japan). Because HSA itself or HSA conjugated with galactose could not retain In^{3+} under the electrophoretic conditions, the amount of In^{3+} bound to GSA represents that of effective chelating sites. The degree of DTPA conjugation is expressed as a molar ratio of DTPA to HSA.

The GSA used in the experiments has a purity of more than 95% as GSA monomer. The majority of impurities are GSA dimers. The coupling molar ratios of galactose and DTPA to HSA are 30–40/1 and 4–7/1, respectively. The UV absorption spectrum of GSA is similar to that of HSA. However, the extinction coefficient at 278 nm is influenced to some extent by the introduction of sugar or DTPA. The extinction coefficient (ε) of GSA (1.0 mg/ml in neutral saline) at the absorption maximum of 278 nm is 0.56 when the GSA contains 30–40 mol of Gal and 4.5–7.0 mol of DTPA per mole of HSA.

Radiolabeling of Galactosyl Human Serum Albumin

Labeling of GSA with ^{99m}Tc is carried out as follows. The concentration of HPLC-purified GSA solution is adjusted to 3 mg/ml with pyrogen-free saline and adjusted by 1 *N* HCl to pH 3.1–3.4. Then the GSA solution is deoxygenated by gently flushing argon gas (flow rate 0.2 liter/min) onto the meniscus while the solution is continuously stirred. This procedure requires a special glass flask equipped with two flushing lines and three gas-tight cap sealings. The oxygen concentration of the solution is monitored by soaked O_2 sensor. When the oxygen level becomes less than 100 ppb, appropriate amounts of $SnCl_2$ and ascorbic acid are dissolved into the solution. The solution is immediately dispensed into glass vials with rubber stoppers through a 0.2-μm membrane filter under a low oxygen atmosphere. The vial should contain 3 mg of GSA, 0.1 μmol of $SnCl_2$, and 0.5 μmol of ascorbic acid in 1 ml. Vials can be kept frozen for a limited period, but preferrably they should be lyophilized and sealed with aluminum caps under argon for storage.

The $^{99m}TcO_4^-$ solution is obtained by eluting the $^{99}Mo/^{99m}Tc$ generator (MEDITEC, Nihon Medi-Physics) with sterile saline. The radioactivity of the $^{99m}TcO_4^-$ solution is calibrated with the dose calibrator (CAPINTEC) and adjusted by dilution to a desired specific radioactivity. The GSA is labeled by introducing 1 ml of $^{99m}TcO_4^-$ solution (1.85–3.70 GBq/ml) into a vial containing lyophilized GSA and swirling gently for a few minutes at room temperature to attain complete dissolution.

The $SnCl_2$ serves as a reducing agent for $^{99m}TcO_4^-$ to initiate the labeling reaction. The amount of components contained in the reaction mixture strongly affects the radiochemical purity (RCP) and stability (RCS) of ^{99m}Tc-GSA. Therefore, different amounts of reducing agent and stabilizer, ascorbic acid, are tested in order to establish the optimal formulation of the additives. The labeling yield is measured by cellulose acetate thin-layer electrophoresis as mentioned above. In the analysis, ^{99m}Tc-GSA migrates about 2 cm from the origin toward the anode while reduced ^{99m}Tc impurities including ^{99m}Tc–Sn colloid remain at the origin and $^{99m}TcO_4^-$ migrates around 5 cm toward anode. Thin-layer chromatography (TLC) could also be used for $^{99m}TcO_4^-$ detection.

Biodistribution Study in Normal Rats and Rabbits

The biodistribution of ^{99m}Tc-GSA is studied in normal Sprague-Dawley rats and normal Japanese white rabbits. The ^{99m}Tc-GSA is injected intravenously into rats under anesthesia at a dose level of 330 μg/kg. Rats are then dissected at 10 min to 48 hr after injection, and the radioactivity in various organs is counted with a single-channel counter.

The dynamic imaging study is performed in rabbits with a γ-camera (GCA-90B, Toshiba, Tokyo) using a low-energy high-resolution collimator. Acquisition is done in the anterior position with 30 sec/frame for 56 min (112 frames total). Regions of interest (ROIs) are set on the heart and liver, and time–activity curves for the ROIs that represent the disappearance and uptake of ^{99m}Tc-GSA, respectively, are obtained. All the ^{99m}Tc-GSA samples used in the animal studies prove to have RCP values higher than 95% and to be free from reduced ^{99m}Tc impurities.

Comments

Synthesis of Galactosyl Human Serum Albumin

Galactosyl human serum albumin has been synthesized by coupling galactose and the bifunctional chelator DTPA to HSA. The molar ratio of galactose to HSA should be 30–40 mol per mole, so that GSA can compete well with naturally existing asialoglycoproteins (ASGP) in the blood pool. It has been reported elsewhere that a synthetic neoglycoprotein having 25 galactoses on bovine serum albumin achieved similar binding affinity to that of asialoorosomucoid, which is known to have the highest binding affinity to the asialoglycoprotein receptor in hepatic parenchymal cells.[7] Because the poor competitiveness of ^{99m}Tc-GSA would result in high background and poor liver uptake, the ^{99m}Tc-GSA for imaging studies requires the 30–40 galactose labeling level.

Conjugation of DTPA with protein should be limited to less than 10 mole per mole of HSA, since HSA coupled with excess DTPA molecules tends to be incorporated nonspecifically into liver (M. Kudo, K. Washino, Y. Yamamichi, and K. Ikekubo, unpublished data, 1989). Therefore, the DTPA coupling ratio was kept below 7 mol per mole of HSA.

Radiolabeling of Galactosyl Human Serum Albumin

Galactosyl human serum albumin has been labeled with ^{99m}Tc using $SnCl_2$ as a reducing agent. The RCP of ^{99m}Tc-GSA in saline solution was greatly influenced by the concentrations of GSA, $SnCl_2$, and ascorbic acid. Ascorbic acid was added in order to protect $SnCl_2$ from the oxidative impurities, thus suppressing regeneration of $^{99m}TcO_4^-$ after labeling. Formulation of lyophilized GSA vial was optimized in terms of RCP and RCS.

Figure 3 shows the fraction of reduced ^{99m}Tc species in ^{99m}Tc-GSA solution that remained unbound to GSA as a function of the GSA/Tc molar ratio. As easily seen, more than a 400-fold excess of GSA was necessary for keeping unbound reduced ^{99m}Tc below 5%.

Figure 4 demonstrates the percentage of $^{99m}TcO_4^-$ in the ^{99m}Tc-GSA solution as a function of the $SnCl_2$/Tc molar ratio. The $^{99m}TcO_4^-$ in the labeling mixture can result from either incomplete reduction initially or regeneration from ^{99m}Tc due to a deficiency of reducing agents. There is a critical Sn/Tc molar ratio for preventing the regenaration of $^{99m}TcO_4^-$, which is governed by several factors such as the number of total Tc atoms, the radioactivity, and the concentration of oxidative impurities such as H_2O_2 created by the radiolysis of water. In the case of the experiment carried out in Fig. 4, the critical $SnCl_2$/Tc ratio was approximately 100.

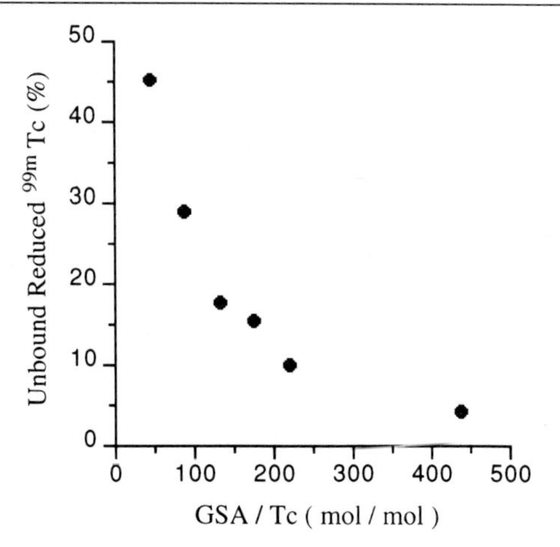

FIG. 3. Labeling of GSA with ^{99m}Tc, showing the effect of GSA concentration on the formation of unbound reduced ^{99m}Tc. Formulation of ^{99m}Tc-GSA solutions was as follows: [GSA] = 1–10 mg/ml in saline, [$SnCl_2$] = 0.15 m*M*, [ascorbic acid] = 0.125 m*M*, [$^{99m}TcO_4^-$] = 3.0 × 10^{-10} mol/1.85 GBq/ml at labeling, pH 3.0, argon saturated. Unbound reduced ^{99m}Tc was measured at 2 hr after labeling by cellulose acetate thin-layer electrophoresis.

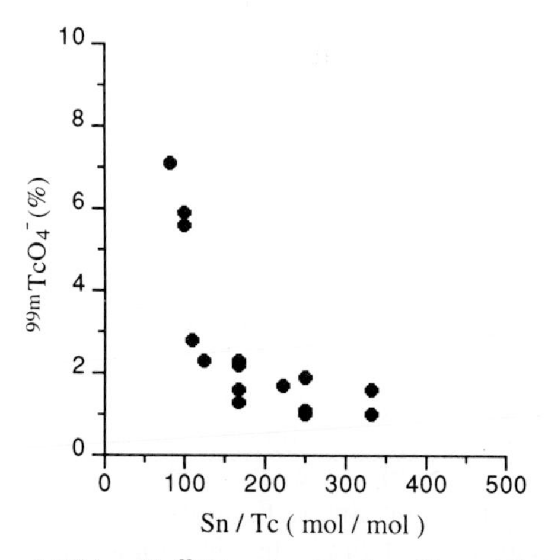

FIG. 4. Labeling of GSA with ^{99m}Tc, showing the effect of $SnCl_2$ concentration on the amount of $^{99m}TcO_4^-$ remaining. Formulation of ^{99m}Tc-GSA solutions was as follows: [GSA] = 5 mg/ml in saline, [$SnCl_2$] = 0.50–0.10 m*M*, [ascorbic acid] = 0.50 m*M*, [$^{99m}TcO_4^-$] = (3.0–6.0) × 10^{-10} mol/1.85–3.7 GBq/ml at labeling, pH 3.4, argon saturated. The $^{99m}TcO_4^-$ was measured at 21 hr after labeling by TLC (Merck silica gel-60 plate, solvent 10% ammonium acetate–methanol, 1 : 3, v/v).

TABLE I

BIODISTRIBUTION OF ^{99m}Tc-GSA IN NORMAL RATS[a]

Organ	% Injected dose/organ (n = 3)							
	10 min	30 min	1 hr	3 hr	6 hr	12 hr	24 hr	48 hr
Whole blood[b]	1.03 ± 0.05	0.65 ± 0.17	0.67 ± 0.23	0.36 ± 0.17	0.17 ± 0.00	0.10 ± 0.03	0.09 ± 0.01	0.07 ± 0.03
Liver	92.4 ± 0.8	86.2 ± 2.6	70.2 ± 4.3	44.2 ± 4.6	34.3 ± 1.1	23.3 ± 1.3	14.0 ± 1.4	8.59 ± 0.75
Heart	0.03 ± 0.01	0.02 ± 0.00	0.02 ± 0.00	0.01 ± 0.01	0.01 ± 0.00	<0.01 ± 0.00	<0.01 ± 0.00	<0.01 ± 0.00
Lung	0.05 ± 0.01	0.04 ± 0.01	0.04 ± 0.01	0.03 ± 0.01	0.01 ± 0.00	0.01 ± 0.00	0.01 ± 0.00	<0.01 ± 0.00
Stomach	0.09 ± 0.02	0.22 ± 0.03	0.81 ± 0.02	0.39 ± 0.01	0.07 ± 0.00	0.06 ± 0.00	0.02 ± 0.00	<0.01 ± 0.00
Spleen	0.09 ± 0.07	0.08 ± 0.26	0.07 ± 0.59	0.04 ± 0.32	0.04 ± 0.02	0.04 ± 0.04	0.03 ± 0.01	0.01 ± 0.01
Small intestines	0.66 ± 0.10	5.72 ± 2.16	19.3 ± 3.4	37.5 ± 4.1	5.43 ± 0.36	3.97 ± 1.73	1.35 ± 0.13	0.44 ± 0.03
Large intestines	0.07 ± 0.02	0.05 ± 0.01	0.06 ± 0.02	0.09 ± 0.06	47.5 ± 1.6	32.6 ± 14.0	8.60 ± 3.28	1.56 ± 0.59
Kidneys	0.81 ± 0.01	0.94 ± 0.10	1.23 ± 0.21	2.06 ± 1.00	2.04 ± 0.07	1.55 ± 0.09	1.60 ± 0.29	1.08 ± 0.18
Testis	0.04 ± 0.02	0.05 ± 0.01	0.03 ± 0.01	0.03 ± 0.01	0.01 ± 0.00	0.01 ± 0.00	0.01 ± 0.00	0.01 ± 0.00
Carcass	3.61 ± 1.25	3.16 ± 1.06	2.71 ± 0.59	2.51 ± 2.11	0.61 ± 0.03	2.29 ± 0.50	0.72 ± 0.60	0.30 ± 0.03
Urine	1.81 ± 0.39	3.22 ± 0.47	5.19 ± 0.67	12.9 ± 4.6	9.88 ± 1.22	8.21 ± 0.82	11.2 ± 2.2	13.2 ± 1.9
Feces	0.00 ± 0.00	0.00 ± 0.00	0.00 ± 0.00	0.00 ± 0.00	0.00 ± 0.00	27.9 ± 15.3	62.4 ± 3.6	74.8 ± 1.6

[a] Normal Sprague-Dawley rats (male, 6 weeks old, 201 ± 3 g), injection dose 330 μg/kg body weight of [^{99m}Tc]GSA.

[b] The % injected dose (ID)/whole blood was calculated from the % ID/ml blood assuming the blood weight to be 6.4% of the body weight.

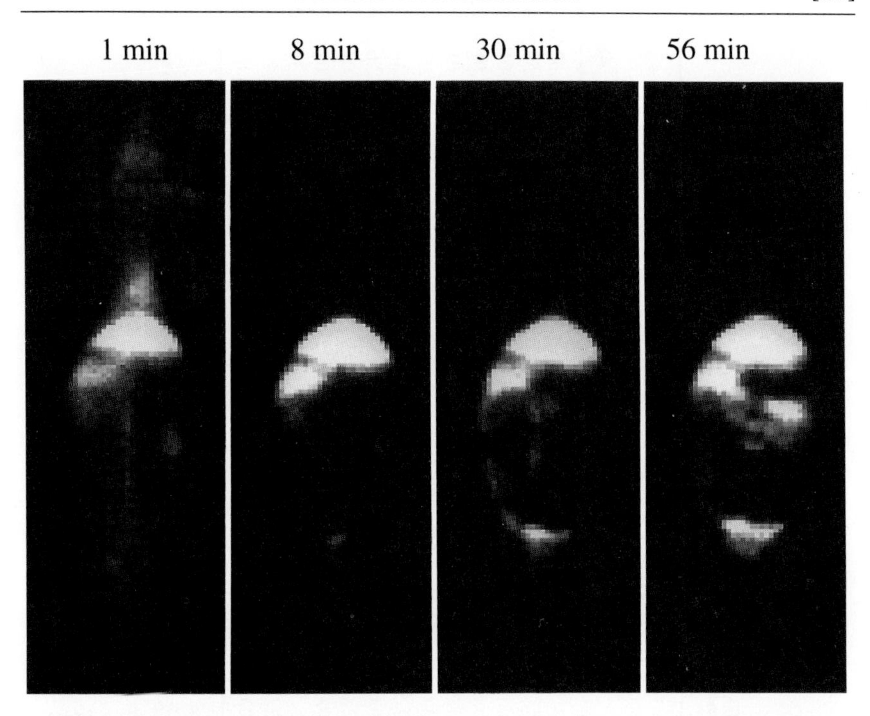

FIG. 5. Whole-body anterior images of a normal Japanese white rabbit (male, 9 weeks old, 2.8 kg) injected with ^{99m}Tc-GSA intravenously under anesthesia with sodium pentobarbital at a dose of 330 μg/kg body weight. Images were obtained using a γ-camera (Toshiba GCA-90B) with a low-energy high-resolution collimator at an acquisition rate of 30 sec/frame from the injection for 56 min (112 frames).

It is easy to prepare stable ^{99m}Tc-GSA using the GSA freeze-dried kit vial (Nihon Medi-Physics, Nishinomiya, Japan) without the formation of reduced ^{99m}Tc species or protein aggregates.

In Vivo Behavior of ^{99m}Tc-Labeled Galactosyl Human Serum Albumin in Normal Animals

The biodistribution of ^{99m}Tc-GSA has been studied in normal male and female Sprague-Dawley rats injected intravenously with 330 μg/kg of ^{99m}Tc-GSA. There was no sex-dependent difference in the biodistribution. Table I summarizes the distribution in male rats. Ten minutes after the injection, 92.4% of the injected radioactivity was found in the liver, and 74.8% of the injected dose was excreted through hepatobiliary into feces at 48 hr after injection. In normal rats, ^{99m}Tc-GSA did not appear to distribute in organs other than liver just after injection.

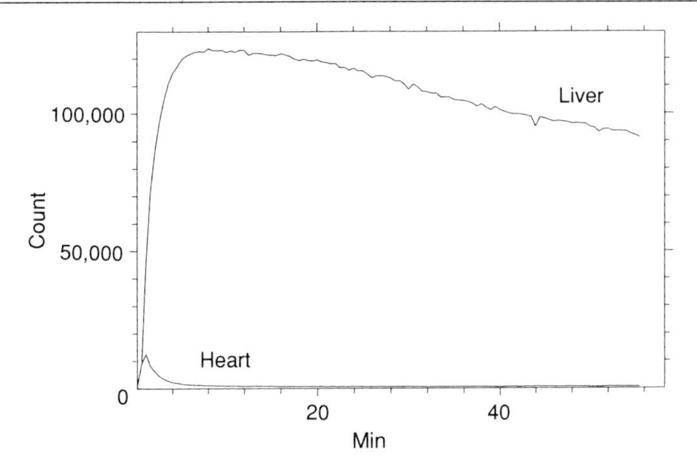

FIG. 6. Time–activity curves for liver uptake and blood clearance of ^{99m}Tc-GSA in normal rabbits. Regions of interest (ROIs) were set on whole liver for uptake and on heart for blood clearance in all frames. The radioactivity in ROIs was plotted against time with decay compensation and smoothing.

The imaging study was carried out in normal male and female Japanese white rabbits with a dose of 330 μg/kg using a γ-camera. Figure 5 shows the whole-body distribution in a male rabbit at 1 to 56 min after injection. No accumulation in spleen or bone marrow is observed throughout the imaging period, which suggests that colloidal or aggregated impurities are absent in the ^{99m}Tc-GSA samples.

Finally, Fig. 6 shows the time-dependent activity curves of heart and liver ROIs. Liver uptake reached a maximum at around 7 min after intravenous injection and remained constant for 7 min thereafter.

Conclusion

Procedures for the synthesis and radiolabeling of GSA have been described. In addition, the radiochemical purity and stability as well as the biodistribution of technetium-99*m*-labeled GSA in animals are presented in this chapter. The radioligand is commercially available in Japan and is actively used for the assessment of liver function, especially hepatic functional reserve[8] by the concurrent use of a kinetic analysis system. It

[8] M. Kudo, A. Todo, K. Ikekubo, K. Yamamoto, D. R. Vera, and R. C. Stadalnik, *Hepatology (Baltimore)* **17,** 814 (1993).

is expected that the usefulness of synthetic asialoglycoprotein analogs will continue to grow, not only for diagnosis of receptor function either by nuclear medicine techniques or by other techniques such as magnetic resonance imaging, but also for therapy of liver diseases.

[28] *In Vitro* Quantification of Asialoglycoprotein Receptor Density from Human Hepatic Microsamples

By DAVID R. VERA, SARA J. TOPCU, and ROBERT C. STADALNIK

Introduction

Technetium-99*m*-labeled galactosyl neoglycoalbumin[1] (Tc-NGA) is a radiopharmaceutical designed for *in vivo* measurement of asialoglycoprotein receptor (ASGP-R) concentration. The measurement is obtained by automated kinetic analysis[2] of time–activity data resulting from 30 min of dynamic liver imaging[3] with standard nuclear medicine instrumentation. Numerous clinical trials[4–7] have confirmed that the measurement of *in vivo* receptor concentration by the kinetic model correlates with the severity of liver disease. However, more critical to the acceptance of the kinetic model is confirmation that *in vivo* receptor concentration measurements accurately reflect hepatic ASGP-R density. As a result, various *in vitro* assays of ASGP-R density[8,9] have been employed to verify[10,11] the accuracy of the *in vivo* measurements.

[1] D. R. Vera, R. C. Stadalnik, M. Kudo, and K. A. Krohn, this volume [29].

[2] D. R. Vera, R. C. Stadalnik, W. L. Trudeau, P. O. Scheibe, and K. A. Krohn, *J. Nucl. Med.* **32,** 1169 (1991).

[3] R. C. Stadalnik, M. Kudo, W. C. Eckelman, and D. R. Vera, *Invest. Radiol.* **28,** 65 (1993).

[4] R. C. Stadalnik, D. R. Vera, E. S. Woodle, W. L. Trudeau, B. A. Porter, R. E. Ward, K. A. Krohn, and L. F. O'Grady, *J. Nucl. Med.* **26,** 1233 (1985).

[5] M. Kudo, D. R. Vera, R. C. Stadalnik, W. L. Trudeau, K. Ikekubo, and A. Todo, *Am. J. Gastroenterol.* **85,** 1142 (1990).

[6] I. Virgolini, C. Müller, J. Höbart, W. Scheithauer, P. Angelberger, H. Bergmann, J. O'Grady, and H. Sinzinger, *Hepatology* **15,** 593 (1992).

[7] M. Kudo, A. Todo, K. Ikekubo, K. Yamamoto, D. R. Vera, and R. C. Stadalnik, *Hepatology* **17,** 814 (1993).

[8] D. R. Vera, K. A. Krohn, R. C. Stadalnik, and P. O. Scheibe, *J. Nucl. Med.* **25,** 779 (1984).

[9] I. Virgolini, P. Angelberger, C. H. Müller, J. O'Grady, and H. Sinzinger, *Br. J. Clin. Pharmacol.* **29,** 207 (1990).

[10] M. Kudo, D. R. Vera, W. L. Trudeau, and R. C. Stadalnik, *J. Nucl. Med.* **32,** 1177 (1991).

[11] I. Virgolini, C. Müller, W. Klepetko, P. Angelberger, H. Bergmann, J. O'Grady, and H. Sinzinger, *Br. J. Cancer* **61,** 937 (1990).

This chapter describes an *in vitro* assay capable of measuring ASGP-R density from small sample sizes (10–15 mg) retrieved by means of percutaneous liver biopsy. Our motivation for development of the assay was 2-fold. The first was to permit access to a wider patient population; previous assays[8,9] required large surgical biopsies (1–2 g). The second was the ability to assay selectively very small regions within the liver for validation of regional *in vivo* concentration measurements via Tc-NGA tomographic imaging[12] and positron emission tomography using gallium-68-labeled deferoxamine-NGA.[13]

Principles of Assay

The ligand used in the assay is the key component which permitted a substantial reduction in sample size. The accuracy of an *in vitro* binding assay is essentially determined by the bound-to-free ratio, with unity being the optimal value. Typically this is controlled by three factors: (1) receptor concentration, (2) initial ligand concentration, and (3) ligand–receptor affinity. Faced with a limited sample size, we elected to maximize the latter two factors. First, we maximized ligand specific activity by employing the short-lived radioisotope technetium-99*m*. Although the 6-hr half-life requires labeling immediately prior to each assay, technetium-99*m* is readily available (~1 hr delivery) and inexpensive. Second, we constructed a neoglycoconjugate of extremely high receptor affinity. The ligand, DTPA-*gal*-polylysine, has three structural features that contribute to an affinity which greatly exceeds that of Tc-NGA: (a) the ligand has an extremely high carbohydrate density imparted by an amino acid backbone composed exclusively of lysine; (b) the coupling reagent, a 3-oxopropyl 1-thiogalactoside,[14] substitutes two carbohydrates per amino group; and (c) tertiary amines maintain a positive charge which increases ligand–receptor affinity.[15] The final result was a binding reaction which comes to equilibrium within 30 min and generates bound-to-free ratios within the 0.5 to 1.5 range. The relative errors for receptor concentration ranged from 10 to 25%, which is comparable to the standard error of the *in vivo* receptor estimates.[16]

[12] I. Virgolini, G. Kornek, J. Höbart, S. R. Li, M. Raolerer, H. Bergmann, W. Scheithauer, T. Pantev, P. Angelberger, H. Sinzinger, and R. Höfer, *Br. J. Cancer* **68,** 549 (1993).

[13] D. R. Vera, *J. Nucl. Med.* **33,** 1160 (1992).

[14] R. T. Lee and Y. C. Lee, *Carbohydr. Res.* **101,** 49 (1992).

[15] R. T. Lee, *Biochemistry* **21,** 1045 (1982).

[16] D. R. Vera, P. O. Scheibe, K. A. Krohn, W. L. Trudeau, and R. C. Stadalnik, *IEEE Trans. Biomed. Eng.* **39,** 356 (1992).

Experimental Methods

Synthesis of DTPA–Polylysine

The mixed anhydride method[17] is used to attach diethylenetriamine-*N,N,N′,N″,N″*-pentaacetic acid (DTPA) to polylysine (Sigma Chemical Co., St. Louis, MO) with an average of 120 lysines per chain. The product, DTPA–polylysine, is separated from low molecular weight components by diafiltration (molecular weight cutoff 3000) against 10 exchange volumes of 0.9% saline (filtered, 0.45 μm) and another 10 exchange volumes of distilled water (filtered, 0.45 μm). After lyophilization a known weight of the product is dissolved in 0.9% saline and assayed for DTPA density by measurement of the primary amine concentration.[18]

Synthesis of DTPA–gal–Polylysine

The compound 2,3,4,6-tetra-*O*-acetyl-α-D-galactopyranosyl bromide[19] is converted to 2-*S*-(2,3,4,6-tetra-*O*-acetyl-β-D-galactopyranosyl)-2-pseudothiourea hydrobromide following the method described by Chipowsky and Lee.[20] The pseudothiourea derivative is converted to 2,3,4,6-tetra-*O*-acetyl-1-thio-β-D-galactopyranose as described by Lee and co-workers.[21] After reaction of the galactopyranose derivative with acrolein according to the method of Lee and Lee,[14] the reaction mixture is fractionated on a Sephadex LH-20 column (5 × 186 cm) in 95% (v/v) ethanol. The major product, 3-oxopropyl 2,3,4,5-tetra-*O*-acetyl-1-thio-β-D-galactopyranoside, a clear viscous oil, is eluted in fractions 440 to 540 (5.2 ml per fraction) and identified with infrared spectrophotometry by the presence of the aldehyde overtone band at 2720 cm^{-1}.

The covalent attachment of galactose to DTPA–polylysine is accomplished by reductive alkylation[14] with the 3-oxopropyl 1-thiogalactoside. After O-deacetylation with 20 m*M* sodium methoxide in methanol (distilled), neutralization with 0.1 *M* acetic acid, and evaporation, the coupling reaction is carried out in 15 ml phosphate buffer (0.2 *M*, pH 7.2) with 0.15 *M* $NaCNBH_3$ at 37° for 12 hr with a 13-fold excess of coupling reagent per lysine (20 m*M*). The neoglycoconjugate is separated from unreacted coupling reagent by diafiltration as previously described. The lyophilized product, DTPA–*gal*–polylysine, is stored at 5°. The galactose density is

[17] G. E. Krejcarek and K. L. Tucker, *Biochem. Biophys. Res. Commun.* **77,** 581 (1977).
[18] R. Fields, this series, Vol. 25, p. 464.
[19] M. Bárczai-Martos and F. Kórösy, *Nature (London)* **165,** 369 (1950).
[20] S. Chipowsky and Y. C. Lee, *Carbohydr. Res.* **31,** 339 (1973).
[21] R. T. Lee, S. Cascio, and Y. C. Lee, *Anal. Biochem.* **95,** 260 (1979).

determined by carbohydrate assay[22] of a known amount of product in 0.9% saline.

Technetium-99m Labeling of DTPA–gal–Polylysine

Radiolabeling is performed immediately prior to the *in vitro* assay. Tin reduction[23] of sodium pertechnetate is employed to label DTPA–gal–polylysine with technetium-99*m*. Labeling is carried out in the following steps. First, 30 μl of a 1.8×10^{-7} mol/ml solution of DTPA-*gal*-polylysine in 0.9% saline is added to a 2-ml glass vial fitted with a rubber septum. Next, a saline solution (0.9%) of sodium pertechnetate (typically 0.15 Ci) is delivered via syringe; the maximum volume is 0.50 ml. The vial is then purged with medical-grade nitrogen for 10 min. Finally, an acidified saline (0.9%) solution (0.10 ml) of tin chloride is rapidly injected into the vial. The tin chloride solution [5.0 mg $SnCl_2$ in 0.175 ml HCl (37%) followed by the addition of 25 ml of 0.9% saline] is stored in a 30-ml multidose vial at 5° and used within 1 week. The final volume of the labeling reaction is 0.60 ml. After standing 2 hr at room temperature, the labeled product is purified via size-exclusion chromatography (Sephadex G-10, 1 × 10 cm, 0.9% saline, 1 ml/min). Labeling yields are typically greater than 95% with specific activities in excess of 10 Ci/μmol.

In Vitro Assay

Equipment and Software

Dounce homogenizer (2 ml; Kontes Glassware, Vineland, NY)
Ultrasonicator (XL2020 with microtip sonicator horn; Heat Systems, Farmingdale, NY)
Centrifuge (High Speeed with HSR-16 rotor; Savant Instruments, Farmingdale, NY)
Dry bath set to 37° with heating block for 13-mm tubes
Micropartition assemblies (MPS-1; Amicon Corp., Danvers, MA)
Personal computer (IBM-compatible running DOS)
LIGAND (Division of Computer Research and Technology, National Institutes of Health, Bethesda, MD)

Disposables

Polypropylene test tubes (12 × 75 mm, snap closures)
Scintillation vials (20 ml, polyethylene)

[22] M. Dubois, K. A. Gilles, J. K. Hamilton, P. A. Rebers, and J. Smith, *Anal. Chem.* **28,** 350 (1956).
[23] W. C. Eckelman, G. Meinken, and P. Richards, *Radiology* **102,** 185 (1972).

Filters (HVLP 013; Millipore Corp., Bedford, MA)
Centrifuge tubes (2 ml, polypropylene, screw-top with O ring seal)

Solutions

Solution A: ^{99m}Tc-DTPA–gal–polylysine (galactose density >1.5 per lysine, specific activity >10 Ci/μmol)
Solution B: Dilution buffer (50 m*M* Tris-HCl, 10 m*M* $MgCl_2$, 10 m*M* $CaCl_2$, 2.0 *M* NaCl, 0.2% HSA, pH 7.5; filtered, 0.45 μm)
Solution C: Homogenization buffer (50 m*M* Tris-HCl, pH 7.5; filtered, 0.45 μm)
Solution D: Assay buffer (50 m*M* Tris-HCl, 5 m*M* $MgCl_2$, 5 m*M* $CaCl_2$, 1.0 *M* NaCl, 0.1% HSA, pH 7.5; filtered, 0.45 μm)
Solution E: Rinsing buffer (50 m*M* Tris-HCl, 5 m*M* $MgCl_2$, 5 m*M* $CaCl_2$, 1.0 *M* NaCl, pH 7.5; filtered, 0.45 μm)

Procedure

1. Prepare six dilutions of ^{99m}Tc-DTPA–*gal*–polylysine. Use solution B and polypropylene test tubes (see Note 1 below). The dilutions, designated *a* through *g*, are 1.0×10^2, 1.0×10^3, 1.0×10^4, 2.0×10^4, 1.0×10^5, and 2.0×10^5.
2. Prepare tissue for homogenization.
   a. Measure weight of liver tissue by placing it within the 2 ml Dounce homogenizer, which was previously tared (see Note 2 below).
   b. Add to the homogenizer 25 μl of solution C per milligram of tissue.
3. Homogenize tissue. Complete 10 strokes each with pestles A and B.
4. Ultrasonicate the homogenized sample by inserting the microtip directly into the homogenizer. Sonicate for 10 sec at power setting 3.
5. Prepare tubes for a blank assay.
   a. Aliquot 20 μl of solution C into two polypropylene centrifuge tubes.
   b. Add 20 μl of dilution *a* to each tube. Seal each tube with a screw top (see Note 3 below).
6. Prepare assay tubes.
   a. Aliquot 20 μl of homogenized sample (0.8 mg tissue) into 12 polypropylene centrifuge tubes.
   b. Add 20 μl of each dilution (*a*–*g*) to two tubes. Seal each tube (see Note 4).

7. Incubate tubes for 1 hr at 37°.
8. Prepare MPS-1 micropartition units.
   a. Soak HVLP filters in solution D for at least 1 hr.
   b. Assemble 13 MPS-1 micropartition units with prepared HVLP filters.
9. Separate bound and free ligand.
   a. After incubation transfer the entire volume of each sample to a filter assembly.
   b. Centrifuge at 3600 *g* for 5 min.
   c. Add 200 μl solution E and centrifuge at 3600 *g* for 5 min.
   d. Repeat step c.
10. Prepare for radioactivity assay.
    a. Disassemble filter units.
    b. Place filters into scintillation vials labeled "bound."
    c. Place filter assembly cup containing filtrate into scintillation vials labeled "free."
11. Assay vials for radioactivity (100–200 keV window) (see Note 5 below).
12. Calculate total and bound ligand concentrations.
    a. Measure the DTPA–*gal*–polylysine concentration of solution A using UV absorbance at 225 nm (see Note 6 below).
    b. Calculate total DTPA–*gal*–polylysine concentration [L] of each dilution.
    c. Calculate the bound concentration [B] for each sample:

$$[B] = [L]*CPM_{bound}/(CPM_{bound} + CPM_{free}) \quad (1)$$

13. Format [L] and [B] data for LIGAND (see Note 7 below).
    a. Set weighing parameters a0, a1, and a4 to zero and a3 to 0.0001.
    b. Set the "varying ligand number" and the "labeled ligand number" to 1.
    c. Set the total hot ligand concentration to the lowest of the [L] values.
    d. The last twelve lines (one line per assay sample) of the data file hold three values per line: $[L]_i$, the total ligand concentration of the *i*th sample; $[B]_i$, the bound ligand concentration of the *i*th sample; and the replication weight, which equals one for all samples.
14. Execute LIGAND software to calculate receptor concentration and affinity.

a. Run program SCAFIT.[24]
b. Enter 1 for the number of binding sites.
c. The program will consider parameters *K11*, *R1*, *N1*, and *C1*.[25]
d. Direct the program to hold parameter *C1* constant at 1.0.
e. Use the initial estimates provided by SCAFIT.
f. Iterate until a minimum sum-of-squares is obtained.
g. SCAFIT will output final estimates for parameters *K11*, *R1*, and *N1* with the relative standard errors expressed as percent coefficient-of-variation.
h. The symbols, representations, and units are as follows: *K11*, the ligand–receptor equilibrium binding constant of association in units of reciprocal concentration; *R1*, receptor concentration; and *N1*, a ratio that represents the amount of nonspecifically bound ligand relative to the amount of free ligand.

15. Calculate receptor density *R* as moles per gram of liver tissue:

$$R = \frac{(40 \times 10^{-6}\ \text{liter})(R1\ \text{mol/liter})}{0.8 \times 10^{-3}\ \text{g}} \qquad (2)$$

Results and Notes

Results

The results from a typical *in vitro* assay are illustrated in Fig. 1 in the form of a Scatchard plot.[26] The sample was obtained from the periphery of a surgical biopsy of a liver tumor. Consequently, this tissue should not be regarded as healthy and may have a diminished receptor density and affinity. The solid triangles represent the measured binding data. The radiolabeled ligand was ^{99m}Tc-DTPA$_{24}$–*gal*$_{185}$–polylysine$_{120}$, which had average DTPA and galactose densities of 24 and 185 mole per mole of polylysine, respectively. Using a one-ligand, one-site model, LIGAND produced a fit (curved line in Fig. 1) to the binding data with the following values for parameters *R1* (receptor concentration), *K11* (equilibrium binding constant of association), and *N1* (nonspecific binding fraction): 0.265 ± 0.029 n*M*, 2.99 ± 0.36 nM^{-1}, and $0.213 + 0.009$, respectively. The asialoglycoprotein receptor density, as calculated by Eq. (2), was $13.2 \pm 1.6 \times 10^{-12}$ mol/g. The total sum-of-squares for the fit was 98.8

[24] P. J. Munson, *in* "A User's Guide to LIGAND." Division of Computer Research and Technology, National Institutes of Health, Bethesda, Maryland, 1992.
[25] P. J. Munson and D. Rodbard, *Anal. Biochem.* **107,** 220 (1980).
[26] G. Scatchard, *Ann. N.Y. Acad. Sci.* **51,** 660 (1949).

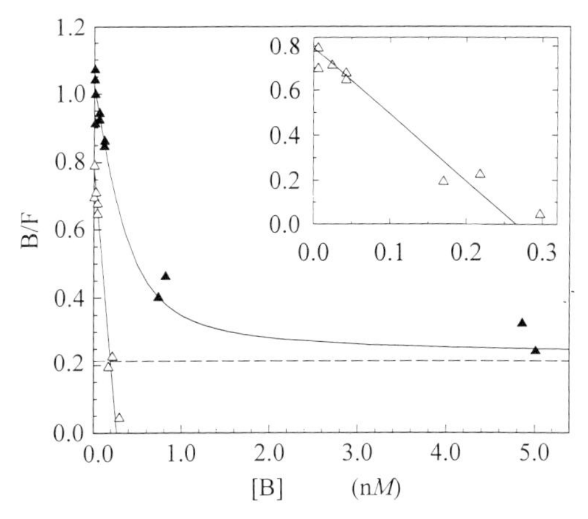

FIG. 1. Typical Scatchard plot resulting from ^{99m}Tc-DTPA$_{24}$–*gal*$_{185}$–polylysine$_{120}$ binding to 0.8 mg of human liver in 40 μl assay buffer. The computer program LIGAND was used to fit the binding data (solid triangles) to a one-ligand, one-binding site model. LIGAND estimated the receptor concentration, the equilibrium binding constant of association, and nonspecific binding to equal 0.265 ± 0.029 n*M*, 2.99 ± 0.36 nM^{-1}, and 0.213 ± 0.009, respectively. The curved line represents total binding. The two straight lines indicate the specific (solid) and nonspecific (dashed) components. The hollow triangles represent data corrected by LIGAND for nonspecific binding (NSB). The inset is an expansion of the specific binding curve with the NSB-corrected data.

with nine degrees of freedom. The two straight lines in Fig. 1 are the specific (solid) and nonspecific (dashed) components of the total binding curve. The hollow triangles represent data corrected by LIGAND for nonspecific binding (NSB). An inset in Fig. 1 displays the NSB-corrected data and specific binding curve more clearly.

Notes

1. DTPA–*gal*–polylysine will not adhere to polypropylene in the presence of 0.1% albumin.
2. Tissue must be stored at −70° in screw-top polypropylene centrifuge tubes with O ring seals and not permitted to thaw.
3. The tubes will provide a measure of nonspecific binding by the filters and are not used to calculate receptor density. Nonspecific binding by the filters should be less than 1% of total ligand.
4. The entire volume of the assay samples must reside at the bottom of each assay tube. If a drop clings to the side of a tube, tap it to the bottom.

5. Count samples in order of descending ligand dilution. Subtract background from each measurement. Calculate the counting background as the mean activity (counts per minute, CPM) of 12 empty scintillation vials counted for 60 min each. Place the empty vials before the dilution *a* and *b* samples. This permits these samples to decay by two half-lives prior to counting.
6. Wait at least 60 hr (ten ^{99m}Tc half-lives) before the absorbance measurement; most spectrophotometers are sensitive to gamma radiation.
7. Data are entered by editing a file with a .CRV extension. An example of the file format is listed in the LIGAND User's Guide (page 37).[24]

Acknowledgments

This work was supported by U.S. Public Health Service Grant RO1AM34768 and U.S. Department of Energy Grant FG03-87ER60553.

[29] Radiopharmaceutical Preparation of Technetium-99*m*-Labeled Galactosyl Neoglycoalbumin

By David R. Vera, Robert C. Stadalnik, Masatoshi Kudo, and Kenneth A. Krohn

Introduction

There exists a stage during the development of a new radiopharmaceutical when a moderate amount of agent is required for initial clinical trials. Such studies usually test the hypothesis that the proposed radiopharmaceutical offers increased diagnostic accuracy over a current agent or test. It is at this point that an injectable form of the radiopharmaceutical is required.

The principal safety features of an injectable solution are sterility and apyrogenicity. A sterile solution does not contain living organisms; an apyrogenic solution does not contain metabolic products of living organisms or debris of dead organisms which can cause a pyretic response on injection. In humans the pyretic response illicits a sudden onset of chills and fever. We will describe a method for the laboratory-scale production of an injectable solution of a radiolabeled neoglycoalbumin. This method is applicable to any radiolabeled macromolecule and is directed toward the academic setting where moderate amounts of injectable material are required for testing "proof-of-principle" or a clinical hypothesis.

Although successful liver imaging was achieved with technetium-99*m*-labeled asialoglycoprotein,[1] we selected galactosyl neoglycoalbumin as a more appropriate pharmacon. When selecting the macromolecular backbone for the synthetic ligand, our primary consideration was the advantages offered by albumin concerning biological safety. It is available at low cost and as a sterile apyrogenic solution. Human serum albumin (HSA), when prepared as an injectable, is heat-treated to destroy viral contaminants. It is also tested for HIV, the human immunodeficiency virus. Additionally, after extensive human use with HSA as a plasma expander, the low immogenicity of the protein has been thoroughly documented. Some other considerations were also in favor of albumin. Its large globular structure minimizes glomerular filtration and interstitial diffusion. As a result, an albumin-based radioligand remains within the vasculature long enough to be readily available for binding by the asialoglycoprotein receptor. Finally, a number of techniques exist by which albumin can be directly labeled with technetium-99*m*[2] or indirectly labeled via chelation to a bifunctional reagent.[3]

Of the several methods[4] available for the covalent attachment of carbohydrate to proteins, we selected amidination using the bifunctional reagent developed by Lee and co-workers.[5] As described by Lee and co-workers,[5] amindine-based thioglycosides have a number of advantages over other bifunctional reagents. The reagent 2-imino-2-methoxyethyl-1-thiogalactoside is specific for primary amino groups at a mild pH. Because the p*K* values of ε-amines and imino hydrogens are similar,[6] the charge distribution of the neoglycoconjugate is maintained. This minimizes alteration in tertiary structure. We wished to avoid any change in conformation of the albumin backbone which could result in immunogenicity or reticuloendothelial uptake. Additionally, the amidine bond and thioglycosides are stable under acidic conditions[7] which is a requirement for successful labeling with technetium. Finally, thioglycosides are also resistant to

[1] D. R. Vera, K. A. Krohn, and R. C. Stadalnik, *in* "Radiopharmaceuticals II" (V. J. Sodd, D. R. Hoogland, D. R. Allen, and R. D. Ice, eds.), p. 565. Society of Nuclear Medicine, New York, 1979.

[2] W. C. Eckelman, G. Meinken, and P. Richards, *J. Nucl. Med.* **12,** 707 (1971).

[3] D. J. Hnatowich, W. W. Layne, and R. L. Childs, *Int. J. Appl. Radiat. Isot.* **33,** 327 (1982).

[4] C. P. Stowell and Y. C. Lee, *Adv. Carbohydr. Chem. Biochem.* **37,** 225 (1980).

[5] Y. C. Lee, C. P. Stowell, and M. J. Krantz, *Biochemistry* **15,** 3956 (1976).

[6] M. J. Hunter and M. L. Ludwig, this series, Vol. 25, p. 585.

[7] D. Horton and D. H. Hutson, *Adv. Carbohydr. Chem.* **18,** 123 (1963).

plasma glycosidases,[8] an important requirement if the neoglycoprotein is to maintain its affinity for the receptor after injection into the plasma volume.

A final, but significant, feature of a synthetic ligand is the ability to control the affinity of the neoglycoprotein via galactose density.[9] Simulations of radiopharmaceutical biokinetics predicted that estimation of receptor quantity required a radiolabeled ligand of moderate affinity.[10] Liver uptake of a high-affinity ligand (more specifically, a high forward-binding rate constant) would be rate-limited by hepatic plasma flow and consequently would not be controlled by receptor quantity.[11] We argued that if the shape of the liver uptake data is not controlled by the number of hepatic receptors, it would not be possible to measure accurately the receptor quantity from the uptake data. We therefore considered the ability to chemically control ligand affinity of galactosyl neoglycoalbumin via carbohydrate density to be of paramount importance for successful kinetic modeling of the radiopharmaceutical.

Procedure

The preparation of ^{99m}Tc-galactosyl neoglycoalbumin[12] (Tc-NGA) is divided into two phases. First is the synthesis of the galactosyl neoglycoalbumin, which we perform once a year in lots of approximately 5 g. The second is the labeling step. Because technetium-99*m* has a half-life of 6 hr, the labeling must be performed within 1 to 2 hr of the imaging study.

Synthesis of Galactosyl Neoglycoalbumin

Synthesis of NGA follows the procedures developed by Chipowsky and Lee.[13] Our early preparations started with the bromination of D-galactose in acetic anhydride,[14] which yields 2,3,4,6-tetra-*O*-acetyl-α-D-galactopyranosyl bromide. This product can now be purchased from Sigma Chemical Co. (St. Louis, MO). Starting with 100 g, the acetobromogalactose is dispensed into eight 50-ml glass centrifuge tubes, quickly

[8] J. Monod, G. Cohen-Bazire, and M. Cohn, *Biochim. Biphys. Acta* **7,** 585 (1951).

[9] D. R. Vera, K. A. Krohn, R. C. Stadalnik, and P. O. Scheibe, *J. Nucl. Med.* **25,** 779 (1984).

[10] K. A. Krohn, D. R. Vera, and R. C. Stadalnik, *in* "Receptor-Binding Radiotracers" (W. C. Eckelman, ed.), Vol. 2, p. 41. CRC Press, Boca Raton, Florida, 1982.

[11] D. R. Vera, K. A. Krohn, P. O. Scheibe, and R. C. Stadalnik, *IEEE Trans. Biomed. Eng.* **32,** 312 (1985).

[12] D. R. Vera, R. C. Stadalnik, and K. A. Krohn, *J. Nucl. Med.* **26,** 1157 (1985).

[13] S. Chipowsky and Y. C. Lee, *Carbohydr. Res.* **31,** 339 (1979).

[14] M. Bárczai-Martos and F. Kórösy, *Nature (London)* **165,** 369 (1950).

dissolved in 100 ml acetone, centrifuged, and decanted into a round-bottomed flask containing 25 g of thiourea. This extraction step is required to remove the barium carbonate used to stabilize the bromo sugar. The product, 2-*S*-(2,3,4,6-tetra-*O*-acetyl-β-D-galactopyranosyl)-2-thiopseudourea hybrobromide, precipitates out during 30 min of reflux. It is then filtered, air-dried, and reacted with chloroacetonitrile to yield cyanomethyl 2,3,4,6-tetra-*O*-acetyl-β-D-galactopyranoside (CNM-thiogalactose), which is recrystallized twice with decolorizing carbon. Typically, 100 g of bromoacetogalactose will yield 50 g of CNM-thiogalactose. We have stored this product at −20° desiccated under nitrogen for up to 5 years.

The final three steps of the synthesis, namely, (1) formation of the imidate reagent 2-imino-2-methoxyethyl-1-thio-β-D-galactopyranoside (IME-thiogalactose), (2) coupling to human serum albumin (HSA), and (3) separation of NGA from uncoupled thiogalactose, are performed under sterile conditions.[9] All solutions use USP-grade[15] water for irrigation or saline for irrigation. The apparatus used in each step is completely disassembled, thoroughly washed with a detergent (1-Stroke Ves-Phene, Vestal Laboratories, St. Louis, MO) solution of fresh, glass-distilled water, rinsed with USP-grade irrigation water, air-dried, packaged in sterile pouches, and gas-sterilized (12% ethylene oxide). This includes glassware for the IME-thiogalactose synthesis, the rotary evaporator glassware with the Teflon seal, and the entire diafiltration system but excluding the ultrafiltration membrane.

The sterile work is performed in a radiopharmacy with controlled entry and work surfaces prepared by thorough cleaning with detergent and alcohol. The individual that constructs the apparatus and handles the reagents wears a face mask, disposable cap, shoe covers, sterile gloves, and a sterile gown. An assistant wears disposable cap and shoe covers, a clean laboratory coat and latex gloves. The entire process should be completed within 72 hr. This precaution and periodic filtration through 0.2-μm sterile filters are designed to prevent bacterial contamination. Of special concern are gram-negative strains which contain endogenous pyrogens within the cell wall. Although 0.2-μm filtration can remove whole cells, cellular debris cannot be filtered and will render a solution pyrogenic.

We synthesize IME-thiogalactose using the base-catalyzed method outlined by Lee and co-workers.[5] We typically react 9 g of CNM-thiogalactose with 0.65 g sodium methoxide in 225 ml dry methanol (Karl Fisher grade, distilled, stored over 4 Å molecular sieves). After 18 hr a sample

[15] The United States Pharmacopeia 1990, USP XXII, United States Pharmacopeial Convention, Inc., Rockville, Maryland, 1989.

of the reaction is aseptically removed for determination[9] of the IME-thiogalactose yield.[16] The following regression equation is then used to determine the amount of HSA required for the desired galactose density[12]:

$$\text{Gal/HSA} = 4.5 \text{ mol/mol} + 0.12(\text{IME/HSA}) \quad (1)$$

where Gal/HSA is the average number of galactose units per albumin molecule and IME/HSA is the molar ratio of the starting reactants, IME-thiogalactose and HSA. Equation (1) is useful when the target ratio is between 5 and 45 Gal/HSA. During removal of the methanol by rotary evaporation a 0.2 *M* solution of Clark's borate buffer (pH 8.5) is prepared using aseptic technique, and the desired amount of USP-grade human serum albumin (Plasbumin, 25%, Cutter Laboratories, Berkeley, CA) is drawn into a sterile syringe. Typically, 4.38 g of HSA and 200 ml of borate buffer are employed and yield a 2% solution of HSA. After evaporation of the IME-thiogalactose solution, the reaction flask (1000 ml) is removed from the rotary evaporator, and the borate buffer is rapidly added to the flask via a sterile 0.2-μm filter. The HSA is immediately added, and the flask is then sealed with a sterile Teflon stopper and set on a gyratory shaker for 3 hr at room temperature.

The diafiltration system consists of a 400-ml stirred ultrafiltration cell (Model 8400; Amicon, Danvers, MA), fitted with an ultrafiltration membrane (YM30; Amicon), a 5-liter reservoir, and a gas/liquid selector valve. A double-door refrigerated cold box fitted with two deck-height shelves is used. The shelves and floor of the cold box are lined with sterile drapes which are packaged with the sterilized ultrafiltration equipment. The ultrafiltration cell is placed on a stirring table which sits on a shelf, and the reservoir is placed on the floor directly below. A stirring hot plate is placed on the opposite shelf, and a large stainless steel bucket is set on the cold box floor.

The system is assembled and rendered apyrogenic during the 3-hr coupling reaction. Initially it is assembled without the ultrafiltration membrane. Connection to the pressure source, a cylinder of nitrogen (medical grade), is via a stainless steel filter assembly holding a gas-sterilized 0.2-μm filter. Silicone tubing is attached to the efflux port of the ultrafiltration cell. A sterile hemostat is used to clamp the tubing when the system is not in operation. The ultrafiltration membrane is placed in a covered beaker equipped with a large stirring bar and is washed with USP-grade water containing 0.025% Clorox for 15 min, four changes of USP-grade water (15 min each), 1% phosphoric acid (30 min), 50 m*M* sodium hydroxide for 1 hr at 80°–90°, and four rinses with USP-grade water. Two factors

[16] H. Brown, *Clin. Chem.* **14,** 967 (1968).

are critical to successful treatment of the membrane: gentle stirring to prevent scarring of the membrane and total emersion—the membrane cannot float atop the liquid, and bubbles at the membrane–liquid interface must be removed. The diafiltration apparatus is also treated with 1% phosphoric acid (5 min), 50 m*M* sodium hydroxide for 1 hr at 80°–90°, and receives at least four rinses with irrigation water. The last rinse is checked for pH, which must match the pH of the irrigation water. The cell and reservoir are then emptied and the reservoir filled with USP-grade saline for irrigation. To complete the construction of the diafiltration apparatus the ultrafiltration membrane is carefully removed from the beaker using fresh sterile gloves and installed into the base of the ultrafiltration cell.

The product of the coupling reaction is processed in the following manner. Using aseptic technique the product is transferred from the reaction flask into the ultrafiltration cell. This is accomplished by withdrawing the NGA solution through a sterile extension tube (K50L; Pharmaseal, Toa Alta, PR) into a large disposable syringe. The catheter is then removed, and a sterile 0.2-μm filter is attached to the syringe. The solution is then filtered into the cell. This is repeated with a fresh filter until the entire volume of product is transferred. The ultrafiltration cell is sealed, and the tubing from the efflux port is unclamped and directed to the waste reservoir. The cell is then pressurized to approximately 30 kPa with the gas/liquid selector set in the gas position. After checking for air and liquid leaks, the cell is pressurized to 200 kPa. Higher pressures result in decreased flows after 4–5 hr of operation. When the cell is fully pressurized, stirring is commenced and the gas/liquid selector switched to liquid. The doors to the cold box are closed, and diafiltration is carried out for at least ten exchange volumes, which is equivalent to a 2^{20} dilution of low molecular weight solutes from the original coupling reaction. At this point, using aseptic technique, the gas/liquid selector is switched to gas and the NGA solution is concentrated to approximately one-fifth of the original volume. The system is depressurized and stirred for approximately 1 hr. Again using aseptic technique, the NGA solution is removed from the cell with a disposable syringe fitted with an extension tubing. The solution is then filtered (0.5 μm) into sterile multidose vials, typically in 5 ml lots.

Samples are obtained for sterility ⟨USP 71⟩,[15] apyrogenicity ⟨USP 151⟩,[15] and analysis of albumin and galactose concentration.[17] The ratio of the two latter measurements yields the carbohydrate density of the product. The solution must pass USP requirements for sterility and apyro-

[17] M. Dubois, K. A. Giles, J. K. Hamilton, P. A. Rebers, and F. Smith, *Anal. Chem.* **28**, 350 (1956).

genicity at five times the scaled human dose. Sterile apyrogenic galactosyl neoglycoalbumin was tested for acute toxicity (mice) 100 and 2500 times the scaled human dose and subacute toxicity (rabbits) at 350 times the scaled human dose. No significant changes in behavior, hematology, or histology were observed.[18] After certification of sterility and apyrogenicity, the NGA is stored in a locked refrigerator for up to 1 year. The expiration period is based on galactose density which diminished during prolonged storage.

Labeling

Radiolabeling of galactosyl neoglycoalbumin with technetium-99*m* uses an electrolytic method developed by Benjamin[19] and later modified for clinical use by Dworkin and Gutkowski.[20] The procedure is divided into two steps: preparation of kit components and radiolabeling prior to each imaging study.

Preparation of Labeling Components. Three components of the labeling kit are prepared in groups of fifty and stored for up to 1 year at 5°. Inside a laminar air flow hood using aseptic technique, the following are prepared:

Reaction vials: Reaction vials are constructed by inserting two acid-cleaned [0.1 : 1 : 1 (v/v/v) HF–water–HNO_3] zirconium rods (26,770-8; Aldrich, Milwaukee, WI) through the rubber septum of a sterile 10-ml multidose vial; each assembly is then inserted into a sterilizing pouch and autoclaved

Acidic saline: Acidic saline is prepared by the addition of 75 ml of hydrochloric acid to a 1-liter bottle of USP-grade saline for irrigation; the solution (pH 1.2) is then dispensed into 10-ml multidose vials via a 0.2-μm syringe filter

Neutralization buffer: Neutralization buffer is prepared by the addition of 42 g $NaHCO_3$ and 20 g NaOH to a 1-liter bottle of USP-grade water for irrigation; This solution is also dispensed into 10-ml multidose vials via a 0.2-μm syringe filter

Radiolabeling Prior to Imaging. Labeling is conducted behind a lead barrier shield and proceeds with the following sequence:

1. Two milliliters of the acidic saline solution is delivered into a vented (22-gauge needle) reaction vial.
2. NGA (25 mg) is transferred to the reaction vial using an insulin syringe (27-gauge, fixed needle).

[18] IND 26,065, U.S. Food and Drug Administration, Rockville, Maryland.

[19] P. P. Benjamin, *Int. J. Appl. Radiat.* **20,** 187 (1969).

[20] J. H. Dworkin and R. F. Gutkowski, *J. Nucl. Med.* **12,** 562 (1969).

3. Three milliliters of injectable saline containing sodium pertechnetate of the desired radioactivity (0.2 mCi/kg) is added.
4. The reaction vial is purged with nitrogen (medical grade, 0.2 μm in-line filter) for 10 min.
5. Current (50 mA) is applied across the electrodes for 42 sec.
6. The solution is allowed to stand for 25 min at room temperature.
7. Then 0.4 ml of the neutralization buffer is added and the pH checked for pH 6–7.
8. The labeled product is filtered (0.2 μm) into a sterile 10-ml multi-dose vial.
9. A 3-ml syringe is used to draw the imaging dose, which is based on the body weight (0.035 ml/kg) of the patient. The maximum administered radioactivity is 8 mCi.

Quality control is conducted prior to injection. Of primary concern is radiochemical quality,[21] which denotes the percentage of radioactivity bound to the pharmacon. Our early imaging studies utilized polyacetate electrophoresis (250 V, 20 min) in barbital buffer (pH 8–6, 25°), followed by radiochromatographic scanning.[22] We later employed high-performance liquid chromatography (HPLC) using a silica-based stationary phase (TSK-3000SW; Beckman Instruments, Fullerton, CA) and saline as the mobile phase (0.9%, 1.0 ml/min). Both methods isolate free unlabeled technetium-99*m*, which typically is less than 2%.[12]

[21] K. A. Krohn and A. L. Jansholt, *Int. J. Appl. Radiat. Isot.* **28,** 213 (1977).
[22] A. L. Jansholt, K. A. Krohn, D. R. Vera, and H. H. Hines, *J. Nucl. Med. Technol.* **8,** 222 (1980).

[30] Culturing Hepatocytes on Lactose-Carrying Polystyrene Layer via Asialoglycoprotein Receptor-Mediated Interactions

By Kazukiyo Kobayashi, Akira Kobayashi, and Toshihiro Akaike

Introduction

Design of biofunctional materials employing specific combinations between ligands on synthetic polymers and receptors on cell surfaces is a promising approach to construct fascinating biomedical systems for cell separations, cell cultures, drug delivery agents, artificial antigens, and so on. Carbohydrate is one of the most important candidates as ligands or recognition signals. This chapter concerns the preparation of a lactose-carrying polystyrene (abbreviated as PVLA) and the culture of hepatocytes using PVLA as substratum which can interact specifically and strongly with asialoglycoprotein receptors on the surface of mammalian liver cells (Fig. 1).[1–7]

Interactions of hepatocytes with PVLA were similar in some aspects to those with asialoglycoproteins, their oligosaccharides, and neoglycoproteins.[8–13] The modified substratum PVLA has the following unique characteristics. The most important property of PVLA is the presence of highly concentrated or clustered β-galactopyranose residues. β-Galactopyranose residues are attached to every repeating unit along the polymer chain. In addition, they tend to gather on the outside of the polymer in water, because the hydrophobic polystyrene main chain is buried inside the molecule to form a hydrophobic core that is sheltered from water. Hence, the numerous multiantennary or clustered galactose terminals of a PVLA molecule may provide regions which can bind strongly to the liver cells through asialoglycoprotein receptors. It is also important that, owing to the amphiphilic character of PVLA, the carbohydrate-carrying polystyrene has a better adhesivity to hydrophobic solid surfaces.

Hepatocytes on PVLA showed a round morphology and functioned well. Epidermal growth factors and insulin stimulated the formation of multicellular aggregates of hepatocytes. Coculture of hepatocytes with

[1] K. Kobayashi, A. Kobayashi, S. Tobe, and T. Akaike, *in* "Neoglycoconjugates" (Y. C. Lee and R. T. Lee, eds.), Academic Press, San Diego, California, **261,** (1994).

[2] K. Kobayashi and T. Akaike, *Trends Glycosci. Glycotech.* **2**(2), 26 (1990).

[3] A. Kobayashi, T. Akaike, K. Kobayashi, and H. Sumitomo, *Makromol. Chem., Rapid Commun.* **7,** 645 (1986).

[4] T. Akaike, A. Kobayashi, K. Kobayashi, and H. Sumitomo, *J. Bioact. Compat. Polym.* **4,** 51 (1989).

[5] A. Kobayashi, K. Kobayashi, S. Tobe, and T. Akaike, *J. Biomater. Sci., Polym. Ed.* **3,** 499 (1992).

[6] T. Kugumiya, A. Yagawa, A. Maeda, H. Nomoto, S. Tobe, K. Kobayashi, T. Matsuda, T. Onishi, and T. Akaike, *J. Bioact. Compat. Polym.* **7,** 337 (1992).

[7] S. Tobe, Y. Takei, K. Kobayashi, and T. Akaike, *Biochem. Biophys. Res. Commun.* **184,** 225 (1992).

[8] G. Ashwell and J. Harford, *Annu. Rev. Biochem.* **51,** 531 (1982).

[9] Y. C. Lee, *Ciba Found. Symp.* **145,** 80 (1989).

[10] P. H. Weigel, R. L. Schnaar, M. S. Kuhlenschmidt, E. Schmell, R. T. Lee, Y. C. Lee, and S. Roseman, *J. Biol. Chem.* **254,** 10830 (1979).

[11] P. H. Weigel, *J. Cell. Biol.* **87,** 855 (1980).

[12] D. T. Connolly, R. R. Townsend, K. Kawaguchi, W. R. Bell, and Y. C. Lee, *J. Biol. Chem.* **257,** 939 (1982).

[13] O. A. Weisz and R. L. Schnaar, *J. Cell Biol.* **115,** 485 (1991).

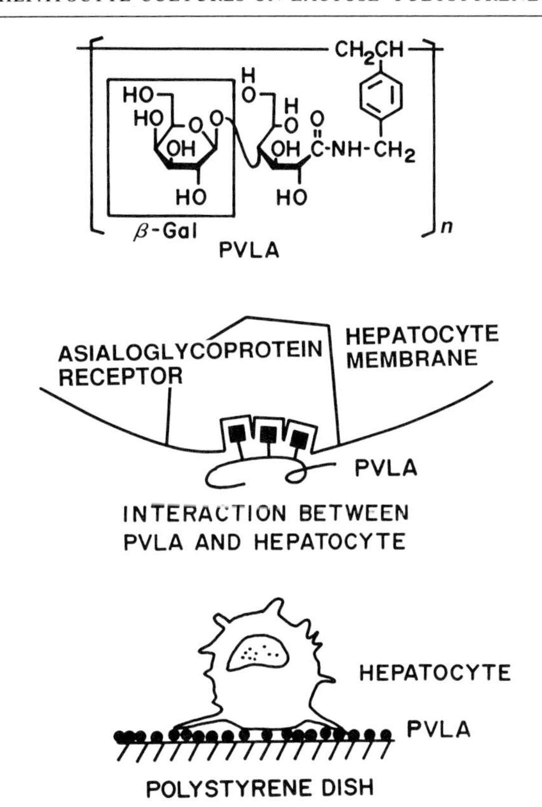

FIG. 1. Adhesion of hepatocytes on lactose-carrying polystyrene via asialoglycoprotein receptor-mediated interactions.

nonparenchymal liver cells induced the formation of tissue structures composed of hepatocytes, Kupffer cells, and endothelial cells. The behaviors of hepatocytes on PVLA were distinct from those on naturally occurring proteins such as collagen, fibronectin, laminin, and proteoglycan substrata.[14–17] The culture system using PVLA substratum is a useful model to investigate the regulation of cell behaviors in high-density cultures.

[14] P. O. Seglen, *Methods Cell Biol.* **13,** 29 (1976).

[15] A. Ichihara, T. Nakamura, and K. Tanaka, *Mol. Cell. Biochem.* **43,** 145 (1982).

[16] S. Shimaoka, T. Nakamura, and A. Ichihara, *Exp. Cell Res.* **172,** 228 (1987).

[17] N. Koide, K. Sakaguchi, K. Koide, T. Asano, M. Kawaguchi, H. Matsushima, T. Takenami, T. Shinji, M. Mori, and T. Tsuji, *Exp. Cell Res.* **186,** 227 (1990).

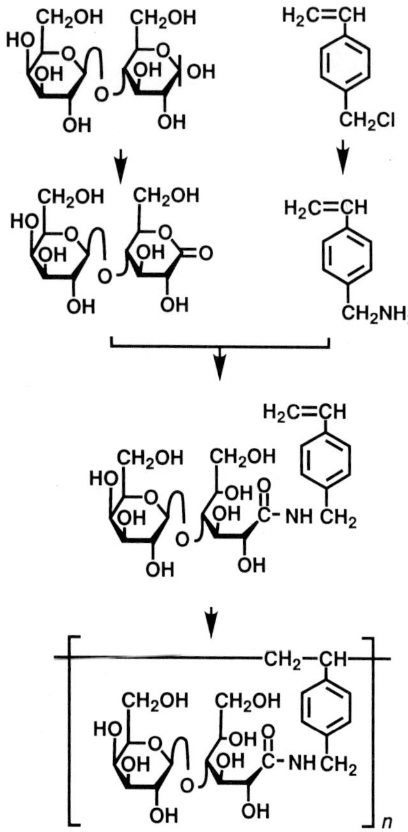

SCHEME 1. Synthesis of lactose-carrying polystyrene.

Materials

The PVLA is prepared according to Scheme 1.[18,19] The simple preparation method requires no protection of the hydroxyl groups of the oligosaccharide, and the yield in each step is high.

p-Vinylbenzylamine. *p*-Vinylbenzyl chloride (15.3 g, 0.10 mol) commercially available from Seimi Chemicals Ltd. (Chigasaki, Kanagawa 253, Japan) and potassium phthalimide (18.5 g, 0.10 mol) are dissolved in 50 ml of *N,N*-dimethylformamide (DMF), and the solution is heated at 50° for 4 hr. The DMF is removed in a vacuum evaporator, and the residue

[18] K. Kobayashi, H. Sumitomo, and Y. Ina, *Polym. J.* **15,** 667 (1983).
[19] K. Kobayashi, H. Sumitomo, and Y. Ina, *Polym. J.* **17,** 567 (1985).

is dissolved in chloroform. The solution is washed with a 0.2 *N* aqueous sodium hydroxide solution and with water, then concentrated. The product is crystallized from methanol to give *N*-*p*-vinylbenzylphthalimide in 22.2 g (84%) yield. Recrystallization from methanol gives pure crystals of mp 107°–108°.

A solution of hydrazine hydrate (80% solution in water, 6.6 g, 0.105 mol) in ethanol (10 ml) is added to a refluxing solution of *N*-*p*-vinylbenzylphthalimide (18.4 g, 70 mmol) in ethanol (50 ml). Immediately, a white stiff crystalline mass starts to precipitate. Refluxing and vigorous mechanical stirring are continued for 90 min. The precipitate is filtered, and the filtrate is concentrated to dryness. The combined solids are treated with an aqueous potassium hydroxide solution (20 g of KOH and 120 ml of water). The resulting *p*-vinylbenzylamine is extracted with ether (once with 140 ml and then 4 times with 70 ml). The combined ether solutions are washed with 2% w/v potassium carbonate solution (4 times, 40 ml) and dried on potassium carbonate. The solvent is then removed, and the residue is distilled under reduced pressure (72°–73°/3 mm Hg), yielding 7.1 g (76%).

N-p-Vinylbenzyl-O-β-D-galactopyranosyl-(1→4)-D-gluconamide. Lactose 1-hydrate (12 g, 33 mmol) is dissolved in water (9 ml), diluted with methanol (25 ml), and added to an iodine (17.1 g, 67 mmol) solution in methanol (240 ml) at 40°. At this temperature, a 4% w/v potassium hydroxide solution in methanol (400 ml) is added dropwise with magnetic stirring for 35 min until the color of iodine disappeared. The solution is cooled externally in an ice bath. The precipitated crystalline product is filtered, washed with cold methanol and then cold ether, and recrystallized from a mixture of methanol and water (12:5, v/v, 500 ml). The yield is 10.7 g (81%).

The resulting potassium lactonate is then converted to the free acid by passing the aqueous solution through a column of Amberlite IR-120. The acidic eluate is collected and concentrated in a rotary evaporator. Repeated evaporation of methanol and ethanol converts the lactonate to lactonolactone containing a small amount of water in a quantitative yield.

Lactonolactone (15.3 g, 45 mmol) is dissolved in refluxing methanol (140 ml), and a *p*-vinylbenzylamine (6.0 g, 45 mmol) solution in methanol (30 ml) is added. The mixed solution is refluxed with magnetic stirring for 120 min and allowed to stand at room temperature to yield white crystals. The crystals are filtered, washed with a small amount of cold methanol, and dried *in vacuo* (18.5 g, 87%). The product is recrystallized from methanol; mp 188°–189°, $[\alpha]_D^{25}$ +29.1°.

Poly[N-p-vinylbenzyl-O-β-D-galactopyranosyl-(1→4)-D-gluconamide]. The monomer (2.37 g, 5 mmol), azobisisobutyronitrile (1.6–4.1 mg, 0.2–0.5 mol% to monomer), and dimethyl sulfoxide (4–5 ml) are charged in a glass ampoule equipped with a three-way stopcock through a silicon rubber adaptor. The resulting solution is frozen by dipping the ampoule in a solid carbon dioxide–methanol bath, degassed, and then returned to solution at room temperature. The procedure is repeated three times. The ampoule is sealed under reduced pressure and maintained in a thermostat at 60° ± 0.05°. The solution is chilled and poured into an excess amount of cold methanol (50 ml) to precipitate the product. The product is repeatedly precipitated from its aqueous solution into methanol five times and freeze-dried from an aqueous solution. The yield is 2.06–2.25 g (87–95%); $[\alpha]_D^{25}$ +35.1 to +36.1°. The polymer is soluble in both water and dimethyl sulfoxide.

Aqueous solutions of PVLA (0.1 mg/ml) are available from Nagase Chemicals Ltd. (Tatsuno, Hyogo 679-41, Japan).

Methods

Methods for preparing PVLA-coated dishes and evaluating PVLA–hepatocyte interactions follow previously published procedures.[3–7]

Preparation of Polymer-Coated Dishes. One-milliliter aliquots of PVLA solution (0.1 mg/ml) are placed in 35-mm diameter polystyrene dishes (Falcon 1008, Becton Dickinson, Lincoln Park, NJ) at room temperature for 10 min to adsorb the polymer on the surface of the dish. The solution is decanted, and the surface of the dish is rinsed with a Dulbecco's phosphate buffer solution three times.

Strong adsorption of the polymer can be confirmed by wettability of the surface with water, surface analysis using fluorescence-labeled PVLA, and X-ray photoelectron spectroscopy. This simple coating method can be applied to hydrophobic polystyrene microsphere beads, which also interact with asialoglycoprotein receptors on cell surfaces and are used to investigate their dynamic behavior.

Preparation and Culture of Hepatocytes. Hepatocytes are isolated from livers of female Sprague-Dawley rats (150–200 g) using the two-step collagenase perfusion technique of Seglen.[14] The cell density is adjusted to 3×10^5 cells/ml in Williams' medium E. Small portions (1.5 ml) of the cell suspension are seeded onto PVLA-coated culture dishes, which are then kept at 37° in a humidified air–CO_2 incubator (95%/5%, v/v). The adhesion increases rapidly within 1 hr and then gradually over 6–12 hr to reach a maximum value (>90%). Hepatocytes maintain a round morphology on the PVLA substratum (Fig. 2) that is quite distinct from the spreading morphology on collagen and fibronectin. In spite of the unique cell morphology, hepatocytes on PVLA are found to exhibit active functions of synthesis of albumin, urea, and bile acid.

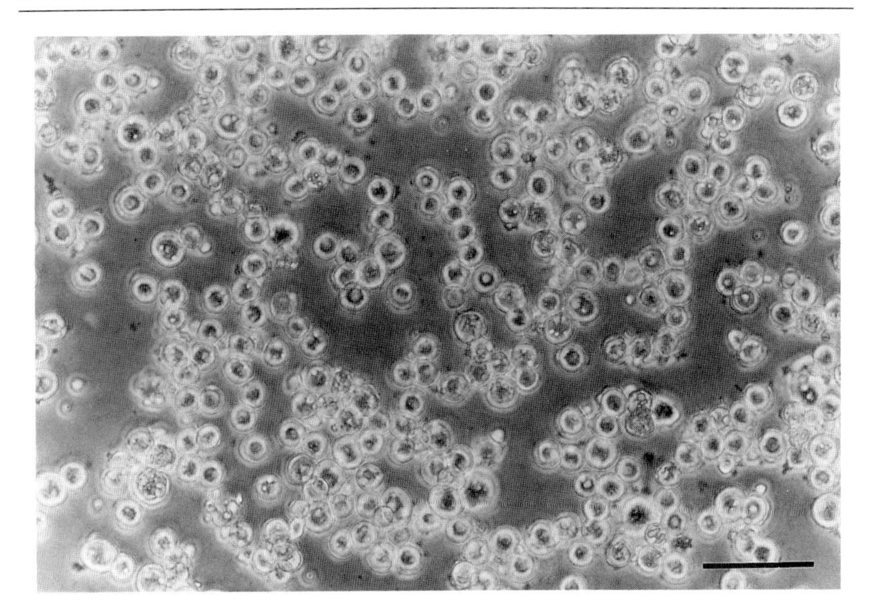

FIG. 2. Phase-contrast micrograph of soft-landed hepatocytes on PVLA. Bar: 100 μm.

Detachment of Hepatocytes by Treatment with EDTA Solution. Calcium ion is required for hepatocytes to adhere to the PVLA substratum, and the soft-landed hepatocytes can be removed from the dish merely by treating the culture with an aqueous solution of a calcium chelator under mild conditions as follows. An EDTA solution (0.02% wt/v, 1 ml) is added to hepatocytes on PVLA-coated dishes and incubated for 15 min. The dish is shaken for a while, and the supernatant is stirred with a pipette and decanted. The surface of the dish is washed with 1 ml of phosphate buffer solution. The recovered cells are counted to estimate the detachment efficiency. The EDTA is removed from the combined hepatocyte suspension by gentle centrifugation (500 rpm for 1.5 min) and the hepatocytes are recovered as a suspension in Williams' medium E. The recovered hepatocytes are still active, and they could be recultured on collagen-coated dishes, where they exhibited the ability to incorporate [^{3}H]thymidine.

Formation of Multilayer Aggregates of Hepatocytes by Treatment with Epidermal Growth Factor. Freshly isolated hepatocytes are incubated on PVLA for 4 hr, and then 50 ng/ml of epidermal growth factor (EGF) is added to the hepatocytes. The cells move gradually and come into contact with one another to form masses. They grow in size to aggregates of about 100 μm wide in 2 days. As the culture is continued, the number of single cells around the aggregates is decreased, and the aggregates are merged to become bigger to a diameter of about 100–150 μm. When the cell density is increased, aggregates of a diameter of 300 μm are also observed in 18 days.

The aggregates have three-dimensional layered structures. A scanning electron micrograph (Fig. 3) shows that several hepatocytes climb up the side of the aggregates and that deep holes are opened along the cell–cell interfaces on the surface of the aggregates. However, the cells directly in contact with the substratum take on a spreading morphology. The surfaces of the aggregates are covered with some extracellular matrix components.

The aggregates could be removed from the substratum with a silicone rubber scraper. When they are transfered onto the collagen layer and incubated for 2 days, larger aggregates of 100 μm diameter adhere well to the collagen layer and retain their three-dimensional structure; however, small-size aggregates lose their spherical shapes and spread rapidly on the substratum to form a monolayer.

Cell Assembly by Coculture of Hepatocytes with Nonparenchymal Liver Cells. When hepatocytes and nonparenchymal liver cells are cocul-

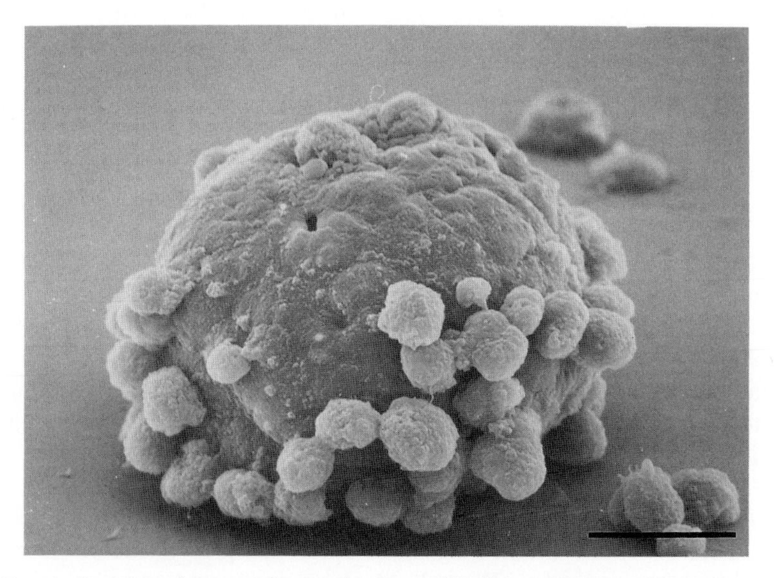

FIG. 3. Scanning electron micrograph of a multicellular aggregate of hepatocytes on PVLA. Bar: 20 μm.

tured on collagen and fibronectin, nonparenchymal liver cells exhibit much greater active proliferating power and hence force out the hepatocytes from the substratum. On PVLA, however, nonparenchymal cells are attached weakly and hepatocytes are attached specifically and strongly. Cell assembly consisting of hepatocytes and nonparenchymal liver cells can be achieved by coculture on PVLA as follows.

Hepatocytes and nonparenchymal liver cells are separated by differential centrifugation.[16] Hepatocytes (1×10^6 cells/dish) are incubated in Williams' medium E containing insulin (10^{-9} *M*), dexamethasone (10^{-9} *M*), and aprotinin (0.12 μg) for 4 hr. Dead hepatocytes and the medium are removed, and nonparenchymal liver cells ($1–4 \times 10^6$) are added and incubated in the medium containing 5% fetal calf serum. Nonparenchymal cells are attached only around the periphery of multicellular aggregates of hepatocytes, and their proliferation is depressed. The diameter of cell aggregates increases to about 100 μm in 2 days. Figure 4 is a representative scanning electron micrograph of the cell assembly with well-developed morphology. Kupffer cells are found to be incorporated into grooves along the surface. This coculture system appears to approach the status of the *in vivo* environment of hepatocytes.

FIG. 4. Scanning electron micrograph of a cell assembly by coculture of hepatocytes and nonparenchymal liver cells. Bar: 10 μm.

Prospects

Functions and survivability of hepatocytes on PVLA were enhanced through the aggregate formation and coculture with nonparenchymal liver cells. The behaviors of hepatocytes on PVLA were distinct from those on commonly used collagen and fibronectin substrata. As the ease of preparation and handing is also an advantage of the artificial material PVLA, the culture system using PVLA is a useful model to investigate the regulations of cell behaviors in high-density culture. Several carbohydrate-carrying polystyrenes could be prepared by applying the simple synthetic method[20,21] to the corresponding oligosaccharides having reducing terminals such as maltose, cellobiose, melibiose, maltopentaose, maltoheptaose, and *N*-acetylchitooligosaccharide as the starting substances and also by applying other synthetic methods.[22]

[20] K. Kobayashi, H. Sumitomo, and T. Itoigawa, *Macromolecules* **20,** 906 (1987).

[21] K. Kobayashi, K. Aoki, H. Sumitomo, and T. Akaike, *Makromol. Chem., Rapid Commun.* **11,** 577 (1990).

[22] K. Kobayashi, T. Akaike, and T. Usui, this series, Vol. 242 [21].

Author Index

Numbers in parentheses are footnote references numbers and indicate that an author's work is referred to although the name is not cited in the text.

C

D

E

F

H

I

J

K

N

O

T

U

V

W

X

Y

Z

Subject Index

A

D

E

F

G

K

L

M

R

S

V

X

ISBN 0-12-182148-X